Global Energy and Water Cycles

The abundance of water in all three phases (solid, liquid, vapor) makes the Earth unique in the Solar System. Knowledge of the fluxes and changes of phase of water are essential for an understanding of weather, climate and, indeed, of life itself.

Global Energy and Water Cycles provides a state-of-the-art treatment of advances in our understanding through improvements in global models, in the representation of the processes included in the models, and in related observations. It deals with fluxes within the atmosphere, at and beneath the land and ocean surface, and the interaction between them. This area of environmental science is developing rapidly and it is important to remain in touch with related developments across the wide range of the meteorological, hydrologic and oceanographic topics involved. In order to provide authoritative coverage, the book draws upon the expertise of many of the world's leading researchers. It provides a comprehensive treatment of a subject which is currently scattered through the literature, and therefore makes it accessible as a coherent whole for the first time.

The book will be of main interest to graduate students and researchers in meteorology, hydrology and oceanography, but it will also appeal to final-year undergraduates in these subjects.

K. A. Browning: Professor in the Department of Meteorology and Director of the Joint Centre for Mesoscale Meteorology at the University of Reading. Formerly Director of Research at the UK Meteorological Office, Chief Meteorological Officer at the Meteorological Office Radar Research Laboratory, and Chief Scientist, (US) National Hail Research Experiment.

R. J. Gurney: Professor and Director of the Environmental Systems Science Centre at the University of Reading. Formerly Head, Hydrological Sciences Branch, NASA Goddard Space Flight Center.

Global Energy and Water Cycles

EDITED BY

K. A. Browning and R. J. Gurney

CAMBRIDGE UNIVERSITY PRESS
Cambridge, New York, Melbourne, Madrid, Cape Town, Singapore, São Paulo

Cambridge University Press
The Edinburgh Building, Cambridge CB2 2RU, UK

Published in the United States of America by Cambridge University Press, New York

www.cambridge.org
Information on this title: www.cambridge.org/9780521560573

First published 1998
This digitally printed first paperback version 2006

A catalogue record for this publication is available from the British Library

Library of Congress Cataloguing in Publication data

Global energy and water cycles / edited by K. A. Browning and R. J. Gurney.
p.   cm.
Includes index.
ISBN 0-521-56057-8 (hb)
1. Energy budget (Geophysics)   2. Hydrologic cycle.
3. Global Energy and Water Cycle Experiment.   I. Title.
QC809.E6B766   1998
551.48–dc21   98–4546   CIP

ISBN-13 978-0-521-56057-3 hardback
ISBN-10 0-521-56057-8 hardback

ISBN-13 978-0-521-03285-8 paperback
ISBN-10 0-521-03285-7 paperback

Contents

Contributors

Jean-Claude André, Météo-France, Centre National de Recherches Météorologiques, F-31057 Toulouse Cédex, France. Present address: Centre Européen de Recherche et de Formation Avancée en Calcul Scientifique (CERFACS), 42 Avenue Gaspard Coriolis, F-31057 Toulouse Cédex, France

Anton C. M. Beljaars, ECMWF, Shinfield Park, Reading RG2 9AX, UK. Also affiliated to The Royal Netherlands Meteorological Institute (KNMI)

Sabine Brinkop, Deutsches Zentrum für Luft- und Raumfahrt e.V., Institut für Physik der Atmosphäre, Oberpfaffenhofen, D-82234 Weßling, Germany

Andrew Brown, Meteorological Office, London Road, Bracknell, RG12 2SY, UK

K. A. Browning, Department of Meteorology, The University of Reading, Earley Gate, PO Box 243, Reading RG6 6BB, UK

Andrew Bushell, Meteorological Office, London Road, Bracknell RG12 2SY, UK

Moustafa T. Chahine, Jet Propulsion Laboratory, Mail Stop 180-904, 4800 Oak Grove Drive, Pasadena, CA 91109-8099, USA

C. G. Collier, Telford Institute of Environmental Systems, University of Salford, UK

J. Dooge, Centre for Water Resources Research, University College, Earlsfort Terrace, IRL – Dublin 2, Ireland

Lydia Dümenil, International CLIVAR Project Office, Max-Planck-Institute for Meteorology, Bundesstrasse 55, D-20146 Hamburg, Germany

Jean-Paul Goutourbe, Météo-France, Centre National de Recherches Météorologiques, F-31057 Toulouse Cédex, France

David Gregory, Meteorological Office, London Road, Bracknell RG12 2SY, UK

P. J. Gregory, Department of Soil Science, The University of Reading, Whiteknights, PO Box 233, Reading RG6 6DW, UK

Christian J. Guillemot, National Center for Atmospheric Research, P. O. Box 3000, Boulder, CO 80307, USA

R. J. Gurney, Environmental Systems Science Centre (ESSC), The University of Reading, Harry Pitt Building, 3 Earley Gate, Whiteknights, PO Box 238, Reading RG6 6AL, UK

Thomas Hauf, Deutsches Zentrum für Luft- und Raumfahrt e.V., Institut für Physik der Atmosphäre, Oberpfaffenhofen, D-82234 Weßling, Germany

Peter V. Hobbs, Department of Atmospheric Sciences, University of Washington, Seattle, Washington 98195, USA

Ming Ji, Environmental Modeling Center, National Centers for Environmental Prediction, NOAA/National Weather Service, Washington DC 20233-9910, USA

Peter Jonas, Department of Physics, UMIST, PO Box 88, Manchester M60 1QD, UK

Eugenia Kalnay, Environmental Modeling Center, National Centers for Environmental Prediction, NOAA/National Weather Service, Washington DC 20233-9910, USA. Present address: School of Meteorology, University of Oklahoma, 100 E. Boyd St., Room 1310 SEC, Norman, OK 73019, USA

Kristina B. Katsaros, Atlantic Oceanographic and Meteorological Laboratory, 4301 Rickenbacker Causeway, Miami, FL 33149, USA

Arun Kumar, Environmental Modeling Center, National Centers for Environmental Prediction, NOAA/National Weather Service, Washington DC 20233-9910, USA

P. C. D. Milly, US Geological Survey, Geophysical Fluid Dynamics Laboratory/NOAA, Princeton, New Jersey, USA

Mitchell W. Moncrieff, Mesoscale and Microscale Meteorology Division, National Center for Atmospheric Research, Box 3000, Boulder, CO 80307-3000, USA

J. L. Monteith, Institute of Terrestrial Ecology, Bush Estate, Pencuik, Midlothian, EH26 0QB, UK

P. Morel, Office of Earth Science, NASA Headquarters, 300 E Street SW, Washington, DC 20546, USA

Joël Noilhan, Météo-France, Centre National de Recherches Météorologiques, F-31057 Toulouse Cédex, France

Taikan Oki, Institute of Industrial Science, University of Tokyo, 7-22-1 Roppongi, Minato-ku, Tokyo 106, Japan

Hua-lu Pan, Environmental Modeling Center, National Centers for Environmental Prediction, NOAA/National Weather Service, Washington DC 20233-9910, USA

Richard D. Rosen, Atmospheric and Environmental Research, Inc., 840 Memorial Drive, Cambridge, MA 02139, USA

D. Rosenfeld, The Institute of Earth Sciences, The Hebrew University of Jerusalem, Israel

Peter Rowntree, Hadley Centre, Meteorological Office, London Road, Bracknell RG12 2SY, UK

Suranjana Saha, Environmental Modeling Center, National Centers for Environmental Prediction, NOAA/National Weather Service, Washington DC 20233-9910, USA

Raymond W. Schmitt, Woods Hole Oceanographic Institution, Woods Hole, MA 02540, USA

Graeme L. Stephens, Colorado State University, Fort Collins, CO 80523-1371, USA

Wei-Kuo Tao, NASA Goddard Space Flight Center, Mesoscale Dynamics and Precipitation Branch, Code 912, Laboratory for Atmospheres, Greenbelt, MD 20771, USA

Peter K. Taylor, James Rennell Division (254/27), Southampton Oceanography Centre, Southampton SO14 3ZH, UK

E. Todini, Dipartimento di Scienze della Terra e Geologico Ambientali, Via Zamboni, 67, I-40126 Bologna, Italy

Kevin E. Trenberth, National Center for Atmospheric Research, P. O. Box 3000, Boulder, CO 80307, USA

Pedro Viterbo, ECMWF, Shinfield Park, Reading, RG2 9AX, UK

Glenn White, Environmental Modeling Center, National Centers for Environmental Prediction, NOAA/National Weather Service, Washington DC 20233-9910, USA

David Williamson, National Center for Atmospheric Research, P. O. Box 3000, Boulder, CO 80307, USA

Eric F. Wood, Department of Civil Engineering, Princeton University, Princeton, NJ 08544, USA

Milija Zupanski, Environmental Modeling Center, National Centers for Environmental Prediction, NOAA/National Weather Service, Washington DC 20233-9910, USA

Preface

There is a growing realization that new global problems, faced both by scientists and by citizens, can only be tackled adequately by new partnerships formed for this purpose. Within science itself, new partnerships are needed to understand the complex scientific realities that underlie these global problems. The development of a comprehensive description of the climate system is an activity that takes the combined efforts of meteorologists, hydrologists, oceanographers and others. A particularly important part of this is the description of global energy and water cycles.

The World Climate Research Programme (WCRP), following the innovative pattern of its predecessor, the Global Atmospheric Research Programme (GARP), is an example of successful partnership and cooperation between the independent scientific community represented by the International Council of Scientific Unions and the intergovernmental community represented by the World Meteorological Organization, a specialized agency of the United Nations.

Part of the WCRP is the Global Energy and Water Cycle Experiment (GEWEX). The problems facing the GEWEX project, the state of our present knowledge, the progress to date, and the key problems that remain – all these were well exposed in the presentations and discussions at the GEWEX Conference held at the Royal Society in London in July 1994. This conference was successful in encouraging real dialogue between experts in different specialisms. This volume is not a volume of proceedings, but is based in part on contributions to that conference, and builds upon it. This book gives the opportunity of extending this important exchange of results and opinions to a wider audience, and brings together the state of our knowledge in this important area. The scientific community are in debt to the editors and contributors for the energy and judgement they have shown throughout this enterprise.

J. Dooge

Foreword

GEWEX: the international context of this book

Striking advances were made in the 1970s in observing the general circulation of the atmosphere and modeling its dynamics, thus developing the relatively new field of *geophysical fluid dynamics* into an effective tool for planetary-scale weather forecasting. However, using the words of Richard P. Feynman, these advances were limited to studying the 'flow of dry water' – a mathematical abstraction in which energy sources and sinks could be largely ignored or very much simplified. The performance of existing general circulation models (GCMs) and computers, as well as the limitations of the observing systems of the time, effectively precluded investigating the diabatic processes which drive the global atmosphere, land surface hydrology and the world ocean circulation.

Nevertheless, environmental scientists understood very well that the great machine which determines the Earth climate and had maintained conditions favorable to life on our planet for billions of years, depends on the operation of a wide range of *physical, geochemical and biological processes*. They recognized that climate was primarily controlled by radiation transfer through the clear air of the stratosphere as well as the cloudy, wet air of the troposphere. They knew that rainfall was controlled by the transport of atmospheric water vapor and the dynamics of clouds. They understood that water resources depended upon rain, groundwater storage and river discharge as well as evapotranspiration and vegetal life.

Indeed, in the 1970s and early 1980s, scientists in all branches of environmental research had already made considerably progress in the study of these various processes individually. For example, a strong radiation science community was actively investigating radiative transfer through clear air and aerosols from satellites or aircraft. Cloud physicists were busy studying the dynamics and microphysical constitution of rain clouds. Agrometeorologists were building the scientific foundations for a quantitative understanding of land surface and vegetation processes in hydrology, while professional hydrologists themselves were far along building 'conceptual models' of water storage and discharge in river catchments.

However, these scientific investigations were largely discipline oriented and piecemeal. It was difficult to see how the patchwork of individual process studies could be transmuted into a comprehensive description of the interactive Earth system. Atmospheric and climate modelers were not interested because their fluid-dynamical models did not allow enough scope for such complicated physics. Observational scientists were not interested because reliable global climatological records of such exotic properties as energy and water fluxes were non-existent.

Yet, the portents of future advances were already in evidence, based as it is often the case upon *technical breakthroughs* such as dramatic advances in computer performance and new concepts for Earth observation from space. From its inception, the

World Climate Research Programme (WCRP) was steered toward a scientific exploitation of these new tools. Climate researchers pressed the atmospheric modelers to consider the thermodynamics as well as the pure dynamics of the atmospheric circulation, and to free themselves from the specified surface boundary conditions and initial value problems that are the hallmark of weather forecasting. Simultaneously, the planners of WCRP sought the support of science-funding and space research agencies to promote essential new instrument technologies and climatologically relevant space missions.

A most significant achievement in this period was the success of the multi-satellite Earth Radiation Budget Experiment (ERBE), which effectively appeared as the crowning achievement of a long series of space research projects to monitor the Earth radiation balance from space. For the first time, it was possible to obtain a quantitative assessment of the contribution of water vapor to the greenhouse effect of clear air, as well as the radiative forcing of clouds. Yet, monitoring radiation fluxes at the top of the atmosphere could not suffice in order to understand climate phenomena and the sensitivity of climate to external forcing: it was clearly necessary to probe into the wet and cloudy troposphere, the deep ocean and the soil on continental surfaces. New active microwave sensors that could penetrate rain clouds, as well as non-precipitating but optically thick clouds, were under development and would obviously constitute a major step in this direction.

At the same time, a new generation of climate models and climate change simulations provided both the means and the incentive to incorporate more realistic formulations of energetic and hydrologic processes. Radiation and clouds, water vapor transport and rain, evaporation from the ocean and land, groundwater storage and river flow, were indeed recognized as components of a single interactive system. The time was right to take one more step in the direction of *integrated earth system sciences* and promote a more active dialogue between scientific disciplines that were close enough and mature enough to benefit from such interactions.

It was also very timely, especially from the perspective of planning space missions and other long-term research initiatives, to formulate a unifying scientific strategy for addressing the global energy balance and hydrologic cycle. The Global Energy and Water Cycle Experiment (GEWEX) is the planetary-scale environmental research program formulated by WCRP in the late 1980s to fulfill these requirements. The present book deals with our current understanding of the related science.

P. Morel

Acknowledgements

This book was stimulated by the European Conference on the Global Energy and Water Cycle held at the Royal Society in 1994 under the auspices of the UK GEWEX Forum, whose members, in addition to the editors, are: M. A. Beran, B. J. Hoskins, P. R. Jonas, D. T. Llewellyn-Jones, A. Lorenc, P. J. Mason, J. L. Monteith and W. B. Wilkinson. The members of this Forum also assisted with the formulation of this book, and their help is gratefully acknowledged.

Each chapter and section of this book has been scientifically peer reviewed by experts in their fields. These reviewers are thanked for their assistance. They include: S. Allen, J.-C. André, N. Arnell, R. Avissar, A. Betts, K. Bryan, H. Charnock, K. Emanuel, D. Entekhabi, K. Georgakakos, D. Gregory, J. Harries, A. Illingworth, P. Jonas, R. Koster, A. Mahfouf, P. Mason, M. Miller, P. Milly, J. Mitchell, M. Moncrieff, J. Monteith, P. Naden, T-E. Nordeng, J. Peixoto, J-L. Redelsberger, G. Rasmussen, R. Rosen, P. Rowntree, D. Salstein, R. Schmitt, K. Shine, L. Simmonds, A. Slingo, J. Slingo, P. Smith, G. Stephens, P. Taylor, E. Todini, K. Trenberth, D. Tsintikidis, H. Wheater, W. Wilkinson and J. Woods, and others who wish to remain anonymous. We have also been greatly helped by assistance from Mrs. J. R. Brookling, particularly in consolidating the sections of the chapters.

In individual sections, the authors would like to acknowledge the following.

André, Noilhan, Goutorbe: Help from Pavel Kabat when preparing the overall structure of the paper.

Beljaars and Viterbo: Permission from the American Meteorological Society to reproduce Figures 6.4 to 6.7.

Oki: Tenure as a visiting scientist at the Laboratory for Atmospheres, NASA Goddard Space Flight Center. The visit was supported by JSPS (Japan Society for the Promotion of Science) Postdoctoral Fellowships for Research Abroad in collaboration with USRA (Universities Space Research Association).

Rosen: Support by the Climate Dynamics Program of the US National Science Foundation under grant ATM-9223164 and the NASA EOS project under grant NAGW-2615.

Schmitt: Support by the US National Oceanic and Atmospheric Administration (NOAA award No. NA47GP0188) and the National Science Foundation (Grant OCE-9520375).

Stephens: Support by the Cooperative Centre for Southern Hemispheric Meteorology, which is in turn supported under the Australian Government Department of Industry, Science and Technology. Also support by ONR Contract No. N00014-91-J-422 P0002, NOAA Contract NA67RJ0152, NASA Grant NAG8-981, and DOE DE-FG03-94ER6l748.

Trenberth & Guillemot: Support by the Tropical Oceans Global Atmosphere Project Office under grant NA97AANRG0208, and NASA under NASA Order No. W-18,077.

K. A. Browning and R. J. Gurney

1 The global energy and water cycles

1.1 The global energy cycle

Richard D. Rosen

Introduction

Given the simple periodic forcing represented by solar radiation at the top of the atmosphere, it is rather remarkable to witness the extraordinary range in fluctuations present in the global climate system, fluctuations that span time and space on scales from millennia to seconds and from global to centimeters. This complexity arises not only from the dynamical nonlinearities inherent in the fluid components of the global system but also from the myriad interactions among those components, the land, and the biosphere. The existence of water in all three phases throughout the system adds further complexity, for water exerts a major influence on the transformation and exchange of the solar energy supplied to the system.

A framework for dealing with the complexities of the Earth system and predicting its response to perturbations comes to us from classical physics and involves tracing the flow of energy through the system, i.e. diagnosing the global energy cycle. Identifying the key processes involved in this cycle offers the promise of understanding how and why the system evolves. Because much of the energy flow in the Earth system is accomplished by atmospheric circulations, the study of the global energy cycle has been central to considerations of large-scale atmospheric behavior since at least the pioneering efforts of Starr (1951) and Lorenz (1955). The subject is well embedded in atmospheric survey texts (e.g. Wallace and Hobbs, 1977; Grotjahn, 1993) and is a focus of recent, extensive monographs (Peixoto and Oort, 1992; Wiin-Nielson and Chen, 1993). In light of this coverage elsewhere, no attempt at a comprehensive treatment of the global energy cycle is made here. Instead, the aim is to provide a brief overview of the subject while at the same time pointing out shortcomings that currently exist in our ability to observe and model the energy cycle.

Diagnoses of the energy cycle have yielded a general appreciation for the fundamental workings of the global atmosphere, but it is sobering to recognize how uncertain the estimates of many of the cycle's main components remain to this day. While such uncertainties may not have prevented a qualitative understanding of the mean state of the global system from being developed, they are a major hindrance to the sort of quantitative assessments needed today as attention turns to anomalies, both short-term and long, in the system. New observational techniques and a new willingness to address global-scale problems may help reduce these uncertainties, and so focusing attention on some of them here seems timely.

The global energy balance

The primary energy source and sink for the Earth are solar (shortwave) and terrestrial (longwave) radiation, respectively. Over the long term these two must balance, given that the planet is observed to be in thermal near-equilibrium. The manner in which the radiant energies are absorbed, emitted, scattered, or reflected is not constrained by this global equilibrium, however, and is vital in determining the response of the global system to the radiative forcing. It has become customary to display the various elements involved in the annual-mean global energy balance schematically; an example is given in Figure 1.1.

Of the 100 units of solar shortwave radiation incident at the top of the atmosphere (representing an irradiance of around 340 W m^{-2}), 31 units are reflected and scattered back to space by clouds, cloud-free air, and the surface, this proportion being the mean albedo of the planet. The remainder is absorbed partly in the atmosphere but, because of the atmosphere's large transparency to solar radiation, primarily by the surface, where it can be transformed into other forms of energy. A combination of longwave (i.e., infrared) radiation and sensible and latent heat fluxes are then returned from the surface to balance the absorbed solar radiation, as shown on the right-hand side of Figure 1.1. Of the longwave radiation emitted by the surface, most is absorbed in the atmosphere

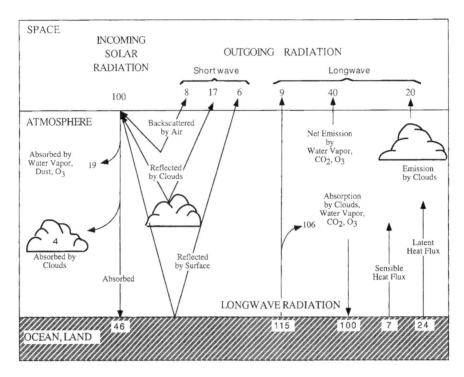

Figure 1.1. Schematic diagram of the annual-mean global energy balance. Units are percent of incoming solar radiation. The solar (shortwave) fluxes are shown on the left-hand side, and the terrestrial (longwave) fluxes are on the right-hand side. (From Mitchell, 1989.)

by clouds, water vapor, and other trace gases, only to be re-emitted out to space or back to the surface, this last process giving rise to the so-called greenhouse effect. Were it not for the additional surface warming caused by the re-radiation of infrared energy by the atmosphere, the mean surface temperature of the planet would be some 33 K colder.

Two points regarding Figure 1.1 are worth highlighting. The first concerns the ubiquitous influence of water in the global energy balance: water vapor and clouds interact with the radiation streaming through the system in critical ways. For example, more than half of the planetary albedo is due to clouds, and most of the infrared radiation emitted by the atmosphere is due to water vapor. Moreover, most of the net energy transferred from the surface to the atmosphere is in the form of latent heat associated with changes in phase of water substance. In combination with the smaller flux of sensible heat from the surface, the latent heat flux is responsible for maintaining a much cooler surface than would exist in the absence of these fluxes: if infrared radiation were the only mechanism to maintain the surface in thermal equilibrium, the surface would need to be more than 50 K hotter! From the viewpoint of the atmosphere, latent heat flux from the surface is the major process compensating for the roughly 2°C day^{-1} cooling that would otherwise result

from the net loss of radiation by the atmosphere in Figure 1.1.

The second aspect of Figure 1.1 deserving comment regards the approximate nature of many of the values assigned to the various processes in the schematic. Although the values are based largely on observations, these may be incomplete in sampling or other respects and are often not direct. Hence, other versions of Figure 1.1 in the literature report different values, and although the discrepancies may appear slight, they can represent large amounts of energy, particularly when compared with the perturbations being studied in connection with global change. For example, according to the estimate reported in the figure, the amount of solar radiation absorbed by clouds and the rest of the atmospheric column is on the order of 80 W m^{-2}, but other estimates are as much as 15 W m^{-2} lower (Kiehl and Trenberth, 1997; Table 1). Some observations (Ramanathan et al., 1995; Cess et al., 1995, 1996) suggest, however, that the true value for solar absorption by the entire cloudy-sky column could be as much as 85 W m^{-2}, indicating that differences of 20 W m^{-2} or so exist in our current understanding of the shortwave radiation budget. Incorporating such differences in climate models can be expected to have a large impact on their climatologies, because of the attendant changes that would

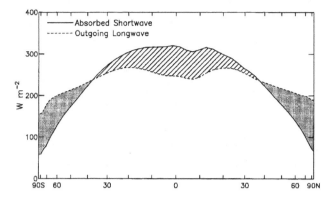

Figure 1.2. Zonally averaged annual mean top-of-the-atmosphere radiation from Earth Radiation Budget Experiment (ERBE) measurements for 1988 as a function of latitude. Shown are the absorbed shortwave (solar) radiation (solid line) and outgoing longwave radiation (dashed line), with their difference, representing the net radiational forcing, highlighted. (From Trenberth and Solomon, 1994.)

occur in other parts of the energy budget, such as the latent heat flux and, therefore, in the models' hydrologic cycles (see, for example, Kiehl *et al.*, 1995). The possibility of enhanced shortwave absorption by clouds is a controversial issue, however (Wiscombe, 1995; Arking *et al.*, 1996; Crisp, 1997), and measurement campaigns are under way to help resolve it. Clearly, effort is needed in placing all the values shown in Figure 1.1 on as firm a foundation as possible.

Flow of energy in the climate system: the role of atmospheric and oceanic circulations

The schematic of Figure 1.1 provides a globally averaged view of the Earth's energy budget, but important consequences for the planet's climate, including the general circulation of the atmosphere, arise from latitudinal differences in this budget. Figure 1.2 displays meridional profiles of the solar radiation absorbed by the Earth–atmosphere–ocean system (i.e., the incoming solar radiation minus that reflected back to space) and of the longwave radiation emitted by the planet out to space. In principle, the integral under each curve should be the same if thermal equilibrium exists, although Trenberth and Solomon (1994) note that the satellite data used to create Figure 1.2 contain a small imbalance of around 4 W m^{-2}. The most striking, and critical, aspect of the figure, though, relates to the very different shapes of the two profiles. Because of the Earth's shape and orbital characteristics, the amount of solar energy reaching the top of the atmosphere varies sharply with latitude: values near the Equator are more than a factor of two larger than those near the poles in the annual mean.

This contrast is even more pronounced for the absorbed solar radiation plotted in the figure, because of the larger albedo in (snow- and ice-covered) polar regions than in the tropics. Outgoing terrestrial radiation, on the other hand, is solely a function of the temperature at which it is emitted and is therefore considerably less dependent on latitude.

The resulting difference between the two curves, i.e., the net radiation, demonstrates that most of the planet is not in local radiative equilibrium. At low latitudes, the amount of radiation absorbed exceeds that emitted; the reverse is true at high latitudes. To achieve a balance in each region, the atmosphere and ocean transport energy from the tropics towards the poles. Within the atmosphere, the mode of transport is governed by the nature of the forces acting to accelerate air parcels, including the pressure gradient force that arises from the meridional contrast in heating. A thermally direct circulation, the 'Hadley cell', develops in the tropics in response to the heating contrast, but the temperature gradient across mid-latitudes is of such a magnitude relative to the effects of the Earth's rotation that large-scale waves become a more efficient means of transporting heat. This combination of a Hadley regime in the tropics and an eddy regime in the extratropics constitutes the general circulation of the atmosphere.

The energy transported by the planet's fluid envelope can assume a number of forms. The main forms in the atmosphere are internal energy I, potential energy Φ, kinetic energy K, and latent energy associated with (liquid-vapor) phase transitions of water L. Expressions for these energy forms per unit mass are:

$$I = c_v T \tag{1.1.1}$$
$$\Phi = gz \tag{1.1.2}$$
$$K = \tfrac{1}{2}(u^2 + v^2 + w^2) \approx \tfrac{1}{2}\boldsymbol{v}\cdot\boldsymbol{v} \tag{1.1.3}$$
$$L = Lq \tag{1.1.4}$$

where c_v is specific heat at constant volume, T temperature, g acceleration due to gravity, z geopotential height above the surface, u zonal wind, v meridional wind, w vertical velocity, $\boldsymbol{v}$ horizontal wind vector, L latent heat of evaporation, and q specific humidity. It can be shown that because the atmosphere is very nearly in hydrostatic equilibrium, the internal and potential energies in a vertical column are proportional to each other, and so it is customary to consider instead their sum, which for a column is given by

$$\int_0^\infty \rho c_p T \, \mathrm{d}z$$

where ρ is density and c_p is specific heat at constant pressure. This integral represents the enthalpy in an atmospheric column, but it is also sometimes referred to as the total potential energy.

An equation governing the balance of energy in the atmosphere can be derived in isobaric coordinates from the horizontal equations of motion, the thermodynamic equation, the mass continuity equation, and the moisture equation. The resulting energy balance equation may be written in differential form as

$$\frac{\partial}{\partial t}(c_pT + Lq + K) + \nabla\cdot(s + Lq + K)\,\boldsymbol{v} + \frac{\partial}{\partial p}(s + Lq + K)\omega = Q \quad (1.1.5)$$

where p is pressure, ω vertical velocity in isobaric coordinates, $s = c_pT + \Phi$ dry static energy, and Q diabatic heating involving transfers of heat due to radiative, sensible, latent, and/or frictional processes. The dry static energy s incorporates mechanical work done by the pressure force (which appears as a flux of potential energy in the equation) and, when combined with the latent energy, forms the moist static energy ($h = s + Lq$) of the atmosphere.

Because the kinetic energy is several orders of magnitude smaller than the total potential energy of the atmosphere, its contribution in the above equation can be neglected. Upon integrating the equation vertically through the depth of the atmosphere, we then obtain

$$\frac{1}{g}\int\frac{\partial}{\partial t}(c_pT + Lq)\mathrm{d}p + \frac{1}{g}\int\nabla\cdot h\boldsymbol{v}\,\mathrm{d}p = F_\mathrm{T} - F_\mathrm{B} \quad (1.1.6)$$

where F_T and F_B are the net downward fluxes of energy at the top and bottom of the atmosphere, respectively. Note that F_T consists simply of the net radiative flux at the top of the atmosphere, but F_B will include radiative, sensible, and latent heat fluxes at the surface. The equation above states that the net vertical flux of energy into an atmospheric column ($F_\mathrm{T} - F_\mathrm{B}$) can act either to change the total atmospheric energy of the column (the first term on the left-hand side) or to induce horizontal transports of moist static energy out of the column. Because the meridional contrast in net radiative heating (Figure 1.2) dominates the shaping of $F_\mathrm{T} - F_\mathrm{B}$, the energy balance equation is often further integrated over a polar cap volume whose lateral boundary is a latitudinal wall, so that meridional transports of energy become an explicit focus:

$$\frac{\partial}{\partial t}\int(c_pT + Lq)\mathrm{d}m = \frac{1}{g}\iint_{\mathrm{wall}} hv\,\mathrm{d}x\mathrm{d}p + (F_\mathrm{T} - F_\mathrm{B}) \quad (1.1.7)$$

where $\mathrm{d}m$ is an element of atmospheric mass and $\mathrm{d}x$ is unit distance in the zonal direction. On an annual-mean basis, changes in energy stored within an atmospheric volume are small, so the left-hand side of the equation becomes zero,

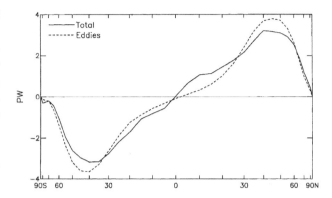

Figure 1.3. Meridional profiles of the annual-mean, total northward transport of moist static energy by the atmosphere (solid line) and of the component due to the sum of transient and stationary eddies (dashed line), derived from analyses of radiosonde observations by Oort and Peixoto (1983) for May 1963 to April 1973.

and a balance is achieved between fluxes of moist static energy across the latitudinal wall of a polar cap volume and the net flux of energy across the volume's upper and lower boundaries.

The annual-mean meridional flux of moist static energy in the atmosphere can be computed from upper-air analyses of wind, temperature, and humidity fields based on observations taken at the global network of radiosonde stations. An example is shown in Figure 1.3 from the analysis of Oort and Peixoto (1983) for the 10-year period from May 1963 through April 1973. In addition to the total flux, the figure also includes the contribution made by eddies to the total. Note that the total flux peaks in mid-latitudes where the region of excess net radiation in the tropics is separated from the region of net radiation deficit in higher latitudes (cf. Figure 1.2). Eddies are clearly the dominant mechanism in the atmosphere for accomplishing the required annual-mean poleward transport of energy. At the latitudes of peak flux, almost half of the transport is in the form of latent energy (Figure 1.4), once again attesting to the importance of water and water vapor to the global energy cycle.

The estimates of meridional energy fluxes shown in Figures 1.3 and 1.4 are subject to numerous uncertainties related to the imperfect sampling of the atmosphere by the radiosonde network. Stations are especially scarce over the oceans and tropical continents, adversely affecting the calculation of net meridional fluxes, particularly in low and southern latitudes. Approaches using the global analyses of operational weather forecast centers, such as the US National Centers for Environmental Prediction or the European Centre for Medium-range Weather Forecasts (ECMWF), have become popular alternatives, because such analyses utilize observations not only

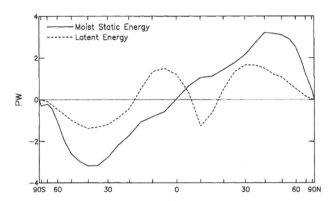

Figure 1.4. Meridional profiles of the annual-mean, total northward transport of moist static energy by the atmosphere (solid line) from Figure 1.3 and of the part due to the total transport of water vapor and its latent energy (dashed line), derived from analyses of radiosonde observations by Oort and Peixoto (1983) for May 1963 to April 1973.

from radiosondes but also from other available platforms, and blend all the observations together in a dynamically consistent manner with a 'first guess' produced by a numerical weather prediction (NWP) model. The output from these operational analyses is available in a convenient, gridded format, allowing calculations to be made readily. Differences in the net meridional flux of moist static energy between radiosonde-based and operational NWP-based analyses can amount to 1 PW (1 PW = 10^{15} W) or more (Michaud and Derome, 1991). Whereas it is tempting to assign most of this discrepancy to errors in the radiosonde-based values, substantial shortcomings also exist in the operational analyses (Trenberth and Solomon, 1994). These include the strong dependence of the analyses upon the verisimilitude of the NWP model in data-sparse regions and the assumptions made in blending values from the model with observations.

Determining the meridional energy flux for the oceans from direct observations is even more problematic than for the atmosphere, because of the general scarcity of ocean measurements. A few direct oceanic calculations do exist, however, such as the value of 2.0 PW across 24°N reported by Bryden (1993). A more common approach has been to infer the ocean transport indirectly from top-of-the-atmosphere radiation budget considerations. This approach takes advantage of the atmospheric energy balance equation written earlier, namely,

$$\frac{1}{g}\int \nabla \cdot h\,\boldsymbol{v}\,\mathrm{d}p = F_{\mathrm{T}} - F_{\mathrm{B}} \qquad (1.1.8)$$

(where the storage term has been disregarded) and the direct measurements of the atmospheric flux to evaluate the left-

hand side of this equation. Satellite measurements provide estimates of F_{T}, so that F_{B} can then be obtained as a residual. Because one may assume that over an annual period the land is in thermal equilibrium and there are no horizontal transports of heat within the land, F_{B} equals zero there. Hence, the value of F_{B} obtained as a residual must apply entirely over the oceans. With this boundary condition and the assumption that thermal equilibrium is maintained within the oceans, the required ocean transport of heat may be inferred, usually on the basis of zonal-mean values.

The indirect estimate of ocean heat transport incorporates uncertainties in the calculation of atmospheric heat transports mentioned above, and it is also sensitive to small biases in the satellite measurements of solar and terrestrial radiation at the top of the atmosphere. Both sources of error can together easily account for a one petawatt or so difference between most indirect and direct estimates of ocean heat transport. Note that because $F_{\mathrm{B}} = 0$ over land, the last equation reduces to

$$\frac{1}{g}\int \nabla \cdot h\,\boldsymbol{v}\,\mathrm{d}p = F_{\mathrm{T}} \text{ (over land)} \qquad (1.1.9)$$

so that a balance should exist between the atmospheric energy flux divergence and F_{T} over land. Trenberth and Solomon (1994) point out, however, that this balance is, in fact, not generally satisfied in the atmospheric data sets normally used to infer the ocean heat flux, and so zonal-mean calculations of this flux are flawed. Instead, Trenberth and Solomon (1994) solve a Poisson equation for the *local* divergence of the ocean heat flux that is forced by F_{B} and is subject to the boundary condition of no flux through the continental boundaries. Although adjustments to the results are still necessary because of errors in the residual F_{B} field, this method leads eventually to zonal-mean meridional ocean heat fluxes that agree better with the direct ocean estimates. Although it is possible that the problem of the 'missing petawatt' can thus be resolved by recognizing the shortcomings present in the atmospheric measurements and analyses, it is important to remember that considerable uncertainties still exist in both the indirect and direct ocean heat transport estimates (Gleckler, 1993; Trenberth and Solomon, 1994).

In summary, the error bars on global energy transports by the atmosphere and ocean remain large, and concerted efforts will be required to reduce them to the point where useful information about the variability of the energy budget on climate-related time scales can be obtained. Nevertheless, the results of Trenberth and Solomon (1994) for an annual mean are illuminating (Figure 1.5). It is clear that both atmosphere and oceans play important roles in effecting the pole-

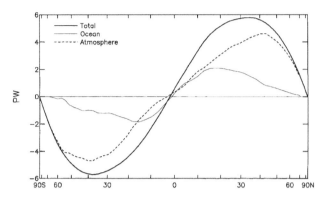

Figure 1.5. Meridional profile of the annual-mean, total northward transport of moist static plus kinetic energy required on the basis of top-of-the-atmosphere (TOA) satellite radiation measurements for 1988 (heavy solid line), along with profiles of the contributions to the total energy flux from the ocean (thin solid line) and the atmosphere (dashed line). The ocean contribution is inferred from the net local fluxes of energy across the surface of the ocean determined using the TOA satellite measurements and ECMWF atmospheric analyses; the total ocean meridional transport so inferred is then adjusted south of 30° S to require it to be zero at 68° S. The atmospheric profile in the figure is simply the residual of the other two curves. Values have been taken from Trenberth and Solomon (1994, portions of Figs. 16 and 17).

ward transport of energy required by the planet's meridional radiation imbalance. Indeed, within the tropics, the two fluxes are comparable in magnitude.

The atmospheric energy cycle: the atmosphere as a heat engine

Although because of its relatively small size the kinetic energy of the atmosphere could be neglected when considering the flow of total energy across the globe, winds and the processes that maintain them are of central importance to the dynamics of the climate system. Aside from shaping life on land and driving the ocean's circulation, winds are the agent, after all, for transporting energy from its source in the tropics to its sink at higher latitudes. Understanding the manner in which the kinetic energy of the winds is maintained against frictional dissipation, on a variety of time scales relevant to climate, is therefore critical.

A now-classical approach to framing this topic is to view the atmosphere as a heat engine in which the large-scale motions of the atmosphere are driven by the potential energy arising from the pole-to-Equator temperature difference. A key to quantifying the workings of this engine is to recognize that only a small portion of the total potential energy in the atmosphere can possibly be converted to kinetic energy, giv-

ing rise to the concept of 'available potential energy'. Lorenz (1955) derived an expression for available potential energy in terms of the variance of pressure on isentropic surfaces, but more importantly from the standpoint of numerous diagnostic studies thereafter, he also provided an approximate formula involving the variance of temperature on isobaric surfaces. Because meteorological observations are commonly made on constant pressure surfaces, the approximate expressions for available potential energy and its generation have seen considerable use. Valid objections to the approximate formulation have been raised, although Siegmund (1994) demonstrates that using the approximate formula has only a small impact on the globally averaged value for the generation of available potential energy. Concerns about the interpretation of individual elements of the Lorenz energy cycle formalism have also been expressed, and alternative approaches have been proposed, but the intuitive appeal of the Lorenz framework outlined below remains compelling.

Because the largest horizontal temperature contrasts are in the meridional direction, the atmospheric energy cycle formalism is normally reckoned in terms of zonal mean quantities; departures from zonal symmetry are also accounted for explicitly to recognize the major role played by large-scale extratropical waves in the energy cycle. With this partitioning in mind, (approximate) expressions for available potential energy (P) and kinetic energy (K) become:

$$P_{M} = c_{p} \frac{\gamma}{2} \int ([\bar{T}]^2 - \tilde{\bar{T}}^2) \, \mathrm{d}m \qquad (1.1.10)$$

$$P_{E} = P_{TE} + P_{SE} = c_{p} \frac{\gamma}{2} \int ([\overline{T'^2}] + [\overline{T^{*2}}]) \, \mathrm{d}m \qquad (1.1.11)$$

$$K_{M} = \frac{1}{2} \int ([\bar{u}]^2 + [\bar{v}]^2) \, \mathrm{d}m \qquad (1.1.12)$$

$$K_{E} = K_{TE} + K_{SE} = \frac{1}{2} \int ([\overline{u'^2} + \overline{v'^2}] + [\bar{u}^{*2} + \bar{v}^{*2}]) \, \mathrm{d}m \qquad (1.1.13)$$

where the subscripts M and E refer to zonal-mean and eddy components, respectively, the latter being further separated in transient eddy (TE) and standing eddy (SE) contributions. Also in the above expressions, an overbar refers to a time mean, a prime is a deviation from the time mean, brackets denote a zonal mean, an asterisk is a departure from the zonal mean, a tilde is an average over an isobaric surface, and γ is a measure of the global mean static stability. Note that the temperature variability over the globe that helps constitute P has been separated into a component measuring the variance between latitudes (P_{M}) and a component measuring the variance within a latitude belt (P_{E}).

The equations governing the balance for these four energy forms can be written symbolically as:

$$\frac{\partial P_M}{\partial t} = G(P_M) - C(P_M, K_M) - C(P_M, P_E) \qquad (1.1.14)$$

$$\frac{\partial P_E}{\partial t} = G(P_E) - C(P_E, K_E) + C(P_M, P_E) \qquad (1.1.15)$$

$$\frac{\partial K_M}{\partial t} = C(P_M, K_M) + C(K_E, K_M) - D(K_M) \qquad (1.1.16)$$

$$\frac{\partial K_E}{\partial t} = C(P_E, K_E) - C(K_E, K_M) - D(K_E) \qquad (1.1.17)$$

where $G(x)$ is the rate of generation of x, $C(x,y)$ the rate of conversion from x into y, and $D(y)$ the rate of dissipation of y. Expressions for each of the terms on the right-hand side of the balance equations have been presented by Peixoto and Oort (1974) and may be written (approximately) as

$$G(P_M) = \int \overline{\gamma(Q - \tilde{Q})(T - \tilde{T})} \, dm \qquad (1.1.18)$$

$$G(P_E) = \int \gamma [\overline{Q' \, T'} + \overline{Q} * \overline{T}*] \, dm \qquad (1.1.19)$$

$$C(P,K) = - \int \overline{\boldsymbol{v} \cdot \nabla \Phi} \, dm = - \int \overline{\omega \alpha} \, dm \qquad (1.1.20)$$

$$C(P_M, P_E) = -c_p \int \gamma [\overline{v' \, T'} + \bar{v} * \bar{T}*] (\partial[\bar{T}]/a\partial\phi) \, dm \qquad (1.1.21)$$

$$C(K_E, K_M) = \int [\overline{v' \, u'} + \bar{v} * \bar{u}*] \cos\phi \, \frac{\partial([\bar{u}] / \cos\phi)}{a\partial\phi} \, dm \qquad (1.1.22)$$

where $\alpha = \rho^{-1}$ is specific volume, ϕ latitude, a Earth's radius, and only the major terms in the last two expressions are included.

From the expressions for $G(P)$, it is clear that adding heat (in the form of radiative, sensible, and/or latent energies) to regions of high temperature and removing heat from regions of low temperature will increase the amount of available potential energy in the atmosphere. It is through this process that the energetics of the atmosphere is seen especially to resemble that of a heat engine, where fuel must be added to sustain the warm furnace and cooling must be provided to maintain the required temperature difference (Peixoto and Oort, 1992a). The expression for the conversion from potential to kinetic energy, $C(P,K)$, reveals that kinetic energy is created only when motions occur in the direction of the pressure-gradient force, i.e. across isobars towards lower pressure. Given the connection between such ageostrophic motions and vertical velocities (through mass continuity and hydrostatic considerations), the second expression for $C(P,K)$ written above is often loosely interpreted as representing the rising of warm air and the sinking of cold air. Such action, which can occur on either zonal mean or eddy scales, results in a lowering of the center of gravity of the atmosphere, thereby reducing its potential energy at the

expense of increasing the kinetic energy of horizontal motions.

The remaining two conversion processes, $C(P_M, P_E)$ and $C(K_E, K_M)$, represent the interaction between zonal mean and eddy scales in the atmosphere. Their expressions exhibit a parallel structure in that each involves an eddy transport relative to the gradient of a zonal mean field. In the case of $C(P_M, P_E)$, a transport of sensible heat by transient and standing large-scale waves down the meridional gradient of zonal-mean temperature results in a conversion of available potential energy from that inherent in the Equator-to-pole temperature contrast to that associated with temperature differences around a latitude circle. Coupled with $C(P_E, K_E)$, $C(P_M, P_E)$ may be pictured as depicting the 'baroclinic instability' of the zonal-mean temperature gradient across mid-latitudes and the subsequent growth of large-scale waves that feed off that instability. In the case of $C(K_E, K_M)$, an eddy transport of angular momentum up the meridional gradient of relative angular velocity results in a conversion of kinetic energy from eddy to zonal-mean form. The fact that this is the direction $C(K_E, K_M)$ is generally observed to proceed in the atmosphere attests to the 'barotropic stability' of the zonal-mean subtropical jet streams and the role that the eddies play in maintaining them against frictional dissipation.

The system of equations governing the budgets of P_M, P_E, K_M, and K_E is generally depicted schematically in the form of a so-called Lorenz energy cycle diagram. An example of one such diagram is given in Figure 1.6 (Kung, 1986) for global, annual-mean conditions based on twice-daily analyses of the state of the atmosphere during December 1978 – November 1979 (the year of the First GARP Global Experiment, FGGE). The diagram indicates that most of the annual-mean energy conversion in the atmosphere is associated with its baroclinicity, proceeding from P_M to P_E to K_E. The bulk of the source of this transformation comes from $G(P_M)$, i.e., from diabatic processes that sustain the mean Equator-to-pole temperature gradient. Most of the energy converted into K_E is then dissipated by surface friction, but some of it is transformed into K_M to maintain the zonal-mean jets against frictional dissipation. Although Figure 1.6 suggests that $C(P_M, K_M)$ is important in the budget for K_M, results reported in the literature for this term are quite varied because of its dependence on the difficult-to-measure mean meridional velocity.

Because diabatic processes in the atmosphere are not routinely measured and analyzed, the values for the generation and dissipation terms in Figure 1.6 were inferred as residuals in the budgets for the various energies. Unfortunately, this budget approach does not allow contributions

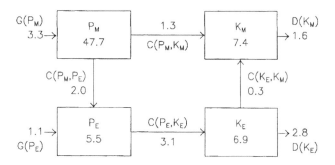

Figure 1.6. Global annual-mean atmospheric energy cycle for the FGGE year (December 1978 to November 1979), from Kung (1986). As defined in the text, available potential (P) and kinetic (K) energies are separated into zonal-mean and eddy components, indicated by the subscripts M and E respectively; units of energy are 10^5 Jm^{-2}. (Note that it is customary in these diagrams to report the values of the various energies themselves rather than their mean time rates of change, as is strictly called for by the balance equations. The latter are typically very small, however, and so can be safely neglected.) Conversions from one energy form to another (symbolized by the letter C), generations of available potential energy (symbolized by G), and frictional dissipations of kinetic energy (symbolized by D) are all reported in units of W m^{-2}.

from different physical processes to be separated for further study. An alternative method which permits such a separation is to utilize the heating fields generated by the 'physics package' of the NWP model used in assimilating the wind and temperature observations. Results from this alternative approach, however, tend to be sensitive to the parameterization schemes used in the physics package and to suffer from sampling and other difficulties as well (Fortelius, 1995). Hence, quantitative assessments of the role of, say, latent heating in generating zonal-mean available potential energy remain highly uncertain, although it does appear (Salstein and Sud, 1994) that this process is the major contributor to the positive, annual-mean value found for $G(P_\mathrm{M})$. This result appears to stand in marked contrast to the role of latent heating within the tropics, where Emanuel *et al.* (1994) suggest that convection acts to damp larger-scale circulations, implying that $G(P_\mathrm{E})$ due to latent heating within the tropics is negative.

To close this section on the atmospheric energy cycle, it is appropriate to raise the issue of how efficient a heat engine the atmosphere is. Lorenz (1967) remarked that 'the determination and explanation of the efficiency [of the atmospheric heat engine] constitute the fundamental observational and theoretical problems of atmospheric energetics', and this perspective remains valid today. The efficiency η of the atmosphere is constrained to be relatively small by the pole-to-

Equator temperature difference: for the ideal case of a Carnot engine,

$$\eta = (T_\mathrm{W} - T_\mathrm{C}) / T_\mathrm{W}$$

where T_W is the temperature of the warm source and T_C that of the cold sink, which would make η less than about 10% for the atmosphere. Other definitions of η are appropriate, however; for example, one may regard η to be the ratio of the rate at which kinetic energy is produced by the atmosphere to the rate at which solar energy reaches the top of the atmosphere, in which case η is on the order of only 1%.

Regardless of precisely how the efficiency is defined, it is intriguing to inquire as to whether the atmosphere is operating as efficiently as it can under present conditions. Phrased differently, one may wonder what factors limit the strength of the winds in the current climate and whether such factors may be altered in a different climate scenario. These questions seem not to be addressed often in the literature, but early works by Schulman (1977) and by Lin (1982) do suggest that the general circulation is operating near maximum efficiency, or at least within a factor of about two of doing so. This conclusion remains tentative, however, and deserves further investigation.

Final remarks

As noted at the outset, the application of energy principles to the study of the global climate system helps place this study on firm physical grounds. It also illuminates, in broad terms, the workings of the general circulation of the atmosphere. Our ability to measure the flow of energy through the system, however, has not advanced to the stage where we can reliably quantify behavior on a variety of time scales. Significant seasonal variations in the energy cycle exist, of course, and are generally understood, at least qualitatively. Space limitations here do not permit a discussion of these seasonal aspects, but such a discussion would emphasize the moderating influence of the oceans on climate. Because of their large heat capacity, the oceans can release to the atmosphere substantial amounts of heat during winter that they accumulated during summer. In the Northern Hemisphere especially, the contrast between the heat capacities of ocean and land imparts a strong seasonal, and non-zonal, signature to the long-term mean circulation. Variations in ocean heat storage on other time scales are also critical in explaining the existence of such notable interannual fluctuations in atmospheric circulation and climate as those associated with the tropical El Niño/Southern Oscillation phenomenon (Philander, 1990).

Our understanding of the energetics of the climate system

and its variability is limited by the fact that transfers of heat within the system associated with diabatic processes remain notoriously difficult to observe and quantify. Uncertainties in the energy generation terms due to these processes can be comparable to the size of the energy conversions which drive the general circulation in Figure 1.6. Techniques for estimating diabatic processes using data assimilation models are emerging, but for the present these estimates are also highly uncertain. The challenge of accurately quantifying anomalies or trends in the global energy cycle is daunting, but with concerted effort, there is reason to believe that it can be met.

1.2 The global water cycle

Taikan Oki

The total volume of water on the Earth is estimated as approximately 1.4×10^{18} m³, and it corresponds to a mass of 1.4×10^{21} kg. Compared with the total mass of the Earth (5.974×10^{24} kg), the mass of water constitutes only 0.02% of the planet, but it is critical for the survival of life on the Earth. There are various forms of water on the Earth's surface. Approximately 70% of its surface is covered with salty water, the oceans. Some of the remaining area (continents) are covered by fresh water (lakes and rivers), solid water (ice and snow), and vegetation (which implies the existence of water). Even though the water content of the atmosphere is comparatively small (approximately 0.3% by mass and 0.5% by volume of the atmosphere), approximately 60% of the Earth is always covered by cloud (Rossow *et al.*, 1993). The Earth is the planet whose surface is dominated by the various phases of water.

Before addressing the water cycle, it is necessary to consider the budget of total water on the Earth. Hot springs and water vapor ejected from volcanoes seem to be the major sources, but most of this is recycled water. The generation of juvenile water from the interior of the solid Earth to its surface has been roughly estimated as 1.0×10^{11} kg year^{-1} (Kuenen, 1963), and is negligible compared with the total mass of water at the Earth's surface. A certain amount of water vapor may be destroyed in the upper atmosphere through photodissociation by solar radiation, but it is also relatively negligible. The H_2O molecule is too heavy to escape from the Earth's gravity. The other source and sink terms are the generation and decomposition of water by respiration and photosynthesis, but they are expected to be balanced. Water is also released by the burning of fossil fuels

$$(CH_2O)_n + nO_2 \rightarrow nCO_2 + nH_2O$$

The same number of water molecules as carbon dioxide molecules is made by this process. With the carbon emissions from fossil fuels at 5×10^{12} kg year^{-1} (Marland and Rotty, 1984), the emission of water should be 7.5×10^{12} kg year^{-1}. This, too, is negligible compared with the total water on the Earth's surface. Altogether, the total amount of water on the Earth can be regarded as constant on the time scale of up to thousands of years that we are concerned with here.

It is also important to consider how much water is associated with each subsystem (reserve) of the water cycle; see Table 1.1 (simplified from a table in Korzun, 1978). The proportion in the ocean is large (96.5%). Other major reserves are solid water on the continent (glaciers and permanent snow

cover) and ground water. Ground water in Table 1.1 includes both gravitational and capillary water. Gravitational water is water in the unsaturated zone (vadose zone) which moves under the influence of gravity. Capillary water is water in the soil above the water table by capillary action, a phenomenon associated with the surface tension of water in soils acting as porous media. Ground water in Antarctica (roughly estimated as 2×10^6 km³) is excluded from Table 1.1. The amount of water stored transiently in a soil layer, in the atmosphere, and in river channels is relatively minute, and the time spent through these subsystems is short, but, of course, they play dominant roles in the global hydrologic cycle.

The objective of this section is to illustrate the global water cycle as quantitatively as possible by means of the latest global data sets. Some of the values and distributions presented here may be different from those in other chapters, emphasizing the uncertainty in our current knowledge of the global water cycle. The precise numbers in the figures and tables will change in the future through the development and use of more dense and comprehensive observing systems, but the framework of the global water cycle that is presented is believed to be valid.

Nature of the water cycles in the climate system

The global hydrologic cycle is one of the key elements in the global environment. Any change in precipitation, evapotranspiration, or runoff may have serious effects for societal activities, and quantitative estimation of the current and future hydrologic cycle is crucial for planning purposes. Climate change, caused either by natural variability of the climate system or by the increases of carbon dioxide and other greenhouse gases by anthropogenic activities, may have important impacts on the water cycles.

The water cycle plays many important roles in the climate system via its various subsystems. Figure 1.7 schematically illustrates the subsystems of the water cycle. Values are taken from Table 1.1 and also calculated from the precipitation estimates by Xie and Arkin (1996). Precipitable water, water vapor transport, and its convergence are estimated using ECMWF (European Centre for Medium-range Weather Forecasts) objective analyses, obtained as 4-year means from 1989 to 1992. The roles of these subsystems in the climate system are now briefly introduced.

Table 1.1. *World water reserves.*

Form of water	Covering area (km²)	Total volume (km³)	Mean depth (m)	Share (%)
World ocean	361 300 000	1 338 000 000	3 700	96.539
Glaciers and permanent snow cover	16 227 500	24 064 100	1 463	1.736
Ground water	134 800 000	23 400 000	174	1.688
Ground ice in zones of permafrost strata	21 000 000	300 000	14	0.0216
Water in lakes	2 058 700	176 400	85.7	0.0127
Soil moisture	82 000 000	16 500	0.2	0.0012
Atmospheric water	510 000 000	12 900	0.025	0.0009
Marsh water	2 682 600	11 470	4.28	0.0008
Water in rivers	148 800 000	2 120	0.014	0.0002
Biological water	510 000 000	1 120	0.002	0.0001
Total water reserves	510 000 000	1 385 984 610	2 718	100.00

Note: Simplified from Table 9 of Korzun (1978).

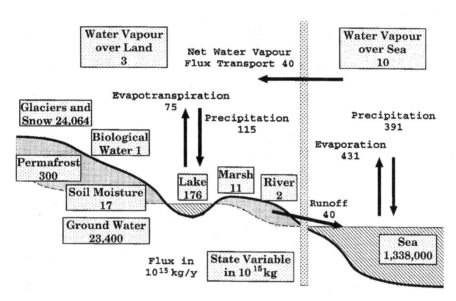

Figure 1.7. Schematic illustration of the water cycle on the Earth. Values are taken from Table 1.1 and calculated from atmospheric water vapor data by ECMWF and precipitation by Xie and Arkin (1996) for 1989–92.

The atmosphere carries *water vapor*, which influences the heat budget as latent heat. Condensation of water releases latent heat, heats up the atmosphere, and affects the atmospheric general circulation. Water vapor is also the major absorber in the atmosphere of both shortwave and longwave radiation.

Liquid water in the atmosphere is another result of condensation. *Clouds* significantly change the radiation in the at-mosphere and at the Earth's surface. *Precipitation* drives the hydrologic cycle on the land surface and changes surface salinity over the ocean.

Snow has special characteristics compared with rainfall. Snow may be accumulated, the albedo of snow is quite high (as high as clouds), and the surface temperature will not rise above 0°C until the completion of snow melt. Consequently the existence of snow changes the surface

11

energy budget enormously. A snow surface typically reduces the aerodynamic roughness, so that it may also have a dynamical effect on the atmospheric circulation.

Evaporation is the return flow of water from the surface to the atmosphere and gives the latent heat flux from the surface. The amount of evaporation is determined by both atmospheric and hydrologic conditions. From the atmospheric point of view, the fraction of incoming solar energy to the surface leading to latent and sensible heat flux is important. Wetness at the surface influences this fraction because the ratio of actual evapotranspiration to the potential evaporation is reduced due to drying stress. The stress is sometimes formulated as a resistance.

Transpiration is the evaporation of water through stomata of leaves. It has two special characteristics different from evaporation from soil surfaces. One is that the resistance of stomata is related not only to the dryness of soil moisture but also to the physiological conditions of the vegetation through the opening and the closing of stomata. Another is that roots can transfer water from deeper soil layers than in the case of evaporation from bare soil. Vegetation also modifies the surface energy and water balance by altering the surface albedo and by intercepting precipitation and evaporating this rain water.

Soil moisture influences the energy balance at the land surface as a lack of available water suppresses evapotranspiration and as it changes surface albedo. Soil moisture also alters the fraction of precipitation partitioned into direct runoff and percolation. The water accounted for in runoff cannot be evaporated from the same place, but the water infiltrated into soil layers may be evaporated again.

Ground water is the subsurface water occupying the saturated zone. It contributes to runoff in its low-flow regime, between floods. Deep ground water may also reflect the long-term climatological situation.

Runoff returns water to the ocean which may have been transported in vapor phase by atmospheric advection far inland. The amount of water mass carried by rivers is smaller than that carried by the atmosphere and oceans, but yet it is not negligible. The runoff into oceans is also important for the freshwater balance and the salinity of the oceans.

Ocean is a giant subsystem of the global water cycle. Even though classical hydrology has traditionally excluded ocean processes, the global hydrologic cycle is never closed without including them. The ocean circulation carries huge amounts of energy and water. The surface ocean currents are driven by surface wind stress, and the atmos-

phere itself is sensitive to the sea surface temperature. Temperature and salinity determine the density of ocean water, and both factors contribute to the overturning and the deep ocean general circulation.

The global water cycle unifies the subsystems consisting of the state variables (precipitable water, soil moisture, etc.) and the fluxes (precipitation, evaporation, etc.).

Water balance requirements

The conservation law of water mass in any arbitrary control volume implies a water balance. In this section, the water balance of land surface, atmosphere, and their combination are presented. Some applications are also introduced.

Water balance at land surface

In the field of hydrology, river basins have commonly been selected for study, and water balance has been estimated using ground observations, such as precipitation, runoff, and storage in lakes and/or ground water.

The water balance at land surface is described as

$$\frac{\partial S}{\partial t} = -\nabla_{\mathrm{H}} \cdot \vec{R}_o - \nabla_{\mathrm{H}} \cdot \vec{R}_u - (E - P) \qquad (1.2.1)$$

where S represents the water storage within the area, $\vec{R}_o$ is surface runoff, $\vec{R}_u$ is the ground water movement, E is evapotranspiration, and P is precipitation. The term $\nabla_{\mathrm{H}} \cdot$ represents the horizontal divergence. S includes snow accumulation in addition to soil moisture, ground water, and surface water storage including retention water. These terms are shown in Figure 1.8(*a*). If the area of water balance is set within an arbitrary boundary, $\nabla_{\mathrm{H}} \cdot \vec{R}_o$ represents the net outflow of water from the region of consideration (i.e., the outflow minus total inflow from surrounding areas). Generally it is not easy to estimate ground water movement $\vec{R}_u$, and the net flux per unit area within a large area is expected to be comparatively small. In this section, all ground water movement is considered to be that observed at the gauging point of a river ($\nabla_{\mathrm{H}} \cdot \vec{R}_u = 0$), and equation (1.2.1) becomes:

$$\frac{\partial S}{\partial t} = -\nabla_{\mathrm{H}} \cdot \vec{R}_o - (E - P) \qquad (1.2.2)$$

Water balance in the atmosphere

It is known in the field of climatology that atmospheric water vapor flux convergence gives water balance information that can complement the traditional hydrologic elements such as

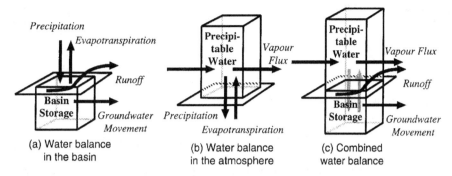

(a) Water balance
in the basin

(b) Water balance
in the atmosphere

(c) Combined
water balance

Figure 1.8. (*a*) terrestrial water balance, (*b*) atmospheric water balance, and (*c*) combined atmosphere–land surface water balance. (*a*), (*b*), and (*c*) correspond to equations (1.2.2), (1.2.3), and (1.2.10), respectively.

precipitation, evapotranspiration, and discharge. The basic concept and an application of using atmospheric data to estimate the terrestrial water balance was presented by Starr and Peixoto (1958).

The atmospheric water balance is described by the equation

$$\frac{\partial W}{\partial t} + \frac{\partial W_c}{\partial t} = -\nabla_{\mathrm{H}} \cdot \vec{Q} - \nabla_{\mathrm{H}} \cdot \vec{Q}_c + (E-P) \qquad (1.2.3)$$

where W represents precipitable water (i.e., column storage of water vapor), W_c is the column storage of liquid and solid water, $\vec{Q}$ is the vertically integrated two-dimensional water vapor flux, and $\vec{Q}_c$ is the vertically integrated two-dimensional water flux in the liquid and solid phases. The water vapor flux vector $\vec{Q}$ may be broken down into two components of Q_λ and Q_ϕ, where λ and ϕ are longitude and latitude, respectively. The components are directed towards east and north

$$Q_\lambda \equiv \int_0^{p_0} qu\frac{\mathrm{d}p}{g} \qquad (1.2.4)$$

$$Q_\phi \equiv \int_0^{p_0} qv\frac{\mathrm{d}p}{g} \qquad (1.2.5)$$

$$W \equiv \int_0^{p_0} q\frac{\mathrm{d}p}{g} \qquad (1.2.6)$$

where q, u, v, g, p, and p_0 represent specific humidity, E–W and N–S wind components, gravitational acceleration, pressure, and pressure at the surface. Water vapor flux divergence ($\nabla_{\mathrm{H}} \cdot \vec{Q}$) may be computed assuming that the Earth is a sphere of radius R_e:

$$\nabla_{\mathrm{H}} \cdot \vec{Q} = \frac{1}{R_e \cos \phi} \left(\frac{\partial Q_\lambda}{\partial \lambda} + \frac{\partial Q_\phi \cos \phi}{\partial \phi} \right) \qquad (1.2.7)$$

Generally, the water content in the atmosphere in the solid and liquid phases is small and will be neglected; then equation (1.2.3) can be simplified as:

$$\frac{\partial W}{\partial t} = -\nabla_{\mathrm{H}} \cdot \vec{Q} + (E-P) \qquad (1.2.8)$$

This balance is schematically illustrated in Figure 1.8(*b*).

Combined atmosphere–river basin water balance

Equations (1.2.2) and (1.2.8) can be combined into:

$$-\frac{\partial W}{\partial t} - \nabla_{\mathrm{H}} \cdot \vec{Q} = (P-E) = \frac{\partial S}{\partial t} + \nabla_{\mathrm{H}} \cdot \vec{R}_o \qquad (1.2.9)$$

Figure 1.8(*c*) illustrates the balance in this equation.

The following further assumptions are often employed in annual water balance computations:

Annual change of atmospheric water vapor storage is negligible $\left(\frac{\partial W}{\partial t} = 0 \right)$.

Annual change of water storage at land is negligible $\left(\frac{\partial S}{\partial t} = 0 \right)$.

With these assumptions, equation (1.2.9) simplifies to:

$$-\nabla_{\mathrm{H}} \cdot \vec{Q} = (P-E) = \nabla_{\mathrm{H}} \cdot \vec{R}_o \qquad (1.2.10)$$

If a river basin is selected as the water balance region, $\nabla_{\mathrm{H}} \cdot \vec{R}_o$ is simply the discharge from the basin. In the simplified equation (1.2.10), the water vapor convergence, 'precipitation-evaporation', and net runoff are equal over the annual period.

Estimation of large-scale evapotranspiration

The equation

$$E = \frac{\partial W}{\partial t} + \nabla_H \cdot \vec{Q} + P \qquad (1.2.11)$$

obtained from equation (1.2.8) is applicable over periods shorter than a year. If atmospheric data with precipitation data are available over short time scales such as months or days, evapotranspiration can be estimated at the corresponding time scales, subject of course to severe limitations imposed by the accuracy of the data. The region over which the evapotranspiration is estimated is not limited to a river basin but depends only on the scales of the available atmospheric and precipitation data.

Estimation of total water storage in a river basin

Equations (1.2.2) and (1.2.8) give

$$\frac{\partial S}{\partial t} = -\frac{\partial W}{\partial t} - \nabla_H \cdot \vec{Q} - \nabla_H \cdot \vec{R}_o \qquad (1.2.12)$$

which indicates that the change of water storage on land can in principle be estimated from atmospheric and runoff data. Although an initial value is required to obtain the absolute value of storage, the atmospheric water balance can be useful in estimating the seasonal change of total water storage in large river basins.

Estimating the zonally averaged net transport of fresh water

The meridional distribution of the zonally averaged annual energy transports by the atmosphere and the oceans have been evaluated, even though there are quantitative problems in estimating such values (Trenberth and Solomon, 1994). However, the corresponding distribution of water transport has not often been studied although the cycles of energy and water are closely related. Wijffels *et. al.* (1992) used values of $-\nabla_H \cdot \vec{Q}$ from Bryan and Oort (1984) and discharge data from Baumgartner and Reichel (1975) to estimate the freshwater transport by oceans and atmosphere, but their results seem to have large uncertainties and they did not present the freshwater transport by rivers.

The annual freshwater transport in the meridional (north–south) direction can be estimated from $-\nabla_H \cdot \vec{Q}$ and river discharge with geographical information such as the location of river mouths and basin boundaries. The governing equations of the transports, crossing at latitude (ϕ_0), can be written separately for over land $R^L(\phi_0)$ and over sea $R^S(\phi_0)$ as:

$$R^L(\phi_0) = -\int_{-\frac{\pi}{2}}^{\phi_0} \oint_{\text{Land}} R_e^2 \cos\phi \nabla_H \cdot \vec{Q} \, d\lambda d\phi - \int_{-\frac{\pi}{2}}^{\phi_0} D(\phi) d\phi \qquad (1.2.13)$$

$$R^S(\phi_0) = -\int_{-\frac{\pi}{2}}^{\phi_0} \oint_{\text{Sea}} R_e^2 \cos\phi \nabla_H \cdot \vec{Q} \, d\lambda d\phi + \int_{-\frac{\pi}{2}}^{\phi_0} D(\phi) d\phi \qquad (1.2.14)$$

where $D(\phi)$ represents the total discharge from continents to the oceans at latitude ϕ, and $\oint_{\text{Land}} d\lambda$ and $\oint_{\text{Sea}} d\lambda$ indicate zonal integration over land and sea, respectively.

Errors and limitations

The procedures for obtaining water balance estimates introduced above may seem attractive; however, one should be careful to recognize the inaccuracies in the estimates. The equations in (1.2.9) are valid, but the equations (1.2.11) to (1.2.14) can give reliable estimates only when accurate input data are available. As Trenberth and Guillemot (1995) have shown, there are large uncertainties in the $\nabla_H \cdot \vec{Q}$ field. Characteristics of atmospheric data, such as spatial and temporal-sampling or vertical resolution, and also the treatment of the lower boundary in the computation, limit the accuracy of estimated $\nabla_H \cdot \vec{Q}$. They showed large discrepancies of $\nabla_H \cdot \vec{Q}$ between the estimates from two sets of data assimilated by ECMWF and NMC (National Meteorological Center, USA; currently called NCEP, National Centers for Environmental Prediction).

Rasmusson (1977) suggested that the method of estimating regional water balance using vapor flux convergence derived from the operational rawinsonde network and current observational schedules, should be useful and relatively accurate for the climatological estimates over areas larger than 10^6 km² and over periods longer than a month. In the case of estimating E as a residual of $-\nabla_H \cdot \vec{Q}$ and P, however, the errors in both terms affect the results. There are many studies in progress that seek to observe or to estimate precipitation over large spatial scales using radar, satellite remote sensing, and surface measurements, but, as explained in Section 4.1, it is not easy to obtain reliable estimates.

In summary, one should be very careful about the confidence limits of the results obtained when applying the atmospheric water balance method, equations (1.2.11) to (1.2.14). However, there may be no other way to estimate E or S on the large scale. In the following subsections, the water in the atmosphere and its transport are presented based on such uncertain atmospheric information. Consequently some presented results could be inaccurate and a few are obviously unrealistic. The results should be regarded as illus-

Table 1.2. *Global water balance* $(-\nabla_{H}\cdot\vec{Q} = P - E = R)$ *estimates (mm year^{-1}) over oceans.*

All oceans	Arctic	Indian	Pacific	Atlantic	
−111	50	−250	91	−384	Baumgartner and Reichel (1975)
−132	263	−97	−56	−333	Korzun (1974)
−18	(in Atlantic)	−53	20	−136	Bryan and Oort (1984)
−66	163	−113	12	−190	Masuda (1988), ECMWF–FGGE
−114	175	−147	14	−345	Masuda 1988, GFDL–FGGE
−78	185	−126	6.3	−236	Oki *et al.* (1995*b*), ECMWF 1985–88
−115	208	−198	−6.7	−299	Oki *et al.* (1995*c*), ECMWF 1989–92

Table 1.3. *Global water balance* $(-\nabla_{H}\cdot\vec{Q} = P - E = R)$ *estimates (mm year^{-1}) over continents.*

All	Asia	Eur.	Afr.	N.A.	S.A.	Au.	Ant.	
256	260	255	113	223	611	267	143	Baumgartner and Reichel (1975)
269	281	273	140	258	578	222	157	Lvovitch (1973)
303	300	273	153	315	678	278	164	Korzun (1978)
42	32	−181	7	162	333	−400	43	Bryan and Oort (1984)
152	94	164	63	227	422	56	100	Masuda (1988), ECMWF–FGGE
260	100	91	333	223	850	211	107	Masuda (1988), GFDL–FGGE
165	235	136	−100	263	415	54	112	Oki *et al.* (1995*b*), ECMWF 1985–88
244	244	197	4	318	773	24	130	Oki *et al.* (1995*c*), ECMWF 1989–92

Note: Mean over continents, Asia, Europe, Africa, North America (N.A.), South America (S.A.), Australia, and Antarctica.

trative of what can be obtained now and as providing an indication of future prospects.

Atmospheric water balance estimation

Twice-daily data sets from ECMWF, analyzed objectively by the 4-dimensional assimilation technique, are used in this subsection. Precipitable water W and water vapor flux convergence $-\nabla_{H}\cdot\vec{Q}$ are calculated by equations (1.2.4), (1.2.5), and (1.2.7) using algorithms described in Oki *et al.* (1995*b*). Monthly and annual mean values are integrated from the twice-daily estimates of W, $\vec{Q}$, and $-\nabla_{H}\cdot\vec{Q}$. The sampling effect on such estimates is discussed by Phillips *et al.* (1992). One problem is the diurnal variation which, especially in tropical areas, is very large (e.g., Oki and Musiake, 1994). Trenberth and Guillemot (1995) compared estimates of monthly means based on twice daily data and four times daily data. They found that the differences of W were regional and went up to 0.5 mm throughout the tropics. Such errors can be neglected. The differences of $-\nabla_{H}\cdot\vec{Q}$ were 10–20 mm month^{-1} over a limited tropical region, and rms differences averaged around latitude circles were 30–60 mm month^{-1} re-

gionally. Thus the diurnal cycle is significant for divergence estimates.

The model used in the ECMWF analyses has changed many times; major changes occurred on 1 May 1985 and 2 May 1989. Therefore 4-year means for the periods 1985 to 1988 and 1989 to 1992 were examined for their accuracy. Specific aspects that were checked were:

the negative areas of 4-year mean vapor flux convergence $(-\nabla_{H}\cdot\vec{Q} < 0)$ over land (refer ahead to Figure 1.10*a*),

the water balance $(-\nabla_{H}\cdot\vec{Q})$ over oceans and continents (see Tables 1.2 and 1.3), and

the degree of quantitative correspondence between $-\nabla_{H}\cdot\vec{Q}$ and river runoff (refer ahead to Figure 1.15).

A region with negative annual water vapor flux convergence is where the annual evaporation exceeds the annual precipitation (see equation (1.2.10)). Such a situation may occur over land at some part of an inland river basin or at the downstream end of a large river; however, most of these estimates are probably erroneous. In addition, one has to be careful in that $P - E$ estimated by the model will not generally agree with $-\nabla_{H}\cdot\vec{Q}$ in the 4-dimensional data assimilation cycle, because such a model has its own bias, and an artificial supply or extraction of water vapor is applied during the assimilation cycle. It is believed that improvements in data and in the treatment of the surface may eliminate the negative annual convergence regions over land on the global scale.

Annual water vapor flux convergence estimates are compared with earlier estimates in Table 1.2 and Table 1.3 based on equation (1.2.10). Baumgartner and Reichel (1975), Lvovitch (1973), and Korzun (1978) estimated runoff by hydrological methods using surface observations. Others listed in these tables used the atmospheric water balance method. Note in the case of Bryan and Oort (1984), the Middle East is included in Europe, not in Asia, and the Arctic Ocean is included in the Atlantic Ocean. The signs of $-\nabla_H \cdot \vec{Q}$ in Table 1.2 are mostly the same among the estimates, but the absolute values vary. In the Pacific Ocean, the precipitation and evaporation approximately balance each other. Evaporation exceeds precipitation in the Atlantic and Indian Oceans, and vice versa in the Arctic Ocean. In Table 1.3, results from the hydrologic method produce broadly similar estimates compared with those by the atmospheric water balance method, although estimates by atmospheric data tend to be generally smaller than the hydrologic estimates. The annual river runoff averaged over the area of a continent is approximately 200–300 mm year^{-1} except for South America, where the estimated annual runoff is 400–800 mm year^{-1}. Continents with high aridity (Africa) and with very cold regions (Antarctica) have small runoff.

The global mean continental discharge (see the first column in Table 1.3) of the 1989–92 mean, 244 mm year^{-1}, is closer to the hydrologic estimates than that of the 1985–88 mean (165 mm year^{-1}). From these comparisons, the 4-year mean of 1989–92 is used here as a good representation of our current knowledge of the global aspects of water and its transport in the atmosphere.

Global aspects of atmospheric water vapor storage, transport, and divergence

The annual mean over the global ocean for precipitable water W is estimated as 28.4 mm using ECMWF data from 1989 to 1992. This value is close to the 28.9 mm estimated by Trenberth and Guillemot (1994) using 4-year data of ECMWF from July 1987 to June 1991, but they found this to be larger than that of Special Sensor Microwave/Imager (SSM/I) satellite estimates of 26.8 mm for the same period for the corresponding global ocean. The SSM/I estimates are regarded as more credible, in which case the ECMWF estimates presented below may be overestimated by 10%. The annual mean W over

the globe before any such correction is 26.1 mm, and 21.1 mm over land.

Zonal means (i.e., averages in an east–west direction) of precipitable water, averaged over (a) land and sea, (b) land only, and (c) sea only are shown in Figure 1.9. Annual, December–January–February (DJF), and June–July–August (JJA) means are shown in each figure, and the length of the horizontal axis is proportional to the area at each latitude $(\cos(\phi))$. The fraction of land at each latitude is shown by bars in Figure 1.9(b). Zonal mean values are of interest because the atmosphere is comparatively more uniform in the east–west direction than in the north–south direction. A particular point of interest is to see how the distribution of zonal means departs from north–south symmetry, because the radiative forcing is nearly equal for both hemispheres in the annual mean, even though the solar irradiance is larger by a few percent in January. One of the major differences between the characteristics of the two hemispheres is in the fraction of sea and land. More than 40% is covered by land in the northern hemisphere, but only 20% in the southern hemisphere.

It is worth considering the water cycle over land and sea separately. Sea water has large heat capacity and large volume. Water over oceans can evaporate as required and the changes of surface temperature are relatively small. However, evapotranspiration over land is suppressed with a deficit of water storage over land, and the surface temperature is determined as a result from land–atmosphere coupling of energy and water. Accordingly, the Bowen ratio (the ratio of sensible heat flux to latent heat flux) may differ over land and sea.

The peak of zonal mean W is situated at 10° N for the overall mean (Figure 1.9a) and for the mean over sea (Figure 1.9c). The zonal mean W over land (Figure 1.9b) is nearly symmetrical about the Equator for the annual mean, and the DJF mean is close to the mirror image of the JJA mean between 40° S and 40° N. The water vapor content in the atmosphere is highly sensitive to the temperature through the saturation vapor pressure (e_s), and the temperature (which determines the e_s via a monotonic exponentially increasing function) decreases toward the upper atmosphere. As a result, more than 50% of the water vapor is concentrated below the 850 hPa surface, and more than 90% is confined to the layer below 500 hPa (Peixoto and Oort, 1992b). Therefore we can consider that the W-field corresponds closely to the surface temperature over the ocean (Stephens, 1990). The fact that the difference in mean W between DJF and JJA is larger over land than sea is due to the large annual difference of surface temperature over continents and the relatively small change of sea surface temperature.

Global distributions of vertically integrated water vapor flux convergences $-\nabla_H \cdot \vec{Q}$ are shown in Figure 1.10. Tropical

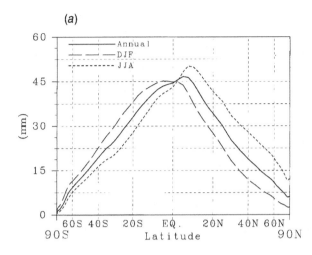

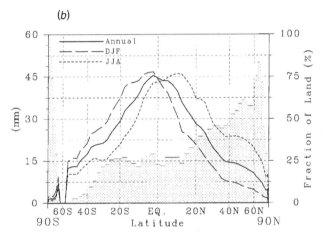

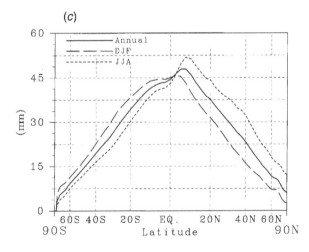

Figure 1.9. Meridional distribution of zonal mean precipitable water W, (a) over land and sea, (b) mean over land only, and (c) mean over sea only. Annual mean, December–January–February (DJF) mean, and June–July–August (JJA) mean for 4 years from 1989 to 1992.

regions are major convergence zones globally. In these areas, precipitation exceeds evapotranspiration. The deficit is made up of water coming from divergence zones, mainly over subtropical oceans. The north–south width of the tropical convergence zone is wider from the Indian Ocean to the western Pacific Ocean than elsewhere. The regions of high convergence generally correspond to the regions of high precipitable water. An exception is the convergence zone at the Pacific coast of North America near 50°N, where there are so-called storm tracks. While subtropical oceans are generally regions of divergence, parts of them include convergence zones, e.g., the South Pacific Convergence Zone (SPCZ) to the northeast of Australia in DJF and the Asian monsoon convergence zone in the western Pacific Ocean in JJA, which corresponds to the Bai-u Front.

The zonal mean $-\nabla_{\mathrm{H}}\cdot\vec{Q}$ (Figure 1.11) shows relatively small seasonal changes at latitudes higher than 40°S and 40°N. Over land, in the regions within 20° of the Equator, the surface stores water from an excess of precipitation over evapotranspiration during the summer season and returns moisture to the atmosphere during the winter season. The situation is reversed between 40°N and 70°N. The values of $-\nabla_{\mathrm{H}}\cdot\vec{Q}$ for DJF and JJA over land are fairly symmetric with positive (summer hemisphere) and negative (winter hemisphere) peaks at roughly 15°. However, over the sea, there are two negative peaks, at 20° for the winter hemisphere and at 25° for the summer hemisphere, but the positive peak of the Intertropical Convergence Zone (ITCZ) stays at 5°N to 10°N for the whole year. The peak is larger in JJA than DJF, which reflects the fact that the activity of the ITCZ is higher in the northern hemisphere's summer.

Global aspects of precipitation and evaporation

The global distributions of water fluxes at the surface are briefly introduced in this subsection. To illustrate the global distribution of precipitation, gridded global precipitation data by Xie and Arkin (1996) for the years 1989 to 1992 were employed. There are other published estimates, but all the long-term averages agree well over land. Large discrepancies exist over the oceans, but it is thought that estimates that include satellite information should be better. Xie and Arkin (1996) merged satellite data with rain-gauge observations and model estimates, and they also discussed the accuracy of their estimates. These are the reasons why their estimates are used here.

Zonal mean precipitation P is shown in Figure 1.12; overall, over land, and over sea. Annual precipitation overall and over sea have the largest peak at 10°N; however, the peak of annual precipitation over land stays at the equator. Zonal

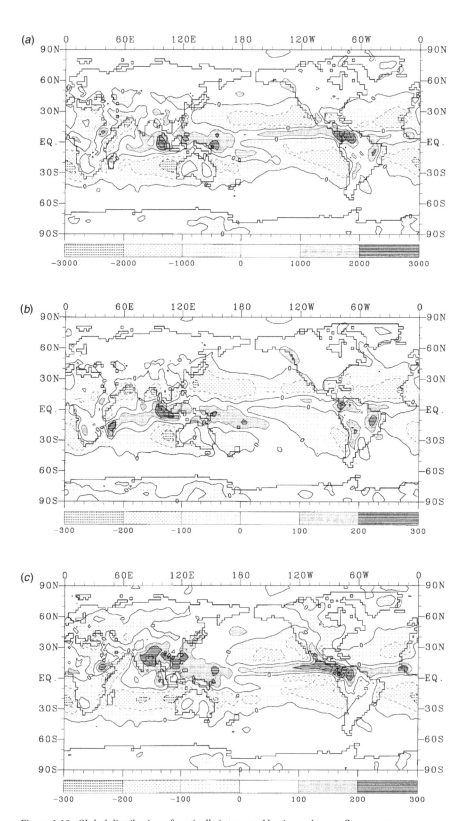

Figure 1.10. Global distribution of vertically integrated horizontal vapor flux convergence $-\nabla_H \cdot \vec{Q}$; (a) annual mean (mm yr⁻¹), (b) DJF mean (mm month⁻¹), and (c) JJA mean (mm month⁻¹). Four-year mean from 1989 to 1992 estimated from ECMWF 4 DDA.

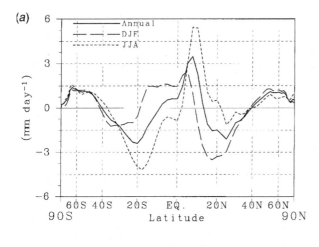

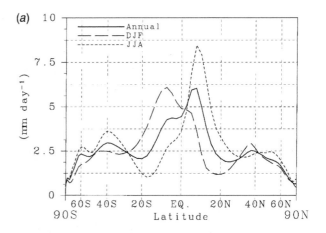

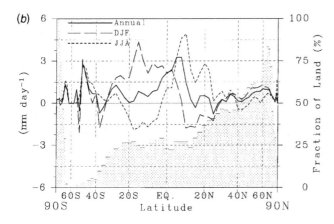

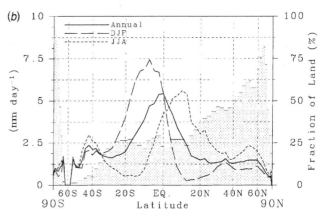

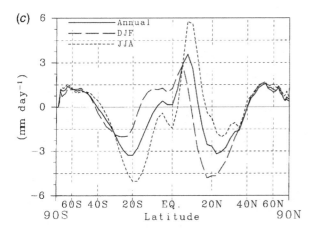

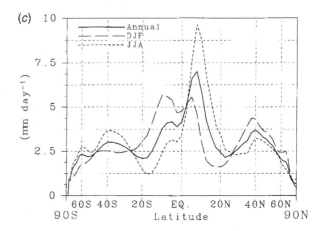

Figure 1.11. Meridional distribution of vertically integrated horizontal vapor flux convergence $-\nabla_H \cdot \vec{Q}$ for (a) over land and sea, (b) mean over land only, and (c) mean over sea only. Annual mean, DJF mean, and JJA mean for 4 years from 1989 to 1992.

Figure 1.12. Meridional distribution of zonal-mean precipitation by Xie and Arkin (1996) for (a) mean over land and sea, (b) mean over land only, and (c) mean over sea only.

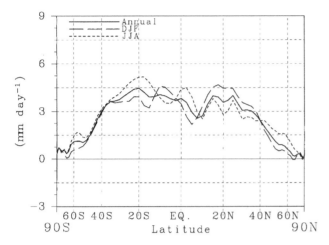

Figure 1.13. Meridional distribution of overall zonal mean evapotranspiration E. Estimated as residuals of $-\nabla_H \cdot \vec{Q}$ and precipitation P.

mean precipitation over sea has secondary peaks at 40°N and 40°S, which are larger in the winter time. However, over land P does not have any dominant secondary peak, and the amount is larger in summer. In the global distribution of annual precipitation (not shown), very limited parts of continents in Southeast Asia and South America have more than 1500 mm year^{-1}. As can be seen from the zonal mean, precipitation along the mid-latitude storm tracks over the North Pacific and Atlantic oceans are stronger in DJF than in JJA. The ITCZ is enhanced in JJA, when the southeastern part of the Asian continent is covered by the southwest monsoon rainfall.

Evapotranspiration E was estimated from equation (1.2.11), and the zonal mean distribution for the globe is shown in Figure 1.13. The distribution of E is less dependent on latitude and has smaller seasonal changes compared with P (see also Trenberth and Guillemot, 1997). The mean evapotranspiration in tropical areas is approximately 4 mm day^{-1}. Because E is calculated as the residual of $-\nabla_H \cdot \vec{Q}$ and P estimates, the reliability may not be high. The zonal mean error of P is estimated as 50–100% by Xie and Arkin (1996), and the error of $-\nabla_H \cdot \vec{Q}$ should be the same order or more. Consequently, E in Figure 1.13 may have serious errors, and the values are subject to change when improved data sets become available.

Monthly water balance of major river basins

For the better understanding of the global water cycle over land, the monthly water balance in major river basins will now be examined. River runoff data observed at gauging stations are obtained from the Global Runoff Data Centre (GRDC, 1992) and UNESCO (1969). Long-term mean values are calculated by averaging over the available data period at each station. Seventy rivers which have basin areas greater than approximately 100 000 km^2 were selected. The continental land mass on 2.5° grid boxes was divided manually into river basins by referring to published atlases. The annual runoff was mapped according to this geographical information with the result shown in Figure 1.14. As discharge is a point measurement, it should be reasonably accurate. According to WMO (1994), the recommended accuracy (uncertainty level) expressed at the 95% confidence limits is 5% for the measurements of discharge.

In most major river basins, the annual runoff R is less than 400 mm year^{-1}, with a few exceptions in Southeast Asia and South America. The distribution of R is basically similar to the global distribution of P. From the annual atmospheric water balance requirement, equation (1.2.10), the runoff distribution (Figure 1.14) should coincide with that of $-\nabla_H \cdot \vec{Q}$ (Figure 1.10a). The differences between them (without regard to sign) are mapped in Figure 1.15. The poorest correspondence occurs in South Africa, Central and Southeast Asia, and South America. These areas have few observational points and/or large amounts of precipitation. However, there are many basins that show differences of less than 100 mm year^{-1} in the mid- and high-latitude regions of the northern hemisphere. This must reflect the higher quality and density of the observations, especially for the atmosphere.

Figure 1.16 illustrates the relationship between error index (%) and the size of the basin. The error index is derived as $|-\nabla_H \cdot \vec{Q} - R|/R \times 100$ (%) in each river basin. From Figure 1.16, it is clear that some river basins have extremely large discrepancies between $-\nabla_H \cdot \vec{Q}$ and R but that in many river basins the error is less than 100%. The error tends to decrease in larger river basins; on continental scales ($\geq 10^7$ km^2), $-\nabla_H \cdot \vec{Q}$ should be accurate to within a few tens of percentage points.

Normalizing for basin area, the mean $-\nabla_H \cdot \vec{Q}$ in these 70 river basins is approximately 300 mm year^{-1}. This value is larger than that of over the entire land area (i.e., 244 mm year^{-1}), but it is smaller than the mean observed runoff of these specific rivers, which is 314 mm year^{-1} according to GRDC, and 365 mm year^{-1} according to UNESCO. These 70 major river basins cover only 55% of the land surface, yet they represent 70% of the total $-\nabla_H \cdot \vec{Q}$ over land and their mean runoff per unit area is larger than the global average. It indicates that the areas where large rivers exist have comparatively high annual runoff. Therefore it is suggested that the global water balance estimated by an extrapolation from the water balances of large rivers may overestimate the true global value for the continents.

Figure 1.17 shows plots of E against P for major river basins.

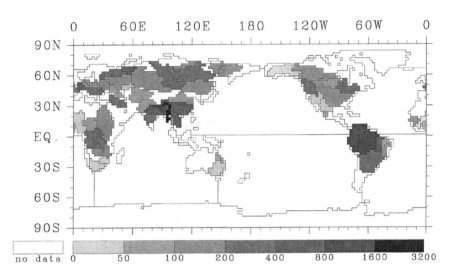

Figure 1.14. Annual runoff $\vec{R}_o$ (mm yr^{-1}) of major rivers. Climatic mean is calculated from GRDC data.

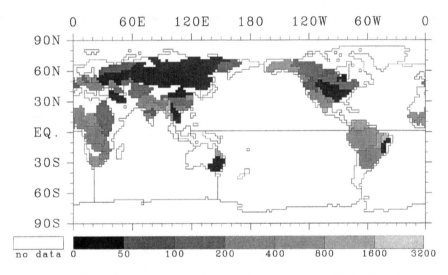

Figure 1.15. Difference between annual water vapor flux convergence estimated from ECMWF objective analysis data (mean from 1989 to 1992) and climatological annual runoff in large river basins, estimated from GRDC data (mm yr^{-1}).

The plotted values are estimated using the geographical information on 2.5° grid boxes. Precipitation is obtained from Xie and Arkin (1996) and E is calculated from P minus the GRDC runoff (i.e., without invoking the atmospheric water balance information of $-\nabla_H \cdot \vec{Q}$). Plots are shown differently for tropical rivers and for other regions, where the radiative forcing is weaker. In the case of mid- and high-latitude river basins, the relation $E \approx 0.7P$ can be observed but with large scatter. In the case of tropical rivers, up to an annual precipitation of 1000 mm year^{-1}, $E \approx P$ and very little runoff is observed. However, E does not increase after P reaches approximately 1000 mm year^{-1}, and excesses above this threshold value become runoff. It roughly reflects the two water balance regimes: water-limited ($P \leq 1000$ mm year^{-1}) and energy-limited ($P \geq 1000$ mm year^{-1}) conditions.

The mean annual $-\nabla_H \cdot \vec{Q}$ does not always match closely with the mean annual runoff, but it should provide a reasonable estimate of total water storage S from equation (1.2.12). Rasmusson (1968) found an accumulated discrepancy of 210 mm within 5 years over continental US and he corrected $-\nabla_H \cdot \vec{Q}$ by the uniform addition of 3.5 mm month^{-1}, and Oki et al. (1991) used a reduction factor of 0.18 for the Chao

21

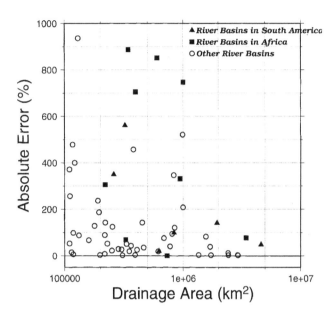

Figure 1.16. Absolute error (%) of annual water vapor flux convergence against climatological annual runoff. Dependency on the basin size is illustrated.

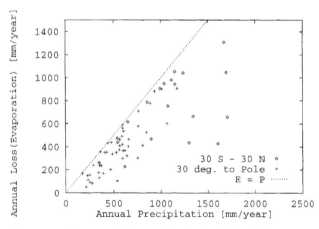

Figure 1.17. Mean water balance of major river basins. P is calculated from Xie and Arkin (1996), and E is calculated as $P - R$ using runoff R from GRDC data.

Phraya River basin in Thailand. Similar adjustments were applied for $-\nabla_H \cdot \vec{Q}$ of the 70 rivers (Oki et al., 1995c). Consequently, from this atmosphere–river basin water balance method, the relative water storage as a function of time of year can be estimated, including the change of snow accumulation, soil moisture, groundwater table, and water storage in river channels.

The monthly water balance of climatological means in five individual river basins are shown in Figures 1.18 through

1.22. P is summarized from Xie and Arkin (1996), E by equation (1.2.11), $\vec{R}_o$ from the Global Runoff Data Centre (GRDC, 1992), and S by equation (1.2.12). The minimum value of S is set to zero. S represents the instantaneous value at the beginning of each month and other values are monthly totals. In the two tropical river basins, the Amazon (Figure 1.18) and the Zaire (Figure 1.19), monthly mean precipitation is above 50 mm month^{-1} throughout the year and the seasonal change of monthly runoff is small. In the case of the Mississippi river basin (Figure 1.20), the characteristics of monthly water balance are similar to that of the tropical rivers, but the excess of E over P from July to September is distinct. This characteristic has also been reported by Roads et al. (1994); however, the total water storage S (noted as W in Roads et al., 1994) is somewhat different. The peak value for S of approximately 75 mm reported by Roads et al. (1994) occurred from March to April, while the peak of 100 mm occurs in June in Figure 1.20. The water balance of the Nile river basin is shown in Figure 1.21, which is relatively dry on average. The rainfall is 20–50 mm month^{-1} and the seasonal change of E, R, and S roughly follows the pattern of P. Finally, as a representative of a high-latitude river basin, the monthly water balance of the Ob river basin is shown in Figure 1.22. S increases until May and suddenly decreases with the increase of R. This corresponds to the snow accumulation and the melting runoff. An excess of E over P is estimated to occur during the warm season from May to July.

The accuracy of the estimates in Figures 1.18 to 1.22 should be questioned, especially the estimates of S, but it is a very difficult problem to compare the total water storage with actual observations on large scales. Field observations are point measurements and their representativeness is a critical issue. Satellite remote sensing can directly measure only shallow soil layers at the surface of nearly bare soil. Even though the realizations of soil moisture or water storage in general circulation models are varied and still uncertain, it may be worthwhile comparing the S obtained by the atmospheric water balance against a GCM result because it gives soil moisture and water storage on a large scale. A numerical experiment under climatological conditions, using the atmospheric general circulation model of the CCSR (Center for Climate System Research, University of Tokyo) and NIES (National Institute for Environmental Studies), produced very good results that correspond well with values of S estimated by the atmospheric water balance method (Oki et al., 1995a). In the case of the Amazon river basin (Figure 1.23), the river water storage calculated in the associated river-routing model has a dominant role in the seasonal change of total water storage (Oki et al., 1996).

The amplitude of the annual change of total water storage

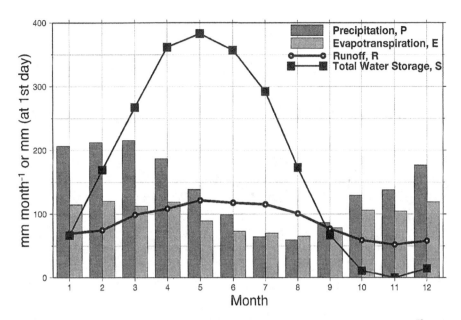

Figure 1.18. Water balance of the Amazon river basin. E and S are estimated using $-\nabla_{H} \cdot \vec{Q}$. P is estimated from Xie and Arkin (1996), and R from GRDC. Catchment areas are 6150 (total) and 4640 (at Obidos) $\times 10^3$ km².

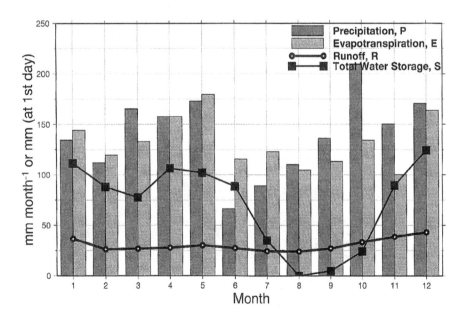

Figure 1.19. Water balance of the Zaire river basin. Catchment areas are 3690 (total) and 3475 (at Kinshasa) $\times 10^3$ km².

in each river basin for which data exist is mapped globally in Figure 1.24. It is large for the Amazon, Yenisey, Lena, and the rivers in Southeast Asia. Although these amplitudes include the change of snow accumulation and storage in river channels etc., roughly speaking, they should correspond to the amplitude of annual soil moisture change; the field capacity may be larger than this range. It is interesting and encouraging that only a few river basins have an annual change of total water storage larger than 200 mm, and the number of 200 mm is comparable to the values of field capacity of soil moisture in the land surface parameterizations of general circulation models.

23

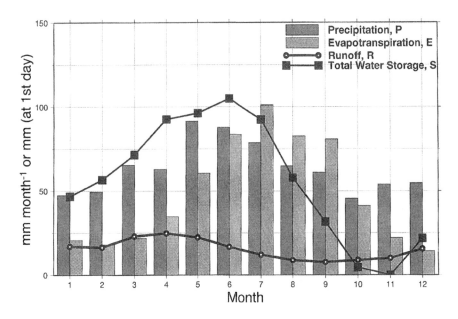

Figure 1.20. Water balance of the Mississippi river basin. Catchment areas are 3248 (total) and 2964 (at Vicksburg) $\times 10^3$ km².

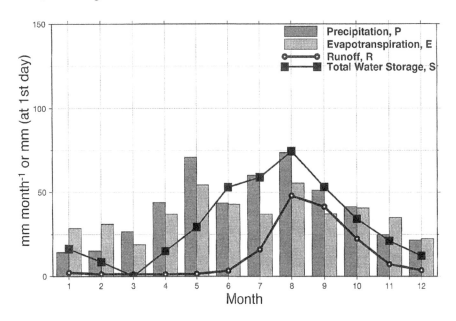

Figure 1.21. Water balance of the Nile river basin. Catchment areas are 3007 (total) and 325 (at Khartoum) $\times 10^3$ km².

Global aspects of freshwater distribution and fluxes into the ocean

The freshwater supply to the ocean has an important effect on the thermohaline circulation because it changes the salinity and thus the density. Annual freshwater transport by rivers and the atmosphere to each ocean is summarized in Table 1.4.

Some part of the water vapor flux convergence remains in the inland basins. There are a few negative values in Table 1.4, suggesting that net freshwater transport occurs from the ocean to the continents. This is physically impossible and is caused by the errors in the source data. Although a detailed discussion of the values in Table 1.4 may not be meaningful, it is nevertheless interesting that such an analysis does make at

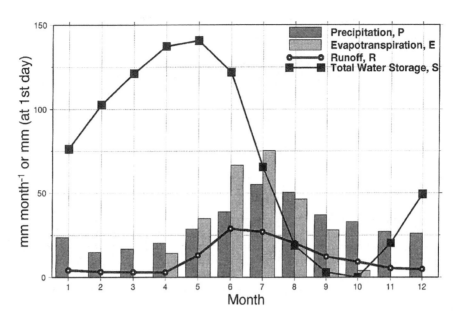

Figure 1.22. Water balance of the Ob river basin. Catchment areas are 2978 (total) and 2950 (at Salekhard) $\times 10^3$ km^2.

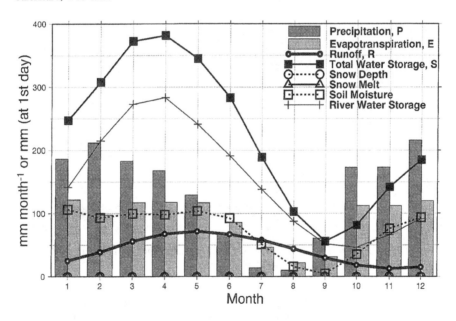

Figure 1.23. Water balance of the Amazon river basin calculated by an atmospheric GCM.

least qualitative sense using the atmospheric water balance method with geographical information on basin boundaries and the location of river mouths. In this analysis, it should be noted that the total amount of freshwater transport into the oceans from the surrounding continents has the same order of magnitude as the freshwater supply that comes directly from the atmosphere, expressed by $-\nabla_H \cdot \vec{Q}$.

The annual freshwater transport in the meridional direction has been estimated from equations (1.2.13) and (1.2.14) with results shown in Figure 1.25. The estimates in Figure 1.25 are the net transport; i.e., in the case of oceans, it is the residual of northward and southward freshwater flux by all ocean currents globally, and it cannot be compared directly with individual ocean currents such as the Kuroshio and the

Table 1.4. *Annual freshwater transport from continents to each ocean (10^{15} kg year^{-1}).*

		N.P.	S.P.	N.At.	S.At.	Indian	Arctic	Inner	Total
From rivers	Asia	4.7	0.4	0.2		3.3	2.7	0.1	11.4
	Europe			1.7		0.0	0.7		2.4
	Africa			− 0.2	0.9	− 0.2		− 0.4	0.1
	N.America	2.9		4.8			1.1		8.8
	S.America	0.5	0.4	5.7	8.3				14.9
	Australia		0.1			0.1			0.2
	Antarctica		1.0		0.1	0.8			1.9
From atmosphere	Total	8.1	1.9	12.2	9.3	4.0	4.5	− 0.3	39.7
	$-\nabla_{\mathrm{H}}\cdot\vec{Q}$	9.9	− 11.1	− 12.7	− 14.0	− 14.0	2.2		− 39.7
	Grand total	18.0	− 9.2	− 0.5	− 4.7	− 10.0	6.7	− 0.3	0.0

Note: 'Inner' indicates the runoff to the inner basin within Asia and Africa.
$-\nabla_{\mathrm{H}}\cdot\vec{Q}$ indicates the direct freshwater supply from the atmosphere to the ocean.
N.P., S.P., N.At., and S.At. represent North Pacific, South Pacific, North Atlantic, and South Atlantic Ocean.

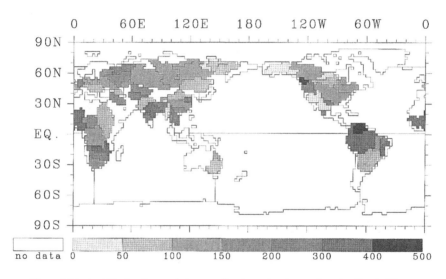

Figure 1.24. Amplitude of annual change of basin water storage (mm) estimated from ECMWF (1989–1992) and mean GRDC data.

Gulf Stream. Transport by the atmosphere and by the ocean have almost the same absolute values at each latitude but with different sign. The transport by rivers is about 10% of these other fluxes globally (this may be an underestimate because $-\nabla_{\mathrm{H}}\cdot\vec{Q}$ tends to be smaller than river discharge observed at land surface). The negative (southward) peak by rivers at 30°S is mainly due to the Parana River in South America, and the peaks at the equator and 10°N are due to rivers in south America, such as the Magdalena and Orinoco. Large Russian rivers, such as the Ob, Yenisey, and Lena, carry the fresh water towards the north between 50–70°N.

These results indicate that the hydrologic processes over land play non-negligible roles in the climate system, not only by the exchange of energy and water at the land surface, but also through the transport of fresh water by rivers which affects the water balance of the oceans and forms a part of the hydrologic circulation on the Earth between the atmosphere, continents, and oceans.

Future prospects

The aim of Section 1.2 was not so much to show authoritative figures concerning the global aspects of the water cycle, but

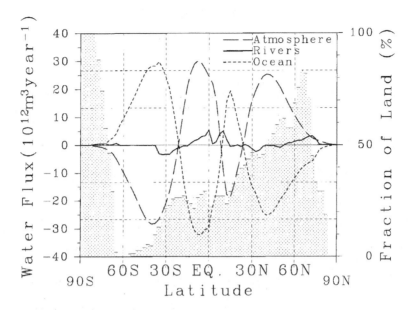

Figure 1.25. The annual freshwater transport in the meridional direction by atmosphere, ocean, and rivers (land) (mean for 1989–1992). Water vapor flux transport of 20×10^{12} m³ year⁻¹ corresponds to approximately 1.6×10^{15} W of latent heat transport.

to introduce a possible methodology for investigating the global water cycle as quantitatively as possible, with tentative results for examples. Nevertheless these examples illustrate the overall characteristics of the global water cycle.

It is expected that the global water cycle will be estimated more accurately in the future, but this will not be accomplished easily. Improvement in physical parameterization and in the horizontal and vertical resolution of the models used in the 4DDA will contribute substantially, as will further developments of observational systems.

It will be worth considering the application of the atmospheric water balance method in future projects that focus on the land surface–atmosphere interactions. This method has the capability to provide the large (macro)-scale constraints for land-surface modeling. For intensive hydrologic field observations, it is desirable to design an observational network that includes Doppler radars and wind profilers as well as rawinsondes, to enclose the region of concern. Then, the atmospheric water balance will provide a better estimate of the areal mean evapotranspiration and the change of total water storage in the area. The required time intervals between atmospheric observations depend on the spatial scale, and the proper resolution of the diurnal cycle is critical. The relationship $v\Delta t \leq \Delta x$ determines the required time increments. Assumption of a wind velocity v of 10 m s⁻¹ and horizontal distance Δx of 100 km leads to a requirement for observations at intervals Δt of 3 hours or less. The present twice daily (i.e., $\Delta t = 12$ hour) data cannot provide adequate estimates for changes on horizontal scales of less than 400–500 km.

References

Arking, A., Chou, M.-D. and Ridgway, W. L. (1996). On estimating the effect of clouds on atmospheric absorption based on flux observations above and below cloud level. *Geophys. Res. Lett.*, **23**, 829–32.

Baumgartner, F. and Reichel, E. (1975). *The World Water Balance: Mean Annual Global, Continental and Maritime Precipitation, Evaporation and Runoff*. Ordenbourg, Munchen, 179 pp.

Bryan, F. and Oort, A. (1984). Seasonal variation of the global water balance based on aerological data. *J. Geophys. Res.*, **89**, 11717–30.

Bryden, H. L. (1993). Ocean heat transport across 24N latitude. Interactions between global climate subsystems, The Legacy of Hann, *Geophys. Monogr.* 75, IUGG 15, Amer. Geophys. Union, 65–75.

Cess, R. D., Zhang, M. H., Minnis, P., *et al.* (1995). Absorption of solar radiation by clouds: observations versus models. *Science*, **267**, 496–9.

Cess, R. D., Zhang, M. H., Zhou, Y., Jing, X. and Dvortsov, V. (1996). Absorption of solar radiation by clouds: interpretations of satellite, surface, and aircraft measurements. *J. Geophys. Res.*, **101**, 23299–309.

Crisp, D. (1997). Absorption of sunlight by water vapor in cloudy conditions: a partial explanation for the cloud absorption anomaly. *Geophys. Res. Lett.*, **24**, 571–4.

Emanuel, K. A., Neelin, J. D. and Bretherton, C. S. (1994). On large-scale circulations in convecting atmospheres. *Quart. J. Roy. Meteor. Soc.*, **120**, 1111–43.

Fortelius, C. (1995). Inferring the diabatic heat and moisture forcing of the atmosphere from assimilated data. *J. Climate*, **8**, 224–39.

Gleckler, P. J. (1993). The partitioning of meridional energy transport between the ocean and the atmosphere. Ph.D. thesis, U. California at Davis, 161 pp.

GRDC (1992). *Second Workshop on the Global Runoff Data Centre. Report 1*. Bundesanstalt fur Gewasserkunde, Federal Institute of Hydrology, Germany, 96 pp.

Grotjahn, R. (1993). *Global Atmospheric Circulations: Observations and Theories*. Oxford Univ. Press, 430 pp.

Kiehl, J. T., Hack, J. J., Zhang, M. H. and Cess, R. D. (1995). Sensitivity of a GCM climate to enhanced shortwave cloud absorption. *J. Climate*, **8**, 2200–12.

Kiehl, J. T. and Trenberth, K. E. (1997). Earth's annual global mean energy budget. *Bull. Amer. Meteor. Soc.*, **78**, 197–208.

Korzun, V. I. (1978). *World Water Balance and Water Resources of the Earth*. Vol. 25 of *Studies and Reports in Hydrology*, UNESCO. English translation; original 1974.

Kuenen, P. H. (1963). *Realms of Water*. John Wiley & Sons, Inc., 327 p.

Kung, E. C. (1986). Spectral energetics of the global circulation during the FGGE year. In *Preprints, National Conf. on Scientific Results of the First GARP Global Experiment, Miami, Florida*, pp. 84–7. Amer. Meteol. Soc.

Lin, C. A. (1982). An extremal principle for a one-dimensional climate model. *Geophys. Res. Lett.*, **9**, 716–18.

Lorenz, E. N. (1955). Available potential energy and the maintenance of the general circulation. *Tellus*, **7**, 157–67.

Lorenz, E. N. (1967). The Nature and Theory of the General Circulation of the Atmosphere. World Meteor. Org., 161 pp.

Lvovitch, M. I. (1973). The global water balance. *Trans. Am. Geophys. Union*, **54**, 28–42.

Marland, G. and Rotty, R. M. (1984). Carbon dioxide emissions from fossil fuels: a procedure for estimation and results for 1950–1982. *Tellus*, **36B**, 232–61.

Masuda, K. (1988). World water balance; analysis of FGGE IIIb data. In *Tropical Rainfall Measurements*, Theon, J. S. and N. Fugono, pp. 51–5. A. Deepak Publ.

Michaud, R. and Derome, J. (1991). On the mean meridional transport of energy in the atmosphere and oceans as derived from six years of ECMWF analyses. *Tellus*, **43A**, 1–14.

Mitchell, J. F. B. (1989). The ''greenhouse'' effect and climate change. *Rev. Geophys.*, **27**, 115–39.

Oki, T., Kanae, S. and Musiake, K. (1996). River routing in the global water cycle. *GEWEX News*, **6**, 4–5.

Oki, T. and Musiake, K. (1994). Seasonal change of the diurnal cycle of precipitation over Japan and Malaysia. *J. Appl. Meteor.*, **33**, 1445–63.

Oki, T., Musiake, K. Emori, S. and Numaguti, A. (1995*a*). Estimation of hydrological cycle and water balance in large river basins by an atmospheric general circulation model. *Annual Journal of Hydraulic Engineering, JSCE*, **39**, 103–8, in Japanese with English abstract.

Oki, T., Musiake, K., Matsuyama, H. and Masuda, K. (1995*b*). Global atmospheric water balance and runoff from large river basins. *Hydrol. Proces.*, **9**, 655–78.

Oki, T., Musiake, K., Matsuyama, H. and Masuda, K. (1995*c*). Global soil moisture extraction using 4DDA and observational runoff data by combined atmospheric–river basin water balance. In *Second International Symposium on Assimilation of Observations in Meteorology and Oceanography I*, No. 651 in WMO/TD, 355–60, WMO.

Oki, T., Musiake, K. and Shiigai, H. (1991). Water balance using atmospheric data: a case study of Chao Phraya river basin, Thailand. *Mitteilungsblatt des Hydrographischen Dienstes in Osterreich*, **65/66**, 226–30, Wien.

Oort, A. H. and Peixoto, J. P. (1983). Global angular momentum and energy balance requirements from observations. *Adv. Geophys.*, **25**, 355–490.

Peixoto, J. P. and Oort, A. H. (1974). The annual distribution of atmospheric energy on a planetary scale. *J. Geophys. Res.*, **79**, 2149–59.

Peixoto, J. P. and Oort, A. H. (1992*a*). *Physics of Climate*. Amer. Inst. of Phys., 520 pp.

Peixoto, J. P. and Oort, A. H. (1992*b*). *Water cycle*. In *Physics of Climate*, 282, American Institute of Physics.

Philander, S.G. (1990). *El Niño, La Niña, and the Southern Oscillation*. Academic Press, 293 pp.

Phillips, T. J., Gates, W. L. and Arpe, K. (1992). The effects of sampling frequency on the climate statistics of the European Centre for Medium-Range Weather Forecasts. *J. Geophys. Res.*, **96**, 20427–36.

Ramanathan, V., Subasilar, B., Zhang, G. J., *et al.* (1995). Warm pool heat budget and shortwave cloud forcing: a missing physics? *Science*, **267**, 499–503.

Rasmusson, E. M. (1968). Atmospheric water vapor transport and the water balance of north America II. Large-scale water balance investigations. *Mon. Wea. Rev.*, **96**, 720–34.

Rasmusson, E. M. (1977). Hydrological application of atmospheric vapor-flux analyses. *Hydrology Rep. 11*, WMO.

Roads, J. O., Chen, S.-C., Guetter, A. K. and Georgakakos, K. P. (1994). Large-scale aspects of the United States hydrologic cycle. *Bull. Amer. Met. Soc.*, **75**, 1589–610.

Rossow, W. B., Walker, A. W. and Garder, L. C. (1993). Comparison of ISCCP and Other Cloud Amounts. *J. Climate*, **6**, 2394–418.

Salstein, D. A. and Sud, Y. C. (1994). Large-scale diabatic heating in a model with downdrafts. *Preprints, Tenth Conf. on Numerical Weather Prediction, Portland, Oregon*, pp. 72–4. Amer. Meteor. Soc.

Schulman, L. L. (1977). A theoretical study of the efficiency of the general circulation. *J. Atmos. Sci.*, **34**, 559–80.

Siegmund, P. (1994). The generation of available potential energy, according to Lorenz' exact and approximate equations. *Tellus*, **46A**, 566–82.

Starr, V. P. (1951). Applications of energy principles to the general circulation. *Compendium of Meteorology*, 568–76. Amer. Meteor. Soc.

Starr, V. P. and Peixoto, J. (1958). On the global balance of water vapor and the hydrology of deserts. *Tellus*, **10**, 189–94.

Stephens, G. L. (1990). On the relationship between water vapor over the oceans and sea surface temperature. *J. Climate*, **3**, 634–45.

Trenberth, K. E. and Guillemot, C. J. (1994). The total mass of the atmosphere. *J. Geophys. Res.*, **99**, 23088–97.

Trenberth, K. E. and Guillemot, C. J. (1995). Evaluation of the global atmospheric moisture budget as seen from analyses. *J. Climate*, **8**, 2255–72.

Trenberth, K. E., and Solomon, A. (1994). The global heat balance: heat transports in the atmosphere and ocean. *Climate Dyn.*, **10**, 107–34.

UNESCO (1969). *Discharge of Selected Rivers of the World*. Vol. I, II, III, UNESCO.

Wallace, J. M. and Hobbs, P. V. (1977). *Atmospheric Science: An Introductory Survey*. Academic Press, 467 pp.

Wiin-Nielsen, A. and Chen, T.-C. (1993). *Fundamentals of Atmospheric Energetics*. Oxford Univ. Press, 376 pp.

Wijffels, S. E., Schmitt, R. W., Bryden, H. L. and Stigebrandt, A. (1992). Transport of freshwater by the oceans. *J. Phys. Oceanogr.*, **22**, 155–62.

Wiscombe, W. J. (1995). An absorbing mystery. *Nature*, **376**, 466–7.

WMO (1994). *Guide to Hydrological Practices*. WMO-No. 168, World Meteorological Organization.

Xie, P. and Arkin, P. A. (1996). Analyses of global monthly precipitation using gauge observations, satellite estimates, and numerical model predictions. *J. Climate*, **9**, 840–58.

2 Global atmospheric models, data assimilation, and applications to the energy and water cycles

In this chapter we present an introductory overview of atmospheric models (Section 2.1), the numerical methods used for global atmospheric modeling (Section 2.2), the methodology of physical parameterizations (Section 2.3), data assimilation and the relative importance of the different observing systems (Section 2.4), the use of data assimilation products to study the hydrologic and energy cycles, especially using the newly available reanalyses (Section 2.5), and the use of coupled atmospheric/ocean models applied to the prediction of climate anomalies (Section 2.6).

2.1 Atmospheric models for weather prediction and climate simulations

Eugenia Kalnay

Atmospheric models used in meteorology are based on the equations of fluid mechanics discretized in space and time, with geophysical parameters appropriate to the Earth's atmosphere. From the beginning, applications of atmospheric models have been oriented towards both numerical weather prediction (Richardson, 1922), and climate simulations (Phillips, 1956). The quasi-geostrophic approximation (Charney, 1948) was widely used until the early 1960s, and then replaced in most models by the 'primitive equations' (Charney, 1962), which are computationally more expensive but more accurate. These equations are conservation laws applied to individual parcels of air: conservation of the 3-dimensional momentum (equations of motion), conservation of energy (first law of thermodynamics), conservation of dry air mass (continuity equation), and equations for the conservation of moisture in all its phases. For models with a horizontal grid size larger than 10 km, it is customary to replace the vertical component of the equation of motion with its hydrostatic approximation. With this approximation, it is convenient to use atmospheric pressure, instead of height, as a vertical coordinate. For a complete derivation of the equations see, for example, Haltiner and Williams (1980).

When these 'conservation' equations are discretized over a given grid size (typically a few kilometers to several hundred km) it is necessary to add 'sources and sinks' terms due to small-scale physical processes that occur at scales that cannot be explicitly resolved by the models. As an example,

the equation for water vapor conservation on pressure coordinates is typically written as

$$\partial q/\partial t + u\,\partial q/\partial x + v\,\partial q/\partial y + \omega\partial q/\partial p = E - C + \partial\,\overline{(\omega' q')}/\partial p \quad (2.1.1)$$

where q is the mixing ratio (ratio between water vapor and dry air), x and y are horizontal coordinates with appropriate map projections, p pressure, t time, u and v are the horizontal air velocity (wind) components, $\omega = dp/dt$ is the vertical velocity in pressure coordinates, and the primed product represents turbulent transports of moisture on scales unresolved by the grid used in the discretization, with the overbar indicating a spatial average over the grid of the model. It is customary to call the left-hand side of the equation, the 'dynamics' of the model, which are computed explicitly. Major progress has been made over the years in developing numerical discretizations that represent accurately the continuous dynamics, and in particular attacking the problems associated with spherical geometry, as discussed in Section 2.2.

The right-hand side represents the so-called 'physics' of the model, i.e. for this equation, the effects of physical processes such as evaporation(E), condensation(C), and turbulent transfers of moisture which take place at small scales that cannot be explicitly resolved by the 'dynamics'. These subgrid-scale processes, which are sources and sinks for the equations, are then 'parameterized' in terms of the variables explicitly represented in the atmospheric dynamics (Section 2.3).

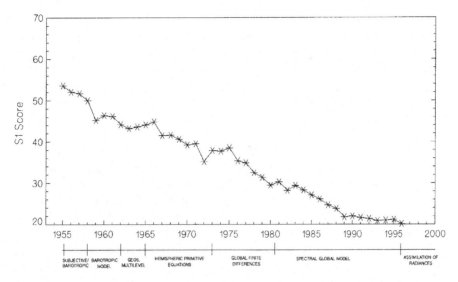

Figure 2.1. Evolution of a long-term skill score for the 500 hPa 36-hour operational forecasts over North America calibrated from the S1 score, which measures the relative error in the forecast of the horizontal pressure gradient. Values of S1 of 0.7 or more have been empirically found to correspond to forecasts which are useless as synoptic guidance, and forecasts with scores of 0.2 or less, to an essentially perfect guidance.

Numerical weather prediction (NWP) provides the basic guidance for modern weather forecasting. Forecasts are performed running (integrating in time) advanced atmospheric models so as to simulate, given today's weather observations, the evolution of the atmosphere in the next few days. The time integration of an atmospheric model is an initial value problem; the ability to make a skillful forecast requires both that the computer model be a realistic representation of the atmosphere, and that the initial conditions represent accurately the state of the atmosphere, a process achieved through data assimilation (Section 2.4). The relative importance of the different components of the observing system is also briefly discussed in this section.

Climate simulations are also performed through a time integration, but the integrations are long enough that the details of the initial conditions are 'forgotten' (Lorenz, 1963). In principle, climate and NWP models are similar, except that climate models are integrated in time much longer (months to years), and therefore they tend to have less spatial resolution than forecast models. At the National Centers for Environmental Prediction (NCEP), for example, the global models currently used for weather forecasting are integrated for two weeks with a horizontal resolution of about 100–200 km, and the coupled ocean–atmosphere models used for seasonal and interannual predictions have a horizontal resolution of 200–300 km (Section 2.6). Additionally, NWP models focus on deterministic predictions, which require accurate initialization of the atmosphere. On the other hand, climate models

deal with atmospheric variability well beyond the time scale of deterministic weather prediction, and focus on atmospheric variability resulting from slowly varying anomalous variations of boundary forcing fields such as sea surface temperature (SST), sea ice, snow cover, and soil moisture. For this reason, for climate models, accurate initialization of the atmosphere is not critical, whereas robust and realistic response to the variations of surface forcing is of central importance.

Another factor that affects both NWP and climate models is their tendency to drift away from the atmospheric climatology, and towards their own model climatology. In NWP forecasts this is reflected as forecast bias (e.g., a model may have excessive westerly winds in the northern hemisphere), and forecast spin-up (e.g., the forecast may take a few days before reaching a level of statistically steady precipitation). For climate models, with very long time integrations, the problem of climate drift is also an important issue. One of the foci of present research is the development of atmospheric models suitable for both weather forecasts and climate simulations.

Over the last decades, the quality of the models and methods for using atmospheric observations has improved continuously, resulting in major forecast improvements. Figure 2.1 shows the longest available record of the skill of numerical weather prediction. The score used measures the percentage error in the horizontal gradient of the height of the constant pressure surface of 500 hPa (in the middle of the atmosphere, since the surface pressure is about 1000 hPa) for

36-hour forecasts over North America. Empirical experience of human forecasters (Shuman, 1989) indicated that a value of the relative error of 70% or more corresponded to a useless forecast, and a score of 20% or less, to an essentially perfect forecast (20% was the average score obtained when comparing analyses hand-made by different experienced forecasters fitting the same observations over the data-rich North America). The figure indicates that current 36-hour 500 hPa forecasts over North America (with a score of just 21% for the 1994 and 1995 annual averages) are able to locate generally very well the position and intensity of the large-scale atmospheric waves, major centers of high and low pressure that determine the general evolution of the weather in the 36-hour forecast. Smaller-scale atmospheric structures, such as fronts, mesoscale convective systems that dominate summer precipitation, etc., are still difficult to forecast in detail, although their prediction has also improved significantly over the years. The model's short range forecasts of precipitation, fluxes of heat, moisture and momentum, vertical velocity, radiation, etc., provide an estimate of many fields that are not directly measured, and can be quite useful in studies of the hydrologic and energy cycles of the atmosphere (Section 2.5).

The improvements in atmospheric models have also resulted in better climate simulations, and in the ability to capture climate anomalies associated with anomalous sea surface temperatures, soil moisture and snow cover. As a result several research efforts were started to extend atmospheric predictions beyond the limit of weather predictability (about two weeks) taking advantage of the long memory of the coupled ocean/land/atmosphere system. Section 2.6 describes research in ocean data assimilation and the ability to predict the El Niño/Southern Oscillation phenomena.

2.2 Numerical approximations for global atmospheric models

David Williamson

In this section we discuss the types of numerical approximations that are used in global atmospheric models today. We emphasize those aspects of the methods that are most important with respect to the spherical geometry of the earth which adds a degree of complexity not found in many disciplines involving numerical approximations. The special characteristics of the water vapor distribution in the atmosphere also introduce difficulties into the modeling arena. For more details about basic numerical methods used in atmospheric models and the concepts mentioned in this section the reader is referred to the textbook by Haltiner and Williams (1980), and the less accessible, but still very valuable GARP Publications Series No 17 (Mesinger and Arakawa, 1976, and Kasahara, 1979) on Numerical Methods Used in Atmospheric Models. For a summary of numerical methods that have been developed for global models and additional references see Williamson (1992).

Atmospheric modeling in general and its water vapor aspects in particular present us with a global problem; the entire global domain must be included in any model integration extending over more than a few days. In time sequences of atmospheric analyses and in long model simulations, coherent tongues of water vapor can be seen moving from the tropics to the polar regions. Even though the turnover time for individual water vapor molecules in the atmosphere is around 10 days, which is less than the time needed for air masses actually to move from the tropics to the polar regions, numerical models must be capable of transporting air masses and maintaining their characteristics for long distances over long periods of time.

Water itself plays a fundamental role in many atmospheric processes. It provides a direct transfer of energy between components of the climate system such as land, sea-ice, ocean and atmosphere. It is the primary greenhouse gas in the atmosphere providing a strong influence on the radiative balance of the overall system and in particular on the radiative heating of the atmosphere. The clouds which form when it is transformed to liquid or ice phase in the atmosphere strongly modulate the radiative heating. There is no hope in modeling the climate of the atmosphere correctly if water vapor is not well modeled.

One important water vapor process that must be well modeled is transport by atmospheric winds. Thus it is important to have accurate estimates of the winds themselves, produced by the approximations to the dynamics of the atmosphere. In addition, given accurate winds, models must also be based on accurate numerical techniques for the transport of water vapor itself. Water vapor is extremely difficult to model accurately because of its large variation in amplitude. It varies by orders of magnitude from Equator to pole and from the Earth's surface to the tropopause. The polar regions, where the amount of water vapor is very small, have been especially difficult to model, yet these regions are extremely important in the climate system.

All global atmospheric models in production use today are based on spherical coordinates (latitude and longitude) and numerical approximations which involve an underlying latitude–longitude grid structure. One such grid is illustrated in Figure 2.2(a). The lines indicate the edges of grid boxes and the grid points can be thought of as being in the centers of the boxes. The type of grid in Figure 2.2(a) is referred to as a uniform latitude–longitude grid. The grid boxes have the same width in longitude everywhere, and the same width in latitude everywhere. The spherical coordinate system has pole points to which the coordinate lines converge, and at which horizontal (relative to the Earth's surface) wind vector co mponents become undefined. These singularities introduce a variety of difficulties into numerical modeling. Collectively they are referred to as the 'pole problem'. The singularity itself is more of an irritant than a major problem and has been successfully dealt with. The so-called 'pole problem' is more economic than technical as will be explained in the following.

Global grids need not be uniform in latitude and longitude. Figure 2.2(b) shows a grid in which the longitudinal grid interval is reasonably constant in terms of physical distance rather than angular distance. Nevertheless, the boxes (and points) are aligned in longitude. Such grids, first introduced by Kurihara (1965) and reintroduced more recently in a different context by Hortal and Simmons (1991), are being used successfully in atmospheric models today with some types of approximations as will also be described below. However, these grids still present difficulties for other types of approximations.

Explicit finite difference approximations

In its simplest form, the transport of a quantity like water vapor may be described by the one-dimensional advection equation

$$\frac{\partial q}{\partial t} + U\frac{\partial q}{\partial x} = 0 \tag{2.2.1}$$

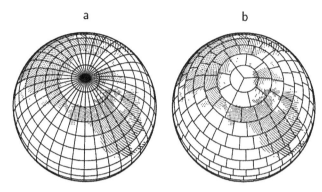

Figure 2.2. (a) Example of a uniform latitude–longitude grid in which all grid boxes have the same width in longitude and the same width in latitude. (b) Example of a reduced grid in which the longitudinal grid interval is reasonably constant in terms of physical distance rather than angular distance.

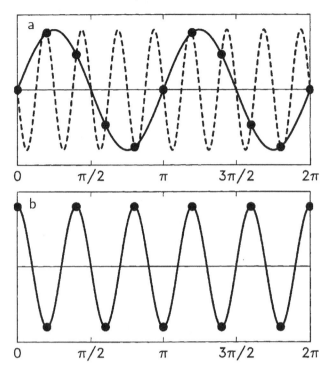

Figure 2.3. (a) Example of a short wave (dashed line) which is not resolved by the gridpoints (circles) and is indistinguishable from the long wave (solid line) at those points. (b) Shortest (2 Δx) wave resolvable by the grid.

where $q(x,t)$ is the transported quantity, x is the independent space coordinate and t is time. For the purposes of discussion we consider the wind U to be constant. The solution of this equation at any future time t is just the initial field shifted downwind a distance Ut, $q(x,t) = q(x - Ut,0)$. The initial structure moves with uniform speed U with no change of shape or amplitude. Of course the atmosphere is more complex than a constant wind field, so we cannot easily use that knowledge to solve the atmospheric situation and must develop approximate methods to solve equation (2.2.1). In order to determine the properties of various approximations, we continue to consider the case of constant U so we can compare approximate solutions to the true solution.

A common approximation to equation (2.2.1) is obtained by replacing the differentials with differences defined over a grid of points such as in Figure 2.2.

$$\frac{q_j^{n+1} - q_j^{n-1}}{2\Delta t} + U\frac{q_{j+1}^n - q_{j-1}^n}{2\Delta x} = 0 \qquad (2.2.2)$$

The shorthand notation q_j^n stands for $q(x_j, n\Delta t)$ where $x_j = (j-1)\Delta x$ are the discrete grid points separated by an interval Δx, and the discrete times are denoted $t^n = n\Delta t$, $n \geq 0$. The approximation (2.2.2) is referred to as centered differences.

Because the solutions are defined at the discrete set of points x_j, small structures whose variation falls between the grid points cannot be distinguished. This is illustrated in Figure 2.3(a) in the case of regular wave-like structures

$$q = Q\sin\left(\frac{2\pi(x - \theta)}{L}\right) \qquad (2.2.3)$$

in which Q is the amplitude, L is the wavelength, the distance

over which the structure goes from 0 to a maximum to 0 to a minimum and back to 0, and θ is the phase, the value of x where the first 0 occurs. In the figure, the short wave drawn with the dashed line is indistinguishable at the discrete grid points, denoted by the circles, from the long wave drawn with the solid line. The shortest wave that can be resolved by the grid has wavelength $L = 2\Delta x$, with a maximum at one grid point, minimum at the next, maximum again at the third, and so on (Figure 2.3b).

The centered approximations (equation 2.2.2) can be solved analytically for the linear case discussed here. From that solution it can be shown that for the amplitude of the wave to remain at its initial and correct value,

$$|U\tfrac{\Delta t}{\Delta x}| \leq 1$$

If

$$|U\tfrac{\Delta t}{\Delta x}| > 1$$

the amplitude will grow exponentially with time. Such erroneous solution behavior is referred to as computational instability because in a short time the amplitude becomes so large that it cannot be represented on a computer and the computation halts. This behavior was first identified by Courant *et*

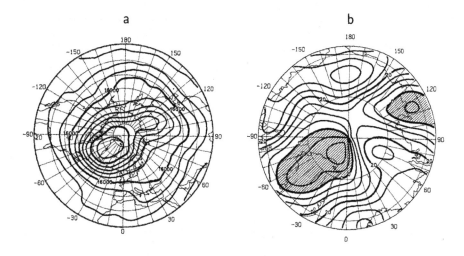

a   b

Figure 2.4. Northern hemisphere 100 mb (*a*) height field and (*b*) zonal wind field for 1200 UCT
3 Jan 1983. The 10° grid boxes indicated by dashed lines are the same in both panels, (*a*) covers
the region poleward of 20 degrees N and (*b*) poleward of 68° N. Contour intervals are 200 m for
height field and 5 m s^{-1} for zonal wind component.

al. (1928) and is referred to as the CFL stability condition for the centered approximations (2.2.2). The combination

$$|U\tfrac{\Delta t}{\Delta x}|$$

is also commonly referred to as the Courant number and is denoted by

$$C = |U\tfrac{\Delta t}{\Delta x}|$$

In solving an equation such as (2.2.1) numerically with approximations (2.2.2) one first defines the grid interval Δx and grid x_j on which the solution will be calculated. The grid interval Δx must be chosen small enough so the phenomena of interest are resolved by the waves with $L \geq 2\Delta x$. The velocity U is determined by the problem and not under our control. Thus, the only remaining parameter Δt must be chosen to satisfy the stability condition.

$$\Delta t \leq \frac{\Delta x}{|U|} \qquad (2.2.4)$$

On a global grid as in Figure 2.2(*a*), the grid distance in the longitudinal (east–west) direction is $\Delta x = \Delta\lambda \cos\varphi$ and $\Delta\lambda$ is uniform, where λ denotes longitude and φ is latitude. The convergence of meridians in a uniform latitude-longitude grid implies that Δx becomes very small near the poles. The zonal wind component u on the other hand, which serves as U in the previous equations, is not necessarily small there. This is illustrated in Figure 2.4 which shows an example of an asymmetric polar vortex. Such structures commonly occur in the atmosphere in the northern hemisphere in winter. The

left side of Figure 2.4 shows the northern hemisphere 100 mb height field for 1200 UCT 3 Jan 1983. The geostrophic relationship implies that the wind is parallel to the height contours with its strength proportional to the distance between contours. Because the center of the vortex is shifted away from the pole the wind is actually blowing across the pole and between 60°E and 90°E is blowing parallel to the 80° latitude circle. The right side of Figure 2.4 shows the analyzed zonal wind u in an enlargement of the polar region (the grid lines are the same in the left and right sides of the figure). The zonal wind u exceeds 25 m s^{-1} at 80° latitude. In addition, at 85° latitude the zonal wind at 75°E is 20 m s^{-1} and is still 15 m s^{-1} at 87.5° latitude.

The effect of such winds near the poles on the maximum allowed time step is illustrated in Figure 2.5 which shows the maximum (over height and longitude) zonal wind speed as a function of latitude for the same case as in Figure 2.4. In both hemispheres the maximum wind is found in mid-latitudes where the jet stream occurs. However, the bound for Δt in the stability condition (equation 2.2.4), $\Delta x/U$, is $\Delta\lambda \cos\varphi/u$ on the global grid. Although u is not excessively large, $\cos\varphi$ becomes so small near the poles that $u/\cos\varphi$ is significantly larger than u. For this particular date, the time step near the poles would have to be seven times smaller than the time step determined for the mid-latitudes where the winds are actually the strongest. Such a time step is too small to be economically viable. This is why we referred earlier to the 'pole problem' as being an economical one. Much effort has been expended to find schemes which do not require such a small time step.

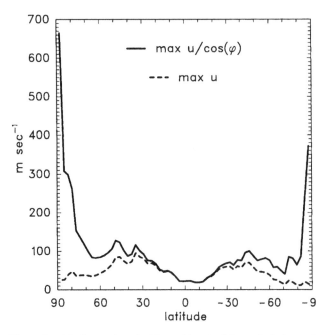

Figure 2.5. Maximum (over height and longitude) zonal wind speed and maximum zonal wind speed divided by cosine of latitude for 1200 UCT 3 Jan 1983.

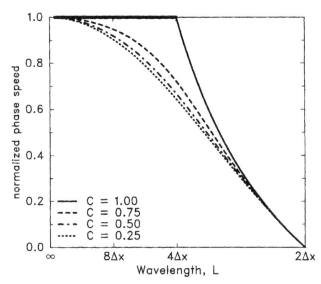

Figure 2.6. Normalized phase speed (approximate divided by true) for explicit centered difference approximations as a function of wavelength for various Courant numbers.

The stability condition (equation 2.2.4) was determined by considering the amplitude of the solution. The speed with which the waves move is another important characteristic of the solutions. This phase speed can also be solved analytically for approximation (2.2.2) with constant U. It is shown in Figure 2.6 for various values of C, normalized by the true speed. The abscissa covers the range of possible wavelengths, from the longest on the left to the shortest, $2\Delta x$, on the right. For $C = 1$, all waves with wavelength $L \geq 4\Delta x$ move with the correct speed, but shorter waves with $2\Delta x \leq L < 4\Delta x$ are slowed and the shortest $2\Delta x$ wave is stationary. In practice one cannot choose Δt such that $C = 1$ because U is unknown beforehand and also generally varies over the domain. Thus Δt must be chosen so that $C < 1$ everywhere. The wind speeds in the atmosphere are 100 m s^{-1} or more in the regions of the jet, but 25 m s^{-1} or less over much of the domain. Thus a Courant number of 0.25 or less is not uncommon, and all waves are slowed down in the approximate numerical solution. As $C \to 0$, the speeds asymptote to a curve slightly below the $C = 0.25$ line in Figure 2.6. Thus the $C = 0.25$ provides a good indication of the worst phase errors of centered differences. The longer the waves are relative to the grid interval, the closer the phase speeds are to the correct values. Therefore, in order to obtain accurate solutions with approximations (2.2.2) we must choose Δx small enough compared to the scale of the phenomena of interest, keeping in mind

the curve for $C = 0.25$ in Figure 2.6, i.e. the smallest structures of interest should be no smaller than 4 to $8\Delta x$.

Implicit finite difference approximations

The approximation (2.2.2) is one member of a class of approximations referred to as explicit. The primary characteristic defining explicit schemes is that the forecast at a grid point (q_j^{n+1}) depends only on the values at past times and not on other future predicted values. Another class of approximations is referred to as implicit. A characteristic of this class is that the forecast at a grid point (q_j^{n+1}) may depend on the forecast at neighboring grid points, e.g. $(q_{j+1}^{n+1}, q_{j-1}^{n+1})$, as well as on the data from the past times. An example of such a scheme applied to equation (2.2.1) is

$$\frac{q_j^{n+1} - q_j^{n-1}}{2\Delta t} + \frac{U}{2}\left[\frac{q_{j+1}^{n+1} - q_{j-1}^{n+1}}{2\Delta x} + \frac{q_{j+1}^{n-1} - q_{j-1}^{n-1}}{2\Delta x}\right] = 0 \qquad (2.2.5)$$

Equation (2.2.5) is a little more complicated to solve than (2.2.2) because each q_j^{n+1} is related to its unknown neighbors $(q_{j+1}^{n+1}, q_{j-1}^{n+1})$ and all values must be determined simultaneously. But it is not extremely complicated and in our simple case can be solved with some straightforward algebra. Unlike (2.2.2) which required a bound on $U\frac{\Delta t}{\Delta x}$ for stable solutions, there is no limit on $U\frac{\Delta t}{\Delta x}$ for (2.2.5) and the scheme is stable for all choices of Δt. Because of the lack of a stability restriction on Δt, at first glance (2.2.5) seems to be an ideal solution

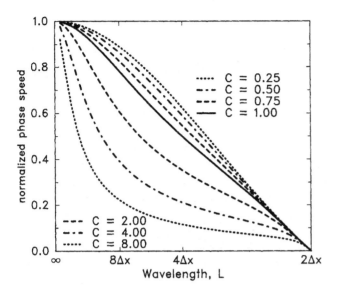

Figure 2.7. Normalized phase speed (approximate divided by true) for implicit centered difference approximations as a function of wavelength for various Courant numbers.

of the pole problem. Consideration of the phase speed of the approximate solution shows this not to be the case. Figure 2.7 shows the normalized phase speed for a variety of Courant numbers C. As $C \rightarrow 0$ the asymptote of the phase speed is the same as that for the explicit differences, and as in that case, the $C = 0.25$ line provides a good indication of it. With the implicit scheme in Figure 2.7, the phase speeds decrease with increasing time step, away from the correct value. The stability of the implicit approximation (2.2.5) is obtained at a price which is a very large phase error even for the longest waves as the Courant number exceeds 1. In Figure 2.5 we see that the Courant number near the poles would be between 4 and 8 if the mid-latitude Courant number were between 0.5 and 1.0. Figure 2.7 shows that in such a case the signal would be significantly slowed down compared with the true solution. The stability of the implicit scheme is obtained by slowing the waves significantly, such that the scheme is not really useful (for large C) for predicting atmospheric flows in which advection is a dominant process.

Semi-implicit approximations

The equations governing atmospheric flow admit other types of motions in addition to advection. These include inertial-gravity waves, which also have stability conditions such as (2.2.4) for explicit approximations, but with the advection speed U replaced by a wave propagation speed. Implicit

schemes are useful for some of these additional components of the complete equations of atmospheric models. For atmospheric type flows the advecting velocity, U, is the order of 10 to 100 m s^{-1}, while the gravity wave speeds can be as high as 300 m s^{-1}, over three times as large as U. This introduces a severe burden to the explicit calculation for this more complex case since Δt must be at least four times smaller than for the simple advection case, and because of this time-step reduction alone a forecast with the complete equations would cost four times as much as with just the advection equations.

In the large-scale global atmospheric models the fast motions associated with gravity waves are of very small amplitude compared with the advection component and do not influence the forecast in any significant way. Even though they are unimportant, they are present in the system of equations and any small perturbations would grow unboundedly and spoil the solution if the time step violated the stability condition. However, because they are unimportant, they can be treated implicitly to decrease their speeds and allow longer time steps without ruining the forecast. Such an approach, introduced by Robert (1969), is referred to as semi-implicit as only the terms that contribute to the fast components are treated implicitly. The advection terms remain explicit. The stability condition associated with the fast waves is eliminated by this procedure and only the explicit advection condition (2.2.4) remains.

Experience has shown that semi-implicit approximations with time steps bounded by (2.2.4) alone produce forecasts which are as accurate as those from an explicit, centered approximation with a time step at least four times shorter. Thus semi-implicit approximations are commonly adopted in atmospheric models. Although the gravity wave components of semi-implicit schemes are stable for long time steps, the advective component continues to impose a restriction on Δt. Thus the 'pole problem' remains for these approximations.

Approaches to the pole problem

We return now to the basic advective stability condition (2.2.4) which becomes excessively restrictive near the poles where Δx becomes small. Equation (2.2.4) presents the most restrictive condition for all waves. The condition for each wave individually is

$$\left| U \frac{\Delta t}{\Delta x} \sin\left(\frac{2\pi \Delta x}{L}\right) \right| \leq 1 \qquad (2.2.6)$$

A second possible way to satisfy the condition, rather than

restricting Δt, is to limit the waves included in the solution in the polar regions so that sin $(2\pi\Delta x/L)$ is small enough for equation (2.2.6) to be satisfied for those waves retained in the solution. Such an approach requires application of a filter in longitude such as a Fourier transform and resynthesis. The filter can be applied to the predicted variables or to just those terms in the equations responsible for the instabilities as in the grid point model of Arakawa and Lamb (1977). However, both approaches are somewhat arbitrary and introduce distortions into the flow. Because of this and the fact that there are other schemes which provide more natural solutions to the 'pole problem', they have not been widely adopted. The major problem is that at each latitude the filters are applied in longitude only, with no account taken of the appropriate latitudinal structures that should be maintained.

A third method that has been attempted to satisfy (2.2.6) is to increase $\Delta\lambda$ approaching the poles so that Δx remains constant; i.e. instead of keeping $\Delta\lambda$ constant as in Figure 2.2(a), $\Delta x = \Delta\lambda\cos\varphi$ is held constant as in Figure 2.2(b). The atmosphere also contains structures called Rossby waves which have a longitudinal wave-like structure, $A(\varphi)$ sin $(2\pi(\lambda+\omega t)/L)$, and propagate with the same longitudinal phase speed ω at all latitudes. In this case the angular propagation velocity is the same at each latitude, and an effective Courant number is $C=\frac{\omega\Delta t}{\Delta\lambda}$ where ω is a constant propagation velocity in degrees longitude per second, rather than meters per second as described earlier. These coherent waves are important in the atmosphere and must be solved for accurately. They tend to have rather long wavelengths, so the longitudinal filtering described above does not affect them. That filtering is applied to shorter waves. In addition, they are representable with the longer $\Delta\lambda$ associated with the type of grid in Figure 2.2(b). However, increasing the longitudinal grid interval $\Delta\lambda$ approaching the poles so that the distance $\Delta x = \Delta\lambda\cos\varphi$ is reasonably constant results in $C=\frac{w\Delta t}{\Delta\lambda}$ decreasing with increasing latitude. From Figure 2.6 we see that this will introduce a differential phase error as a function of latitude leading to a latitudinal tilt in the wave structures (Kreiss and Oliger, 1973) which will produce spurious poleward transport of heat, momentum and moisture.

Spectral transform method

The most common numerical method employed in global atmospheric models today is the spectral transform method. Machenhauer (1979) provides a thorough review of this approach. In this method at various stages in a time step the predicted variables are represented by grid-point values or by a series of linearly independent spectral functions such as sines. In the one-dimensional case the series can be written as

$$q(x_j,t) = \sum_{k=0}^{K} q_k(t)\,\sin\left(\frac{2\pi(x_j-\theta_k(t))}{L_k}\right) \quad (2.2.7)$$

Since this representation is also associated with an underlying grid, the waves included are restricted to have wavelengths $L_k \geq 2\Delta x$. The grid-point representation $q(x_j,t)$ is equivalent to the (amplitude, phase) representation $(q_k(t), \theta_k(t))$. The transform method takes advantage of the best of both representations. Spatial derivatives are approximated by the analytic derivative of the individual terms of the series and thus are extremely accurate. More complicated calculations such as nonlinear products of fields and the physical parameterizations (not included in our simple example but discussed in the following section) are calculated from grid-point values. The solution is determined by the time evolution of the amplitudes and phases (q_k, θ_k) of the terms of the series. The spectral transform method commonly adopts centered time differences as in (2.2.2) or (2.2.5). In this case, in order to have stable solutions

$$\left|U\frac{\Delta t}{\Delta x}\right| \leq \frac{1}{\pi} \quad (2.2.8)$$

which is more restrictive than the condition for the centered spatial differences (2.2.4). Although the time step is more restricted, the solution is more accurate for a given wavenumber and therefore fewer waves need to be retained in the solution, allowing a coarser grid (larger Δx). Figure 2.8 shows the phase speed of the discrete solution as a function of wavenumber. In this spectral case the waves actually move faster than the true speed, due to the centered time differences. The acceleration effect of the centered time differences is also present in the grid-point scheme (2.2.2) but it is overwhelmed by a retardation introduced by the spatial differences. This acceleration explains why the phase error is smaller with larger time steps in the centered difference scheme. The longer the time step, the larger the acceleration which counteracts the deceleration of the spatial differences.

The final advantage of the spectral transform approach is that it provides a natural solution to the spherical 'pole problem'. In the two-dimensional spherical case the natural representation is Fourier series in longitude, and associated Legendre functions in latitude

$$q(\lambda,\varphi,t) = \sum_{n=0}^{N}\sum_{k=0}^{n} q_n^k(t)P_n^k(\sin\varphi)\,\sin\left(\frac{2\pi(x_j-\theta_k(t))}{L_k}\right) \quad (2.2.9)$$

The latitudinal structures $P_n^k(\sin\varphi)$ have the useful property

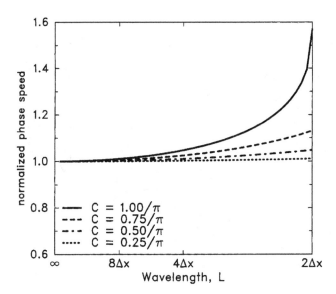

Figure 2.8. Normalized phase speed (approximate divided by true) for spectral transform method as a function of wavelength for various Courant numbers.

that for small longitudinal wavelengths their amplitudes are essentially zero near the poles. Thus a condition like (2.2.8) or (2.2.6) for individual waves does not have to be satisfied for the smaller Ls near the poles. The time step is not excessively restricted by the small longitudinal grid distance near the poles and the more natural limit associated with mid-latitude flow holds. Figure 2.9 shows the latitudinal structure of selected $P_n^k(\sin \varphi)$ for a few wavenumbers, k. The wavenumber is related to the wavelength by $k = 2\pi/L$ and is a more natural way to characterize the functions. The figure illustrates that the amplitudes are indeed zero near the poles for the smaller wavelengths (larger wavenumbers).

Although the spectral transform method has many advantages and seems ideal for spherical domains, it also has some serious disadvantages. The method is formally equivalent to a least square approximation and minimizes the mean square error over the global domain. This implies that the size of the error is likely to be the same everywhere, which is not a problem for a field like temperature where, say, a one degree error is no more serious in the polar regions than in equatorial regions. However, it is a serious problem for a field like water vapor, usually treated in atmospheric models as the mixing ratio (the ratio of the mass of water vapor to the mass of dry air containing the vapor). At the Earth's surface the mixing ratio varies from close to 20 g kg⁻¹ in the equatorial regions to 1 g kg⁻¹ or less in the polar regions. An error of 2 g kg⁻¹ may only be 10% in equatorial regions and have little effect, but such an error in polar regions completely changes the character of the field. It can even make the mixing ratio

negative which, of course, is physically impossible. Yet just such a case commonly occurs in spectral models (Rasch and Williamson, 1990). In addition, the error can lead to significantly larger values than should be present and create spurious supersaturation which in turn leads to spurious rain (Williamson, 1990).

Semi-Lagrangian method

One of the reasons for the introduction of semi-Lagrangian methods into global models was to reduce the problem of spurious rain in the polar regions that occurs with the spectral transform method. This approach has other advantages as well which will be discussed in the following. Staniforth and Côté (1991) provide an excellent review of this approach. The method is based on the Lagrange form of the equations rather than the Euler form. Instead of (2.2.1), the approximations are based on the pair of equations

$$\frac{dq}{dt} = 0 \tag{2.2.10}$$

$$\frac{dx}{dt} = U \tag{2.2.11}$$

Equation (2.2.10) states that q is unchanged along a trajectory (in the absence of sources and sinks). The trajectory is given by (2.2.11). Once again we seek a forecast at a specified set of points such as the grid in Figure 2.10. (The grid points here are given by the intersections of the grid lines.) For each point we approximate the trajectory that would arrive at that point at time $(n+1)\Delta t$ using the wind U at time n. We refer to the point at which the forecast is to be made as the arrival point (A) and the point from which the trajectory departed at time $(n-1)\Delta t$ the departure point (D). The point in the middle of the trajectory at time $n\Delta t$ is referred to as the midpoint (M). The departure point is determined by approximating (2.2.11) with

$$x_D = x_A - 2\Delta t U(x_M) \tag{2.2.12}$$

The midpoint, at which U is needed in (2.2.12) is obtained by a similar expression

$$x_M = x_A - \Delta t U(x_M) \tag{2.2.13}$$

which is implicit since U at x_M cannot be determined until x_M is known. Thus x_M is generally found by an iterative process in which the next guess of x_M on the left is based on U at x_M from the previous iteration. In atmospheric applications, only a few iterations are required. Once x_M is found, the departure point x_D is calculated from (2.2.12) and (2.2.10) yields

$$q(x_A)^{n+1} = q(x_D)^{n-1} \tag{2.2.14}$$

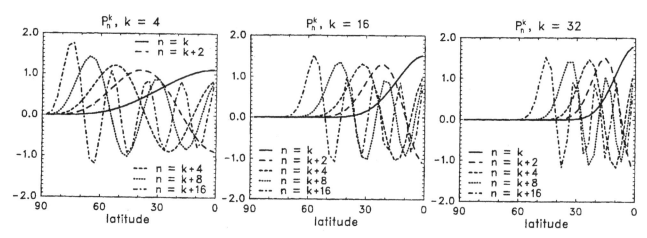

Figure 2.9. Latitudinal structure of selected $P_n^k (\sin \varphi)$.

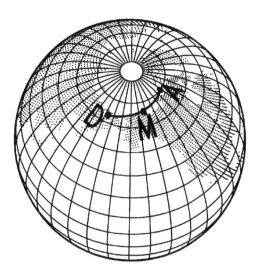

Figure 2.10. Illustration of a trajectory near the poles.

There is no reason to expect that the departure point x_D or midpoint x_M will be grid points. Thus values of q at x_D and U at x_M are obtained by interpolation from the surrounding grid points. The scheme is stable as long as the interpolation is based on data points surrounding the departure point (or midpoint) and it has no time step restriction. Thus, near the poles, the departure point may be many grid intervals away from the arrival point as illustrated in Figure 2.10 without being unstable, and the convergence of meridians does not introduce a stability restriction. The scheme is also accurate and does not suffer from the phase errors seen earlier with finite differences.

The semi-Lagrangian method has an additional advantage in that the interpolation used to obtain $q(x_D)^{n-1}$ in equation (2.2.14) can incorporate some physical constraints. One that has been used for water vapor prediction is monotonicity, i.e. the interpolated value q_D is no larger or smaller than the surrounding grid-point values. Thus negative values cannot be introduced by the scheme if the field is positive to start with, and spurious supersaturation and its accompanying spurious rain will not be produced by overshooting.

As mentioned above, advection is only one component making up the equations representing atmospheric motions. Semi-Lagrangian advection approximations have been successfully combined with finite difference, spectral, and finite element based approximations for the remaining components. An example of a spectral, semi-Lagrangian model is Ritchie *et al.* (1995). Chen and Bates (1996) developed a finite difference, semi-Lagrangian model applied to both numerical weather prediction and climate simulation. Côté *et al.* (1997) developed a variable resolution, finite element, semi-Lagrangian model for the sphere which also has the capability of concentrating resolution over an area of interest, while retaining a global domain.

Finite element approximations

The finite element method is based on principles similar to those of the spectral method. Both methods represent the unknown variables by a finite series of linearly independent basis functions with a prescribed spatial structure. The coefficients of the series are the unknowns which must be determined as a function of time. One difference between the spectral and finite element approaches is that the spectral method uses global basis functions, i.e. functions that tend to be non-zero almost everywhere. The finite element method uses basis functions that are zero except in limited regions, usually spanning the equivalent of only a few grid cells, where they are piecewise linear or low order polynomials. Finite element methods applied over uniform latitude–longitude grids as in Figure 2.2(a) suffer from the same Courant number restriction as finite difference schemes because the elements are local and do not possess the global filtering properties of the spherical harmonic basis functions of equation (2.2.9). However, coupled with semi-Lagrangian and semi-implicit approximations, finite element schemes lose this restriction and are very suitable for spherical geometry (Côté *et al.* 1997).

Reduced grids

In the uniform latitude–longitude grid illustrated in Figure 2.2(a), the longitudinal grid points converge approaching the poles. As described earlier this convergence results in a restrictive permissible time step for finite difference and finite element schemes. The spectral transform method avoids this restriction by excluding from its representation the shortest longitudinal waves associated with this grid near the poles. A grid more like that in Figure 2.2(b), referred to today as a reduced grid, is consistent with these spectral basis functions (Hortal and Simmons, 1991) and in fact is now used in some spectral transform models. The semi-Lagrangian method does not have a time step restriction and thus can be used with the grid in Figure 2.2(a). However, that grid is wasteful computationally since the grid points are very close together in the longitudinal direction in the polar region. Semi-Lagrangian schemes can also easily be based on grids like Figure 2.2(b). As mentioned earlier, finite difference based schemes are less amenable to application on grids such as that in Figure 2.2(b) because of the potential introduction of differential phase errors. The variability of the longitudinal and latitudinal unit vectors over the grid boxes also introduces severe difficulties. Finite difference approximations often involve an assumption that a grid point value represents a grid box average. However, the unit vectors defining the spherical vector coordinates vary significantly over a single grid box near the poles, invalidating any assumption that the wind over the box is representable in that box by a single value of the components.

Vertical aspects

So far we have considered only the horizontal aspects of numerical approximations for atmospheric models. The vertical dimension adds an additional degree of complexity out of proportion to just being a third dimension. This is in part because of the complex shape of the mountainous Earth's surface, because of the preferential direction of gravity toward the center of the Earth coupled with the shallowness of the atmosphere relative to the radius of the Earth, and because of the several orders of magnitude variation in the water vapor mixing ratio between the Earth's surface and the tropopause. We continue to concentrate here on global models designed to forecast or simulate synoptic-scale and large-scale atmospheric motions for which hydrostatic equilibrium is a valid assumption.

Mathematically, the most natural vertical coordinate for global models is the radial distance from the center of the Earth or its linear transformation, height above mean sea level, denoted z. The hydrostatic assumption, however, allows us to choose variables other than geometric height for the vertical coordinate. In all coordinate systems, the vertical terms involve integrals, which can be approximated by well known methods such as the trapezoidal rule, and differentials, which can be approximated by the methods described in the previous sections. Finite difference, finite element and semi-Lagrangian approximations are all being used in the vertical in models today. Because of the nature of the top and bottom boundary conditions, the spectral method has been less successful in the vertical. Constraints are often applied to the vertical approximations to ensure that various energy and angular momentum conservation properties of the continuous equations are satisfied by the approximate equations (Simmons and Burridge, 1981; Arakawa and Suarez, 1983). The issues of accuracy and stability described in the previous sections with regard to the horizontal approximations are equally valid for the vertical coordinate. We will not repeat them here but rather concentrate on the choice of vertical coordinate. For more details on vertical coordinates see Haltiner and Williams (1980) and Sundqvist (1979).

In the z vertical coordinate system, the equation for the horizontal velocity **V** (assuming no forcing terms) is

$$\frac{d\mathbf{V}}{dt} + f\hat{\mathbf{k}} \times \mathbf{V} = -\frac{RT}{p}\nabla p \qquad (2.2.15)$$

where p is pressure, T is temperature, R is the gas constant, $f = 2\Omega \cos \varphi$ is the Coriolis parameter, Ω is the rotation rate of the Earth and $\hat{\mathbf{k}}$ is the outward radial unit vector. Application of these equations is made difficult by the fact that the mountains stick up into the domain and thus cut holes in the horizontal coordinate surfaces (levels of constant z). Lateral boundary conditions must be applied at these irregular boundaries, and although the pressure gradient term is a simple gradient operator, its approximation at these boundaries is difficult. In addition because the horizontal domain is incomplete, it is very difficult to apply the spectral transform method in the horizontal dimensions. Because of these and other difficulties the z system has not proven popular for global hydrostatic atmospheric models.

The hydrostatic approximation

$$\frac{\partial p}{\partial z} = -\frac{gp}{RT} \tag{2.2.16}$$

where g is gravity, allows the choice of variables other than geometric height for the independent vertical coordinate. Pressure at first appears to be an attractive choice because the continuity equation is particularly simple in that system and because of the tradition of analyzing meteorological variables on pressure surfaces. However, as in the z system, mountains stick up into the domain and intersect surfaces of constant pressure. In addition, in the p system the location of these lateral boundaries is a function of time since the pressure in the atmosphere varies with time.

The most common vertical coordinate in use in global models today is a normalized pressure coordinate first introduced by Phillips (1957). This is usually referred to as the σ system after Phillips' original definition of $\sigma = p/p_s$ where p_s denotes the surface pressure. In this system the Earth's surface is a coordinate surface and the difficulties of the z and p systems associated with mountains intersecting 'horizontal' coordinate surfaces do not arise. Nevertheless, this advantage does not come without cost. In the σ system the equation for the horizontal velocity is

$$\frac{d\mathbf{V}}{dt} + f\hat{\mathbf{k}} \times \mathbf{V} = -\nabla\phi - RT\nabla \ln p_s \tag{2.2.17}$$

where $\phi = gz$ is the geopotential. The pressure gradient is the sum of the two terms on the right-hand side of equation (2.2.17). In the vicinity of mountains these two terms are of opposite sign and are significantly larger than the pressure gradient itself. Thus any errors in either term are amplified in a relative sense when the difference is taken. A common problem observed in atmospheric models is excessive precipitation locked to the orography. For example, almost all models show excessive precipitation over the Andes in January. This is likely caused by a pressure gradient error driving spurious moisture convergence over the mountain, leading to condensation. The error is not necessarily reduced by

increasing vertical resolution as it may be due to the last term in equation (2.2.17) alone. The release of latent heat associated with the spurious condensation drives additional convergence, exacerbating the problem. A hybrid system introduced by Simmons and Burridge (1981) is currently gaining favor in global models. This system looks like the σ system at the Earth's surface, but becomes a pressure system around the tropopause. Thus the pressure gradient error discussed above does not occur in the upper troposphere or stratosphere, but it does remain in the lower troposphere. Since most of the water vapor is concentrated in the lower troposphere, the orographically locked precipitation problem remains in this hybrid system.

To address this pressure gradient force problem in the region of steep terrain, Mesinger et al. (1988) introduced a step-mountain coordinate which achieves approximately horizontal coordinate surfaces. Conceptually, the mountains are constructed from three-dimensional blocks (grid boxes), and above these blocks a σ type system is used. This system does once again introduce horizontal boundaries into the problem while reducing the pressure gradient error. It is proving successful for regional models (Mesinger et al., 1988) but has not yet been adopted in any production global model.

Potential temperature (θ) surfaces have always been appealing as a model vertical coordinate in spite of the difficulties introduced by these coordinate surfaces intersecting the ground and by the fact that isentropes can become nearly vertical or unstable. Some of the potential advantages are: for adiabatic flow, isentropic surfaces are material surfaces and motions are 'horizontal' in such a coordinate system; the pressure gradient force is a single term avoiding the need to difference two large terms as in the σ system; and fronts tend to be parallel to isentropes and thus the large gradients associated with them in other coordinate systems are avoided in the θ system and coarser grids can be used to provide equivalent accuracy.

Although isentropic coordinates have been used for various applications since the beginning of atmospheric modeling, they never became common in global models because of their inherent difficulties. However, currently there is a resurgence of activity in the development of such approximations. Arakawa et al. (1992) have developed a model in which artificial massless layers are introduced to simplify the lower boundary condition. They report difficulty, however, in defining the temperature at the surface because the number of degrees of freedom available to define the distribution of surface temperature is determined by the number of coordinate surfaces intersecting the ground. Zapotocny et al. (1994) have developed a global model using a composite mesh approach with a σ system near the surface and θ above.

Variables are interpolated between the two systems in a region of overlap. Zhu *et al.* (1992) have derived a vertical discretization based on a general coordinate that smoothly transitions from a terrain-following σ system at the ground to an isentropic system at higher levels. For a global model, however, the pure θ system cannot be attained much below the tropopause, so this system appears more useful for models which include a detailed stratosphere. Because of the problems with normalized vertical coordinates such as σ, because θ appears to offer many advantages, and because significant progress has been made in the development of local approximations with nonlinear constraints for transport and nonlinear fluid flow, there is a resurgence of developmental activity devoted to overcoming the remaining problems with the θ system in global models. Undoubtedly, isentrope-based coordinates will be much more common in global models in the future.

Future directions

For all its advantages, the semi-Lagrangian method as currently implemented is not perfect. One of its major detractions is that it is difficult to achieve conservation without the introduction of *ad hoc* corrections. There is currently a flurry of activity devoted toward developing new numerical schemes for atmospheric models. One path of activity is directed toward developing conservative semi-Lagrangian schemes (e.g. Leslie and Purser, 1995; Rančić, 1995). Also in the semi-Lagrangian arena is the development of schemes based on the advection of potential vorticity, a more naturally conserved quantity (Bates *et al.*, 1995). Another path of activity is directed toward developing schemes on grids which are not latitude–longitude based, but rather which provide more

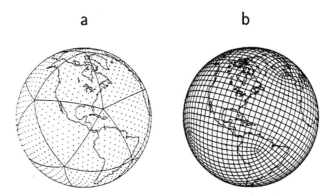

Figure 2.11. Examples of global grids based on (*a*) subdividing the triangle sides of an icosahedron into smaller triangles and (*b*) subdividing the square sides of a cube into smaller squares.

homogeneous distributions of points over the sphere without the clustering near the poles. Two examples of such grids are shown in Figure 2.11. Part (*a*) illustrates a grid based on subdividing the triangles of an icosahedron inscribed in a sphere into smaller triangles. Part (*b*) illustrates a grid based on subdividing the square sides of a cube inscribed in a sphere into smaller squares. The triangular grids are being pursued by, among others, Heikes and Randall (1995) and the cube based grids by Rančić *et al.* (1996). Approximations for both these grids were developed 25 years ago with only limited success. However, since then much progress has been made in developing numerical approximations for nonlinear fluid flow in Cartesian geometry. There is every reason to expect that this work can be extended to spherical geometry and adapted to grids like those in Figure 2.11. Ten years from now atmospheric models are likely to look very different from those in use today.

2.3 Parameterization of subgrid-scale processes

Hua-lu Pan

As indicated in the introduction, computer resources limit the resolution of atmospheric models, and it is not possible to predict the fluid behavior on the scales smaller than those resolved by the model grids. Following Reynolds (1895), we develop equations governing mean quantities, such as the mean specific humidity or the water vapor mixing ratio. To emphasize this, we now write the prediction equation for the specific humidity as

$$\frac{\partial \bar{q}}{\partial t} = -\bar{V}\cdot\nabla\bar{q} - \overline{V'\cdot\nabla q'} + S/\rho \qquad (2.3.1)$$

Here the specific humidity has been decomposed into a mean ($\bar{q}$) and a fluctuation (q') such that

$$q = \bar{q} + q' \qquad (2.3.2)$$

While the traditional Reynolds average is considered a temporal average, in NWP applications it is generally interpreted as a grid box average. The terms in equation (2.3.1), then, represent: (a) the local time rate of change of the grid averaged specific humidity, (b) the advection of the mean specific humidity by the grid-scale wind, (c) the advection of the subgrid-scale humidity fluctuation by the subgrid-scale velocity fluctuations, and (d) the source/sink term. Since the subgrid-scale quantities are not resolved by the model, we cannot predict the behavior of these quantities without additional equations. In many atmospheric situations, the effect of the subgrid-scale fluctuations can exert a strong influence on the grid-mean quantities, so that it is necessary to include these effects. Instead of trying to predict the subgrid-scale fluctuations, it is common to emulate their effects using grid-scale variables. This is the essence of the parameterization of the subgrid-scale processes.

A good example of an important process that takes place at a subgrid-scale is the turbulent mixing in the planetary boundary layer. During the daytime, the solar heating at the Earth surface not only warms the soil but also causes the plants to transpire and wet soil to evaporate, thus transporting moisture into the atmosphere. The surface heating leads to turbulent motion that is on the scale of a few meters to a few hundred meters. On a grid size of 50 km to 100 km, NWP global models cannot resolve this motion. Yet the transport of the heat and moisture into the boundary layer is crucial to the development of afternoon thunderstorms and a host of other phenomena that are important to the resolvable atmospheric fields. Another important example is tropical cumulus con-

vection. The cumulus clouds in the tropics are known to be extremely important to the global energy balance, yet each cloud typically occupies only a few kilometers of space horizontally and vertically.

In addition to the problem of resolution, some of the processes are not well enough known to be properly formulated on any scale. Transpiration is one such example. It is known biologically that plants try to maximize the absorption of sunlight for photosynthesis and minimize the release of water vapor (transpiration). To organize this into a mathematical formula that takes into account the condition of the plants, the condition of the soil, and the weather condition is a science that is only in the developing stage. This makes the problem of parameterization for NWP models even more difficult.

We will use the parameterization schemes that are currently employed in the NCEP models as examples to illustrate the procedures and the philosophy of parameterization. We will start with the surface layer. It is well known that the lowest 10–20 m layer of the atmosphere is an important layer for momentum dissipation and for heat and moisture transport from the ground. Through the work of Monin and Obukhov (1954), we have come to understand that the profiles of wind and temperature in this very turbulent layer can be described by a set of equations that depends on only a few parameters (the surface roughness length is one of them), and by the hypothesis of similarity, based on many observational studies, that suggest that the fluxes of momentum and heat are nearly constant with height in the surface layer. According to the similarity theory there exist, near the ground, a velocity u^*, a length L, and a temperature T^* that are essentially invariant with height within the surface layer. When the atmospheric variables are scaled by these invariants to become non-dimensional variables, the vertical profiles of these new variables are similar for many different ground surfaces and many different atmospheric conditions. These so-called similarity profiles are used in the models to provide us with the exchange coefficients for surface stress, sensible and latent heat fluxes. Without these similarity profiles, we would need to have model vertical resolutions in the range of a centimeter in order to try to resolve the turbulence. The surface layer similarity theory is one of the most successful parameterization schemes in the NWP field. At NCEP, the global spectral model uses similarity profiles derived from field experiments, as do most other NWP models.

Above the surface layer, turbulence can still be important

when either the solar heating or the wind shear are strong. This is the layer commonly called the planetary boundary layer (PBL) with a typical depth of 1 km. Over the ocean, when the wind is steady (e.g. over the trade wind regime), the boundary layer is relatively well mixed so that potential temperature and moisture are constant with height, since turbulence has a sufficiently long time to smooth out the gradients. Over land, the boundary layer grows deeper as the sun rises and collapses after sunset. The nocturnal boundary layer can grow when there are strong wind gusts. Earlier GCMs (general circulation models) tended to use the so-called mixed-layer parameterization to model the PBL. This assumes that the PBL is always well-mixed and we need to estimate only the depth of the layer. Potential temperature and moisture are assumed constant within the PBL and are easily estimated. This works very well over most of the oceanic regions, is very easy to implement, and computationally inexpensive. From the Reynolds-averaging point of view, this is called the 'zero-th order closure' as only the mean variables are estimated.

Similar to the procedure to obtain equation (2.3.1), we can derive an equation for the flux of the subgrid-scale moisture by the subgrid-scale wind (e.g. $u'q'$). This equation would contain terms that are cubic in the subgrid-scale variables. In principle, one can go on to form predictive equations for the n-th order subgrid-scale variables and it would contain $n + 1$-th order terms. Mellor and Yamada (1974) provided a hierarchy of closure models that were ordered in terms of a succession of simplifying assumptions. In general, one has to parameterize the $n + 1$-th order terms as functions of the lower order terms. This would be called the $n + 1$-th order closure. When we parameterize the second term on the right-hand side of equation (2.3.1) using grid-averaged variables, we are dealing with a first order closure. This order of closure takes the form of a diffusion equation, as we normally parameterize the subgrid-scale fluxes by the gradient of the grid-averaged variables. For example, the turbulent transport of the horizontal momentum by the vertical wind component of the turbulence (the wind stress) is normally parameterized as

$$-\rho \overline{u' w'} = \tau_x = \rho K \frac{\partial \bar{u}}{\partial z} \tag{2.3.3}$$

Experience with this type of scheme has taught us that a simple constant coefficient of diffusivity (K) is not sufficient to simulate the diurnally varying PBL over land. In the current NCEP models, various methods are utilized to formulate this coefficient better to simulate the turbulent mixing. In the Eta regional model (Janjic, 1994), the turbulent kinetic energy is

predicted and then used to determine K (sometimes called the M-Y level 2.5 scheme, as the zero-th order scheme is also called the M-Y level one scheme and the first order scheme is called the level two scheme by Mellor and Yamada). In the global NCEP model, the coefficient is specified based on a simple model of turbulence deduced from Large-Eddy Simulation model simulations of turbulence (Hong and Pan, 1996). The NCEP Eta model also uses a surface layer parameterization that is a level two scheme of Mellor and Yamada.

In a way, cumulus convection can also be considered a form of turbulence. The major differences are: (a) there is a phase change as water vapor condenses into liquid (or ice), and (b) the vertical scale is an order of magnitude larger than in surface turbulence. Because of these differences, the usual approach of diffusion does a poor job of simulating the phenomenon. Over vast regions of the tropical ocean, the atmosphere is stable to perturbations that send air parcels upward or downward. Because of the decrease of pressure with height in the atmosphere, the rising parcels of air will expand and cool (a process called 'adiabatic cooling'). The atmosphere is normally in a statistically stable state such that the rising parcels find the environment to be warmer and the parcels will sink back downward. When the air parcel becomes saturated due to adiabatic cooling, the release of latent heat slows down the cooling and as a result the parcel may become warmer than the environment (called conditional instability) and the parcels will continue to move upward. This is the basic way cumulus clouds form. Since the air is rarely saturated at the grid scale, the grid-scale precipitation processes do not operate. If we do not parameterize the convection except for the removal of supersaturation, there will be situations when the atmosphere eventually becomes saturated. When, in addition, there is upward motion, the model atmosphere becomes absolutely unstable (rising air parcels will continually rise up to the tropopause) and can also become numerically unstable. The first type of parameterization designed to deal with this situation applied an adjustment of the vertical profile of temperature and moisture (the moist convective adjustment, Manabe et al., 1965) that removes the instability and converts the excess heat into precipitation. Recognizing that convection probably starts long before the grid-averaged atmosphere becomes saturated, several schemes were later developed (e.g. Kuo, 1965, 1974; Betts and Miller, 1986) that perform adjustments to the model temperature and moisture fields over a period of time when the scheme determines that convection should be occurring. Each scheme has its own adjustment profile and its own triggering mechanism that determines the onset of convection and the amount of adjustment. The NCEP eta model uses a modified Betts and Miller scheme (Janjic, 1994) and has proven to be very skillful in predicting convective precipitation in the short range forecast.

Primarily through the work of Arakawa (1969), Yanai *et al.* (1973), and Ooyama (1971), another view of the effect of cumulus convection began to take form. This culminated in a new type of cumulus parameterization scheme (Arakawa and Schubert, 1974). The key concept involved is that the cumulus updraft is a result of the release of latent heat but the compensating subsidence is the mechanism that warms and dries the atmosphere. In addition, they proposed that an ensemble of clouds is involved in the heating and drying of the grid-averaged atmosphere. Their scheme, implemented by Lord (1978), formulated the adjustment of the atmosphere as a process to balance between the large scale atmosphere destabilization, through moisture convergence and thermal advection, and the stabilization by the convective clouds (the quasi-equilibrium assumption). This elegant treatment of the cumulus brought the art and science of parameterization to a higher level by using a better understanding of the phenomenon to build a better algorithm. The complexity of the scheme, however, prevented widespread use of the scheme. In addition, the triggering mechanism is based on a mixed-layer PBL assumption and can become a problem for convection over land. In recent years, research work on mesoscale circulations such as the mesoscale convective complexes has resulted in new parameterization schemes (e.g. Kane and Fritsch, 1990) that placed major emphasis on the sub-cloud layer structure and on the onset of convection. In addition, we now know that strong downdrafts can develop near regions of updraft and exert significant influence on the continued development of intense thunderstorms. The current NCEP MRF model uses a scheme that is based on Grell (1993) which is a simplified version of the Arakawa and Schubert scheme with major modifications on the cloud trigger mechanism and the addition of downdrafts (Pan and Wu, 1995).

The processes through which the Earth surface exchanges heat and moisture with the atmosphere are extremely complicated but very important to all scales of the atmospheric circulation. Over the ocean, because the sea surface temperature (SST) varies relatively slowly compared with atmospheric activity, NWP models usually assume the sea surface temperature to be constant. Daily SST observations from satellites are used in operational centers to update the model boundary condition. For seasonal predictions, to assume that the SST anomaly remains constant with time, as is generally done in NWP, is clearly inappropriate and ocean circulation models are needed to provide the anomalous surface boundary condition. To estimate the surface heat fluxes, the condition of the sea is also an important ingredient. In principle,

we need to know the wave energy of the ocean surface in addition to the atmospheric variables. At present, most of the NWP models use a parameterization that relates an effective roughness length to the surface stress (Charnock, 1955). In an iterative procedure, we can deduce both the stress and the roughness length as is done in the NCEP MRF model. Together with the surface-layer similarity profiles, we can then use bulk-aerodynamical formulae to deduce the sensible and latent heat fluxes.

Over land, the situation is more complicated. Daytime solar heating will quickly warm up the top layer of the soil and plant leaf surfaces. This triggers many processes that remove the excess heat. The Earth radiates upward in the infrared range at the near-surface temperature. Turbulent air motion transfers sensible heat to the lower atmosphere and larger-scale eddies transfer the heat further up into the PBL. Atmospheric circulations eventually advect the heat away. There is also a heat flux downward, warming the soil. When the soil is moist or when there is vegetation, latent heat fluxes through evaporation and transpiration are also important processes within the surface energy balance. One of the key variables that determines the amount of latent heat flux is the extent of the wetness of the soil. When the soil is dry, plants will slow down photosynthesis to conserve water. On the other hand, over the wet tropical rainforest, most of the net radiative energy is converted to latent heat flux. These processes clearly require a model of the land surface hydrologic budget.

The surface hydrologic balance starts with precipitation and melted snow. Part of this goes into the soil and part becomes surface runoff that feeds into streams and rivers. Within the soil, conduction of heat and water reduces the gradients while the deep soil drainage can also become runoff. The water that is near the surface can evaporate and be replenished from water in the deeper soil layer. Plants can extract water from the root zone and transpire. In the area of a typical grid box, all of these mechanisms are operating simultaneously. In addition, precipitation does not fall uniformly within a grid box, and trees do not cover the entire grid box. The existence of sloping terrain within the grid box is very important to the surface runoff amount. In addition to trying to parameterize all these processes as a grid box budget, we also have to consider the subgrid-scale irregularities. When the snow cover in winter is also considered, the number of processes that we need to include is quite large.

In early days, the latent heat flux over land was parameterized by assuming the underlying surface is like a bucket with a surface temperature. The latent heat flux was parameterized with a bulk-aerodynamic formula (Manabe, 1969). Over the last decade, it has become apparent that the bucket formulation often produces large and unrealistic latent heat fluxes in NWP models. As the land surface cumulus convection often depends critically on the differential heating and moistening between a dry region and a wet one, the accuracy of the prediction of sensible and latent heat fluxes is important to the success of mesoscale forecasts. On longer time scales, the correct climate response of the land surface region to changes in the SST or global warming also depends critically on a realistic representation of the effects of soil moisture on evaporation. It is not surprising that there is increasing attention devoted to this topic.

At NCEP, we have taken a simple approach to improve the land surface prediction by the incorporation of a simple soil hydrology model and a soil thermodynamic model (Mahrt and Pan, 1984; Pan and Mahrt, 1987). Diffusion equations are used to model the flow of heat and water in the soil. The hydraulic diffusivity and conductivity are functions of the soil moisture and can vary over several orders of magnitude between dry and wet soil. Evaporation is parameterized as the sum of three components: direct evaporation from bare soil, transpiration from plants, and re-evaporation of rain that is intercepted by the leaf canopy. Runoff is currently crudely parameterized as drainage and surface runoff when the top soil layer becomes saturated. There are other models of the transpiration processes (e.g., Dickinson, 1984 and Sellers *et al.*, 1986) that try to represent the physical process of transpiration as part of the photosynthesis process. Many hydrologists are interested in improving the runoff process modeling for possible flash flood and longer term river flow predictions. This is an area where we can anticipate major improvement of models within the next few years.

There are two other portions of the model parameterization effort that are in the early stages of development and are only crudely parameterized in the NCEP models. One is the cloud specification and/or prediction. The presence of clouds has a tremendous impact on the convection and PBL development over land as the clouds will shield the sun and slow the warming and moistening of the surface layer. Early NWP models specified the amount of low, middle, and high clouds by latitude (Manabe *et al.*, 1965). Next came the use of relative humidity to specify clouds (Slingo, 1987). The current NCEP global model still uses a simple relative humidity function to specify the cloud amount (Campana, 1995). The relationship is derived from actual observations, and so the model roughly predicts the same amount of cloud as observations. In the eta model, cloud and rain water are predicted using budget equations and deduce clouds from the amount of cloud water (Zhao *et al.*, 1995). This is potentially a more accurate way to provide cloud amount for the radiation calculation.

The other area where only partial understanding of the phenomena is available is the effect of subgrid-scale mountains. Despite the obvious importance of the underlying terrain, the effect of the subgrid-scale inhomogeneity of the terrain is not well known. Some NWP modellers utilize an additional mountain height based on the variance of the subgrid-scale terrain and have found such procedures to provide better synoptic forecasts in the short (1–2 days) and medium (6–10 days) ranges. In general, the blocking effect of the mountains, the gravity wave drag and the planetary boundary layer over complex terrain all interact to make this a very complex phenomenon to parameterize. In the 1–2 day forecast, the NCEP models have a tendency to form lee-side cyclones too far to the north and too strong as synoptic wave troughs move over the Rockies. On the 6–10 day time frame, the ECMWF experience demonstrates that there is potential benefit in the reduction of model biases by the inclusion of the subgrid-scale mountains (Lott, 1995). Recently, Lott and Miller (1997) formulated a new parameterization using recent developments in the nonlinear theory of stratified flows around obstacles, paying special attention to the parameterization of the blocked flow when the effective height of the subgrid-scale orography is high enough. They showed that this method can duplicate the effect of the so-called envelope orography used in the ECMWF model to simulate the effects of subgrid-scale orography.

One of the lessons learned through the last few decades is that we can improve parameterizations only by getting to know better the phenomena we want to parameterize. Two tools are very important in this effort. The first one is provided by the field experiments. Only by getting the actual subgrid-scale observations can we begin to understand the physics. A second powerful tool is the use of very high resolution models commonly called the large-eddy simulation (LES) models (e.g. Deardorff, 1974) and cloud-resolving models (in this book see Section 5.2). While field experiments provide the best data, they are also very expensive. The use of LES and CRM can provide many more situations in a controlled environment and is much cheaper. We are seeing both approaches used in the quest for better parameterizations.

2.4 Principles of data assimilation

Milija Zupanski and Eugenia Kalnay

In this section we give an overview of the problem of data assimilation, review briefly the characteristics of the data assimilation methods most commonly used at the present in operations (optimal interpolation, or OI, and three-dimensional variational data assimilation, or 3DVAR), and discuss the main two methods currently under development for future operational use (4DVAR and Kalman Filtering or KF). Finally we give an overview of the relative importance of different types of observations.

General data assimilation problem

As discussed in Section 2.1, NWP is an initial value problem, and in order to make a forecast, initial conditions have to be specified. Since NWP models are designed to simulate the evolution of the atmosphere, the initial conditions should represent well the current atmospheric state. Data assimilation is defined here as a method used for creating an accurate estimation of the present atmospheric state (also called *analysis*). An optimal estimation is given by the initial conditions that produce the 'best' subsequent forecast. Note that the quotation marks are used here to emphasize that the meaning of this word varies according to the adopted choice of measure.

The complexity of a data assimilation system depends directly on the complexity of the NWP model. By complexity we mean the range of scales and phenomena that the NWP model simulates, as well as the mathematical equations used in the model. For the state-of-the-art NWP models, with high horizontal and vertical resolution, there is a substantial sensitivity of the forecast to the wide spectrum of atmospheric scales and phenomena that are simulated. For example, initial conditions which are inconsistent with the quasi-geostrophic balance of the atmosphere will give rise to spurious inertia-gravity waves, and then initialization has to be introduced to filter out their (unphysical) contribution to the forecast. Thus, some kind of balancing of the initial conditions needs to be included in the data assimilation as well. As NWP models have grown in complexity over the past decades, so has the data assimilation. A historical overview of data assimilation methods may be found in Daley (1991).

In everyday operational practice data assimilation is applied continuously, in so-called *data assimilation cycles*. In each cycle a new analysis is produced, based on two essential ingredients: data (observations) and the first-guess (short-range model forecast). Usually, the period of the data

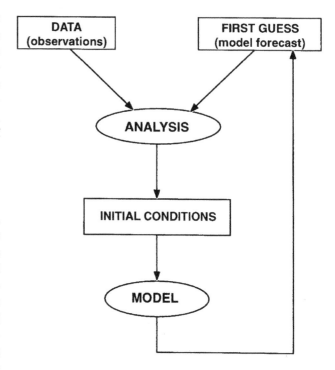

Figure 2.12. A flow diagram of the data assimilation cycle. Input into analysis are observations and the first guess, usually provided by the model forecast. The analysis creates the initial conditions. The consequent short-range forecast is used as a first guess in the next assimilation cycle.

assimilation cycle, given by the length of the forecast, is 1 to 24 hours. The flow diagram above (Figure 2.12) shows a typical data assimilation cycle. As implied in this figure, the data assimilation cycle has to take into account several important factors, as outlined below.

Observations

Intuitively, it is clear that the optimal initial state has to be based on atmospheric observations and fit them well. The number, geographical coverage, and diversity of measuring instruments, has increased dramatically in recent decades. Global coverage maps of some operationally employed observation types are shown in Figure 2.13. It should be noted that:

1. Observations are spatially irregular. Hence, a technique used to assimilate observations has to incorporate some

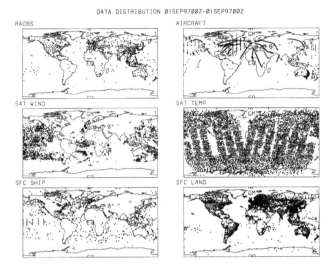

Figure 2.13. Typical global data coverage map for a period of 6 hours. Shown are rawinsonde balloons (top left), aircraft wind speed and direction (top right), satellite cloud-tracked winds (center left), satellite temperature soundings (center right), surface data from ships (bottom left), and surface data from land stations (bottom right). (Courtesy of R. Kistler.)

number, and data processing from the observational location to weather centers may cause many additional errors that need to be detected, and if possible corrected (e.g., Collins and Gandin, 1990).

Background state (first guess)

The problem of creating a complete initial state from an insufficient number of observations requires additional independent information about the atmospheric state. This is usually done by choosing a background field or first guess analysis, such as a climatological analysis, a persistence forecast, or a model forecast. The background state has to be the best available estimate of the present atmospheric state, if no observations are available. Since the NWP models are best suited to transport atmospheric information in space and in time, the best knowledge of the present atmospheric state, prior to including the observations, comes from the use of NWP models. Currently all operational centers use short-range forecasts from a NWP model as the background state.

kind of interpolation between the analysis and observations.

2. Observations are temporally irregular. Some observations are taken almost continuously (e.g., satellites), some are taken at regular time intervals (e.g., radiosondes), or in irregular time intervals (e.g., aircraft data). If all these observations are to be used, a practical method has to be found in order to include a temporal interpolation to the present time. The simplest approach is to take all observations from a certain time interval (say ± 3 hours), and assume that all observations are taken at the middle of the interval.

3. The number of observations is typically two orders of magnitude smaller than the dimension of the analysis (the number of degrees of freedom of the model). This means that there is an indeterminacy problem if only observations are used to create the initial state: the amount of information contained in data is insufficient to produce uniquely and optimally the initial state. This problem can be partially reduced by using observations from the past in creating the present inital state. This is done mainly through the use of the model forecast (background state), as will be discussed below.

4. There is a need for quality control, which rejects, or even corrects bad observations. The variety of data types and observational instruments and platforms, their large

Statistical information

Whenever a measurement (observation) is taken, there will be associated observational errors. Also, NWP models are not perfect, mostly due to the error caused by the limited spatial resolution (Section 2.2), and errors in physical parameterizations (Section 2.3). Since both the observations and the model forecast (background field) are used to create the analysis, they both contribute to the analysis error. It is desirable that a data assimilation algorithm should be able to estimate the analysis error, providing a level of confidence in the resulting analysis. The available information about the observational and model errors is introduced into the analysis by considering their error statistics. In essence, the statistical information about the observational and model errors is used to determine how close the analysis should be to either the observations or the model forecast. Smaller errors imply a closer analysis fit, and vice versa. The specification of these statistics is a difficult problem, due to the the lack of knowledge of the true state of the atmosphere, the ever changing design of NWP models and observational platforms, and to the spatial and temporal correlations between these errors. In addition, there is a computer storage problem due to the large number of observational types and instruments, and to the large number of degrees of freedom of the forecast model.

Data assimilation methods

A good data assimilation method has to pay attention to all available information about the current atmospheric state. This means it has to provide simultaneously a close fit to the observations, to the background state, and to additional dynamical constraints, such as quasi-geostrophic balance, according to their relevance. Although historically approached from different points of view, modern data assimilation methods are related to each other (e.g., Lorenc, 1986). This should be of no surprise, since they all try to solve the same general problem. The difference comes from embedded approximations. Some of the most important aspects of the practical approximations are related to: (a) computational demands, (b) nonlinearity (including physics) of NWP models and observations, and (c) statistical information used (observational and model errors, background state error).

Current operational data assimilation methods

Data assimilation methods used in the major operational centers are the most sophisticated data assimilation techniques that can be run on on present-day computers. Most of the current operational data assimilation systems are based on the optimal interpolation (OI) method, although some NWP centers have implemented a more advanced, three-dimensional variational (3DVAR) data assimilation system. The 3DVAR was implemented operationally at the National Centers for Environmental Prediction (NCEP) in Washington in 1991 (Parrish and Derber, 1992), and in early 1996 at the European Centre for Medium-range Weather Forecasts (ECMWF) (Courtier *et al.*, 1994). Lorenc (1986) showed that in both assimilation methods (OI and 3DVAR) the analysis is obtained by minimizing a cost function defined to measure the distance between the analysis and both the first guess and the observations:

$$J = \frac{1}{2}(H(\boldsymbol{x}) - \boldsymbol{y})^{\mathrm{T}}\boldsymbol{R}^{-1}(H(\boldsymbol{x}) - \boldsymbol{y}) + \frac{1}{2}(\boldsymbol{x} - \boldsymbol{x}^{\mathrm{b}})^{\mathrm{T}}\boldsymbol{B}^{-1}(\boldsymbol{x} - \boldsymbol{x}^{\mathrm{b}}) \qquad (2.4.1)$$

The first term on the right-hand side of equation (2.4.1) measures the misfit (distance) of the analysis to the observations, and the second term is the misfit of the analysis to the background state. The vector $\boldsymbol{x}$ represents the model initial conditions (analysis), and $\boldsymbol{y}$ the observations. The superscript 'T' denotes a transpose, and the superscript 'b' denotes the background state. The matrix $\boldsymbol{R}$ is the observational error covariance, $\boldsymbol{B}$ is the background error covariance, and H is the transformation from the regular (model) space to obser-

vational locations and to observed parameters. For example, in order to use satellite radiances, temperature analyses ($\boldsymbol{x}$) are converted into satellite radiances ($\boldsymbol{y}$) by applying the transformation H. This conversion from analysis to observed variable is a major source of nonlinearity in the above cost function.

Optimal interpolation

First developed by Gandin (1963), the OI method has evolved into an efficient and widely used assimilation method. In OI, the solution to the linearized optimization problem (2.4.1) is explicitly found as

$$\boldsymbol{x}^{\mathrm{a}} = \boldsymbol{x}^{\mathrm{B}} + \boldsymbol{B}\boldsymbol{H}^{\mathrm{T}}(\boldsymbol{H}\boldsymbol{B}\boldsymbol{H}^{\mathrm{T}} + \boldsymbol{R})^{-1}(\boldsymbol{y} - \boldsymbol{H}(\boldsymbol{x}^{\mathrm{b}})) \qquad (2.4.2)$$

where $\boldsymbol{H}$ is a linearization of the transformation H. The superscript 'a' denotes the analysis. The background vector $\boldsymbol{x}^{\mathrm{b}}$ is usually defined as a 6-hour forecast. From a computational point of view, a major obstacle is the inversion of the matrix $\boldsymbol{H}\boldsymbol{B}\boldsymbol{H}^{\mathrm{T}} + \boldsymbol{R}$. In practice, this is solved by representing the background (forecast) error covariance $\boldsymbol{B}$ as a simple function of distance. For example, a function $\exp(-r_{ij}^2/L^2)$ is often used for the correlation of errors in the forecast heights between two points, with L being a typical length scale, and r_{ij} the distance between two points i and j. In addition, only a limited number of observations surrounding an analysis grid point or volume are taken into account (i.e., generating a localized solution), further reducing the computational cost. These simplifications allow the inversion calculation.

Although computationally feasible, this approach implies a forecast error covariance matrix $\boldsymbol{B}$ constant in time, which is certainly inadequate considering that the weather systems, and therefore the forecast errors, change from day to day. Another deficiency of the OI method, being only a linear approximation, is its inability to account properly for the nonlinearity of the transformation H. This is important especially for the assimilation of non-standard observational types, such as satellite radiances, Doppler radar observations, etc., since the transformation H may be very nonlinear in those cases.

In OI, there is also a problem related to balancing the initial conditions created by the analysis. This balance is usually done after the analysis (generally through a method denoted 'nonlinear normal mode initialization'), which modifies the analysis, degrading the fit to the observations and forecast previously achieved, and tends to enhance the problem of 'spin-up' of the hydrologic cycle (Section 2.5).

Three-dimensional variational data assimilation

In the 3DVAR method, a different approach is taken: instead of finding an explicit solution to the minimization problem (equation 2.4.1), the full nonlinear problem is solved in an iterative manner. In this way, the linearity assumption, necessary for the explicit solution (2.4.2), is avoided. As a result of its more general approach, the 3DVAR method has several advantages over OI, including the ability to overcome the linear constraint between the observations and the analysis, and to include all observations simultaneously (i.e. creating a single global solution) rather than a succession of spacially localized solutions. In 3DVAR it is not necessary to assume a specific functional dependence of the forecast error covariance B on distance, as in OI, although it is in current applications still assumed to be constant in time, independent of the current forecast. There is also some flexibility added to the 3DVAR method, since it can include efficiently a number of additional penalty constraints. For example, by including in the cost function (2.4.1) an additional term that penalizes the presence of gravity waves, the fit to observations may be achieved within 3DVAR while simultaneously balancing the initial conditions. As a result, nonlinear normal mode initialization is not needed within 3DVAR, resulting in a considerable amelioration of the problem of spin-up (Section 2.5).

Employment of an iterative minimization algorithm (e.g., conjugate-gradient, quasi-Newton; see Gill *et al.*, 1981) is essential in the 3DVAR technique. In each iteration of minimization, a new (updated) analysis is obtained from

$$x^a_i = x^a_{i-1} + \alpha_i d_i \qquad (2.4.3)$$

where the index 'i' denotes the iteration number, α is a step-length (a constant), and d is a descent direction. The descent direction is obtained from the gradient of the cost function J with respect to the initial conditions (analysis) x, by employing a minimization algorithm. The gradient itself is calculated using an adjoint (conjugate-transpose) operator of the transformation H. The use of the adjoint allows the 3DVAR method to assimilate all types of observations, because it provides the means to transfer the misfit between the first guess and observed parameter back to the analysis space variable. In the current NCEP operational application, the number of minimization iterations is 30–50, which could be further improved by providing a better estimate of the Hessian matrix (second derivative of the cost function J with respect to initial conditions).

Derber *et al.* (1991) and Parrish and Derber (1992) reported that the implementation of the 3DVAR analysis resulted in major improvements with respect to the OI analysis. The 3DVAR analyses were smoother, they fitted the data better, the spin-up of the hydrological cycle was much reduced, there was no need for a nonlinear normal mode initialization, and the consequent forecast skill was also significantly improved. In addition, the computational cost was found to be no larger than the OI.

As suggested earlier, an important consequence of accounting for the nonlinearity of transformation H is that many new types of direct observations can be correctly assimilated. For example, 3DVAR data assimilation allows the three-dimensional assimilation of satellite radiances, rather than the retrieved fields of temperature and humidity used in OI. The benefit of using direct observations instead of retrieved fields is two-fold: first, the observational error statistics are easier to determine, and second, the actual observational errors are smaller because there is no additional error introduced by a retrieval algorithm. This means that the radiance observation bears more significant information than the retrieved fields. This is confirmed by recent results at NCEP (Derber and Wu, 1996), where a substantial improvement in the global NWP forecast has been achieved by assimilating directly the satellite radiance observations in the 3DVAR system. Figure 2.14 shows the evolution of the 5-day forecast skill in the northern hemisphere summer season for the NCEP operational global spectral model (Kanamitsu, 1989; Kalnay *et al.*, 1990). It also shows the skill of the forecasts obtained during parallel testing using 3DVAR of TOVS satellite radiances, compared with the operational assimilation of retrievals. In both the northern and the southern hemispheres the positive impact of the use of the radiances is larger than any other individual improvement made over the last decade.

Future data assimilation methods

A common deficiency of current operational methods is the neglect of the model dynamics in defining the forecast error covariance matrix. As the weather systems constantly change, so does the error of the forecast used to create the first guess. Future data assimilation methods may be viewed as generalizations of current operational methods, in the sense that the current forecast error ('error of the day') is taken into account. This generalization, however, introduces a substantial increase in the computational cost of the analysis method. The two methods for future data assimilation that have been most studied are the Kalman-Bucy filter (Jazwinski, 1970; Ghil *et al.*, 1981) and the four-dimensional variational (4DVAR) data assimilation (e.g., Lewis and Derber, 1985; LeDimet and Talagrand, 1986; Courtier *et al.*, 1994). These two techniques are algebraically equivalent in

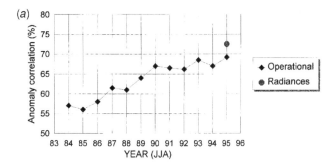

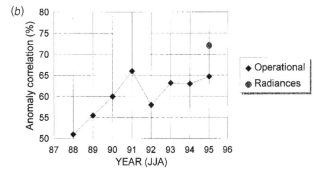

Figure 2.14. Anomaly correlations for the 5 day 500 hPa forecasts averaged over June, July and August months: (*a*) northern hemisphere, and (*b*) southern hemisphere. Solid diamonds are operational scores assimilating infrared and microwave retrievals, and the grey circle for 1995 is from the direct variational assimilation of radiance observations.

the case of a perfect linear NWP model, provided they use the same initial analysis and the associated analysis error covariance matrix (e.g., Jazwinski, 1970, example 7.11).

The Kalman-Bucy filter

The Kalman-Bucy filter (KF) is the optimal data assimilation method for a linear NWP model. The KF is a sequential method, which produces simultaneously the analysis and the analysis error covariance by marching forward in time, from one time level in the data assimilation cycle to the next. The KF algorithm consists of two steps: (a) *a forecast step*, in which the forecast error covariance is estimated, and (b) *an analysis step*, in which the observations are assimilated using the first guess forecast, thus producing the final analysis and the analysis error covariance. In the linear framework, the forecast step is

$$P_f = M P_a M^T + Q \tag{2.4.4}$$

where P_a and P_f are the old (previous cycle) analysis error covariance and the current forecast error covariance, respectively. The matrix Q denotes the model error covariance,

and M denotes a (linearized) forecast model. The analysis step is similar to the general OI solution (equation 2.4.2), except that the constant background error covariance is substituted by a full forecast error covariance (2.4.4). The analysis is given by

$$x^a = x^b + P_f H^T (H P_f H^T + R)^{-1}(y - H(x^b)) \tag{2.4.5a}$$

and

$$P^a = (I - P_f H^T (H P_f H^T + R)^{-1} H) P_f \tag{2.4.5b}$$

In this way, the major advantage of the KF with respect to the OI and 3DVAR methods is the use of a forecast error covariance P_f associated with the first guess x^b instead of a constant background forecast error covariance B.

In principle, the forecast of the error covariance in the KF is equivalent to n model integrations, where n is the number of degrees of freedom of the model. In addition, since there are no specific assumptions made to facilitate the inversion of large matrices in equation 2.4.5, the application of the KF using sophisticated NWP models remains exceedingly costly. The analysis error covariance matrix is of the order of $10^7 \times 10^7$ for state-of-the-art global NWP models, too large to store even on the today's supercomputers. Furthermore, the non-linearity of NWP models may be accounted for only by increasing significantly the computational burden. To date, the Kalman filter has never been applied to the assimilation of atmospheric data in a complete form, using state-of-the-art NWP models with physics. There have been, however, important applications of the Kalman filter using less sophisticated, low-dimensional models such as shallow water and quasi-geostrophic models (e.g., Ghil *et al.*, 1981; Cohn and Parrish, 1991; Todling and Cohn, 1994).

4DVAR data assimilation

This technique, which is a generalization of 3DVAR , offers a practical advantage over the KF: the computational cost is relatively smaller due to the iterative character of the method. Instead of storing the elements of large matrices, the 4DVAR method works with vectors and operators only, thus significantly reducing the memory requirements. Consequently, it can be applied using sophisticated forecast models (e.g., Thepaut and Courtier, 1991; Navon *et al.*, 1992; Zupanski, 1993; Zupanski and Mesinger, 1995). The optimal analysis in the 4DVAR data assimilation method is obtained by fitting the model forecast to the observations and the first guess during the chosen time interval (assimilation period), i.e., by minimizing the cost function defined over the length of assimilation period or cycle

$$J = \frac{1}{2} \sum_{TIME} (H[M\boldsymbol{x}_0, \phi)] - \boldsymbol{y})^{\mathrm{T}} \boldsymbol{R}^{-1} (H[M(\boldsymbol{x}_0, \phi)] - \boldsymbol{y})$$
$$+ \frac{1}{2} \sum_{TIME} \phi^{\mathrm{T}} \boldsymbol{Q}^{-1} \phi + \frac{1}{2} (\boldsymbol{x}_0 - \boldsymbol{x}_0^{\mathrm{B}})^{\mathrm{T}} \boldsymbol{B}^{-1} (\boldsymbol{x}_0 - \boldsymbol{x}_0^{\mathrm{B}}) \qquad (2.4.6)$$

where M is a (nonlinear) NWP model operator, the background error covariance $\boldsymbol{B}$ is defined at the beginning of the data assimilation period, and *TIME* denotes the length of assimilation period. The model error vector is denoted ϕ, and it is a function of both space and time. The subscript '0' is used to indicate the model state at the beginning of assimilation period. The first term on the right-hand side of (equation 2.4.6) represents the distance between the model forecast and observations during the period of assimilation. The second term represents the error introduced by the use of an imperfect forecast model in assimilation, and the third term is the distance between the initial conditions and the background state. The cost function (2.4.6) is similar to the cost function (2.4.1), except for the introduction of time as an additional dimension, and the cost minimization is also performed by iterations. This implies that all observations during the chosen assimilation period are included in assimilation, as well as the forecast model integration.

The iterative character of the 4DVAR method facilitates the inclusion of nonlinearities present in the forecast model M and in the transformation H, in a manner equivalent to the 3DVAR method (equation 2.4.3). In order to apply the method, it is important to define the *control* variable, defined as a vector, or a parameter, which is altered during the cost minimization. This vector includes the initial conditions at the beginning of data assimilation period, the model error, physical parameters, etc. Also, the implementation of the 4DVAR method requires the adjoint of the (linearized) NWP model, in addition to the adjoint of the linearized transformation H. The optimal present atmospheric state (e.g., analysis) in the 4DVAR method is obtained as a short-range forecast, valid at the end of the assimilation period.

The full 4DVAR data assimilation method is being tested at NCEP in a realistic, operational environment, using the operational limited-area Eta model (Mesinger *et al.*, 1988; Black, 1994). In this application the cost function (2.4.6) is defined including an additional gravity wave penalty term to ensure balanced initial conditions. A random model error component ϕ is defined using a coarse time resolution (3 hours) (D. Zupanski, 1996, pers. comm.). An example of the 4DVAR assimilation (after 10 minimization iterations) is shown in Figure 2.15, for a 12-hour assimilation period. The wind forecasts starting from both the OI and 4DVAR analyses are compared against observations every 12 hours of

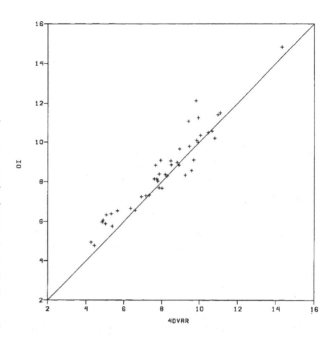

Figure 2.15. A scatter diagram of the total wind rms forecast errors (m s[-1]), averaged over all vertical levels. The 4DVAR based forecasts are compared against the OI based forecasts, every 12 hours. There are nine 4DVAR and OI assimilation experiments shown, randomly chosen during 1994 and 1995. Each point represents a specific forecast time, and a specific assimilation experiment. The points above the middle line show a smaller rms error in the 4DVAR case, hence a better forecast.

the forecast, up to the 48-hour forecast. Preliminary results from nine synoptic situations show a very encouraging improvement of the forecast based on the 4DVAR method. Further work is needed before the full potential of the method can be achieved. A useful estimate of the associated analysis error is difficult to make, due to the computational requirements. Also, there is a need for a better model and observational error statistics, use of all available direct observations etc.

A practical problem with both the KF and the 4DVAR methods is the limited knowledge of model error statistics for sophisticated forecast models. Yet, the inclusion of the model error statistics should be the strongest advantage of the two future methods over current operational methods (OI and 3DVAR). There are still unresolved problems associated with the next generation assimilation methods, but due to the strong relationship between the KF and 4DVAR methods, it can be expected that research will lead towards a merging of the two methods.

53

Relative importance of different types of data

Figure 2.13 shows the typical coverage of data available in four assimilation cycles of 6 hours. The available data are of different types. The 'in situ', or conventional data base, most of which has been available for many decades, is composed of (a) rawinsondes, which provide vertical soundings of temperature, moisture, and wind; (b) aircraft observations of wind and temperature, which are available at flight level; and (c) surface observations from land stations, ships, and moored and drifting buoys, which provide observations of pressure, temperature and humidity at the surface. Since the Global Weather Experiment of 1979 (GWE, also known as FGGE, or First GARP Global Experiment), much use has been made of satellite data that provide additional extensive coverage. The satellite data include: (a) TOVS vertical sounders from the NOAA polar orbiting satellites, and from which vertical retrievals of temperature and moisture are operationally generated by NOAA/NESDIS; (b) cloud-tracked winds from the geostationary weather satellites (GOES from the US, METEOSAT from Europe and GMS from Japan); (c) surface winds from scatterometer instruments available from several polar orbiters such as ERS (Europe), ADEOS (Japan), and surface wind speed from SSM/I (Defense Meteorological Satellite Program, DMSP).

The relative importance and usefulness of this complex data base depends on several factors: whether the data are in the form of a vertical profile or a single level source of data, and the accuracy of the data. In addition, their importance depends on the extent to which the data contribute to defining the value of potential vorticity, a fundamental conserved dynamical quantity of the atmosphere, since potential vorticity, together with a balance constraint, can be used to determine the initial conditions of the atmosphere.

It is necessary to simplify the discussion here for the sake of space, but from arguments based on a theoretical basis (e.g.,

potential vorticity and geostrophic balance) as well as from experience, it can be stated that, in general:

1. Vertical profile data are more useful than single level data. As a result, the rawinsonde network, which provides wind, mass (temperature and pressure) and moisture profiles, is the backbone of the present observing system.

2. Because of its strong influence on potential vorticity, wind data tend to be somewhat more important than mass (temperature and pressure) data. This is particularly true in the tropics.

3. Moisture data are most useful for very-short-range forecasts (i.e., 12 hours or so). For forecasts longer than one day, forecast models generally tend to reach their own hydrologic equilibrium, and basically 'ignore' moisture observations. Recently, however, surface observations of temperature and moisture have been also use to initialize soil moisture (e.g., Mahfouf, 1991).

4. Of the satellite observations, the temperature and moisture sounders have been of vital importance in the southern hemisphere. It is remarkable that southern hemisphere forecast skill, fundamentally dependent on satellite data, especially TOVS, is currently at a level comparable to that of the northern hemisphere (NH) only about 5 years ago, even though the NH has much more abundant observations. In the NH, the impact of satellite data has been generally small (e.g., Mo *et al.*, 1995), except for the large positive impact obtained with the introduction of 3DVAR direct assimilation of satellite radiances (Derber and Wu, 1996).

5. We can expect that in the next decade there will be a major development in the area of mesoscale modeling and data assimilation, and that Doppler radar data, which can provide high resolution wind and precipitation data, will be essential for the mesoscale models' initialization.

2.5 Estimation of the global energy and water cycle from global data assimilation

Glenn White and Suranjana Saha

As discussed in previous sections, modern atmospheric data assimilation systems used in numerical weather prediction (NWP) include models with physical parameterizations that attempt to reproduce all important atmospheric processes, all exchanges of radiation with space, and all exchanges of momentum, water and energy with the Earth's surface (Section 2.3). Precipitation and fluxes from data assimilation are generally accumulated during the 6-hour forecasts that are used in the analysis cycle as the first guess for the next analysis. Such data assimilation systems produce analyses every six hours from a combination of accurate short-range forecasts and all available observations, as described in Section 2.4, and the analyses are carefully monitored by a variety of users. Both the spatial and temporal resolution of NWP analyses continue to increase over the years.

Since NWP analyses use many observations from diverse sources to provide estimates of the global distribution of precipitation and surface fluxes a few hours after observation time, and also offer complete global coverage on a regular grid of precipitation and surface fluxes on a near-real time basis, NWP precipitation and fluxes are potentially very valuable estimates of the actual values and offer an extremely valuable resource in understanding the global energy and water cycles. The accuracy of these NWP products needs to be carefully assessed.

Several studies have examined precipitation and surface fluxes from NWP analyses and forecasts. Arpe (1991), Arpe *et al.* (1988) and Arpe and Esbensen (1989) evaluated fluxes from the European Centre for Medium-range Weather Forecasts (ECMWF) and other NWP systems. Foreman *et al.* (1994) assessed fluxes from ECMWF and the United Kingdom Meteorological Office (UKMO). Janowiak (1992) compared tropical rainfall from ECMWF, the National Centers for Environmental Prediction (NCEP, then known as the National Meteorological Center) and satellite estimates. Da Silva and White (1996) compared surface fluxes from the NCEP/NCAR (National Center for Atmospheric Research) and NASA/Goddard Space Flight Center (GSFC) reanalyses with fluxes estimated from ship observations and from satellite-derived estimates; Reynolds and Leetma (1996) compared surface fluxes from the NCEP/NCAR reanalysis to surface fluxes estimated by Cayan (1992) from ship observations. Bony *et al.* (1997) compared surface and atmospheric fields related to atmospheric hydrology and radiation from the NASA/GSFC and NCAR/NCEP reanalyses with estimates from satellite data over the tropical oceans. White (1995) compared pre-

cipitation, surface fluxes and top-of-the-atmosphere (TOA) outgoing longwave radiation (OLR) from four operational global analysis/forecast systems (ECMWF, the Japan Meteorological Agency (JMA), NCEP and UKMO) to each other and to independent estimates based on ship reports, satellite data and station data.

These studies have found considerable agreement between the NWP fields and independent estimates, especially in terms of large-scale patterns. In many respects the NWP estimates agree with the independent estimates to within the uncertainty in the independent estimates. Such comparisons have found and continue to find specific errors in the NWP precipitation and surface fluxes and as a result have led to improvements in both analysis systems and forecast models.

One problem with NWP-derived precipitation and fluxes has been the frequency of changes in the analysis/forecast systems over the years. While the changes have produced great improvements in weather forecasts, they have also caused changes in the analyzed fields, especially in precipitation and surface fluxes, that can exceed and mask interannual variability. To remedy this problem of perceived climate changes due to analysis or model changes, several centers are conducting extensive 'reanalyses' of past atmospheric data, using a modern, frozen analysis/forecast system. The NCEP/NCAR 40-Year Reanalysis will cover the period 1957–1996 and should be completed in 1997 (Kalnay *et al.*, 1996). It may also be extended back to the establishment of the northern hemisphere radiosonde network in the late 1940s. Fields from the NCEP/NCAR Reanalysis are available on a CD-ROM in the March 1996 issue of the *Bulletin of the American Meteorological Society*. The ECMWF reanalysis includes the 15-year period 1979–1993 and has been completed. The centers plan to perform reanalyses again every 5–10 years with current state-of-the-art data assimilation and modeling systems.

Another problem is that numerical models generally undergo a 'spin-up' process during which considerable changes occur in the hydrologic cycle and in other physical processes. The spin-up takes place during the initial stages of a forecast, reflecting an imbalance between initial conditions of the analyzed atmosphere, influenced by both the forecast model and the observations, and the atmosphere consistent with the forecast model's physics and dynamics. The use of 3-D variational data assimilation (Section 2.4) instead of the more traditional optimal interpolation (OI) method followed by normal mode initialization, has resulted in an amelioration of the problem of spin-up. With OI, the model showed very strong spin-up at low resolution, and strong spin-down at high resolution, with changes in the global precipitation as large as 50–100%. Figure 2.16 shows the change in global mean precipitation over the first five forecast days averaged for May 1–20, 1993 in two versions of the NCEP global model, showing relatively little spin-up: the global rainfall in a version of the model with 18 levels and Kuo parameterization of convection increases by 13% from the analysis cycle to the 5-day forecast (spin-up), whereas the version of the model with 28 levels and simplified Arakawa–Schubert convection (much like the model used in reanalysis, but at higher horizontal resolution, implemented operationally in January 1995) shows a decrease of 4% in global precipitation over the same interval (spin-down). The version of the operational global model implemented at NCEP in October 1995, has an improved bulk parameterization of the boundary layer (Section 2.3), and responds better than the reanalysis version to SST anomalies (Section 2.6), but has a slightly larger spin-down in the global precipitation. Over smaller regions, the spin-up or spin-down can be considerably larger. Since the effect of the spin-up is strongest during the first few hours of the forecast, the ECMWF reanalysis, which uses OI and normal model initialization instead of the 3DVAR used in the NCEP/NCAR reanalysis, stores the 12–36 hr forecasts of precipitation rather than the 6-hr forecast from the analysis cycle.

In the rest of this section we present comparisons of fields characteristic of the energy and water cycles obtained from the NCEP/NCAR reanalysis CD-ROM discussed above and other observational estimates.

Global estimates of precipitation and evaporation

The level of uncertainty in our current knowledge of rainfall is shown in Figure 2.17, which compares three different estimates of the zonal mean distribution of precipitation for July 1987: (a) from the NCEP/NCAR reanalysis, (b) from estimates derived from the MSU satellite sounder blended with station reports by Schemm *et al.* (1992) recently modified by an upgrade to the MSU algorithm (J. Schemm, 1996, pers. comm.), and (c) an estimate by Xie and Arkin (1996) blending two satellite products and station reports. The three estimates display very similar large-scale patterns that are consistent with July rainfall climatologies. Reanalysis amounts near 50°N and 60°N over land and over the southeast United States are larger than the other estimates. The reanalysis precipitation is the lowest in the deep tropics; in mid-latitudes the differences between Xie and Arkin (1996) and

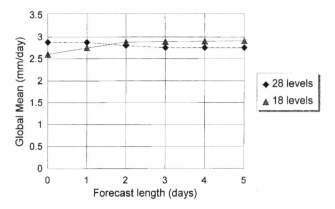

Figure 2.16. Global mean precipitation (mm day⁻¹) over the first five days of the forecasts averaged over the period May 1–20, 1993, in two versions of the NCEP global analysis/forecast system.

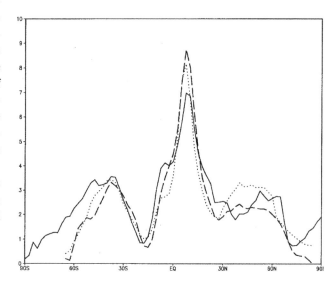

Figure 2.17. Zonal mean precipitation (mm day⁻¹) for July 1987 from Schemm *et al.* (1992) (dotted line), the NCEP/NCAR reanalysis (solid line) and Xie and Arkin (1996) (dashed line).

Schemm (1996), both based on satellite observations, can be as large as the differences between Xie and Arkin (1996) and the reanalysis. It should be noted that the magnitude of rainfall estimates using the same MSU satellite observations was reduced by about 30% in response to modifications in the algorithm used to estimate rainfall. It is not clear which estimate of rainfall is the most accurate.

The interannual variability in precipitation estimated by Schemm (1996), Xie and Arkin (1996) and the NCEP/NCAR reanalysis for July 1987 to July 1988, is shown in Figure 2.18. The three estimates display generally similar patterns, for example in the Pacific and Indian oceans. However, there are

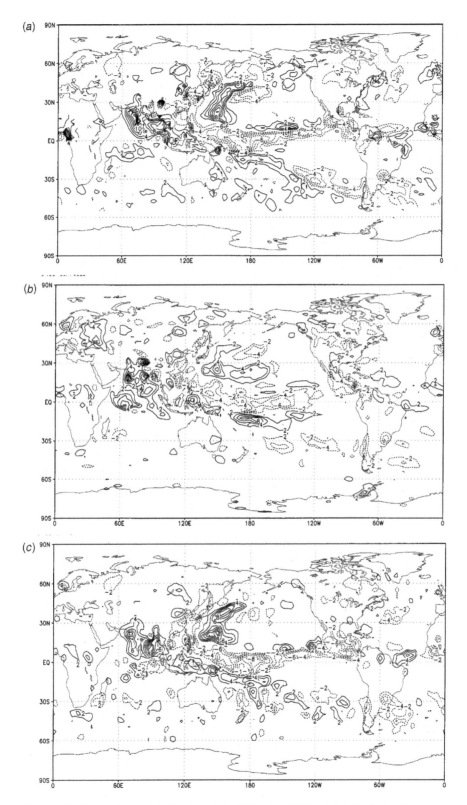

Figure 2.18. Change in precipitation from July 1987 to July 1988 as estimated by (*a*) Schemm *et al.* (1992), (*b*) the NCEP/NCAR reanalysis and (*c*) Xie and Arkin (1996). Contours −6, −4, −2, 2, 4, 6 mm day⁻¹. Dashed lines indicate decreases from 1987 to 1988.

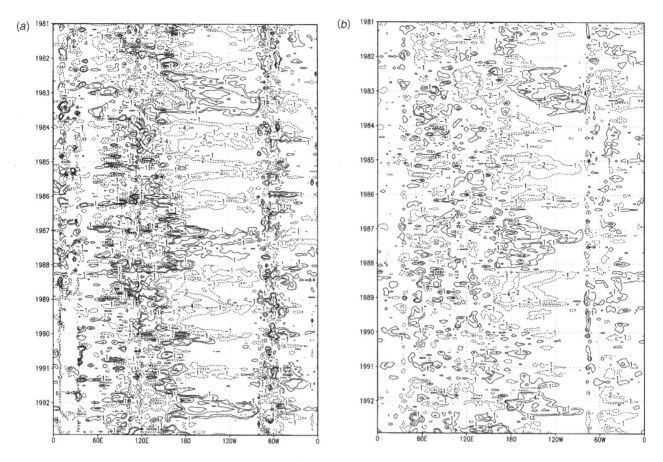

Figure 2.19. Longitude–time diagrams of precipitation anomalies averaged over 5 °N–5 °S from (*a*) Schemm *et al.* (1992) and (*b*) the NCAR/NCEP reanalysis for 1981–1992. Anomalies are from each estimate's own climatology. Contours −8, −4, −2 −1, 1, 2, 4, 8 mm day⁻¹; dashed contours indicates less precipitation than normal.

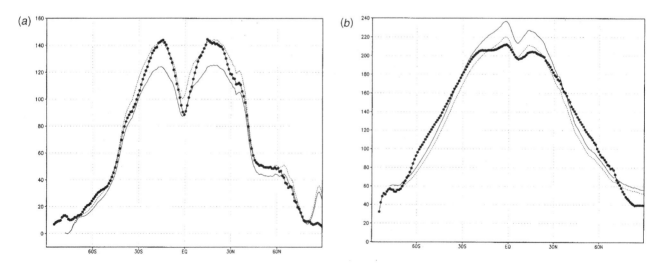

Figure 2.20. (*a*) Annual mean oceanic zonal mean evaporation and (*b*) surface net shortwave radiation from several years of reanalysis (large dots), da Silva *et al.* (1994) unconstrained (solid) and da Silva *et al.* (1994) constrained (small dots).

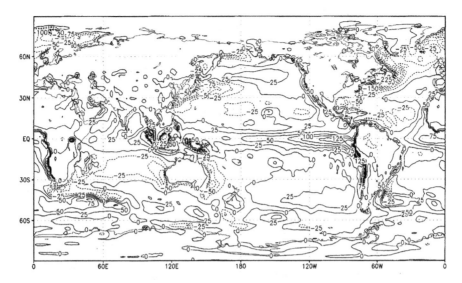

Figure 2.21. Annual mean net surface heat flux from several years of NCAR/NCEP reanalysis. Contours at intervals of 25 W m⁻². Dashed upward fluxes.

also significant differences: the reanalysis has smaller changes in precipitation in the ITCZ (Intertropical Convergence Zone) near South America than the satellite estimates, whereas in other areas one of the satellite estimates is closer to the reanalysis than to the other satellite-derived field.

Figure 2.19 compares monthly anomalies in precipitation averaged over 5°N–5°S from Schemm (1996) and from reanalysis for 1981–1992. Both show similar patterns in variability, reflecting largely the El Niño Southern Oscillation, but the anomalies from reanalysis are smaller than the anomalies from Schemm (1996). Anomalies for the same region from Xie and Arkin (1996) for a shorter period are also larger than the reanalysis anomalies.

Surface fluxes of energy

Figure 2.20 compares zonal mean latent heat flux (*a*) and surface net shortwave radiation (*b*) from several years of reanalysis to climatological estimates from da Silva *et al.* (1994) based on 45 years of ship observations. Da Silva *et al.* (1994) carefully corrected for biases in the ship observations, then calculated the net heat flux and the inferred oceanic heat transport. They found that the heat transport was unrealistic and constrained the fluxes mathematically to obtain a reasonable zonal mean oceanic heat transport. The adjustment needed can be thought of as a lower bound on the uncertainty in da Silva *et al.* (1994)'s climatology. Reanalysis values of evaporation tend to lie between the two da Silva *et al.* (1994) estimates; net shortwave in the tropics is closer to the constrained than the unconstrained estimate. Reanalysis

fluxes tend to be closer to da Silva *et al.* (1994)'s 'corrected' fluxes than to their unconstrained fluxes. Net shortwave in higher latitudes, however, exceeds either estimate by da Silva *et al.* (1994).

A major source of uncertainty in surface fluxes, both in NWP systems and in reality, is cloudiness. Net shortwave radiation at the surface and TOA OLR (top of atmosphere outgoing longwave radiation) from the NCEP/NCAR reanalysis are considerably closer to estimates based on satellite data than in earlier versions of the NCEP global analysis/forecast system, such as studied by White (1995); however, A. da Silva and W. Ebisuzaki (private communication) found that the ocean surface albedo is too high. A new shortwave radiation code being tested at NCEP corrects this problem.

Figure 2.21 shows the regional distribution of annual mean net surface heat flux from several years of reanalysis. Heat enters the ocean in the tropics, particularly in the eastern oceans and leaves the ocean in the subtropics and northern hemisphere mid-latitudes. Strong heat flux out of the ocean is seen over the Kuroshio and Gulf Stream. A surprising feature (also seen in the ECMWF reanalysis, but for different reasons) is heat flux *into* the ocean near 50° S.

Figure 2.22 compares zonal mean annual mean surface net heat flux from several years of reanalysis to da Silva *et al.* (1994)'s unconstrained and constrained estimates. In the tropics and subtropics the net heat flux from reanalysis agrees well with da Silva *et al.* (1994)'s constrained net heat flux; in higher latitudes the reanalysis is closer to their unconstrained estimates.

Figure 2.23 shows the oceanic heat flux implied by the net

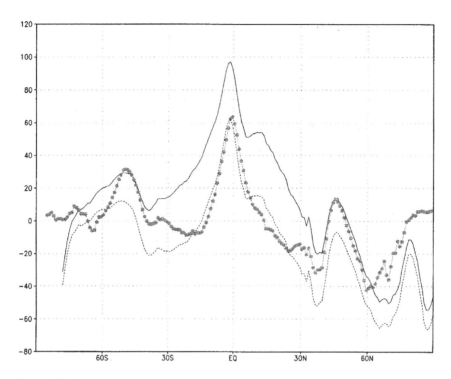

Figure 2.22. Annual mean oceanic zonal mean net surface heat flux from (large dots) several years of reanalysis, (solid) da Silva *et al.* (1994) unconstrained and (small dots) da Silva *et al.* (1994) constrained.

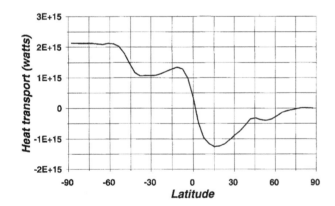

Figure 2.23. Zonal mean oceanic heat transport integrated from surface net heat flux starting at North Pole based on several years of NCAR/NCEP reanalysis.

surface heat flux, integrating from the North Pole. The pattern is reasonable in the northern hemisphere and in the southern hemisphere up to 40°S, but then fails to go to zero at the South Pole, reflecting a global mean net heat flux into the ocean of slightly less than 6 W m^{-2}. (The ECMWF reanalysis shows a slightly larger imbalance in its 0–6 hour forecasts, but a smaller imbalance in its 12–36 hour forecasts.) Maximum magnitudes in the subtropics are somewhat less than independent estimates. No effort was made to adjust the fluxes to achieve a better oceanic heat transport. The uncertainties in the components of the net heat flux clearly exceed the global imbalance.

Fluxes from the NCEP/NCAR reanalysis have some specific problems, as pointed out above, but overall appear to be quite realistic and to agree fairly well with fluxes estimated from ship observations, both in climatological fields and in interannual variability. Comparison with independent estimates of fluxes reveals some specific problems in NWP fluxes; in many respects, however, NWP fluxes agree with fluxes estimated from ship observations to within the uncertainty in these fluxes. More detailed comparisons of NWP fluxes to flux estimates from satellites and field experiments are needed, with careful consideration to the uncertainties in these estimates.

2.6 Coupled atmosphere–ocean modeling for climate simulation and prediction

Ming Ji and Arun Kumar

Ocean data assimilation and the prediction of interannual SST variability

In this section we discuss applications of global atmosphere model and data assimilation techniques in simulation and prediction of short-term climate anomalies. The short-term climate anomalies of the atmosphere are a shift of the mean atmospheric conditions from climatology resulting in part from anomalies in the boundary forcing, primarily the sea surface temperature (SST) changes on a seasonal to interannual time scale. Although the time scale for the short-term climate variability is well beyond the limit of *weather* predictability of the atmosphere (about 2 weeks), prediction of climate anomalies may be possible if the changes in the boundary conditions such as SST can be predicted. Furthermore, since the time scale for prediction of climate variability is beyond the limit of predictability, the atmospheric models used for climate emphasize the robust response to variability of boundary forcing rather than initialization of the atmosphere. However, the climate signals reside in the ocean because of its much greater heat capacity compared with the atmosphere. Therefore, initialization of the ocean is critical for short-term climate prediction. One example is the prediction of the tropical Pacific SST on a seasonal to interannual time scale. It is essentially an initial value problem for the tropical Pacific ocean analogous to the medium range numerical weather prediction; hence accurate initialization of the ocean is crucial for achieving skillful predictions (Chen *et al.*, 1995; Ji and Leetmaa, 1997).

Climate prediction requires coupled ocean–atmosphere models because the slow variations in the ocean drive the climate variations of the atmosphere through coupled ocean–atmosphere interactions. One application of coupled ocean–atmosphere models is the prediction of El Niño-Southern Oscillation (ENSO) variability for the tropical Pacific.

ENSO is the major source of short-term climate variability on a seasonal to interannual time scale in tropical convection and precipitation, and has significant impact on global climate anomalies such as temperature and precipitation (Rasmusson and Wallace, 1986; Ropelewski and Halpert, 1987). Simulation and prediction of changes in tropical convection patterns and their impact on global precipitation and temperature depend on skillful prediction of seasonal to interannual tropical Pacific SST anomalies associated with ENSO. The ENSO variability is the result of coupled interaction between the tropical Pacific ocean and global atmosphere on a basin-wide spatial scale and a seasonal to interannual time scale (Bjerknes, 1969; Wyrtki, 1975). Since the oceanic response to the change in atmospheric wind forcing is much slower than the atmospheric response to changes in SST, evolutions in the ocean dominate the timing and strength of the coupled ocean–atmosphere system's variability on a seasonal to interannual time scale. Numerous studies using coupled models have shown that the predictability for the tropical Pacific SST variability associated with ENSO can be one year or longer (see review by Latif *et al.*, 1994).

Since the tropical ocean circulations depend mainly on the past history of atmospheric wind forcing, initialization of the ocean primarily requires high-quality historical atmospheric wind stress. Unfortunately, the available wind products, both those based mainly on sparse *in situ* observations, and those produced by operational atmospheric analyses, contain large errors, and therefore these wind products alone are not sufficient to produce accurate oceanic initial conditions (Ji and Smith, 1995).

Ocean data assimilation (primarily subsurface temperature observations into the ocean model, similar to the atmospheric data assimilation described in Section 2.4), is used to achieve accurate oceanic initial conditions for ENSO prediction. Although similar in concept and in the way that the data assimilation is carried out, there is an important difference in the objective of ocean data assimilation compared with the atmospheric data assimilation: a main objective of ocean data assimilation is to compensate for deficiencies in the wind forcing field and errors in ocean models in order to achieve better ocean initialization. As the ocean model and the quality of wind forcing improve, through an improved observational network over the open oceans and improved operational atmospheric analyses, the dependence on ocean data assimilation for coupled model initialization probably will decrease, and there will be an improved balance between the oceanic initial conditions and the coupled forecast model. There is already some evidence that such improved balance could lead to improved forecast skill and predictability for ENSO (Chen *et al.*, 1995).

Ocean data assimilation has been an active area of research since the late 1970s, partially inspired by the success in atmospheric data assimilation and partially due to advances in ocean data collection and modeling. A comprehensive review for many of the early works in the field of ocean data assimilation in support of tropical ocean circulation studies is found in the review by Busalacchi (1996).

A decade-long international research program, Tropical Ocean and Global Atmosphere (TOGA, 1985–1994) which focused on understanding the variability and predictability of the coupled tropical ocean–atmosphere system, mainly the ENSO, resulted in the implementation of the TOGA observing system. The TOGA observing system consists of drifting buoys for observations of SST and currents; moored buoys, i.e. the tropical atmosphere ocean (TAO) array (McPhaden, 1993), which is primarily for observing surface winds and subsurface temperatures in the equatorial Pacific; and an expanded voluntary opportunity ship (VOS) network for observing surface winds, SST and subsurface temperatures using expendable bathythermographs (XBT). The implementation of the TOGA observing system brought vast improvement in data quantity and spatial and temporal coverage, both for the surface and subsurface, for the tropical Pacific. This provided an unprecedented opportunity for ocean modelers and data assimilators to develop improved ocean general circulation models and ocean data assimilation systems to analyze and diagnose the tropical Pacific ocean and to initialize coupled models for ENSO monitoring and prediction. A review of the TOGA observing system at the end of TOGA is given by McPhaden et al. (1996).

With the much improved data coverage during the TOGA program, routine basin-scale analyses for monitoring low-frequency climate variability in the tropical Pacific became possible. A number of operational centers are routinely producing monthly analyses for the Pacific and global oceans; however, the focus is in the tropical Pacific. These include the ocean analysis systems developed at the Bureau of Meteorology Research Center (BMRC) of Australia (Smith, 1995), the El Niño monitoring center of the Japan Meteorological Agency (Kimoto et al., 1996), and the National Centers for Environmental Prediction (NCEP, formerly the National Meteorological Center, or NMC) of the USA (Ji et al., 1995).

The NCEP ocean analysis system is based on the system developed by Derber and Rosati (1989). This data assimilation system is based on a variational method (Section 2.4). In this system, assimilation is done continuously during the model integration, and the corrections to the model are spread over a long period; thus changes to the model temperature field during each model time step are very small, and this significantly reduces the impact of the data assimilation to the dynamical balances of the model fields, and also keeps them from drifting back towards the model's own climate. This strategy also compensates for a main drawback to the sequential initialization method: that is, the insertion of data often introduces a strong shock when corrections are applied to the model fields, as discussed in Moore (1990). Further, this system keeps an observation in the model for a

long period of time (2–4 weeks), weighted down by the difference between the model time and the observation time during each assimilation time step, thus significantly increasing the influence of observations to compensate for the lack of spatial and temporal data coverage in many areas. The disadvantage of this method over long periods is that it tends to limit the analyses to resolving only large spatial scales and low frequency phenomena (Halpern and Ji, 1993). A similar data assimilation system is also in use at NOAA's Geophysical Fluid Dynamical Laboratory (GFDL, Rosati et al., 1995).

Shown in Figure 2.24 is an example of a retrospective subsurface ocean analysis for 1980–1995 produced at the NCEP (left panel). The historical monthly pseudo-stress analyses produced at the Florida State University (FSU), (Goldenberg and O'Brien, 1981) were used to force the ocean model. The FSU winds are mainly based on ship board observations. Ji and Smith (1995) found that this data set provides relatively high quality, consistent long-term wind forcing. Shown in the figure is the time history of the depth of 20°C isotherm anomalies along the Equator in the Pacific. The 20°C isotherm lies in the middle of the thermocline and its variation is often used as a proxy for the thermocline variation. For comparison, 20°C anomalies produced from a model simulation, using the same ocean model and the same wind forcing but without assimilating observed subsurface temperature data, are shown in the right panel. The thermocline anomalies produced by the ocean analyses show stronger variability in the central and western Pacific compared with that produced by the model forced with the FSU winds without data assimilation. Comparisons with in situ observations of moorings and tide gauges (not shown) suggest that the model-based analyses are of higher accuracy than the wind-forced simulation (FSU) results (Ji and Smith, 1995). The comparison shows that even when using a high-quality wind-stress forcing and a state-of-the-art ocean general circulation model, ocean data assimilation can still further improve the quality of analyses by compensating for errors in the forcing field and in the model.

The assimilation of observations obtained from the TOGA observing system provides not only the means to produce much improved ocean analyses, but also a great opportunity for improving definition of the initial ocean fields for the prediction of ENSO using coupled models. As indicated earlier, the memory of the climate signals resides in the ocean because of the greater heat capacity of the ocean compared with the atmosphere. Therefore, skillful prediction of SST variations associated with ENSO depends strongly on the accuracy of the ocean initial conditions. For coupled general circulation models, two different methods are presently used for the initialization of the ocean in ENSO predictions. The

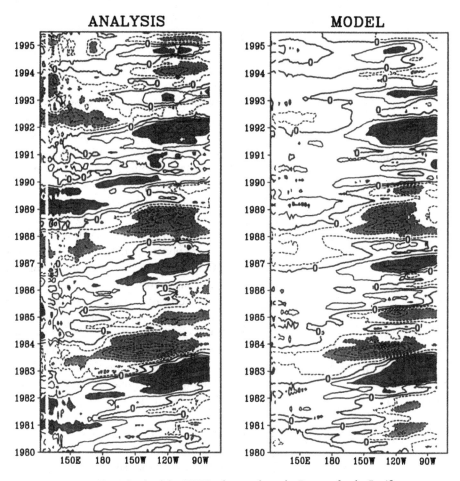

Figure 2.24. Anomalous depth of the 20°C isotherms along the Equator for the Pacific produced from NCEP retrospective ocean analysis (left panel) and from an ocean model simulation forced with monthly surface wind analyses from FSU (right). The contour interval is 10 m. Anomalies greater (less) than 20 m (– 20 m) are indicated by dark (light) shading.

first method uses assimilation of subsurface temperature data together with the best available surface wind forcing to achieve initialization of subsurface ocean states. An example of this initialization method is the NCEP coupled model (Ji *et al.*, 1994). A second method, developed by Kirtman *et al.* (1997), uses a statistical method to produce surface-stress forcing to the ocean model based on near-surface winds from the atmospheric model. This approach essentially uses a statistical method to compensate for errors in the atmospheric general circulation model (GCM) boundary layer parameterization which results in errors in surface stress (Section 2.3).

As shown in Figure 2.24, by assimilating observed subsurface temperature data, the NCEP ocean analysis system produced ocean initial conditions that exhibited stronger thermocline variability on seasonal to interannual time scale compared with wind forced simulations. This indicates that

the data assimilation captured signals in the oceanic initial conditions which are either missing or too weak in the wind forcing field. Ji and Leetmaa (1997) showed impacts of subsurface data assimilation on forecast skill for ENSO.

Shown in Figure 2.25 are temporal anomaly correlation coefficients (ACC) and root mean square (RMS) errors as a function of forecast lead times for area-averaged SST anomalies between forecasts and observations for an eastern equatorial Pacific region (170°W–120°W, 5°S–5°N). The predictions were launched once a month for the period of 1983–1993. Solid curves in the figure represent ACC and RMS errors for forecasts initiated from ocean initial conditions produced with assimilation of surface and subsurface temperature data, whereas dashed curves are for forecasts initiated from ocean initial conditions produced without subsurface data assimilation. Since the only difference between the two sets of forecast experiments is in the generation of ocean initial

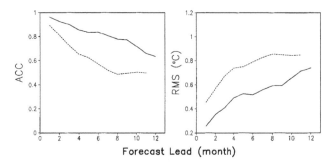

Figure 2.25. Anomaly correlation coefficients (left) and RMS errors (right) between forecasts and observations for area-averaged SST anomalies in the eastern equatorial Pacific region (Niño 3.4), 170° W–120° W and 5° S–5° N. Solid (dashed) curves are for forecasts initiated from ocean initial conditions produced with (without) subsurface data assimilation.

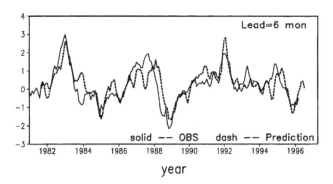

Figure 2.26. Six-month lead predictions of Niño 3.4 (170° W–120° W, 5° S–5° N) SST anomalies for the period of 1981–1995 using the NCEP coupled forecast model. Predicted/observed SST anomalies are shown in dashed/solid curves. Anomalies for predictions are departures from the model climatology.

conditions, this figure demonstrates that data assimilation has a significant positive impact on improving ENSO forecast skill.

Shown in Figure 2.26 are time histories of observations and 6-month lead predictions of area-averaged SST anomalies for the Niño-3 region (150°W–90°W, 5°S–5°N) for the 1982–1995 period. The predictions were made using the NCEP coupled forecast model and ocean initial conditions produced with assimilation of subsurface temperature data. As shown in this figure, the interannual time-scale variability of the eastern equatorial Pacific SST was well predicted by the NCEP coupled model.

An additional note regarding the results shown in Figure 2.25 is that for the forecasts initiated without using data assimilation, there is still useful skill in the ENSO prediction.

This demonstrates that in nature, the history of atmospheric wind forcing is the primary source containing information for ocean initialization. Coupled models of less complexity, such as Cane and Zebiaks' (Cane *et al.*, 1986) and Kleeman's coupled models (Kleeman, 1993), have demonstrated significant predictive skill for ENSO with lead time of up to one and a half years or longer using only FSU wind forcing as the primary source of observed information for model initialization. These evidences support our earlier statement that the ocean data assimilation is a tool rather than an end objective for achieving improved ocean initialization. The goal for ocean initialization is a systematic improvement in the wind forcing field and in ocean models.

Before concluding this subsection, we note that assimilation of subsurface data into ocean models inevitably introduces noise into the ocean initial conditions. This noise is incompatible with the coupled model, and the adjustment during forecasts, or the initialization shock, could result in degradation of predictions. In the study by Chen *et al.* (1995), initial conditions for the Cane and Zebiak model are generated by running the coupled model itself and blending the FSU winds with the model winds. Initial conditions obtained in such a manner were more consistent and better balanced with the coupled model and thus reduced initialization shocks when the model was run subsequently in forecast mode. With this procedure, the coupled model acts as a dynamic filter to the oceanic initial conditions. This example indicates that all aspects of ocean initialization, which include quality of wind forcing fields, accuracy of data assimilation, and balance between coupled model and ocean initial conditions, must be considered together in order to achieve the best possible initialization for coupled forecast models.

Simulation of interannual precipitation variability

The largest interannual variations in the sea surface temperatures occur in the tropical Pacific. These variations are also responsible for interannual variations in tropical convection, diabatic heating, surface momentum and latent heat fluxes and other tropical circulation features. A direct impact of change in the tropical convection is to generate a region of low-level convergence and upper-level divergence over the warm SSTs and thus alter the climatological distribution of the upper-level source of the Rossby wave which can propagate to remote extratropical latitudes. Such global teleconnection relationships between the tropical Pacific and global atmosphere are a major atmospheric climatic perturbation and are collectively referred to as the El Niño-Southern Oscillation (ENSO) phenomenon.

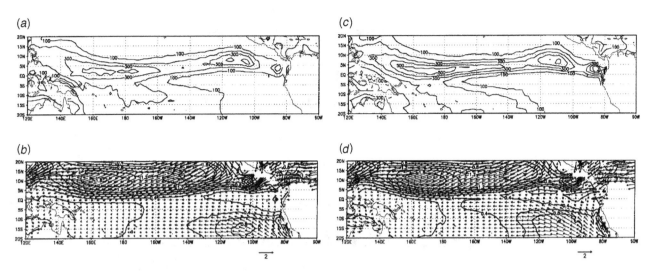

Figure 2.27. January climatological rainfall (mm/month and contour interval of 100 mm/month) and surface wind stress (in N m⁻² and contour interval of 0.2 N m⁻²) for operational (*a* and *b*) and reanalysis (*c* and *d*) models at NCEP. Climatologies are generated by integrating the GCMs under perpetual January conditions and with climatological SSTs. Integrations are 30 month duration.

ENSO events through their impact on the global atmospheric circulation also lead to large-scale changes in the precipitation all over the globe and have considerable impact on the regional water budget, for example producing floods over some regions and droughts over others (Ropelewski and Halpert, 1987). Changes in the availability of the water for the local evaporative processes can also feed back on to the atmospheric circulation, thus further modulating the remote teleconnection links.

The simulation and long-range prediction of ENSO-related interannual variability requires (a) proper simulation of coupled air–sea interactions, and (b) an ability to communicate the impact of anomalous SSTs to the global atmosphere. GCM parameterization schemes play a crucial role in the fulfilment of these requirements. For example, it is well established that the evolution of tropical SSTs is very sensitive to the specification of surface wind stresses, which in GCMs are governed by boundary-layer parameterization schemes and their interaction with the convective and other physical processes (Section 2.3). Further, parameterization schemes determine the GCMs' climatological flow features and hence the characteristic property of Rossby wave energy propagation and instability properties of the latitudinally and longitudinally varying basic states. In this way they also have an effect on the GCMs' ability to simulate the remote atmospheric response to the tropical SST anomalies.

Changes in the GCMs' convective parameterization scheme can change the simulation of surface wind stresses. Weaker-than-observed tropical Pacific wind stresses are a common feature among many different GCMs (Reynolds *et*

al., 1989). Such characteristic behavior can be linked to the GCMs' tropical precipitation field which at times has spatially diffused character. The diabatic heating resulting from such precipitation distribution, due to lack of horizontal gradients and associated equatorward wind acceleration, results in weaker-than-observed tropical surface winds. Ocean simulations forced with such wind stresses fail to maintain equatorial east–west oceanic pressure gradients, often resulting in the collapse of the simulated equatorial Pacific cold tongue. Modifications in and differences amongst convective parameterization schemes can result in substantial changes in the surface wind stresses leading to improvements in the simulation and prediction of coupled interannual SST variability (Ji *et al.*, 1994).

An example of such dependence of the surface wind stresses on the GCMs' parameterization schemes is shown in Figure 2.27, where simulated tropical rainfall anomaly and surface wind stress for two different versions of a GCM are shown. The GCMs are versions of the global model used operationally at NCEP during 1995 and 1996, respectively (Sections 2.3 and 2.5). Both GCMs have the same (T62 spectral) resolution and are integrated under perpetual January conditions with climatological distribution of SSTs. The two versions of the GCM differ in the closure assumption of the convective parameterization scheme among others, and are used as the reanalysis model and the 1996 operational model at NCEP (hereafter referred to as RENGCM and OPEGCM, respectively). The precipitation and the surface wind stresses in Figure 2.27 are the result of averaging over 30 simulated perpetual Januaries. Due to long time averag-

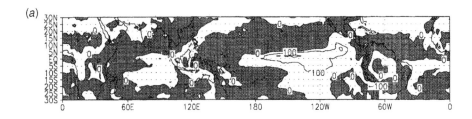

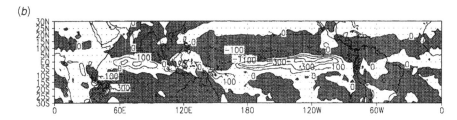

Figure 2.28. Remote rainfall response (mm month^{-1} and contour interval of 100 mm month^{-1}) for (a) operational and (b) reanalysis models at NCEP for the warm tropical Pacific SST anomaly confined within 160° E–60° W and 20° S–20° N. The response is generated by integrating the GCMs under perpetual January conditions with anomalous SSTs for 30 months and differencing the averaged field from the climatological SST integration.

ing and the low tropical internal variability, the differences between two simulations in the tropics are statistically significant.

North of the Equator, OPEGCM has enhanced rainfall as compared with RENGCM. Also for OPEGCM, the dry region in the eastern Pacific is better defined where the region of minimum rainfall (as evidenced by the 100 mm month^{-1} rainfall contour) now penetrates further westwards and closer to the dateline. Consistent with the changes in the rainfall, the surface wind stress climatology for OPEGCM has increased stresses over both the equatorial central and far eastern Pacific. Based on previous studies it is expected that OPEGCM will be a model better suited for simulating the tropical interannual SST variability.

The differences in convective parameterization schemes also result in different model sensitivity to interannual vari-

ations in the SSTs. An example of this for RENGCM and OPEGCM, for an idealized warm SST anomaly in the eastern Pacific is shown in Figure 2.28. The precipitation anomalies are once again based on averaging over 30 simulated perpetual Januaries. Consistent with the stronger climatological rainfall in the equatorial Pacific (Figure 2.27), OPEGCM has more intense anomalous rainfall. Similar differences in the surface wind stress anomalies also exist (not shown). Further, in Figure 2.28 remote teleconnection response to localized SST anomalies is evident. Differences in the direct local response of the SST anomalies to the diabatic heating, with corresponding differences in the upper-level divergent circulation, lead to differences in the teleconnection responses between the two models. Further examples of the sensitivity of the remote teleconnection response to GCM parameterization schemes are discussed in Kumar et al. (1996).

References

Arakawa, A. (1969). Parameterization of cumulus convection. *Proc. WMO/IUGG Symposium on Numerical Prediction in Tokyo*, **IV 8**, 1–6.

Arakawa, A. and Lamb, V. R. (1977). Computational design of the basic dynamical processes of the UCLA general circulation model. *Methods in Computational Physics*, Vol. 17, 174–265, Academic Press.

Arakawa, A., Mechoso, C. R. and Konor, C. S. (1992). An isentropic vertical coordinate model: Design and application to atmospheric frontogenesis studies. *Meteorol. Atmos. Phys.*, **50**, 31–45.

Arakawa, A. and Schubert, W. H. (1974). Interaction of a cumulus cloud ensemble with the large-scale environment, Part I. *J. Atmos. Sci.*, **31**, 674–701.

Arakawa, A. and Suarez, M. J. (1983). Vertical differencing of the primitive equations in sigma coordinates. *Mon. Wea. Rev.*, **111**, 34–45.

Arpe, K. (1991). The hydrological cycle in the ECMWF short-range forecasts. *Dynamics of Atmospheres and Oceans*, **16**, 33–59.

Arpe, K., Burridge, D. and Gilchrist, A. (1988). A comparison of surface stresses and latent heat fluxes over oceans in short range forecasts by ECMWF, UKMO and NMC. *Proceedings of the 13th Annual Climate Diagnostic Workshop*, 31 Oct.–4 Nov. 1988. U.S. Dept. of Commerce, Washington, D.C., 389–97.

Arpe, K. and Esbensen, S. K. (1989). Surface stresses and latent heat fluxes over oceans in short range forecasts: their annual and interannual variability and comparison with climatological estimates. *Annalen der Meteorologie*, **26**, 128–30.

Bates, J. R., Li, Y., Brandt, A., McCormick, S. F. and Ruge, J. (1995). A global shallow-water numerical model based on the semi-Lagrangian advection of potential vorticity. *Quart. J. Roy. Meteor. Soc.*, **121**, 1981–2005.

Betts, A. K. and Miller, M. J. (1986). A new convective adjustment scheme. Part II: Single column tests using GATE wave, BOMEX, ATEX and arctic airmass data sets. *Quart. J. Roy. Meteor. Soc.*, **112**, 692–709.

Bjerknes, J. 1969: Atmospheric teleconnections from the equatorial Pacific. *Mon. Wea. Rev.*, **97**, 163–72.

Black, T. L., (1994). The new NMC mesoscale eta model: Description and forecast examples. *Wea. Forecasting*, **9**, 265–78.

Bony, S., Sud, Y., Lau, K. M., Susskind, J. and Saha, S. (1997). Comparison and satellite assessment of NASA/DAO and NCEP/NCAR reanalyses over tropical ocean: atmospheric hydrology and radiation. *J. Climate*, **10**, 1441–62.

Busalacchi, A. J., (1996). Data assimilation in support of tropical ocean circulation studies. In *Modern Approaches to Data Assimilation in Ocean Modeling*, ed. P. Malanotte-Rizzolli. Elsevier.

Campana, K. A. (1995). Use of cloud analysis to validate and improve model-diagnostic clouds at CNMC. In *Proc. ECMWF Workshop on Modelling, Validation and Assimilation of Clouds, 31 October – 4 November 1994*, pp. 207–31. ECMWF, Reading, UK.

Cane, M. A., Zebiak, S. E. and Dolan, S. C. (1986). Experimental forecasts of El Niño. *Nature*, **321**, 827–32.

Cayan, D. (1992). Variability of latent and sensible heat fluxes estimated using bulk formulae. *Atmosphere-Ocean*, **30(1)**, 1–42.

Charney, J. G. (1948). On the scale of atmospheric motions. *Geofys. Public.*, **17(2)**, 17 pp.

Charney, J. G. (1962). Integration of the primitive and balance equations. In *Proc. Intnl. Symp. on Numerical Weather Prediction*, in Tokyo, 7–13 November 1960, ed. S. Syono, pp. 131–52. Meteorological Society of Japan, 656 pp.

Charnock, H. (1955). Wind stress on the water surface. *Quart. J. Roy. Meteor. Soc.*, **81**, 639–40.

Chen, D., Zebiak, S. E., Busalacchi, A. J. and Cane, M. A. (1995). An improved procedure for El Niño forecasting. *Science*, **269**, 1699–1702.

Chen, M. and Bates, J. R. (1996). A comparison of climate simulations from a semi-Lagrangian and an Eulerian GCM. *J. Climate*, **9**, 1126–49.

Cohn, S. E. and Parrish, D. F. (1991). The behaviour of forecast error covariances for a Kalman filter in two dimensions. *Mon. Wea. Rev.*, **119**, 1757–85.

Collins, W. G. and Gandin, L. S. (1990). Comprehensive hydrostatic quality control at the National Meteorological Center. *Mon. Wea. Rev.*, **118**, 2752–67.

Côté, J., Méthot, A., Patoine, A., Roch M. and Staniforth, A. (1997). Preliminary results from a dry global variable-resolution PE model. *Atmosphere-Ocean*, André J. Robert Memorial Symposium issue (in press).

Courant, R., Friedrichs, K. O. and Lewy, H. (1928). Uber die partiellen differenzengleichungen der mathematischen physik. *Math. Annalen*, **100**, 32–74.

Courtier, P., Thépaut, J.-N. and Hollingsworth, A. (1994). A strategy for operational implementation using an incremental approach. *Quart. J. Roy. Meteor. Soc.*, **120**, 1367–87.

Daley, R. (1991). *Atmospheric Data Analysis*. Cambridge Univ. Press, Cambridge, New York, Port Chester, Melbourne, Sydney, 457 pp.

da Silva, A., Young, C. C. and Levitus, S. (1994). *Atlas of Surface Marine Data 1994 Vol. 1: Algorithms and Procedures*. NOAA Atlas NESDIS 6, U.S. Dept. of Commerce, Washington, D.C., 83 pp.

da Silva, A. and White G. (1996). Intercomparison of surface marine fluxes from GEOS-1/DAS and NCEP/NCAR re-analyses. *Proceedings of the WCRP Workshop on Air-Sea Fluxes for Forcing Ocean Models and Validating GCMs*, 24–27 Oct. 1995. World Meteorological Organization, Geneva, Switzerland, (in press).

Deardorff, J. W. (1974). Three-dimensional numerical study of turbulence in an entraining mixing layer. *Bound. Layer Meteor.*, **7**, 199–226.

Derber, J. C., Parrish, D. F. and Lord, S. J. (1991). The new global operational analysis system at the National Meteorological Center. *Wea. Forecasting*, **6**, 538–47

Derber, J. C. and Rosati, A. (1989). A global oceanic data assimilation system. *J. Phys. Oceanogr.*, **19**, 1333–47.

Derber, J. C. and Wu, Wan-shu (1996). The use of cloud cleared radiances in the NCEP SSI analysis system. Proceedings of the 11th Conference on Numerical Weather Prediction. American Meteorological Society, Norfolk, VA, August 19–23 1996.

Dickinson, R. E., (1984). Modeling evapotranspiration for three-dimensional global climate models: climate processes and climate sensitivity. *Geophys. Monogr.*, **29**, 58–72.

Foreman, S. J., Alves, J. O. S. and Brooks, N. P. J. (1994). *Assessment of Surface Fluxes from Numerical Weather Prediction Systems*. Tech. Report 104, Forecasting Research Division, Meteorological Office, Bracknell, U.K.

Gandin, L. S. (1963). Objective analysis of meteorological fields. *Gidrometeorologicheskoe Izdatelstvo*, Leningrad. English translation by: Israeli Program for Scientific Translations, Jerusalem, 1965, 242 pp.

Ghil, M., Cohn, S., Tavantzis, J., Bube, K. and Isaacson, E. (1981). Applications of estimation theory to numerical weather prediction. In *Dynamic Meteorology: Data Assimilation Methods*, ed. L. Bengtsson, M. Ghil, and E. Kallen. Springer-Verlag, 330 pp.

Gill, P. E., Murray, W. and Wright, M. H. (1981). *Practical Optimization*. Academic Press, London, 401 pp.

Goldenberg, S. B. and O'Brien, J. J. (1981). Time and space variability of tropical Pacific wind stress. *Mon. Wea. Rev.*, **109**, 1190–207.

Grell, G. A. (1993). Prognostic evaluation of assumptions used by cumulus parameterizations. *Mon. Wea. Rev.*, **121**, 764–87.

Halpern, D. and Ji, M. (1993). An evaluation of the National Meteorological Center weekly hindcast of upper-ocean temperature along the eastern Pacific equator in January 1992. *J. Climate*, **6**, 1221–6.

Haltiner, G. J. and Williams, R. T. (1980). *Numerical Prediction and Dynamic Meteorology*. John Wiley & Sons, New York, 477 pp.

Heikes, R. and Randall, D. A. (1995). Numerical integration of the shallow-water equations on a twisted icosahedral grid. Part I: Basic design and results of tests. *Mon. Wea. Rev.*, **123**, 1862–80.

Hong, Song-you and Pan, Hua-lu (1996). Non-local boundary layer vertical diffusion in a medium-range forecast model. *Mon. Wea. Rev.*, **124**, 2322–39.

Hortal, M. and Simmons, A. J. (1991). Use of reduced Gaussian grids in spectral models. *Mon. Wea. Rev.*, **119**, 1057–74.

Janjic, Z. I. (1994). The Step-Mountain Eta Coordinate Model: further developments of the convection, viscous sub-layer and turbulence closure schemes. *Mon. Wea. Rev.*, **122**, 927–45.

Janowiak, J. E. (1992). Tropical rainfall: a comparison of satellite-derived rainfall estimates with model precipitation forecasts, climatologies and observations. *Mon. Wea. Rev.*, **120**, 448–62.

Jazwinski, A. H. (1970). *Stochastic Processes and Filtering Theory*. Academic Press, 376 pp.

Ji, M., Kumar, A. and Leetmaa, A. (1994). An experimental coupled forecast system at the national meteorological center: some early results. *Tellus*, **46A**, 398–418.

Ji, M. and Leetmaa, A. (1997). Impact of data assimilation on ocean initialization and El Niño Prediction. *Mon. Wea. Rev.*, **125**, 742–53.

Ji, M., Leetmaa, A. and Derber, J. (1995). An ocean analysis system for seasonal to interannual climate studies. *Mon. Wea. Rev.*, **123**, 460–81.

Ji, M. and Smith, T. M. (1995). Ocean model responses to temperature data assimilation and varying surface wind stress: Intercomparisons and implications for climate forecast. *Mon. Wea. Rev.*, **123**, 1811–21.

Kalnay, E., Kanamitsu, M. and Baker, W. E. (1990). Global numerical weather prediction at the National Meteorological Center. *Bull. Amer. Meteor. Soc.*, **71**, 1410–28.

Kalnay, E., Kanamitsu, M., Kistler, R., Collins, W., Deaven, D., Gandin, L., Iredell, M., Saha, S., White, G., Woolen, J., Zhu, Y., Chelliah, M., Ebisuzaki, W., Higgins, W., Janowiak, J., Mo, K. C., Ropelewski, C., Wang, J., Leetma, A., Reynolds, R., Jenne, R. and Joseph, D. (1996). The NCEP/NCAR 40-year reanalysis project. *Bull. Amer. Meteor. Soc.*, **77**, 437–71.

Kanamitsu, M. (1989). Description of the NMC global data assimilation and forecast system. *Wea. Forecasting*, **4**, 335–42.

Kane, J. S. and Fritsch, J. M. (1990). A one-dimensional entraining/detraining plume model and its application in convective parameterization. *J. Atmos. Sci.*, **47**, 2784, 2802.

Kasahara, A. (ed.) (1979). *Numerical Methods Used in Atmospheric Models*, Vol. 2, GARP Publications Series No 17, WMO and ICSU, Geneva, 499 pp.

Kimoto, M., Yoshikawa, I. and Ishii, M. (1997). An ocean data assimilation system for climate monitoring. *J. Meteor. Soc. of Japan*, **75**, 471–87.

Kirtman, B. P., Shukla, J., Huang, B., Zhu, Z. and Schneider, E. K. (1997). Multiseasonal predictions with a coupled Tropical Ocean Global Atmosphere system. *Mon. Wea. Rev.*, **125**, 789–808.

Kleeman, R. (1993). On the dependence of hindcast skill in a coupled ocean-atmosphere model on ocean thermodynamics. *J. Climate*, **6**, 2012–33.

Kreiss, H. and Oliger, J. (1973). *Methods for the Approximate Solution of Time Dependent Problems*. GARP Publications Series No 10, WMO and ICSU, Geneva, 107 pp.

Kumar, A., Hoerling, M. P., Ji, M., Leetmaa, A. and Sardeshmukh, P. (1996). Assessing a GCM's suitability for making seasonal predictions. *J. Climate*, **9**, 115–29.

Kuo, H. L., (1965). On formation and intensification of tropical cyclones through latent heat release by cumulus convection. *J. Atmos. Sci.*, **22**, 40–63.

Kuo, H. L. (1974). Further studies of the parameterization of the influence of cumulus convection on large scale flow. *J. Atmos. Sci.*, **31**, 1232–40.

Kurihara, Y. (1965). Numerical integration of the primitive equations on a spherical grid. *Mon. Wea. Rev.*, **93**, 399–415.

Latif, M., Barnett, T. P., Cane, M. A., Flugel, M., Graham, N. E., von Storch, H., Xu, J.-S. and Zebiak, S. E. (1994). A review of ENSO prediction studies. *Climate Dyn.*, **9**, 167–79.

LeDimet, F. X. and Talagrand, O. (1986). Variational algorithms for analysis and assimilation of meteorological observations: theoretical aspects. *Tellus*, **38A**, 91–110.

Leslie, L. M. and Purser, R. J. (1995). Three-dimensional mass-conserving semi-Lagrangian scheme employing forward trajectories. *Mon. Wea. Rev.*, **123**, 2551–655.

Lewis, J. M. and Derber, J. C. (1985). The use of adjoint equations to solve a variational adjustment problem with advective constraints. *Tellus*, **37A**, 309–22.

Lord, S. J. (1978). Development and observational verification of a cumulus cloud parameterization. Ph. D. Thesis, UCLA, 359 pp.

Lorenc, A. (1986). Analysis methods for numerical weather prediction. *Quart. J. Roy. Meteor. Soc.*, **112**, 1177–94.

Lorenz, E. N. (1963). Deterministic non-periodic flow. *J. Atmos. Sci.*, **20**, 130–41.

Lott, F. (1995). The significance of subgrid-scale orography and problems in their representation in GCMs. *Proc. ECMWF Workshop on Parameterization of Sub-grid Scale Physical Processes*, pp. 277–303. ECMWF, Reading, UK.

Lott, F. and Miller, M. J. (1997). A new sub-grid scale orography drag parameterization: its formulation and testing. *Quart J. Roy Meteor. Soc*, **123**, 101–27.

McPhaden, M. J. (1993). TOGA-TAO and the 1991–93 El Niño-Southern Oscillation event. *Oceanography*, **6**, 36–44.

McPhaden, M. J., Busalacchi, A. J., Cheney, R., Donguy, J-R., Gage, K. S., Halpern, D., Ji, M., Julian, P., Meyers, G., Mitchum, G. T., Niiler, P. P., Picaut, J., Reynolds, R. W., Smith, N. and Takeuchi, K. (1996). The Tropical Ocean Global Atmosphere (TOGA) observing system: a decade of progress. *J. Geophy. Res.* (accepted).

Machenhauer, B. (1979). The spectral method. In *Numerical Methods Used in Atmospheric Models*, Vol. 2, GARP Publications Series No 17, pp. 121–275. WMO and ICSU, Geneva.

Mahfouf, J.-F. (1991). Analysis of soil moisture from near surface parameters: a feasibility study. *J. Appl. Meteor.*, **30**, 1534–47.

Mahrt, L. and Pan, H.-L. (1984). A two-layer model of soil hydrology. *Bound. Layer Meteor.*, **29**, 1–20.

Manabe, S., (1969). Climate and the ocean circulation, 1. The atmospheric circulation and the hydrology of the earth's surface. *Mon. Wea. Rev.*, **97**, 739–74.

Manabe, S., Smagorinski, J. and Strickler, R. F. (1965). Simulated climatology of a general circulation model with a hydrologic cycle. *Mon. Wea. Rev.*, **93**, 769–98.

Mellor, G. L. and Yamada, T. (1974). A hierarchy of turbulence closure models for planetary boundary layers. *J .Atmos. Sci.*, **31**, 1791–806.

Mesinger, F. and Arakawa, A. (1976). *Numerical Methods Used in Atmospheric Models*, Vol. 1, GARP Publications Series No 17, WMO and ICSU, Geneva, 64 pp.

Mesinger, F., Janji, Z. I., Nickovic, S., Gavrilov, D. and Deaven, D. G. (1988). The step-mountain coordinate: model description and performance for cases of Alpine lee cyclogenesis and for a case of an Appalachian redevelopment. *Mon. Wea. Rev.*, **116**, 1494–517.

Mo, K. C., Wang, X.L., Kistler, R., Kanamitsu, M. and Kalnay, E. (1995). Impact of satellite data on the CDAS-Reanalysis system. *Mon. Wea. Rev.*, **123**, 124–39.

Monin, A. S. and Obukhov, A. M. (1954). Basic laws of turbulent mixing in the ground layer of the atmosphere. *Akad. Nauk SSSR Geofiz. Inst. Tr.*, **151**, 163–87.

Moore, A. M. (1990). Aspects of geostrophic adjustment during tropical ocean data assimilation. *J. Phys. Oceanogr.*, **19**, 435–61.

Navon, I. M., Zou, X., Derber, J. and Sela, J. (1992). Variational data assimilation with an adiabatic version of the NMC spectral model. *Mon. Wea. Rev.*, **120**, 1433–46.

Ooyama, K., (1971). A theory on parameterization of cumulus convection. *J. Meteor. Soc. Japan*, **49**, Special Issue, 744–56.

Pan, H.-L. and Mahrt, L. (1987). Interaction between soil hydrology and boundary-layer development. *Bound. Layer Meteor.*, **38**, 185–202.

Pan, H.-L. and Wu, W.-S. (1995). Implementing a mass flux convection parameterization package for the NMC Medium-Range Forecast model. NMC Office Note, No. 409, 40 pp. [Available from NCEP, 5200 Auth Road, Washington DC 20233.]

Parrish, D. F. and Derber, J. C. (1992). The National Meteorological Center's spectral statistical-interpolation analysis system. *Mon. Wea. Rev.*, **120**, 1747–63.

Phillips, N. A. (1956). The general circulation of the atmosphere, a numerical experiment. *Quart. J. Roy. Meteor Soc.*, **82**, 123–64.

Phillips, N. A. (1957). A coordinate system having some special advantages for numerical forecasting. *J. Meteor.*, **14**, 184–5.

Rančić, M. (1995). An efficient conservative, monotonic remapping as a semi-Lagrangian transport algorithm. *Mon. Wea. Rev.*, **123**, 1213–17.

Rančić, M., Purser, R. J. and Mesinger, F. (1996). A global shallow-water model using an expanded spherical cube: gnomic versus conformal coordinates. *Quart. J. Roy. Meteor. Soc.*, **122**, 959–82.

Rasch, P. J. and Williamson, D. L. (1990). Computational aspects of moisture transport in global models of the atmosphere. *Quart. J. Roy. Meteor. Soc.*, **116**, 1071–90.

Rasmusson, E. and Wallace, J. M. (1986). Meteorological aspects of the El Niño/Southern Oscillation. *Science*, **222**, 1195–202.

Reynolds, O. (1895). On the dynamical theory of incompressible viscous fluids and the determination of the criterion. *Phil. Trans. Roy. Soc., London*, **186**, 124–64.

Reynolds, R. W., Arpe, K., Gordon, C., Hayes, S. P., Leetmaa A. and McPhaden, M. J. (1989). A comparison of Tropical Pacific surface wind analyses. *J. Climate*, **2**, 105–11.

Reynolds, R. W. and Leetmaa, A. (1996). Midlatitude monthly surface heat fluxes from models and data. *Proceedings of the WCRP Workshop on Air-Sea Fluxes for Forcing Ocean Models and Validating GCMs*, 24–27 Oct. 1995. World Meteorological Organization, Geneva, Switzerland.

Richardson, L. F. (1922). *Weather Prediction by Numerical Process*. Cambridge University Press., 236 pp. Reprinted by Dover.

Ritchie, H., Temperton, C., Simmons, A., Hortal, M., Davies, T., Dent, D. and Hamrud, M. (1995). Implementation of the semi-Lagrangian method in a high-resolution version of the ECMWF forecast model. *Mon. Wea. Rev.*, **123**, 489–514.

Robert, A. (1969). The integration of a spectral model of the atmosphere by the implicit method. *Proc. WMO/IUGG Symposium on NWP*, pp. 19–24. Tokyo, Japan.

Ropelewski, C. and Halpert, M. (1987). Global and regional scale precipitation patterns associated with the El Niño/Southern Oscillation. *Mon. Wea. Rev.*, **114**, 2352–62.

Rosati, A., Budgel, R. and Miyakoda, K. (1995). Decadal analysis produced from an ocean data assimilation system. *Mon. Wea. Rev.*, **123**, 2206–28.

Schemm, J., Schubert, S., Terry, J. and Bloom, S. (1992). *Estimates of Monthly Mean Soil Moisture for 1979–89*. NASA Tech. Mem. 104571, Oct. 1992, NASA/Goddard Space Flight Ctr., Greenbelt, MD 20771.

Sellers, P. J., Mintz, Y., Sud, Y. C. and Dalcher, A. (1986). The design of a Simple Biosphere model (SiB) for use within general circulation models. *J. Atmos. Sci.*, **43**, 505–31.

Shuman, F. G. (1989). History of numerical weather prediction at the National Meteorological Centre. *Weather and Forecasting*, **4**, 286–96.

Simmons, A. J. and Burridge, D. M. (1981). An energy and angular momentum conserving vertical finite-difference scheme and hybrid vertical coordinates. *Mon. Wea. Rev.*, **109**, 758–66.

Slingo, J. M. (1987). The development and verification of a cloud prediction model for the ECMWF model. *Quart. J. Roy. Meteor. Soc.*, **113**, 899–927.

Smith, N. R. (1995). The BMRC ocean thermal analysis system, *Aust. Met. Mag.*, **44**, 93–110.

Staniforth, A. and Coté, J. (1991). Semi-Lagrangian integration schemes for atmospheric models: a review. *Mon. Wea. Rev.*, **119**, 2206–23.

Sundqvist, H., (1979). Vertical coordinates and related discretization. In *Numerical Methods Used in Atmospheric Models*, Vol. 2, GARP Publications Series No 17, pp. 1–50. WMO and ICSU, Geneva.

Thepaut, J-N. and Courtier, P. (1991). Four-dimensional variational data assimilation using the adjoint of a multilevel primitive equation model. *Quart. J. Roy. Meteor. Soc.*, **117**, 1225–54.

Todling, R. and Cohn, S. E. (1994). Suboptimal schemes for atmospheric data assimilation based on the Kalman filter. *Mon. Wea. Rev.*, **122**, 2530–57.

White, G. H. (1995). *An Intercomparison of Precipitation and Surface Fluxes from Operational NWP Analysis/Forecast Systems*. World Meteorological Organization, Geneva, Switzerland, WMO/TD-723, 33 pp + fig.

Williamson, D. L. (1990). Semi-Lagrangian moisture transport in the NMC spectral model. *Tellus*, **42A**, 413–28.

Williamson, D. L. (1992). Review of numerical approaches for modeling global transport. In *Air Pollution Modeling and its Application*, IX, ed. H. van Dop and G. Kallos, pp. 377–94. Plenum Press, NY.

Wyrtki, K. (1975). El Niño: the dynamical response of the equatorial Pacific to atmospheric forcing. *J. Phys. Ocean.*, **5**, 572–84.

Xie, P.-P. and Arkin, P. A. (1996). Analyses of global monthly precipitation using gauge observations, satellite estimates and numerical model predictions. *J. Climate*, **9**, 840–58.

Yanai, M., Esbensen, S. K. and Chu, J.-H. (1973). Determination of bulk properties of tropical cloud clusters from large-scale heat and moisture budgets. *J. Atmos. Sci.*, **30**, 611–27.

Zapotocny, T. H., Johnson, D. R. and Reames, F. M. (1994). Development and initial test of the University of Wisconsin global isentropic-sigma model. *Mon. Wea. Rev.*, **122**, 2160–78.

Zhao, Q., Black, T. L. and Baldwin, M. E. (1995). Cloud prediction scheme in the Eta model at NCEP. In *Proc. ECMWF Workshop on Modelling, Validation and Assimilation of Clouds*, 31 October–4 November 1994, pp. 233–51. ECMWF, Reading, UK.

Zhu, Z., Thuburn, J., Hoskins, B. and Haynes, P. (1992). A vertical finite-difference scheme based on a hybrid sigma-theta-p coordinate. *Mon. Wea. Rev.*, **120**, 851–62.

Zupanski, D. and Mesinger, F. (1995). Four-dimensional variational assimilation of precipitation data. *Mon. Wea. Rev.*, **123**, 1112–27.

Zupanski, M. (1993). Regional four-dimensional variational data assimilation in a quasi-operational forecasting environment. *Mon. Wea. Rev.*, **121**, 2396–408.

3 Atmospheric processes and their large-scale effects

3.1 Radiative effects of clouds and water vapor

Graeme L. Stephens

Introductory remarks

The broad relevance of radiative transfer to the energy budget of the planet is evident in the global-annual view of the energy exchanges in the Earth's climate system. The budget of radiation leaving the planet at the top of the atmosphere (TOA), referred to as the Earth's Radiation Budget (ERB), represents the integration of all energy exchanges that occur below. Although we currently measure the relevant components of the ERB, namely the albedo and longwave emission of the planet, the lack of precision of these measurements and the limited spatial and temporal coverage make them less than ideally suited for detection of global change (Goody *et al.*, 1995). From the meridional distribution of TOA net fluxes, however, the meridional energy transports required by the atmosphere and oceans to bring the N–S gradients of the ERB into balance can be deduced (e.g., Oort and Vonder Haar, 1976).

It is not yet possible to observe the distribution of radiative fluxes at the surface or within the atmosphere over the globe and we must resort to indirect methods to infer them (e.g., Rossow and Zhang, 1995; Ellingson *et al.*, 1994; Li *et al.*, 1997; Whitlock *et al.*, 1995). Determining both the uncertainties of these methods as well as ways to improve them are topics of current research. Until progress is made, these observational limitations on the global energy budget impede progress towards understanding and modeling of the global climate system and impede subsequent progress toward the ultimate prediction of climate. It is now well recognized that such progress requires an advanced understanding of the quantitative links between hydrologic processes and the 4-dimensional distribution of energy fluxes, including the fluxes of radiation.

A simple demonstration of the importance of these links lies in the example of the global mean radiation budget of the atmosphere. In a global average sense, the atmosphere constantly loses energy by radiation exchange at a rate of approximately 100 W m^{-2}. This loss is commonly expressed as the rate

of cooling of the atmosphere and is compensated by energy transfer from the surface via convective and turbulent transfer, principally through excess latent heat release characterized by the precipitation falling from the atmosphere to the surface. Global scale changes in this cooling imply compensating changes in latent heating and thus fundamental changes in the Earth's hydrologic cycle. It is appropriate to consider this cooling as a rudimentary measure of the activity of the Earth's greenhouse effect and in turn an indirect measure of the gross activity of the hydrologic cycle in heating the atmosphere.

This discussion underscores two important aspects of the radiation budget of the climate system.

1. The hydrologic cycle and the radiative heating of the atmosphere are intimately linked and it is necessary to provide a measure of the hydrologic variables that characterize this distribution of heating in the atmosphere and the fluxes at the boundaries of the atmosphere.
2. It is important to go beyond a simple global-annual averaged view of the budget typically described in rudimentary texts of climate. What is crucial is understanding the importance of the 4D distribution of these energy fluxes. Gradients of the fluxes and heating drive the circulation and underpin feedbacks between radiation and the hydrologic factors that govern radiation.

These links are explored below in relation to the effects of clouds and water vapor on the radiative budget of the atmosphere.

Effects of water vapor and clouds on the Earth's Radiation Budget

The ERB is defined as

$$F_{net} = \frac{Q_\otimes}{4}(1-\alpha) - F_\infty \qquad (3.1.1)$$

where F_{net} is the net TOA radiation imbalance, $Q_\otimes$ is the solar insolation at the TOA, α is the albedo of the planet and F_∞ is the outgoing emitted longwave radiation. Since the annual and global mean of F_{net} is zero (at least to the accuracy of present-day satellite measurements),

$$\frac{Q_\otimes}{4}(1 - \bar{\alpha}) = F_\infty \qquad (3.1.2)$$

where the overbar emphasizes the global-annual average of the specified quantities.

Water vapor influence on the ERB

Water vapor directly affects the ERB in at least two distinct and important ways:

1. Through absorption of solar radiation leading to a reduction in the planetary albedo. Although this absorption is relatively straightforward to determine, uncertainty in its estimation is an important factor producing disparities in different solar radiative transfer models (e.g., Fouquart *et al.*, 1991). This is a topic that will not be considered further although it continues to be relevant to the current debate surrounding the solar radiation absorbed by clouds (Stephens and Tsay, 1990; Cess *et al.*, 1995).

2. Through the absorption and emission of infrared (IR) radiation. These processes are the principal controls of F_∞ and thus of the so-called greenhouse effect – a term defined later. The association between the infrared emission by the atmosphere (to the surface and to space) defined largely by the water vapor content of the atmosphere is thought to produce a positive feedback on surface temperature through the control of the latter on water vapor (e.g., Manabe and Wetherald, 1967; Stephens, 1990). According to climate models, this feedback contributes to the major portion of the global warming associated with increasing concentrations of atmospheric CO_2; however, the water vapor feedback loop on Earth is interfered with by transport of vapor and by its condensation into clouds. These clouds, in turn, impart a substantial influence of their own on the greenhouse effect. The actual way these interferences take place and the specific connection between water vapor, cloudiness, SST and the greenhouse effect on Earth are still inadequately understood, poorly observed and largely unexplored. We return to certain key aspects of the problem below.

A further complicating factor in understanding the water vapor feedback arises through the disproportionate influence of the small amount of upper tropospheric water vapor on the radiation emitted by the planet to space. A 10% change in upper tropospheric absolute humidity (above 500 hPa and by

definition assuming no change in temperature) has an approximately threefold larger effect on F_∞ than does an equivalent percentage change (but significantly larger absolute change) in lower tropospheric water vapor. This is due to the nonlinear effects of the curve of growth of emissivity (absorption) as a function of layer water vapor amount where small changes in path in an existing dry layer leads to a larger change in emissivity compared with an increase in the path of a moist layer by the same percentage amount. The enhanced growth of absorption in dry layers of the upper troposphere occurs in the spectral regions occupied by the strong absorption bands of water vapor (the rotational band at wavelengths longer than about 13 µm and the vibrational band centered at 6.3 µm).

The actual change in F_∞ induced by any changes in upper tropospheric humidity is also equally influenced by changes in the temperature in these layers. Spectral F_∞ in these strong absorption bands of water vapor is directly related to the relative humidity of the upper troposphere (UTH) – a relation that serves as the basis for satellite retrievals of UTH using spectral emission to space at 6.3 µm (Soden and Bretherton, 1993; Stephens *et al.*, 1996; Schmetz *et al.*, 1995). The general lack of validated measurements of upper tropospheric moisture is a major observational shortcoming that complicates our understanding of the water vapor feedback and the importance of upper tropospheric water vapor to this feedback. In the absence of these measurements, Slingo and Webb (1992) developed a simulation system, the Simulation and Analysis of Measurements from Satellites using Operational aNalyses (SAMSON), that uses initialized analyses from the operational archive at the European Centre for Medium-range Weather Forecasts (ECMWF). SAMSON is constructed around a high spectral resolution radiative transfer model, and comparisons of clear-sky TOA outgoing longwave radiation with the Earth Radiation Budget Experiment (ERBE) were used by Slingo and Webb to identify problems in both ERBE clear-sky data and in the analyses of ECMWF. SAMSON was also used by Stephens *et al.* (1994) as a basis for the development of a satellite algorithm for estimating the downward longwave flux to the surface. Application of SAMSON continues using ECMWF reanalyses data (Slingo and Webb, 1997).

The influence of clouds on the ERB

Clouds affect the ERB in two competing ways:

1. They reduce the net solar input by reflecting more solar radiation to space. This is sometimes referred to as a cooling or albedo effect.

2. They reduce the longwave emission to space such that relative to clear skies the altitude of emission is effectively raised to heights characterized by colder temperatures.

Analyses of data from ERBE in terms of the difference between clear- and cloudy-sky TOA fluxes (see e.g., Harrison *et al.*, 1990) have demonstrated how the shortwave effect dominates over the longwave effect globally. The strongest cooling effect is associated with the persistent maritime stratus off the west coast of continents and storm-track clouds in the summer hemisphere over the mid- and high-latitude Atlantic and Pacific oceans.

This conclusion was reached through analyses of clear-minus-cloudy-sky flux difference quantities. For longwave fluxes, the difference

$$C_{LW} = F_{clr} - F_{obs} \qquad (3.1.3a)$$

is a measure of the reduction from the clear-sky longwave radiation emitted to space by clouds and hence is a measure of the greenhouse effect of clouds. The highest, coldest clouds, such as occur as part of monsoonal cloud systems or of tropical deep convective systems, produce the largest greenhouse effect as highlighted in Figure 3.1(a) in which C_{LW} is shown as a function of SST.

The shortwave flux difference,

$$C_{SW} = \frac{Q_{\otimes}}{4}(\alpha_{clr} - \alpha_{obs}) \qquad (3.1.3b)$$

is generally negative because clouds reflect more shortwave solar radiation than do the adjacent clear skies. When C_{SW} is presented as a function of SST, as shown in Figures 3.1(b) and (c), the separate influences of the two defining factors, $Q_{\otimes}$ and α, become apparent. In the winter hemisphere, C_{SW} decreases poleward since the available sunlight $(Q_{\otimes})$ decreases poleward. In the summer hemisphere, significant reflection of solar radiation occurs over the illuminated clouds in the mid-latitude storm tracks and C_{SW} increases with decreasing SST by virtue of the increasing cloud albedo with decreasing SST.

The net impact of clouds on the ERB, defined as the sum of equations (3.1.3a) and (3.1.3b)

$$C_{net} = C_{LW} + C_{SW} \qquad (3.1.4)$$

is a residual of two terms of (generally) opposite sign.

Descriptions of the global distributions of the three flux quantities are described elsewhere (e.g. Harrison *et al.*, 1990). Values of C_{net} range regionally from about -140 W m^{-2} to 40 W m^{-2} with a negative global average of approximately -20 W m^{-2}. We also know that the magnitude of C_{SW} is almost as large as the longwave flux difference over tropical cloud systems, producing small values of net differences in these regions.

The extent to which these individual components cancel in tropical regions is perplexing and not yet understood despite recent attempts to simplify the issue (e.g., Kiehl, 1995).

Two broad classes of cloud properties determine both C_{LW} and C_{SW}. These may be introduced through the following approximation of the cloudy-sky flux:

$$F_{obs} = (1 - N) \, F_{clr} + N \, F_{cld} \qquad (3.1.5)$$

where N is the fractional cloud amount and F_{cld} is the flux associated with overcast skies. Simple rearrangement of this expression applied to long and shortwave fluxes gives

$$C_{LW} = N(F_{clr} - F_{cld}) \qquad (3.1.6)$$

and

$$C_{SW} = N\frac{Q_{\otimes}}{4} \, (\alpha_{clr} - \alpha_{cld}) \qquad (3.1.7)$$

respectively. Thus, we might think of changes in both C_{LW} and C_{SW} occurring by macroscopic changes in cloudiness through changes in cloud cover N and/or through bulk changes of the albedo and emittance of the cloud itself (such as through changes in cloud thickness) and secondly through microscopic changes (such as microphysics of clouds) that alter α_{cld} and cloud emissivity.

Ockert-Bell and Hartmann (1992) have used ERBE data and global cloud data from the International Cloud Climatology Project (ISCCP, refer to discussion of ISCCP below) in an attempt to identify the effects of different cloud types on $C_{LW,SW,net}$. This represents an emerging area of research in which data on different parameters are beginning to be correlated to explore relationships between them. A further example of this is given below where the albedo of clouds derived from ERBE is correlated with liquid water path data.

Partitioning of radiative energy in the column

Both clouds and water vapor affect longwave radiative transfer within the atmosphere in such a way that the fluxes at the TOA are to some degree decoupled from the fluxes to the surface. Clouds decouple the longwave fluxes when cloud base and cloud top temperatures are uncorrelated. This has been realized for some time (e.g., Stephens and Webster, 1984) but the extent to which water vapor absorption decouples the longwave fluxes at the atmospheric boundaries has not been appreciated.

Water vapor

Emission by water vapor decouples changes in TOA fluxes from changes in surface fluxes as illustrated in the context of

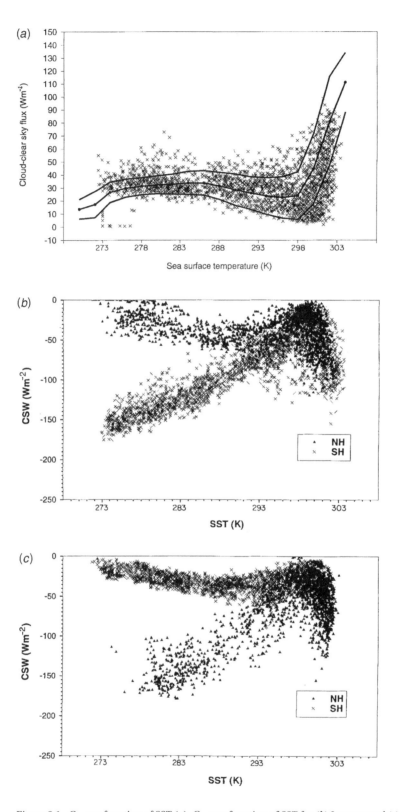

Figure 3.1. C_{LW} as a function of SST (a). C_{SW} as a function of SST for (b) January and (c) July. C_{SW} is separated by hemisphere to illustrate the effects of the seasonal change in insolation on C_{SW}.

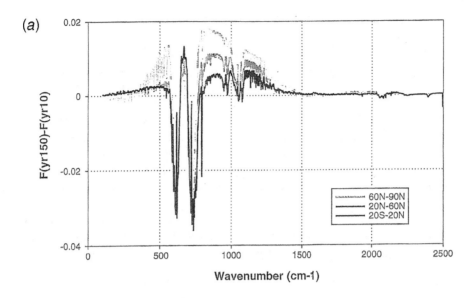

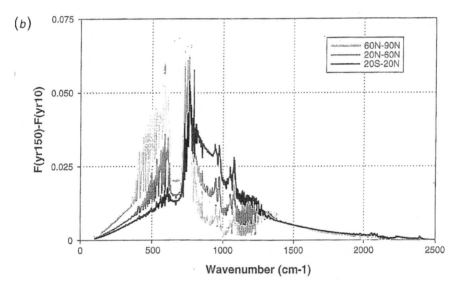

Figure 3.2. (a) Year 150 minus year 10 annual mean clear-sky spectral OLR (units Wm⁻² (cm⁻¹)⁻¹) derived from a transient simulation of a coupled ocean – atmosphere GCM. (b) As in (a) but for clear-sky downward IR flux. Differences are shown for averages over the regions indicated by the latitudes.

changes that occur in response to a forced climate change. Simulations of climate change using a coupled model (Gordon and O'Farrell, 1997) forced by increases in CO_2 at a rate of 1% per year and projected 180 years into the future are used for this purpose. Clear-sky IR, spectral F_∞ and spectral surface fluxes were calculated using the simulated profiles of temperature and moisture as input into a radiation model. The annual mean F_∞ spectra derived for model year 150 into the simulation minus the equivalent spectra of year 10 of the

simulation are shown in Figure 3.2(a). The simulated difference in global mean surface temperature between these years is approximately 2.5 degrees which is not reflected in the simulated differences of global mean broadband F_∞ of 0.2 W m⁻² (Garratt *et al.*, 1996). Figure 3.2(a) does show small shifts in the spectral TOA emission in the window region between 800 to 1300 cm⁻¹ associated with the surface warming and also indicates stratospheric cooling in the CO_2 band between approximately 600 and 800 cm⁻¹. Except for the polar regions where water vapor feedbacks are strong in this model and the surface warming greatest, the total change in emission to space is small. The spectral change is also small in regions of strong water vapor absorption where upper tropospheric water vapor most influences the emission to space. This small difference is a result of the competing effects of increasing temperature which tends to *increase F_∞* and increases in water vapor which produces compensating *decreases* in F_∞ by elevating the effective level of emission to space (e.g., Webb *et al.*, 1993). These competing effects are realized through the relative humidity of the upper layers of the troposphere which are only slightly changed in year 150 from year 10.

The combined effects of moisture and temperature increases on the emission of longwave radiation to the surface are very different from the combined effects on F_∞ as shown in Figure 3.2(a). Whether or not the climate change of the model could have been deduced from spectral measurements similar to the simulated spectra of Figure 3.2(a) remains an issue of some relevance not only to the topic of monitoring and detection of climate change (e.g., Goody *et al.*, 1995) but also to the topic of testing climate models and their predictions (e.g., Haskins *et al.*, 1997). In Figure 3.2(b), the spectral difference of the downward longwave flux to the surface is shown for the same latitudinal regions as those of Figure 3.2(a). The spectral nature of this increase itself conveys information about the nature of the warming and water vapor feedback in the regions indicated. The implication is that the water vapor increase in polar regions occurs more within spectral regions of strong water vapor absorption (notably the rotation band) whereas the lower-latitude increase, including that in the moist tropics, occurs through emission in the atmospheric window region. The change in global mean longwave flux to the surface between these two simulation years is 18.5 W m^{-2} suggesting a significantly enhanced sensitivity to increases in water vapor than was noted for the F_∞. The different sensitivities between fluxes at the TOA and fluxes at the surface profoundly affect the interpretation of flux data in terms of the water vapor feedback and measures of climate change (e.g., Garratt *et al.*, 1996).

Clouds

While the net effect of clouds on the TOA radiation balance arises through compensating effects of the individual longwave and shortwave fluxes, a different picture emerges when the radiation budgets of the atmosphere and surface are considered (Figure 3.3). According to Figure 3.3(c), clouds radiatively heat the atmosphere primarily through the IR absorption of high ice clouds associated with tropical deep convection (Stephens *et al.*, 1994). The radiative heating of the atmosphere by clouds is masked in the TOA budget by an almost equivalent amount of surface cooling principally associated with the reduction of solar radiation at the surface (e.g., Wielicki *et al.*, 1995). Clouds vertically redistribute the absorbed energy of the atmospheric column, depositing radiative energy in the atmosphere in cloud layers largely at the expense of energy absorbed at the surface. We presently are unable to deduce the quantitative details of this partition of energy from observations and must resort to simulations from General Circulation Models (GCMs) to study this effect. The extent that this energy redistribution occurs in the real climate system and our ability to model this redistribution are topics of critical importance, active research and, in part, a topic of continuing controversy.

Observed relationships

It is not yet possible to diagnose the vertical distribution of radiative fluxes and the associated vertical profiles of heating and cooling that derive from these flux profiles using only spaceborne measurements. However, recent research suggests that it may be possible to estimate the IR radiative cooling for the entire clear-sky column (Stephens *et al.*, 1994) or even for a few broad layers with limited success (e.g., Ellingson *et al.*, 1994). Stephens *et al.* (1994) showed that the following expression for the column clear-sky IR cooling rate

$$\frac{\mathrm{d}T}{\mathrm{d}t} = \frac{gF_\infty}{c_p p_s}\left[\frac{\sigma T_s^4}{F_\infty} - \frac{F_g}{F_\infty} - 1\right] \tag{3.1.8}$$

relates to the column water vapor w through the relationships

$$\frac{\sigma T_s^4}{F_\infty} = a + cw \tag{3.1.9a}$$

$$\frac{F_g}{F_\infty} = dw \tag{3.1.9b}$$

where the TOA outgoing clear-sky flux is F_∞ and the downward longwave radiation at the surface is F_g. The coefficients were derived using suitable satellite measures of the SST, F_∞ and w and model estimates of F_g. Properties of these relationships and their bases are explored in detail in Webb *et al.* (1993) and Stephens *et al.* (1994).

The association between the column-averaged heating rate and w is shown in Figure 3.4. According to equations (3.1.6a) and (3.1.6b), we expect the cooling rate to increase in an approximately linear way with increasing w (Figures 3.4(a) and (b)) but as an almost exponential function of SST in Figures 3.4(c) and (d) through the nonlinear relation between

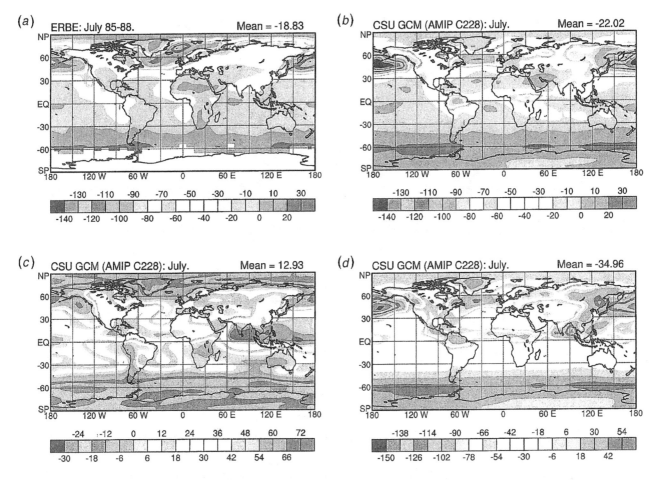

Figure 3.3. The difference between cloudy and clear-sky net radiative fluxes (in W m⁻²) (*a*) at the TOA from ERBE; (*b*) at the TOA from simulations using the Colorado State University (CSU) GCM (*c*) within the atmosphere from the CSU GCM; and (*d*) at the surface from the CSU GCM (D. Randall, personal communication).

w and SST. When plotted as a function of SST, the cooling exhibits winter–summer hemispheric differences. The characteristics of the relation between the column cooling rate and SST, especially the increased rate of cooling with increasing SST, is a result of the enhanced emission from the atmosphere to the surface associated with the increasing water vapor with SST at these temperatures.

Water vapor effects on radiative transfer

The effects of water vapor on radiative transfer are explored here in terms of its effect on IR cooling rates. A convenient way of visualizing the process of cooling is in terms of the exchange between layers as shown schematically in Figure 3.5. Individual layers radiatively heat or cool depending on whether or not the net exchange of radiation between layers is positive or negative. For a layer high in the atmosphere and in a part of the spectrum where absorption is strong (such as the previously

mentioned water vapor rotation band) radiation emitted to space exceeds the radiation gained through the exchange between the layer and the surrounding atmosphere (primarily the atmosphere below the reference layer). This layer loses energy at these wavelengths and is said to *cool to space* (e.g., Rodgers and Walshaw, 1966). Layers lower in the atmosphere are characterized by cooling to space at more transparent wavelengths as well as exchanges both from above and below at wavelengths of stronger absorption. The net exchanges are comparatively weak compared with cooling to space.

Cooling to space is dominant throughout much of the troposphere and its properties shift significantly with wavelength and altitude. Strong cooling to space takes place in the lower atmosphere through emission in the window region. As absorption increases, the cooling to space shifts to higher layers. This picture is conveyed in the spectral distribution of the infrared cooling rate shown in Figure 3.6(*a*) with the total cooling profile highlighted to the left in Figure 3.6(*b*) (modified

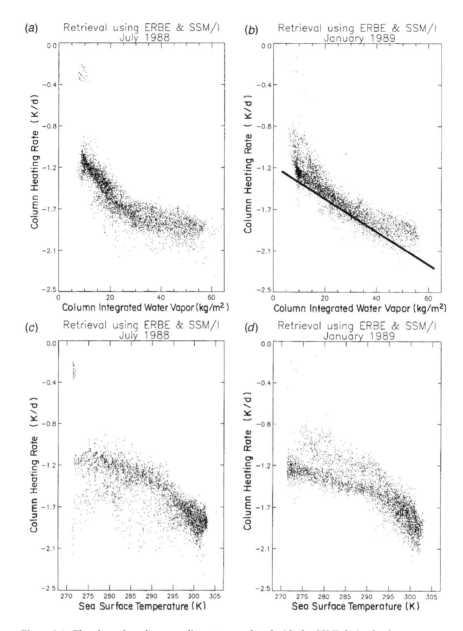

Figure 3.4. The clear-sky column cooling rate correlated with the SSMI derived column water vapor obtained from the data of (*a*) July 1988 and (*b*) January 1989. The clear-sky column cooling rate correlated with SST for (*c*) July 1988 and (*d*) January 1989 (adapted from Stephens *et al.*, 1994).

from Clough *et al.*, 1992). This diagram indicates how the cooling of the layer shifts in its spectral properties from a maximum in the window (associated with the continuum absorption) for layers low in the atmosphere to the stronger absorption regions of the rotation band for layers located in the upper troposphere. This implies that lower tropospheric water vapor (i.e. the majority of the water vapor mass) contributes to lower tropospheric cooling whereas the cooling of the

upper troposphere is governed by upper tropospheric water vapor. Here is direct evidence of the importance of the vertical distribution of water vapor to the vertical profile of IR cooling.

It is also appropriate to comment on the continuum absorption in the window region of the spectrum. The Intercomparison of Radiation Codes for Climate Models (ICRCCM, refer to Ellingson *et al.*, 1991) indicated how many radiation models agree within a few percent of the highest

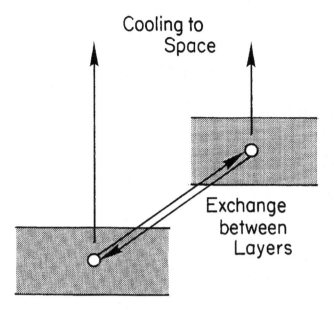

Cooling to Space

Exchange between Layers

Figure 3.5. A schematic of the exchanges of radiation between layers with emphasis on the cooling to space that takes place in the 'upper' layers. The position of these upper layers varies spectrally as suggested in Figure 3.6.

resolution line-by-line models when only line absorption is assumed. The effect of the continuum absorption, which is the dominant component of absorption in the more transparent spectral regions, creates a spread in the models of several W m^{-2} and proper treatment of this absorption is required in these models.

Cloud effects on radiative transfer: general considerations

The properties of clouds that define how much radiation is absorbed in the atmosphere and how much radiation escapes through the boundaries of the atmosphere include:

1. The three-dimensional distribution of cloud. This influence is usually thought of in terms of separate vertical and horizontal effects. Vertical variability is dealt with in models by introducing cloud overlap assumptions (e.g., Geleyn and Hollingsworth, 1979). Horizontal variability is dealt with using assumptions involving cloud amount (Stephens, 1988). Both are empirical and a critical assessment of the uncertainties associated with the assumptions is lacking. Research is now beginning to demonstrate how the 3D nature of clouds is perhaps the most significant factor that influences the transfer of solar radiation through clouds. Hereafter the macroscopic cloud properties that determine this influence are referred to as *extrinsic optical properties* (EOP).

2. The internal optical properties of the cloud. These properties are intrinsically defined by cloud microphysics (such as size and shape of particles). In the case of ice clouds, our understanding of the relationship between optical properties to ice particle microphysics is qualitative. Global climate models and most cloud-resolving models do not predict the relevant microphysical properties even of water clouds. Many of the parameterizations are then carried out in terms of the predictable water or ice mass and a specified microphysical parameter (such as effective particle size). The optical properties of clouds defined by the intrinsic microphysics of clouds will hereafter be referred to as *intrinsic optical properties* (IOP). The single scatter albedo is an example of an IOP. As shown below, cloud optical depth is defined both by macroscopic properties (e.g. cloud depth) and microphysical properties (such as particle size) and thus overlaps the definition of IOPs and EOPs.

Intrinsic optical properties and solar radiative transfer

The impact of IOPs on IR radiative properties of clouds is generally thought to be negligible based on assumptions of the applicability of Rayleigh scatter at these wavelengths (Stephens, 1984). The influence of cloud IOP on solar radiative transfer, however, is germane to a number of cloud-radiation issues including those associated with the indirect effect of aerosol on radiation through the so-called Twomey effect (see Section 3.2).

The relation between IOPs and solar transfer can be explored, as Stephens and Tsay (1990) show, in the context of the limits of cloud optical depth $\tau^* \to 0$ and $\tau^* \to \infty$. For optically thick clouds with $\tau^* \to \infty$, it follows that

$$\mathscr{R}_\infty = \frac{1}{\mu_o}\left[Z_+(\mu_o) - Z_-(\mu_o)\frac{g-}{g+} \right] \qquad (3.1.10a)$$

$$\mathscr{A}_\infty = 1 - \mathscr{R}_\infty \qquad (3.1.10b)$$

where $\mathscr{R}_\infty$ and $\mathscr{A}_\infty$ are respectively the albedo and absorption of this 'semi-infinite' cloud and where $Z\pm$ are functions of the cosine of the solar zenith angle μ_o (see Stephens and Tsay, 1990). According to these simple relationships both the albedo and absorption approach fixed asymptotic limits as τ' increases. These upper (theoretical) limits are defined solely by IOPs of the cloud, namely single scatter albedo, $\tilde\omega_o$ and scattering properties characterizing the nature of the scattering phase function. Stephens and Tsay introduce the basic relationship between $\mathscr{A}_\infty$ (and thus $\mathscr{R}_\infty$) and the droplet absorption factor $(1 - \tilde\omega_o)$ in the form

$$\mathscr{A}_\infty \approx constant \times (1 - \tilde\omega_o)^{0.4} \qquad (3.1.11)$$

which is also similar to the result of Twomey and Bohren (1980).

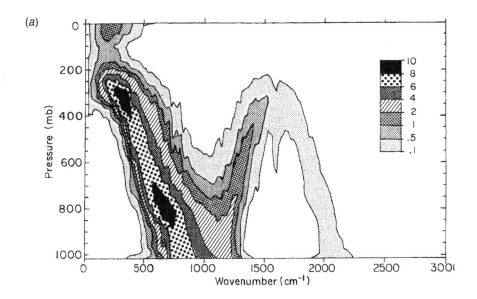

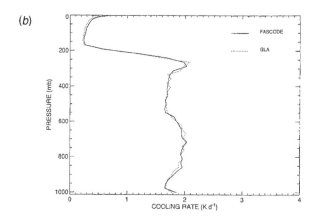

Figure 3.6. (*a*) Spectral cooling rate for water vapor for a hypothetical mid-latitude summer atmosphere. Key is in units of K day^{-1}(cm^{-1})$^{-1}$. (*b*) The total cooling (spectral integral of the data in *a*) as a function of pressure (Clough *et al.*, 1992).

For optically thin clouds with $\tau^* \rightarrow 0$, it can also be shown that (Stephens and Tsay, 1990):

$$\mathscr{R}_o = \frac{\tau^*}{\mu_o} \tilde{\omega}_o b_o \qquad (3.1.12a)$$

$$\mathscr{T}_o = 1 - \frac{\tau^*}{\mu_o}(1 - \tilde{\omega}_o f_o) \qquad (3.1.12b)$$

and

$$\mathscr{A}_o = \frac{\tau^*}{\mu_o}(1 - \tilde{\omega}_o) \qquad (3.1.12c)$$

where b_o and f_o are the backscatter and forward scatter of the solar beam. Thus both the albedo and absorption of thin cloud vary linearly with optical thickness and, as expected, respectively depend on IOPs that define the backscatter and absorption properties of the individual cloud particles.

These relations provide us with a way of deducing the effects of particle size on the albedo and absorption of solar radiation. Two factors are important.

The first is a reciprocal dependence of optical depth on particle size (e.g. Stephens, 1978)

$$\tau = \frac{3W}{2\rho_{\text{water}} r_{\text{e}}} \qquad\qquad (3.1.13a)$$

where W is the liquid water path (LWP, an EOP), r_{e} is the effective radius (the ratio of volume to area of the distribution, an IOP) and ρ_{water} is the density of water. The same sort of reciprocal relationship follows for ice clouds (e.g. Fu and Liou, 1993). An alternative relation is

$$\tau = 2\pi\, h\, N_{\text{o}}^{\frac{1}{3}} \lambda^{\frac{2}{3}} \qquad\qquad (3.1.13b)$$

where h is the cloud thickness, N_{o} is the number density of particles and λ is the liquid water content ($W \approx \lambda h$). Both relationships predict that for fixed liquid water content or path, the optical depth increases through increases in N_{o} or equivalently through decreases in r_{e}. Such an increase in optical depth implies increased albedo of clouds but not necessarily an increase in the albedo of thick clouds since the reflection of these clouds is largely insensitive to any changes in optical depth if deep enough.

The second factor involves the relationship between $(1 - \tilde{\omega}_{\text{o}})$ and r_{e} which Ackerman and Stephens (1987) simplify to

$$1 - \tilde{\omega}_{\text{o}} \approx constant \times \kappa\, r_{\text{e}}^{\,p} \qquad\qquad (3.1.14)$$

where κ is the bulk absorption by water and $p \leq 1$.

From the relationships (3.1.13a), (3.1.14) and the expressions for albedo (3.1.10b), (3.1.11) and (3.1.12a), it follows that the albedo of clouds increases as particle size decreases (Figure 3.7a) through a combination of both a decreased absorption (predicted from equation (3.1.14)) and associated increases of optical depth (3.1.13a). The relationship between absorption and particle size as highlighted in Figure 3.7(b) is complex. The dependence on particle size is such that the absorption of thin clouds (i.e., small LWP) actually decreases with decreasing particle size while the reverse applies for thick clouds (or large LWP). This thick-cloud dependence has been mistakenly interpreted to imply that increases of particle size enhance absorption in clouds. Absorption in marine boundary layer clouds typified by intermediate values of LWP (and optical depth) is only weakly dependent on r_{e}.

In summary, we deduce that the albedo of clouds is sensitive to both particle size and LWP (or ice water path, IWP) and varies in a systematic way with changes in these parameters. By contrast, the solar absorption depends on these parameters in a complex way.

A series of aircraft experiments that seek to confirm the relation between r_{e} and cloud albedo are those of the South-ern Ocean Cloud Experiments (SOCEX) described by Boers *et al.* (1996). They find significant seasonal variations in r_{e} characteristic of marine layered cloud between summer and winter (Figure 3.8) with composite mean values of $r_{\text{e}} = 19\ \mu\text{m}$ in winter and $r_{\text{e}} = 13\ \mu\text{m}$ in summer. Since these measurements were carried out in baseline air (free of continental effects), these results are consistent with the seasonal variations of dimethylsulfoxide (DMS) and the subsequent influence of DMS on cloud condensation nuclei (CCNs) and thus cloud microphysics. These large measured seasonal changes in r_{e} translate to significant percentage changes in cloud albedo (refer to Figure 3.7b). In Figure 3.8 the profile of effective radius is shown when the measured drizzle component is added to the measured profiles of Figure 3.8. Drizzle contributes significantly to the particle size at cloud base but its direct effect on the albedo of clouds has not yet been determined. The indirect effects of drizzle on albedo through its effect on cloud evolution is thought to be significant.

Extrinsic cloud optical properties

The bulk absorption of IR radiation by water and ice is so strong that scattering, it is argued, is negligible. The infrared properties of clouds are thus usually expressed in terms of absorption properties (cloud emissivity for example); however, scattering of infrared radiation is not negligible especially in the case of high clouds (e.g., Stephens, 1980; Edwards and Slingo, 1995). With scattering effects ignored, the IR radiative properties of clouds depend on a combination of the optical depth of the cloud, and the contrast in temperature between clouds and the effective emission temperature of the surrounding atmosphere above and below (this is another way of thinking about the exchange processes discussed in relation to Figure 3.5). It is generally assumed that the optical depth depends on the liquid water or ice path of the cloud (Stephens, 1984).

The most dominant control on the infrared radiative transfer through clouds is the contrast between radiation emitted from the atmosphere to the cloud from above and below and the radiation emitted from the cloud itself. This is readily apparent in Figures 3.9a and b showing the spectral infrared heating of a cloud located in a tropical and subarctic winter atmosphere, respectively. High tropical clouds predominantly heat because the difference between emission at cloud base and absorption of radiation from below leads to a gain in radiative energy and thus a heating of high tropical clouds. This heating occurs principally in the more transparent regions of the spectrum where these differences are largest. The details of this heating and how it penetrates into the cloud

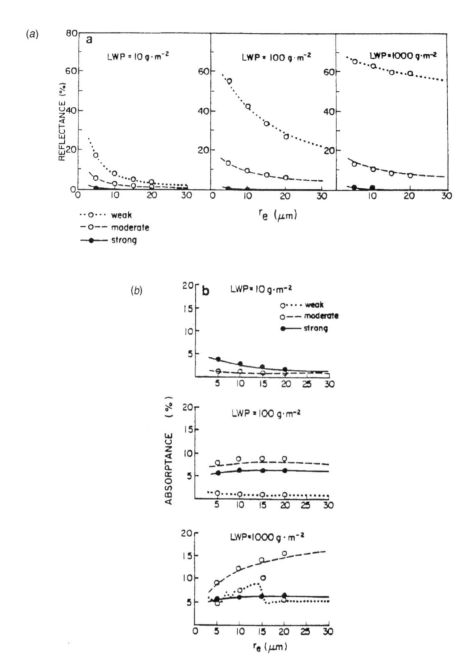

Figure 3.7. Contributions to the (a) albedo and (b) absorptance by the three spectral absorption regimes introduced by Ackerman and Stephens (1987) as a function of r_e for specified values of LWP. The symbols refer to calculations using scattering properties from Lorenz–Mie theory (ignore differences between symbols and curves).

also depend on the optical depth of the cloud which, under the Rayleigh assumptions, depends on the ice water path. For the lower cloud in the subarctic atmosphere, the lack of a distinct contrast between the emission from cloud base and the upwelling radiation from below leads to a much reduced heating at cloud base. These lower clouds predominantly cool at cloud top at most wavelengths through the cooling to space process that is dominant under clear sky conditions.

The important point to be drawn from Figure 3.9 is that the IR radiative properties of clouds, specifically the extent and magnitude of IR heating (recall discussion of Figure 3.3), is strongly dependent on the properties of the environment

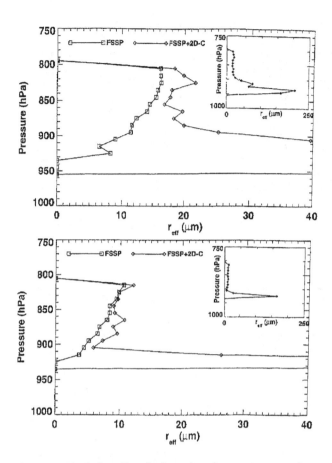

Figure 3.8. Vertical profiles of r_e from aircraft measurements of clouds sampled in winter and summer off the west coast of Tasmania during SOCEX I and II (upper panel; Boers *et al.*, 1996) where r_e is derived from the combination of FSSP and 2-D probes (lower panel; R. Boers, personal communication).

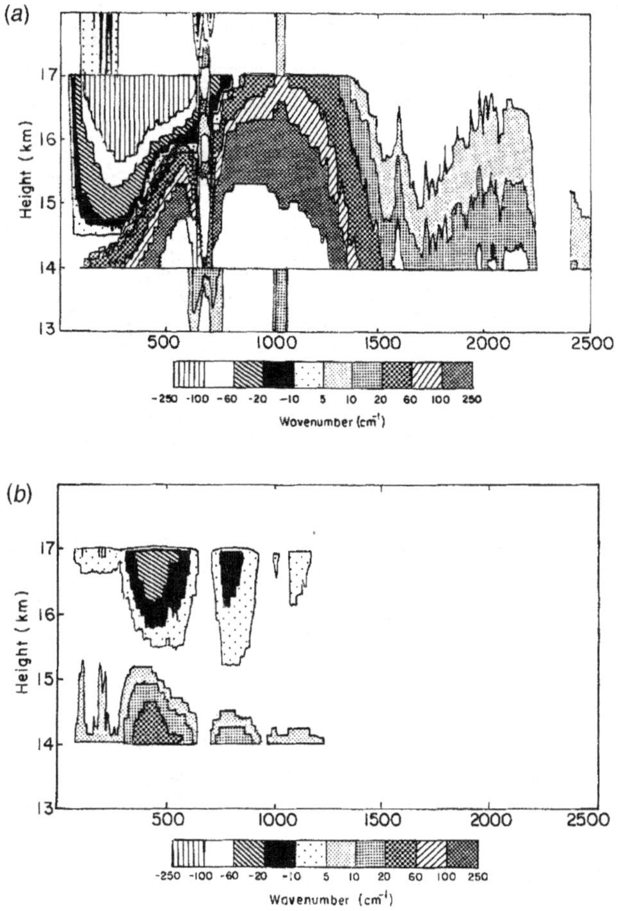

Figure 3.9. The vertical profile of the spectral IR cooling rate for (a) an ice cloud located in a model tropical atmosphere and (b) the same in a subarctic winter atmosphere (Edwards and Slingo, 1995).

around the cloud (such as the temperature contrast and the amount of water vapor above and below the cloud). What is not emphasized in these diagrams but is of equal importance is the effect of optical depth (or IWP) on the radiative heating of the cloud. We expect that the macroscopic influences on the IR radiative properties of clouds noted above are readily handled by proper specification of the atmospheric temperature and water vapor profile (and optical depth) as well as proper specification of cloud base and top heights.

These properties influence the IR heating of the atmosphere. There is another class of macroscopic influence that is also as important but more so to fluxes at the boundaries of the atmosphere, and the appropriate way to treat these influences on radiative transfer is far more uncertain. The effects of complex 3-D geometry of clouds on the bulk radiative properties of clouds is demonstrably large for solar radiation (it is perhaps also not negligible for IR transfer; Evans, 1993).

Tiedtke (1995) provides a clue as to the global importance of these effects using an empirical technique to account for 3-D effects in a global model. Figure 3.10 expresses the difference between the estimated surface radiation budget with and without these effects included. The effects treated include a relationship between albedo of cloud and optical depth that depends on cloud type. In this example, Tiedtke simply scales the optical depth of convective clouds by

$$\tau_{\text{eff}} = \beta\tau \tag{3.1.15}$$

where β is a factor that differs from unity and attempts to account for cloud heterogeneity. A value of $\beta = 0.7$ is assumed in the study and is based on an assessment of aircraft data. With this effect in the model, Tiedtke shows how the TOA net fluxes alter in some regions in excess of 30 W m^{-2} compared with simulations without this effect (Figure 3.10). Research continues on how these effects should actually be incorpor-

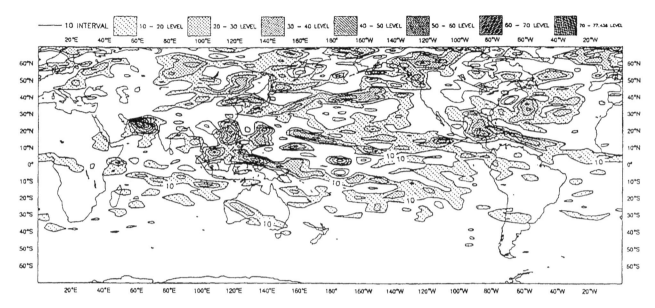

Figure 3.10. A 30-day mean difference of model-generated net shortwave radiation at the top of the atmosphere between forecast experiments with and without inhomogeneous cloud effects. Contours are in W m^{-2}. Values above 10 W m^{-2} are shaded (Tiedtke, 1995).

ated into radiative transfer. The theoretical basis for the scaling of the type introduced by Tiedtke was introduced in Stephens (1988) where it is shown that the scaling approach is a form of subgrid-scale closure and the idea has been confirmed for low-level clouds using both measurements and Monte Carlo models by Cahalan *et al.* (1994, 1995).

Existing global observations

Water vapor

The two major sources of global atmospheric water vapor data include global (i.e. land based) radiosonde data and, more recently, satellite radiance data. The radiosonde data lack spatial coverage, with large data gaps over the oceans, and even over significant land areas (such as Africa). Satellite data, by contrast are near global in nature but the information contained in these data is of variable and often uncertain quality. Examples of presently available large-scale data sets include satellite microwave retrievals from the Defense Meteorological Satellite Program (DMSP) Special Sensor Microwave Imager (SSM/I) data over ocean (e.g., Jackson and Stephens, 1995), TOVS infrared retrievals over land and ocean (Rossow and Kachmar, 1988; Wittmeyer and Vonder Haar, 1994), upper tropospheric relative humidity from geostationary satellites (Schmetz *et al.*, 1995; Soden and Bretherton, 1993; Stephens *et al.*, 1996), upper tropospheric and stratospheric water vapor from SAGE (Rind *et al.*, 1993) and UARS MLS (Read *et al.*, 1995), and a number of data sets using

special radiosonde measurements for research purposes on limited time and space scales. Also available are model-analyzed 4-D data assimilated (in many cases operational analyses) global humidity fields from the European Centre for Medium-range Weather Forecasts (ECMWF), the National Meteorological Center (NMC) (Trenberth and Olson, 1988) and the Goddard Earth Observing System – Data Assimilation System (GOES-DAS) (Schubert *et al.*, 1993).

Since the individual sources of data have certain limitations, it follows that a most useful global moisture data set should be derived from a combination of these measurement systems (Randel *et al.*, 1996). An example of the characteristics of a radiosonde – satellite blended data set in Figure 3.11 highlights the availability of satellite and radiosonde data for one day (July 10, 1989). Data of each type have gaps which when blended produce an improved distribution of water vapor more suitable for climatology. A climatology of water vapor derived from a blending of the three sources of data indicated in Figure 3.11 has been compiled by Randel *et al.*, for the five-year period 1988–1992.

The five-year climatology of total-column water vapor is shown in Figure 3.12. The averaged global distribution is shown in Figure 3.12(*a*) and agrees well with the radiosonde-based climatology of Peixoto and Oort (1991). Figure 3.12(*b*) presents the difference of 1988 minus 1992 column water vapor and indicates the clear fingerprint of the 1992 El Niño year, with drying in the subtropics and mid-latitudes and enhanced atmospheric moisture only in areas near the Equator associated with enhanced SSTs. The interannual

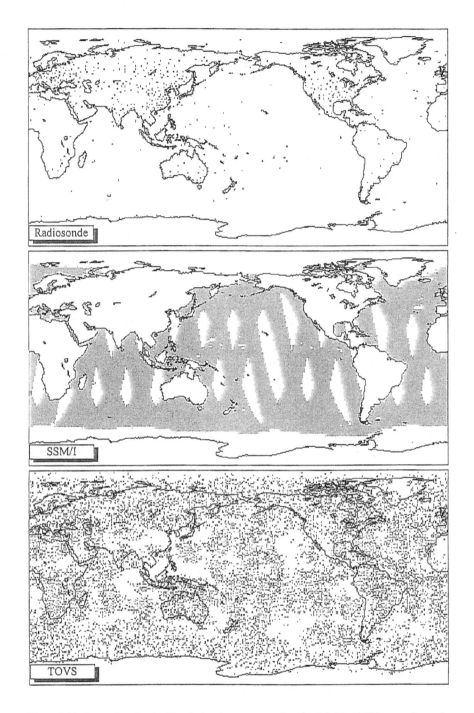

Figure 3.11. Input data for the blended water vapor product for July 10, 1989 from radiosondes, SSM/I and TOVS. These data are gridded on 1° × 1° global grids and all gridpoints containing data are shaded.

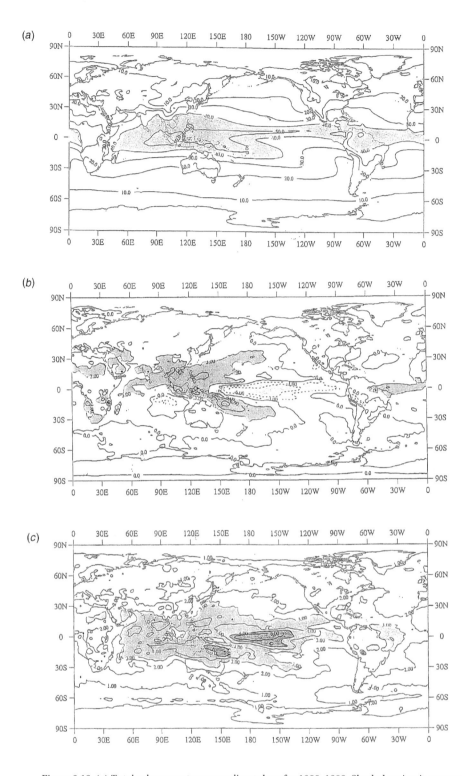

Figure 3.12. (*a*) Total column water vapor climatology for 1988–1992. Shaded region is greater than 40 mm. (*b*) 1988 minus 1992 total column water vapor. Shaded areas are 1988 greater than 1992, light shade 1992 greater than 1988. (*c*) Variability of the total column water vapor with annual cycle removed for 1988 through 1992, expressed as standard deviation (mm). Areas of greater than 3 mm are shaded lightly, greater than 5 mm are shaded darkly.

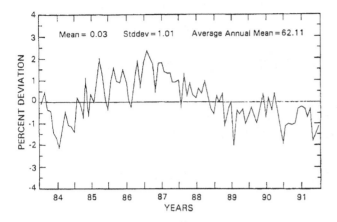

Figure 3.13. Global monthly mean cloud cover anomaly (relative to mean annual cycle) determined from ISCCP analysis for the period July 1983 through June 1991 (Raschke and Rossow, 1995).

variability of the water vapor after removing the annual cycle, is shown in Figure 3.12(c). The values are expressed as a mean monthly standard deviation. The most striking areas of high variability during the five years are due to changes in the tropical circulation patterns caused by the ENSO events.

The International Satellite Cloud Climatology Project (ISCCP)

The ISCCP was established in 1982 (Schiffer and Rossow, 1983) and has been a vital source of information on global cloudiness and cloudiness variability for more than a decade (Figure 3.13). The data are now being applied in a number of ways such as in the estimation of the surface radiation budget (Rossow and Zhang, 1995; Whitlock *et al.*, 1995) and the analysis of cloudiness associated with synoptic scale disturbances (Lau and Crane, 1995). ISCCP has compiled a uniform satellite radiance climatology using sensors on both geostationary and polar orbiting satellites. The principal data are visible (VIS, centered at 0.65 μm) and infrared (IR, 11 μm) calibrated radiances which are used to differentiate clear from cloudy scenes using the ISCCP cloud detection method based on a radiance time–space variational approach (Rossow and Garder, 1993*a*). These data have been validated with information from field programs and other data sources (Rossow and Garder, 1993*b*) and compared with other cloud-amount climatologies (Rossow *et al.*, 1993).

Important information derived from ISCCP radiances include cloud optical depth, cloud pressure and cloud amount. Cloud amount is one of the EOPs required to describe the radiative transfer. Identification of other EOP information of relevance is a topic of research (an example of this other

information in the context of Tiedtke's (1995) work is his β factor).

Global ISCCP radiances are also used to produce information about IOPs. In addition to optical depth, use of near-infrared radiances provides a way of retrieving cloud particle size and, through the relation in equation (3.1.13*a*), cloud liquid water path. Han *et al.* (1994), using ISCCP radiance data, augmented by 3.7 μm radiances from the AVHRR, produced geographic distributions of the effective radius (Figure 3.14) which they analyze to obtain some insight into the nature of hemispheric, land and ocean and seasonal variabilities of r_e. While the variations found were consistent with expected variations in CCN concentrations, with continental clouds characterized by smaller values of r_e than oceanic clouds, they do not reflect the extent of the seasonal variations observed during the SOCEX series of experiments. The importance of r_e in the definition of cloud-radiation effects dictates the need for a concerted validation effort and further refinements to those methods used to derive this information from satellite radiances.

Other global information about clouds

The simple analyses used to produce the results of Figure 3.7 serve to remind us of the importance of the cloud water (and ice path) to the bulk transfer of solar radiation in clouds. Global cloud liquid water information is presently derived from the microwave radiance data obtained from the SSMI operational instrument (e.g., Greenwald *et al.*, 1993) over oceans, and combinations of sensors on different satellites are beginning to show promise (Greenwald *et al.*, 1997). From these data, we hope to establish a better understanding of the links between liquid water path, temperature and radiative properties of clouds. For example, Figure 3.15(*a*) presents the results of the correlation between the LWP and atmospheric temperature much in the way cloud optical depth and temperature were correlated in the study of Tselioudis *et al.* (1992). Figure 3.15(*a*) presents the parameter

$$f = \frac{\mathrm{d}\ln W}{\mathrm{d}T} \qquad (3.1.16)$$

derived from gridded LWP data corresponding to ISCCP-defined low cloud correlated with the mean temperature of the layer between the surface and 680 hPa. These data apply to the region from 60°N to 60°S and show that in the warmest regions of the globe a decrease in W is correlated with an increase of temperature. This result is similar to the optical-depth sensitivities deduced by Tselioudis *et al.* (1992) who

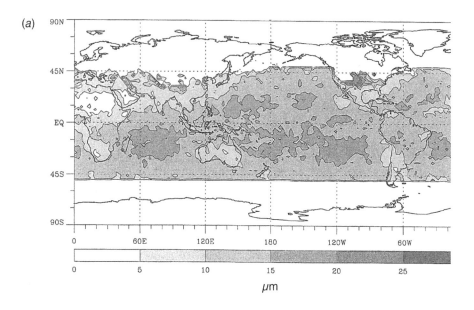

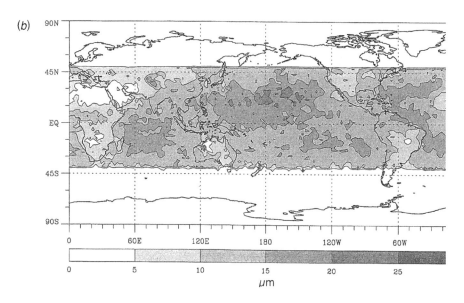

Figure 3.14. Geographic distribution of water cloud droplet effective radii (in μm) derived from the NOAA-9 AVHRR radiances for (*a*) January and (*b*) July, 1987 (Han *et al.*, 1994).

argue that specific regional changes in the optical depth–temperature correlation is more complex than one simply defined by thermodynamical considerations.

The relationship between LWP and cloud albedo can be examined using global information derived from satellite microwave radiance data (SSM/I) and cloud albedo available from ERBE. While this is an important task, it has been difficult to find enough coincident data to carry out correlations between albedo and LWP – a problem rectified with the launch of NASA's Tropical Rainfall Measurement Mission (TRMM) in

1997. Nevertheless a limited match of SSM/I and ERBE was presented in the study of Greenwald *et al.* (1995) in which the albedo of low overcast cloud as determined by ISCCP is presented as a function of LWP (Figure 3.15(*b*)). The curves shown represent relations derived from theory assuming different values of r_e. Possible reasons for the differences between theory and observation need to be explored; they vary from biases introduced in sampling the different data to macroscopic effects that dramatically alter the intrinsic relationship between albedo and LWP.

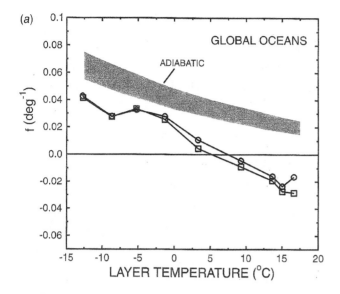

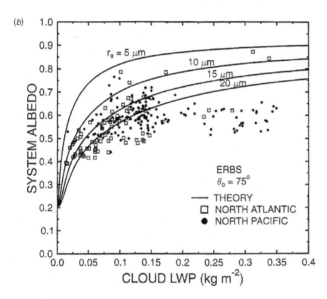

Figure 3.15. (*a*) Global observations of *f* versus temperature (°C) of the atmospheric layer between the surface and 680 hPa. Observations using composite ISCCP and microwave liquid water are given by symbols and the shading indicates the range of the relationship derived from adiabatic assumptions for clouds of varying thickness and height. (*b*) Scatter diagram of the instantaneous albedo measurements from ERBs at a solar zenith angle of 75° versus coincident SSM/I LWP data for low clouds over the Northern Pacific and Atlantic during June and July 1988. Also shown are relationships based on parameterized theory for different values of r_e (Greenwald *et al.*, 1995).

Summary

The broad impacts of water vapor and clouds on the ERB have been discussed and specific effects on the radiative transfer within the atmosphere explored. The main characteristics of these effects include:

Water vapor

1. The chief influence of water vapor on the ERB is through its control of F_∞. The association between the infrared emission defined largely by the water vapor content of the atmosphere and the (sea) surface temperature of the planet is complicated by a number of factors. These include the disproportionate influence of the small amount of upper-tropospheric water vapor on the radiation emitted by the planet to space, largely in the spectral regions of the rotational absorption band of water vapor. Global measurements of the emission in this important spectral region neither exist nor are presently planned for future research satellites.

2. In gross terms, the clear-sky column IR cooling is proportional to the column water vapor. Since this cooling is the dominant component of the radiative budget of the atmosphere, and since the deficit of the budget is largely balanced by latent heat release, it is appropriate to consider this cooling as a rudimentary measure of the activity of the Earth's greenhouse effect and in turn as an indirect measure of the gross activity of the hydrologic cycle in heating the atmosphere.

3. The profile of clear-sky IR cooling is governed to a significant degree by the vertical distribution of water vapor. Lower-tropospheric water vapor (i.e. the majority of the water vapor mass) contributes to lower-tropospheric cooling whereas the cooling of the upper troposphere is governed by upper-tropospheric water vapor. The contributions to cooling in different layers varies spectrally from low-level cooling in the transparent window regions of the IR spectrum to upper-tropospheric cooling from emission in the strong rotational absorption band.

4. Emission by water vapor decouples changes in TOA fluxes from changes in surface fluxes. Increasing atmospheric temperature tends to *increase* TOA F_∞ whereas increases in water vapor produce compensating *decreases* in F_∞ by elevating the effective level of emission to space. These competing effects are realized through the relative humidity of the upper layers of the troposphere. When the relative humidity is unchanged, so too is the F_∞. In contrast, the combined effects of increases of moisture and temperature on the emission of longwave radiation to the

surface add to produce a large sensitivity of flux to changes induced by increases in greenhouse gases. Whether or not it is possible to deduce the climate change, driven by surface changes from spectral measurements at the TOA, remains an issue of some relevance to the topic of monitoring and detection of climate change. The different sensitivities between fluxes at the TOA and fluxes at the surface profoundly affect the interpretation of flux data in terms of the water vapor feedback. This requires a better understanding than is available at present.

Clouds

5. The impact of clouds on the ERB is conveniently described in terms of clear-to-cloudy sky flux differences which quantify the residual effects of longwave and shortwave fluxes on the ERB. On the whole, the magnitude of the shortwave flux effect is slightly in excess of the longwave effect, largely as a result of the highly reflecting summertime mid-latitude clouds. It is often incorrectly assumed, however, that these negative values of global C_{net} necessarily imply a negative cloud feedback to global warming. It is also incorrectly assumed that it is the change ΔC_{net} associated with a change in climate that provides a measure of cloud feedback (e.g. Wielicki *et al.*, 1995). This is not so as ΔC_{net} is merely a measure of the response and provides no direct measure of feedback. Tropical cloud systems in contrast to mid-latitude systems produce a near-zero net forcing and whether or not this perplexing cancellation is fortuitous is not well understood. Because the effects of clouds on C_{LW} and C_{SW} are largely reciprocal, processes that affect one component by a disproportionate amount offer greatest potential for significantly influencing the ERB and thus the Earth's climate. The proposed aerosol influence on cloud albedo is one such example.

6. While the net effect of clouds on the TOA radiation balance arises through compensating effects of the individual longwave and shortwave fluxes, clouds significantly redistribute the absorbed energy of the atmospheric column, depositing radiative energy in the atmosphere largely at the expense of energy absorbed at the surface. The extent to which this energy redistribution occurs in the real climate system, and our ability to represent this redistribution, are topics of critical importance and active research.

7. Clouds affect the radiative transfer in a way that can be described in terms of two broad classes of cloud properties. These include properties determined by the macrophysical characteristics of the clouds (referred to as extensive optical properties, EOP) and properties determined by the microphysics of clouds (referred to as intensive optical properties, IOP). The effects of both properties have been explored in this section and it is shown how cloud particle size affects both the albedo of clouds and the absorption of solar radiation in clouds. The effects of EOPs on cloud radiative properties are large and their treatment is largely empirical. The fundamental role of the ice and liquid water contents, inasmuch as these define the cloud optical depth, is also emphasized.

8. Relative to clear skies, high clouds tend to warm the atmosphere and low clouds cool the atmosphere. The extent of this warming/cooling depends on the cloud top and base heights and the details of the surrounding profile of water vapor. In order to represent this important effect, information about the vertical distribution of clouds is crucial.

9. Present global observations of cloud have advanced our knowledge about clouds and their effects on the planet's radiation budget. Despite this progress, we lack crucial data to validate existing global properties (such as r_e and LWP). We lack information on the 4D distribution of radiative fluxes and the related 4D distribution of cloud optical properties that govern these fluxes. Gaping deficiencies exist in global observations of the vertical distribution of clouds, the ice water path of clouds and the liquid water path of clouds over land regions.

3.2 Effects of aerosols on clouds and radiation

Peter V. Hobbs

Introductory remarks

Small particles in the air (called *aerosols*) originate from both natural and anthropogenic sources. Aerosols play a role in determining the microstructures of clouds and whether or not a cloud precipitates. They also affect the global energy balance. For example, it is apparent that aerosols over large cities reduce the intensity of the radiation from the sun that reaches the ground. This diminution is due primarily to aerosols scattering and, to some extent, absorbing solar radiation. Since aerosols are very variable in their distribution and properties, so are their effects on solar radiation. A topic of considerable current interest and debate is the effect of anthropogenic aerosols on the Earth's radiation balance, and the extent to which these aerosols may offset predictions of global warming due to increases in anthropogenic greenhouse gases.

In addition to their *direct effects* on the Earth's radiation balance, aerosols can also affect solar radiation reaching the ground by changing cloud microstructures (*indirect effect* of aerosols on radiative forcing). As in the case of the direct effect of aerosols, the indirect effect is believed to lower temperatures at the Earth's surface.

Aerosols have very short residence times in the atmosphere compared with long-lived anthropogenic greenhouse gases (such as CO_2 and CH_4). Even if CO_2 emissions, for example, were maintained at today's levels, the effects on the atmosphere of past increases in emissions would be experienced for at least two centuries. By contrast, if the emissions of anthropogenic aerosols were suddenly reduced, their effects on the atmosphere would disappear in a few weeks.

In this section a brief account is given of ways in which atmospheric aerosols may affect cloud structures, precipitation, and the Earth's radiation balance. For more detailed discussions of these topics, and many other aspects of aerosol–cloud–climate interactions, the reader is referred to Hobbs (1993), Charlson and Heintzenberg (1995), Andreae (1995), and Houghton *et al.* (1996).

Sources of atmospheric aerosols

Atmospheric aerosols range in size from hundredths of a micrometer to many tens of micrometers. They are produced by direct injection of particles into the air (*primary* aerosols) and by the conversion of gases into liquid and solid particles (*secondary* aerosols). Primary aerosols dominate atmospheric particles with diameters greater than about 1 μm,

while secondary aerosols comprise most of the smaller particles (Figure 3.16*a,b*).

Aerosols derive from many sources: the Earth's surface, oceans, volcanoes, the biosphere, industries, power plants, automobiles, etc. As can be seen from Table 3.1, anthropogenic sources contribute about one-third to the total aerosol mass flux into the atmosphere and well over one-half of the flux of secondary particles. Consequently, any discussion of the effects of aerosols on the atmosphere must consider both natural and anthropogenic sources.

The large amounts of dust from the major deserts (e.g., the Sahara and the Gobi) are readily seen from satellites. Even though many of the larger particles are not transported far, dust from the Sahara is regularly observed in Barbados. Also, a significant fraction of marine sediments consist of wind-blown dust. Desertification, deforestation and poor agricultural practices have increased dust in the atmosphere.

Although sea salt (produced by bursting air bubbles and winds) is the largest natural flux of aerosol mass into the atmosphere, most of this mass is in large particles that are not transported very far.

Volcanic eruptions can inject large amounts of particles and gases into the atmosphere over relatively short time periods. If these materials are blasted into the stratosphere, where they are removed much more slowly than in the troposphere, they can cause significant climatic effects (see below).

The formation of sulfates from sulfur gases is a very important source of secondary aerosols (Table 3.1). Prior to the industrial revolution, sulfur in the atmosphere was dominated by the emissions of gases (primarily dimethylsulfide, DMS) from marine phytoplankton and continental biota (Andreae, 1990). However, by the early 1960s, anthropogenic emissions of sulfur into the atmosphere (primarily from the burning of fossil fuels) began to exceed the emissions of sulfur from natural sources (Houghton *et al.*, 1996).

If significant amounts of alkaline materials (e.g., soil dust) are present in aerosols, they may absorb nitric acid. Also, in the presence of high concentrations of ammonia, ammonium nitrate can condense onto existing particles. However, the amount of nitrate aerosols in the atmosphere is small compared with sulfate (Table 3.1).

The biosphere provides a large source of organics to the atmosphere in the form of organic particles and gaseous hydrocarbons (e.g., terpenes), some of which condense to form aerosols. Anthropogenic volatile organic compounds (VOC) can also condense to form aerosols. Recent measure-

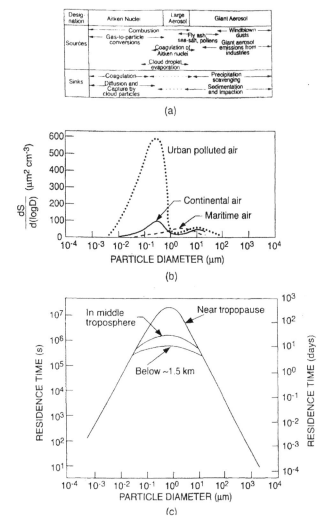

(a)

(b)

(c)

Figure 3.16. (*a*) Sources and sinks of atmospheric aerosols. (*b*) Typical plots of aerosol surface areas (*S*) versus aerosol diameter. (*c*) Estimates of the residence time of atmospheric aerosols. (Adapted from Jaenicke, 1978.)

this size. Figure 3.16(*c*) shows the residence times of aerosols; compared with most greenhouse gases (although water vapor is a notable exception), the lifetimes of aerosols are very short (about 1 week). This, together with the spatial variability of aerosol sources, produces the highly variable geographical distribution of aerosols, particularly those that derive from anthropogenic sources. Consequently, the effects of aerosols on climate are expected to vary greatly with geographical location.

Effects of aerosols on the microstructures of clouds and on precipitation

Some atmospheric aerosols serve as particles on which cloud droplets form. These aerosols, which are called *cloud condensation nuclei* (CCN), tend to be greater than a few hundredths of a micrometer in diameter and soluble in water.

There is no doubt that local sources of aerosols can increase CCN and cloud droplet concentrations (Hobbs *et al.*, 1970; Eagen *et al.*, 1974; see also the discussion of ship tracks below).

On much larger scales, the concentrations of CCN over land are generally about ten times greater than over oceans. Consequently, otherwise similar cloud types (e.g., stratus or cumulus) over land tend to have larger droplet concentrations than they do over the oceans. This has several important consequences. First, since the liquid water contents of continental and maritime clouds do not differ greatly, the larger concentrations of droplets in continental clouds produce smaller average droplet sizes and narrower droplet size spectra than in maritime clouds. This, in turn, generally makes it more difficult for droplets in continental clouds to increase to precipitation size by colliding with each other. Consequently, as far as microphysical processes are concerned, similar cloud types with similar dimensions are less likely to precipitate in a continental air mass than in a maritime air mass. These effects of aerosols on cloud structures and precipitation affect the radiative properties of clouds (see below). However, increasing CCN concentrations does not always reduce the precipitation efficiency of a cloud; injection of an appropriate mix of CCN can sometimes enhance precipitation (e.g., Hobbs *et al.*, 1970). Also, the effects of CCN on cloud microstructures may often be masked by fluctuations in updraft velocity and the entrainment of ambient air in the clouds.

A very small fraction of the aerosols in the air (called *ice nuclei*) can nucleate ice in clouds, which promotes the development of precipitation (see Section 3.4). Many ice nuclei have crystallographic structures similar to that of ice. Sources of ice nuclei include deserts and arid regions (silicate materials), surface soils (e.g., clay particles), decaying vegetation (such as tree leaves), and organic materials from the ocean.

ments on the US East Coast suggest that aerosols in urban/industrial regions contain more organics than previously believed, sometimes rivaling sulfates (Novakov *et al.*, 1997).

Biomass burning in the form of forest and savanna/cerrado fires, the burning of agricultural wastes, and for the production of domestic energy, produces both primary and secondary aerosols. Biomass burning also emits black carbon, which strongly absorbs visible radiation.

Figure 3.16(*b*) highlights typical differences in aerosol amounts in urban, continental and maritime air. Note the peak in particle surface area at particles a few tenths of a micrometer in diameter, particularly in urban polluted air; this reflects the strong sources but weak sinks for particles of

Table 3.1. *Estimates of current mass fluxes, lifetimes, global and column burdens, mass extinction coefficients and global mean optical depths of aerosols from natural and anthropogenic sources.*

Source	Flux (Tg yr⁻¹)			Lifetime (days)	Global burden (Tg)	Column burden (mg m⁻²)	Mass extinction efficiency (hydrated) (m² g⁻¹)	Optical depth*
	Low	High	Best estimate					
(a) Natural sources								
Primary								
Soil dust (mineral aerosol)	500	1,500	750	4	8.2	16.1	0.8	0.013
Sea salt	1,000	10,000	1300	1	3.6	7.0	1.5	0.010
Volcanic dust	4	10,000	33	4	0.4	0.7	2.0	0.001
Biological debris	26	80	50	4	0.5	1.1	2.0	0.002
Secondary								
Sulfates from biogenic gases	60	110	90	5	1.2	2.4	5.1	0.012
Sulfates from volcanic SO_2	4	45	12	5	0.16	0.3	5.1	0.002
Organic matter from biogenic VOC	40	200	55	7	1.1	2.1	5.0	0.010
Nitrates from NO_x	10	40	22	4	0.24	0.5	2.0	0.001
Total natural	1,644	21,975	2,312		15.4	30.2		0.051
(b) Anthropogenic sources								
Primary								
Soil dust mineral aerosol	500	1,500	750	4	8.2	16.1	0.8	0.013
Industrial dust, etc.	40	130	100	4	1.1	2.1	2.0	0.004
Black carbon (soot and charcoal)	10	30	20	6	0.3	0.6	10.0	0.006
Secondary								
Sulfates from SO_2	120	180	140	5	1.9	3.8	5.1	0.019
Biomass burning (excluding black carbon)	50	140	80	8	1.8	3.4	5.1	0.017
Nitrates from NO_x	20	50	36	4	0.4	0.8	2.0	0.002
Organics from anthropogenic VOC	5	25	10	7	0.19	0.4	5.0	0.002
Total anthropogenic	745	2,055	1,136		13.9	27.2		0.063
Total natural and anthropogenic			3,448		29	57.4		0.114
Anthropogenic percentage of total aerosol			33		47	47		55

Note: * The optical depth τ between height z and the top of the atmosphere is defined by the relation $E_z = E \exp(-\tau)$, where E_z and E are the irradiances (units: W m⁻²) due to the direct solar beam at height z and at the top of the atmosphere, respectively. Thus, for a unit optical depth $E_z = E \exp(-1) = 0.37 E$. Note that τ is a dimensionless parameter.

Source: Based on private communication from M. O. Andreae, March 1996.

The connection between ice nuclei and ice in clouds is complicated by *secondary ice producing processes* that do not depend directly on ice nuclei (for a review of this topic see Beard, 1992). For example, the formation of copious ice in clouds by secondary ice producing processes is correlated with the breadth of the droplet size distribution (Rangno and Hobbs, 1994), which, as we have seen, is determined in part by the CCN concentrations. Consequently, maritime clouds, with their lower CCN concentrations and broader droplet size distributions, generally contain more ice than corresponding continental clouds; this is another reason why maritime clouds precipitate more readily.

Direct radiative effects of aerosols

As mentioned above, aerosols scatter and absorb solar radiation and thereby deplete the intensity of the solar beam reaching the Earth's surface. The extent of this depletion is measured by a dimensionless parameter called the *optical depth* (τ) of the aerosol layer, which is given by:

$$\tau = \sec \phi \int \sigma_e dz \qquad (3.2.1)$$

where ϕ is the zenith angle of the sun, σ_e is the *aerosol extinction coefficient* (units: m^{-1}), and the integration extends over the depth of the aerosol layer. The aerosol extinction coefficient is given by:

$$\sigma_e = \sigma_s + \sigma_a \qquad (3.2.2)$$

where, σ_s and σ_a are the aerosol scattering and absorption coefficients. When σ_a is small compared with σ_s, equation (3.2.1) can be written as:

$$\tau \approx \tau_s = \sec \phi \int \sigma_s dz \qquad (3.2.3)$$

The aerosol scattering coefficient (and therefore τ) depends on the wavelength (λ) of the radiation:

$$\sigma_s \propto \lambda^{-\alpha} \qquad (3.2.4)$$

where α (which varies from about 0.5 to 2 for atmospheric aerosols) is called the *Angström exponent*. For aerosol particles that are very small compared with λ the scattering is in the Rayleigh region, and $\alpha = 4$. If the particles are very large compared with λ (e.g. cloud droplets scattering sunlight), geometrical optics applies and $\alpha = 0$.

The direct effects of aerosols on the radiative balance of the Earth–atmosphere system (hereafter called the planet) can be conveniently discussed in terms of the direct aerosol radiative forcing (DARF), which is defined as:

$$DARF = A_{SW} + A_{LW} \qquad (3.2.5)$$

where A_{SW} and A_{LW} are the perturbations in the shortwave and longwave fluxes at the top of the atmosphere (or at the level of the tropopause) by aerosols. By convention, positive and negative values of DARF indicate perturbations that, on average, cause warming and cooling, respectively, of the planet. Since aerosols scatter a portion of the solar radiation back into space, A_{SW} is generally negative. However, A_{LW} is generally positive because aerosols absorb longwave radiation from the Earth and re-emit it at a lower temperature. Carbonaceous particles absorb significant amounts of solar radiation. In this case, a portion of the solar radiation that would otherwise reach the Earth's surface is absorbed by the aerosol, thereby causing local warming aloft. Longwave radiation is then emitted in all directions by the warmed aerosol. Thus, the net effect of an absorbing aerosol on DARF depends on the ratio σ_s / σ_e (or τ_s / τ_e), which is called the *single-scattering albedo* (ω).

For particles that are small compared with λ and for which $\sigma_s \gg \sigma_a$, equations (3.2.3) and (3.2.4) show that τ decreases rapidly with increasing λ. Consequently, such aerosols attenuate solar radiation (λ small) much more than they do longwave radiation (i.e. A_{SW} in equation (3.2.5) dominates over A_{LW}). Since there are strong water vapor absorption bands in the near-infrared solar spectrum, the dominant effect of tropospheric aerosols on DARF is in the window regions of the solar spectrum, mainly at visible wavelengths.

A first-order estimate of the magnitude of DARF produced by the addition of a thin aerosol layer spread uniformly over the globe can be obtained as follows. The reflectance R of an optically thin aerosol layer is given by:

$$R = 2\,\omega\,\beta\,\tau = 2\,\beta\,\tau_s \qquad (3.2.6)$$

where β is the average fraction of solar radiation that is backscattered to space, and the factor 2 comes from the fact that the global mean value of the cosine of the solar zenith angle is $\frac{1}{2}$. The absorptance A of the layer is:

$$A = 2\,(1 - \omega)\tau = 2\,\tau_a \qquad (3.2.7)$$

If $R \ll 1$, the increase in the reflectance when an aerosol layer is placed over a ground surface with reflectance R_s, taking into account multiple reflections, is given approximately by:

$$R(1 - R_s)^2 - 2\,A\,R_s \qquad (3.2.8)$$

If the addition of an aerosol layer is to produce cooling of the Earth's surface, it must increase the reflectance. From equations (3.2.6), (3.2.7) and (3.2.8), we see that this will be the case provided that:

$$\frac{(1 - R_s)^2}{2\,R_s} > \frac{(1 - \omega)}{\beta\,\omega} \qquad (3.2.9)$$

This condition is most readily met for an aerosol that has large values for the single-scattering albedo and the backscatter fraction, and which overlies a poorly reflecting surface. Since most atmospheric aerosols satisfy equation (3.2.9), they tend, on average, to cool the planet.

Since the optical depths of clouds (about 10) is very much greater than that of aerosols (about 0.1), aerosol layers overlying clouds have negligible effects on the reflectivity of the planet. Therefore, from equations (3.2.6) and (3.2.8) the change in the solar reflectivity produced by an optically thin, non-absorbing aerosol layer is:

$$\Delta R_p \approx T_{atm}^2 (1 - N_c)(1 - R_s)^2 2\beta\tau \qquad (3.2.10)$$

where T_{atm} is the fraction of the solar radiation transmitted through the atmosphere above the aerosol layer and N_c is the fractional horizontal extent of the aerosol layer that is overlapped by cloud.

If S is the irradiance of solar energy incident upon the planet (1365 W m^{-2}), the global mean irradiation is $S/4$, because the effective solar-energy collecting area of the Earth is almost exactly one-fourth of the total surface area of the Earth:

$$\frac{\pi r^2}{4\pi r^2} = \frac{1}{4}$$

Hence, if an aerosol layer results in an additional fraction ΔR_p of the incident solar energy being reflected back into space, the change in the mean global insolation of the planet due to the aerosol, which is the globally averaged shortwave DARF, is given by:

$$\text{DARF} = -\Delta R_p \frac{S}{4} = -\frac{S}{2}T_{\text{atm}}^2(1-N_c)(1-R_s)^2\beta\tau \qquad (3.2.11)$$

The optical depth τ in equation (3.2.11) can be derived from (3.2.1) if the height dependence of σ_e is known. Alternatively (and more commonly), it is derived from:

$$\tau = B_a\,\alpha_a\,f(\text{RH}) \qquad (3.2.12)$$

where B_a is the dry aerosol column burden (units: kg m^{-2}), α_a is the dry aerosol mass extinction efficiency (defined as the aerosol extinction coefficient per mass of aerosol in a unit volume of air; units: $\text{m}^2\,\text{kg}^{-1}$), and $f(\text{RH})$ is the relative increase in aerosol light scattering due to increases in aerosol size with increasing relative humidity (RH).

Listed in Table 3.1 are estimates of column burdens (B_a) and the hydrated mass extinction efficiencies $[\alpha_a\,f(\text{RH})]$ for various types of atmospheric aerosols. From (3.2.12), the product of these two quantities gives the optical depth, values for which are also given in Table 3.1. If we use the following as global mean values, $T_{\text{atm}} = 0.71$, $N_c = 0.6$, $R_s = 0.15$ and $\beta \simeq 0.21 - 0.28$, equation (3.2.11) becomes:

$$|\text{DARF}| \simeq 21\,\tau \text{ to } 28\,\tau \qquad (3.2.13)$$

where DARF is positive or negative depending on whether the aerosol causes warming or cooling of the planet. Equation (3.2.13) can be used to obtain rough estimates of the globally averaged DARF from estimates of τ. For example, the estimate given in Table 3.1 for the globally averaged optical depth due to anthropogenic sulfate particles is 0.019; therefore (3.2.13) gives a rough upper limit to the contribution of these particles to DARF of -0.5 W m^{-2}.

More accurate calculations of DARF require the use of more sophisticated numerical models. Most effort has been expended on determinations of DARF due to anthropogenic sulfate (Table 3.2). For example, Langner and Rodhe (1991) used a multiple-box model to simulate the emission, transport, removal and distribution of sulfate in the atmosphere. Charlson et al. (1991) combined these sulfate distributions with specified aerosol radiative properties to estimate regional and global DARFs produced by anthropogenic sulfate. Kiehl and Briegleb (1993) carried out more sophisticated calculations of DARF due to anthropogenic sulfate, by using a multispectral radiation code and including more rigorously the effects of clouds, water vapor and ozone. Haywood and

Shine (1995) modified Charlson et al.'s method, to provide a 'half-way house' between the two previous approaches, and they included absorption as well as scattering by the aerosol.

Aerosol effects on cloud radiative properties

Photographs of the Earth from satellites show that the brightest objects, and therefore the major reflectors of the sun's radiation, are clouds and snow-covered surfaces. Therefore, if atmospheric aerosols modify the radiative properties of clouds, they have the potential for large effects on the Earth's radiation balance.

The optical thickness of a cloud of depth h containing a uniform number concentration $n(r)$ of droplets of radius r is given by:

$$\tau = \pi h \int_0^\infty Q_e r^2 n(r)\,\mathrm{d}r \qquad (3.2.14)$$

where Q_e is the extinction efficiency. At visible wavelengths, and when $\lambda \ll r$, Q_e has a value close to 2. Therefore, if we assume a narrow droplet-size spectrum with a mean radius $\bar{r}$, (3.2.14) simplifies to:

$$\tau = 2\pi h(\bar{r})^2 N \qquad (3.2.15)$$

where N is the total number concentration of droplets. The total liquid water content (W) of the cloud is:

$$W = \frac{4}{3}\pi\rho_L\int_0^\infty r^3 n(r)\mathrm{d}r = \frac{4}{3}\pi\rho_L(\bar{r})^3 N \qquad (3.2.16)$$

where ρ_L is the density of liquid water. From (3.2.15) and (3.2.16),

$$\tau = 2.4\left(\frac{W}{\rho_L}\right)^{\frac{2}{3}}hN^{\frac{1}{3}} \qquad (3.2.17)$$

It follows from (3.2.17) that if W and h are constant:

$$\frac{\Delta\tau}{\tau} = \frac{1}{3}\frac{\Delta N}{N} \qquad (3.2.18)$$

The reflectance (or albedo) of a cloud is the fraction of the incident radiation that is reflected by the cloud integrated over the hemisphere of backscattering. The large areas of rather thin stratiform clouds that cover an appreciable fraction of the Earth's surface have an albedo of about 0.50. Therefore, the energy balance of the Earth is rather sensitive to the albedo of these clouds.

For the global average solar zenith angle of 60°, the albedo A of a cloud is given to a good approximation by:

$$A = \frac{(1-g)\tau}{1+(1-g)\tau} \qquad (3.2.19)$$

Table 3.2. *Some estimates of direct aerosol radiative forcing by anthropogenic aerosols.*

Aerosol type	Global mean forcing (W m^{-2})	Regional maximum forcing (W m^{-2})	Reference and comments
Anthropogenic sulfate	-1.6	$-$	Charlson *et al.* (1990). Single-box model.
	-0.6	-4 (eastern Mediterranean Sea) -2 (eastern USA)	Charlson *et al.* (1991). Global 3D chemical/radiative model, based on Langner and Rodhe (1991) with slow SO$_2$ oxidation rate.
	-1.3	$-$	Charlson *et al.* (1992). Single-box model.
	-0.3	-4.2 (Europe) -3 (eastern USA)	Kiehl and Briegleb (1993). 3D chemical model and GCM.
	-0.66	-11 (central Europe) -5 (eastern USA)	Kiehl and Rodhe (1995). Chemical model.
	-0.95	$-$ (Europe) -3 (eastern USA)	Taylor and Penner (1994).
	-0.56 to -0.94	$-$	Penner (1995). Single-box model.
	-0.36 to -0.79	$-$	Haywood and Shine (1995).
Biomass burning	-0.8	$-$	Penner *et al.* (1992). Box model.
	-0.3	$-$	Hobbs *et al.* (1997). Same model as Penner *et al.* (1992) but using measured radiative properties of smoke from biomass burning in Brazil.
Soot	$+0.35$	$-$	Penner (1995).
	$+0.05$ to $+0.27$	$-$	Haywood and Shine (1995).
	$+0.1$	$-$	Houghton *et al.* (1996). Uncertainty of at least a factor of 3.

Source: Adapted with permission from National Research Council (1996), *Aerosol Radiative Forcing and Climate Change.* Copyright 1996 by the National Academy of Sciences. Courtesy of the National Academy Press, Washington, D.C.

where g is the scattering asymmetry factor, which is the average value of the cosine of the scattering angle. For the scattering of solar radiation by clouds, $g \simeq 0.85$. Hence, (3.2.19) becomes:

$$A \simeq \frac{\tau}{\tau + 6.7} \qquad (3.2.20)$$

Provided W and h are constant, (3.2.18) and (3.2.20) yield:

$$\frac{\Delta A}{\Delta N} = \frac{A(1-A)}{3N} \qquad (3.2.21)$$

It follows from (3.2.21) that for a given N, $\frac{\Delta A}{\Delta N}$ has a maximum value when $A = 0.5$, although the curve is rather flat for a range of values on either side of $A = 0.5$ (Figure 3.17). For a fixed value of A, $\frac{\Delta A}{\Delta N}$ is inversely proportional to N. Hence, as

can be seen from Figure 3.17, the albedo A of a cloud is most sensitive to changes in N when A has values from about 0.25 to 0.75 and N is small (i.e. in air with low CCN concentrations). The maximum value of $\frac{\Delta A}{\Delta N}$ approaches 1% per additional cloud drop per cubic centimeter of air! As Twomey (1991) has pointed out, to produce a change in CCN of 1 cm^{-3} from the surface of the Earth up to a height of 1 km over the whole globe would require only about 50 tonnes of material. Assuming a CCN residence time of just 2 days, this perturbation in CCN concentration would require an injection rate of 1–10 kilotonnes of material per year. Since this is very small compared with anthropogenic aerosol emissions, anthropogenic emissions should be affecting cloud albedos, particularly over the oceans and other remote regions of the world.

Over the remote oceans the main source of sulfate particles and CCN appears to be the oxidation of DMS. Thus, DMS from

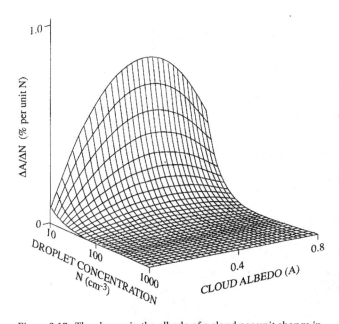

Figure 3.17. The change in the albedo of a cloud per unit change in the droplet concentration ($\Delta A/\Delta N$) as a function of the cloud albedo (A) and the droplet concentration (N), for a cloud with a constant liquid-water content. (Adapted from *Atmos. Environ.*, **25A**, Twomey, S., Aerosols, clouds and radiation, 2435–2, Copyright 1991, with permission from Elsevier Science Ltd, The Boulevard, Langford Lane, Kidlington OX5 1GB, UK.)

the oceans may determine the concentrations and size spectra of cloud droplets, and therefore the cloud albedo, over large regions of the oceans. These ideas are the basis for the DMS–cloud–climate hypothesis, according to which DMS from the oceans enters the atmosphere and is subsequently oxidized to sulfate particles, which serve as CCN in marine stratiform clouds (Shaw, 1983; Nguyen *et al.*, 1983; Charlson *et al.*, 1987). An increase in DMS emissions (due, for example, to global warming) could then lead to an increase in CCN and therefore cloud droplet concentrations. This, in turn, would increase the amount of solar radiation reflected back into space by marine stratiform cloud, thereby tending to offset the initial global warming. However, the sensitivity of cloud microstructures to DMS emissions is much reduced by the ubiquity of anthropogenically derived CCN, even over the remote oceans.

Equations (3.2.17) and (3.2.20) show that the optical thickness (τ) and the albedo (A) of a cloud are most sensitive to perturbations in the number concentration of cloud droplets when N is small. Hence, thin marine stratiform clouds should be particularly susceptible to modification by CCN. So-called *ship tracks* in clouds (i.e., relatively narrow tracks in marine stratiform clouds that appear brighter in satellite imagery) provide an excellent demonstration of this effect (Figure 3.18). Ship tracks are due to CCN from ships increasing the

concentration of cloud droplets and therefore increasing the albedo of marine stratiform clouds (Coakley *et al.*, 1987). Radke *et al.* (1989) and King *et al.* (1993) described a case study of two ship tracks, based on airborne and remote sensing measurements. Droplet concentrations in the ship tracks were about 70–100 cm^{-3} greater than in adjacent cloud regions. The upwelling intensity of radiation at a wavelength of 744 nm was ~110 W m^{-2} μm^{-1} sr^{-1} in the ship tracks compared with 40 W m^{-2} μm^{-1} sr^{-1} in the adjacent clouds. While ship tracks themselves do not have any significant effect on the global radiation budget, they illustrate the potential for increasing aerosol concentrations, produced by anthropogenic activities worldwide, to decrease surface temperatures by increasing the reflectivity of clouds.

There is another way in which the addition of CCN might increase the reflectivity of a cloud. As we have seen above, increasing the concentrations of CCN generally decreases the efficiency of precipitation production. Since precipitation is a sink for cloud water, lowering of the precipitation efficiency will increase the cloud water content and therefore the reflection of solar radiation by the cloud (Albrecht, 1989).

Because clouds are already optically thick at infrared wavelengths, changes in the amounts of terrestrial radiation they absorb due to increases in droplet concentration are negligible. However, increasing anthropogenic emissions of light-absorbing particles (e.g., carbon), which can become incorporated into cloud droplets, will increase the absorption of solar radiation by clouds and therefore tend to warm the Earth's surface. Twomey *et al.* (1984) concluded that on a global scale the increased reflectivity of clouds produced by increasing aerosol concentrations (i.e., the cooling effect) is probably dominant.

Falkowski *et al.* (1992) concluded from an analysis of sulfate data that, although anthropogenic sulfur emissions enhance cloud albedo adjacent to the east coast of the United States, over the central North Atlantic Ocean an anthropogenic effect is not currently discernible.

Kaufman and Nakajima (1993) used NOAA-AVHRR satellite images taken over the Amazon Basin during the biomass burning season to study the effects of smoke aerosol on the properties of low cumulus and stratocumulus clouds. The reflectance at 3750 nm was studied for tens of thousands of clouds. They showed that the presence of dense smoke (with optical thickness of 2.0) reduced the (remotely sensed) size of the droplets in continental clouds from 15 μm to 9 μm. The cloud reflectance at 640 nm was *reduced* by 0.03 in the presence of dense smoke. This reduction could be due to the high initial reflectance of clouds in the visible part of the spectrum and the presence of large quantities of graphitic carbon. Kaufman *et al.* (1991) reviewed the characteristics of the

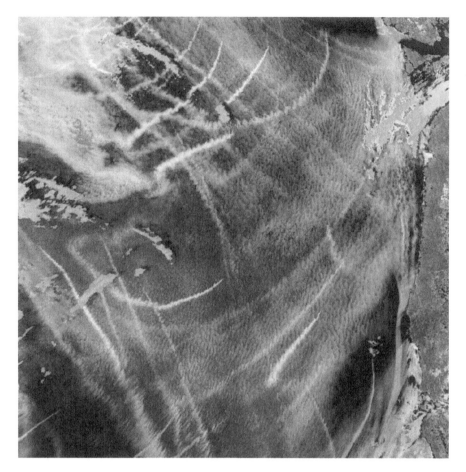

Figure 3.18. A spectacular example of ship tracks in marine stratiform clouds off the west coast of the United States detected at a wavelength of 3700 nm with the NOAA-9 AVHRR satellite. (Courtesy of P. Durkee, Naval Postgraduate School, Monterey.)

cooling effect due to aerosol-induced increases in cloud albedo, and they applied Twomey's theory to check whether fossil fuel or biomass burning produces heating or cooling of the planet. They showed that although coal and oil emit 120 times as many CO_2 molecules as SO_2 molecules, each SO_2 molecule could be 50 – 1100 times more effective in cooling the atmosphere (through the effects of aerosol on cloud albedo) than a CO_2 molecule is in heating it. Kaufman *et al.* concluded that the cooling effect of coal and oil burning may presently range from 0.4 to 8 times their heating effect. However, this analysis did not consider the direct effects of aerosol backscattering on solar radiation.

While there is no doubt that local and regional aerosol sources can increase the reflectivity of clouds, the question remains whether increasing atmospheric aerosol concentrations are having any significant global effects. Schwartz (1988) argued that they are not, because he found no evidence for any systematic differences in cloud albedo or surface temperatures between the northern and southern hemi-

spheres, even though anthropogenic emissions of aerosols are much greater in the northern hemisphere. However, in a survey of satellite measurements, Han *et al.* (1994) found that the mean droplet radius in the northern hemisphere was 11 μm, which was 0.7 μm lower than in the southern hemisphere.

Jones *et al.* (1994) used a climate model to calculate indirect radiative forcing due to sulfate aerosols modifying cloud droplet sizes. They used Langer and Rodhe's sulfate distributions and an empirical relationship between the effective radius of cloud droplets and aerosol concentrations. The global mean indirect-radiative forcing due to anthropogenic sulfate was found to be − 1.3 W m⁻², and the forcing in the northern hemisphere was 1.6 times greater than in the southern hemisphere. Boucher and Lohmann (1995) used aerosol sulfate distributions from a chemical model as inputs to GCMs with rudimentary cloud microphysics to explore the sensitivity of the global radiation balance to the indirect effects of anthropogenic sulfate. They obtained estimates ranging from -0.5 to -1.5 W m^{-2}. Although the uncertainties associated with these various estimates are very large, they indicate the potential importance of indirect radiative forcing due to anthropogenic particles.

Stratospheric aerosols

Stratospheric aerosols, and their effects, are better understood than tropospheric aerosols. Sulfuric acid droplets, which derive from sulfur (COS, SO_2, volcanic SO_2) injections from the troposphere, comprise about 75% of the mass of stratospheric aerosols. Most of these aerosols are located at heights between about 17 and 20 km (the *stratospheric sulfate layer*). These aerosols cool the Earth's surface by scattering solar radiation back into space.

During quiescent periods between major volcanic eruptions, stratospheric aerosols have a monomodal size distribution with a number mode radius near 0.08 m. For periods of several years following a major volcanic eruption, the formation of new particles and the growth of existing particles by gas-to-particle conversion in the stratosphere produces a bimodal aerosol distribution with radii at about 0.05–0.15 μm and 0.3–0.5 μm. The perturbation in radiative forcing produced by such disturbances can be much larger locally than the average tropospheric DARF (e.g., Hansen *et al.*, 1992). For example, the violent eruption of Mt. Pinatubo in the Philippines in June 1991 emitted massive amounts of SO_2 and aerosols into the stratosphere, which gave rise to a mean global radiative forcing of -3 to -4 W m^{-2} for a few months after the eruption but which virtually disappeared by the end of 1994. The ability of climate models to predict observed temperature changes at the surface produced by major volcanic eruptions, about -0.5°C (Hansen *et al.*, 1992), lends credence to these models.

Stratospheric aerosols may also influence the stratospheric chemistry and Earth's radiation balance through their effects on polar stratospheric clouds (Poole and McCormick, 1988).

Problems and prospects

We now have significant insights into the various ways in which aerosols might affect clouds, precipitation, and the Earth's radiation balance. However, our knowledge is still not adequate to quantify these effects with satisfactory precision. For example, simple cloud models suggest that the addition of an appropriate number of giant CCN to some warm clouds, or of ice nucleating material to supercooled clouds, should enhance precipitation. But fifty years of experimentation has failed to demonstrate that precipitation can be in-creased by cloud seeding. Radiative forcing due to increases in anthropogenic greenhouse gases is known with high precision ($+2.45 \pm 0.37$ W m^{-2}). By contrast, direct radiative forcing due to anthropogenic aerosols, which might possibly offset global warming, is not known to better than a factor of 2; indirect radiative forcing due to aerosols is even more uncertain. Very little quantitative information is available on the radiative effects of mineral dusts, or of organic materials in the atmosphere.

To improve estimates of direct aerosol radiative forcing, careful measurements are needed of the optical properties of various types of aerosols (sulfates, silicates, organics, soot). Derivations of aerosol optical depths over the oceans from satellite remote sensing measurements need to be verified against *in situ* and ground-based measurements. Methods need to be developed for measuring aerosol optical depths over land from satellites.

The recently completed SCAR-B and TARFOX field experiments, in Brazil and on the US East Coast, respectively, in which detailed airborne *in situ* measurements of aerosol properties were obtained simultaneously with remote sensing measurements, exemplify the types of studies needed in other parts of the world.

With appropriate knowledge of the physical and chemical properties of various aerosol types, climate models could be used to infer the direct effects of aerosols on global and regional energy budgets, and to predict the effects of aerosol perturbations on climate.

Many of the processes involved in determining the effects of aerosols on cloud structures, precipitation, and indirect radiative forcing are not understood well enough to be incorporated with confidence into climate models. For example, there is no simple relationship between aerosols and CCN, or between CCN and cloud droplet concentrations. The relationship between cloud droplet concentration and sizes and cloud albedo is also complex. Virtually nothing is known about the effects of aerosols on the radiative properties of ice clouds.

Important information on the sensitivity of the albedo (A) of a cloud to perturbations in aerosol number concentration (N) is provided by the quantity $\frac{\Delta A}{\Delta N}$ (called the *cloud susceptibility*). Recently, airborne measurements of $\frac{\Delta A}{\Delta N}$ have been obtained for arctic stratus (Hegg *et al.*, 1996) and for marine stratus (Platnick *et al.*, 1997). Similar measurements are needed for other cloud systems and in other regions of the world. These cloud susceptibilities could then be incorporated into climate models to explore the effects of various types of aerosol perturbation on the Earth's radiation balance and climate.

3.3 Vertical transport processes

T. Hauf and S. Brinkop

On the importance of vertical transport

Vertical motions play an important role in the atmosphere. Nearly all severe weather phenomena like thunderstorms, hurricanes and fronts are accompanied by strong vertical motions. They occur on a wide range of scales: millimeter-sized motions in the laminar surface layer, cumulus updrafts on the kilometer scale, and rising and sinking motion of the hemispheric meridional circulation. Vertical velocities range between the extreme values of $+60$ m s^{-1} and -25 m s^{-1} found in thunderstorm up- and downdrafts. Typical values, for synoptic scale motions, however, are several centimeters per second. Vertical motions cause vertical transport of heat, mass and momentum. These transport processes are an important element in the global water and energy cycle. A complete treatment of all vertical transport processes is beyond the scope of this section. We therefore focus on processes which are of principal relevance for the global energy and water cycle and discuss to a lesser degree the momentum transport, which is nevertheless significant. The selected processes are listed in Table 3.3 together with their location and vertical extent. Figure 3.19 sketches these processes. We describe these processes, study their impact on the large-scale flow and then explain how they are represented in global weather and climate models. For more detailed studies the interested reader is referred to the general textbooks of Dutton (1976) for atmospheric dynamics, of Stull (1988) for boundary layer processes, of Emanuel (1994) for convective processes and to other sections of this book for cloud physics.

The role of vertical transport in a global perspective

One dominant feature of the atmosphere is its layered structure due to the combined action of gravity and radiation (Dutton, 1976). Radiative processes in a motionless atmosphere, however, yield a dynamically unstable temperature field with air parcels close to the surface being lighter than those immediately above. In order to stabilize the lower atmosphere, vertical convective energy exchange processes must be allowed for. They lower the near-surface temperature and heat the free atmosphere until a statistical quasi-equilibrium is achieved where radiative and convective energy exchanges are balanced. Energy exchange by dynamic processes is thus a necessary element in the atmospheric energy budget (Ramanathan *et al.*, 1989).

Vertical convective transports, in general, influence the large-scale atmospheric flow in basically three ways: through heating due to latent heat release, by vertical turbulent transport of heat, moisture and momentum and through the interaction of cumulus clouds with radiation. Moist convection transports four times more energy than dry convection. Both processes together have a value of about 30% of the incoming solar energy flux. They are stronger than the longwave radiative flux from the Earth's surface (about 20%) together with which they balance the absorbed solar energy (about 50%) and transport the energy away from the Earth's surface.

The most important vertical transport processes

Molecular diffusion in the lowest laminar layer

Immediately above the water surface or the soil surface there exists a thin and laminar layer of about 1 mm depth where molecular diffusion and conduction processes prevail (Foken, 1978). Thickness varies in a complex manner with wind speed, friction and surface roughness. All exchange processes between the atmosphere and the underlying surface have to pass through this layer. As the coefficients of molecular heat conduction and viscosity are both about 10^{-5} m^2 s^{-1}, a temperature gradient of 2 K mm^{-1} builds up in order to maintain, for example, a heat flux of 50 W m^{-2} from the surface to the atmosphere. These extraordinary values are found nowhere else in the free atmosphere. In practice in a parameterization scheme, the molecular transports are either subsumed in surface exchange processes or in the turbulent transport processes. Here, they are emphasized because of their importance and the magnitude of gradients. More details on molecular diffusion can be found in Csanady (1980).

Turbulence in the atmospheric boundary layer

The atmospheric boundary layer (ABL) is that part of the atmosphere that feels the frictional effect of the surface. The wind close to the ground is decelerated and the resulting vertical shear together with the surface heating continually lead to the development of turbulent eddies. Turbulence, therefore, is typically a boundary-layer phenomenon (Stull, 1988). Turbulent eddies are very effective mixing agents. They shuffle heat and water vapour away from, and momentum towards, the Earth's surface at diffusion rates six orders

Table 3.3. *Vertical processes considered in this section, their location and vertical extent.*

Transport process	Location	Vertical extent from	to
Molecular diffusion	Close to the Earth's surface	0	1 mm
Turbulence	Atmospheric boundary layer	0	*c.* 1–3 km
	Frontal zones	0	several km
	Jetstream	*c.* 10	*c.* 12 km
	In breaking gravity waves		depth $\approx$ 1 km
Shallow dry convection	Convective boundary layer	0	1–3 km
Shallow moist convection	Lower troposphere	0	1–4 km
Deep moist convection	Troposphere	0	tropopause (TP)
	Plus occasionally parts of the lower stratosphere	TP	TP + 2 km
Frontal circulation	Extratropical cyclones	0	TP
Slantwise convection	Mainly extratropical cyclones tropical cyclones	0	several km

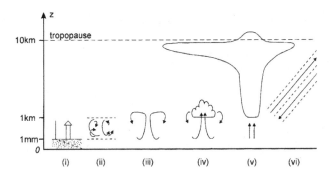

Figure 3.19. Cartoon of vertical transport processes considered: (i) molecular diffusion in the lowermost layer; (ii) turbulence in the atmospheric boundary layer, (iii) shallow dry convection, (iv) shallow moist convection, (v) deep convection, (vi) slantwise convection.

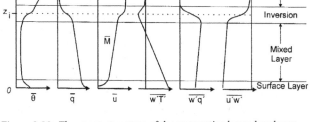

Figure 3.20. The mean structure of the convective boundary layer with: $\bar{\theta}$ potential temperature, $\bar{q}$ humidity, $\bar{M}$ momentum, $\overline{w'T'}$ heat flux, $\overline{w'q'}$ moisture flux, $\overline{u'w'}$ momentum flux. Shown on the right are the lowest superadiabatic surface layer, the mixed layer with nearly constant profiles of mean quantities, the entrainment zone with a strong inversion and the free atmosphere aloft. (From Stull, 1988, after Driedonks and Tennekes, 1984.)

of magnitude higher than those of molecular diffusion. They tend to smooth out contrasts in the field of meteorological parameters and cause typical mean ABL-profiles (Figure 3.20).

Depth of the ABL ranges from about 30 m in conditions of large static stability to more than 3 km in highly convective conditions. Usually, a strong temperature inversion is found at the top of the boundary layer. This stable inversion layer suppresses vertical eddy motions. They, however, still penetrate into it (Figure 3.21), and mix boundary layer from beneath and warm and dry air from aloft. As a result, the inversion is eroded with time, boundary-layer depth increases and local gradients are sharpened, contrary to the

general smearing out within the ABL. More generally, this holds whenever turbulent kinetic energy is advected into a non-turbulent environment. Turbulent eddies vary in size from millimeter-sized ones to those covering the full boundary-layer depth. Turbulence is basically a three-dimensional phenomenon. With decreasing spatial scales, turbulence becomes isotropic and the turbulent eddies lose their transport capability.

Turbulence shows a strong dependency on temperature stratification. During a typical summer day over land when the surface of the Earth is heated by the sun, the temperature stratification close to the ground becomes unstable, because lighter fluid rests under heavier fluid. This leads to

Atmospheric Backscattering   28.6.1991.

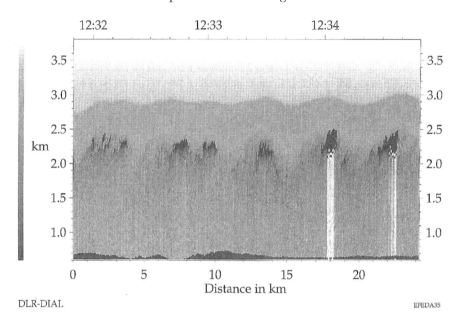

DLR-DIAL   EFEDA35

Figure 3.21. Aerosol backscatter in a vertical plane from an airborne lidar system flying over a convective boundary layer with clouds. The grey-scale (left) is a measure of backscatter. Convective motions penetrate into the stable inversion layer and initiate gravity waves there. Compare with Figures 11.31 and 11.32 in Stull (1988). Clouds are seen on the right with strong backscatter and shadowing the lidar signal. (From Kiemle *et al.*, 1995.)

overturning convective motions which try to restore a near neutral stratification. A common physical picture of this dry boundary-layer convection assumes the latter to comprise individual thermals or rising buoyant air parcels. Thermals increase in size with height and may organize themselves to form cells or rolls (Figure 3.22) which are quite distinct from the unorganized turbulent eddies.

Dry convective motions cover the whole boundary layer, ranging from the lowermost surface layer through the mixed layer up to the inversion layer. Complementary to the rising positively-buoyant thermals, compensating downward motion occurs in between over a much larger area and, consequently, at lower speeds. However, upward- and downward-directed transport can be treated separately as 'top down/bottom up' diffusion (Wyngaard and Brost, 1984). In the convective boundary layer local gradients almost vanish. This is one fundamental result found already in laboratory convection studies (Emanuel, 1994). Despite their difference in appearance, turbulence and dry convection within the ABL are similar in their transport properties and, therefore, they are considered together.

In a statically stable atmosphere, bouyancy forces tend to suppress vertical motions and turbulence is caused solely by dynamical shear instability. With increasing static stability, turbulence becomes, therefore, more two-dimensional and vertical mixing less efficient.

One general feature of the ABL in the presence of wind shear across the top of the boundary layer is the excitation of gravity waves. Figure 3.21 shows the initial phase of this excitation mechanism. Gravity waves propagate in the up-shear direction and usually extend vertically to the tropopause (Clark *et al.*, 1986). Gravity waves carry momentum, simply because convective thermals act like obstacles to the flow. Momentum flux values are comparable to those found for gentle hills (Hauf and Clark, 1989).

In the free atmosphere, turbulence is found in frontal zones (see below), in shallow layers with breaking gravity waves, in mid-tropospheric convectively unstable layers, and in the shear zone of the jet stream.

Moist convection

Energetics of shallow convection change dramatically when latent heat is released. This occurs when a rising air parcel becomes saturated and vapour condenses to form clouds. The released heat increases the parcel's buoyancy and it

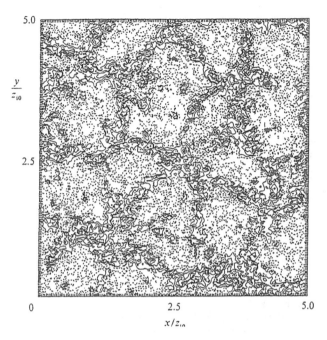

Figure 3.22. Horizontal cross-section of vertical wind velocity field within the convective boundary layer at $z = 0.25\, z_{i0}$, with z_{i0} inversion height. Horizontal coordinates are normalized with z_{i0}. Dashed contour lines indicate negative and solid lines positive velocities. Results were obtained from large-eddy simulations. Updrafts concentrate in narrow bands and reveal a polygonal structure. Downdrafts are slower but cover a wider area. (From Schmidt and Schumann, 1989.)

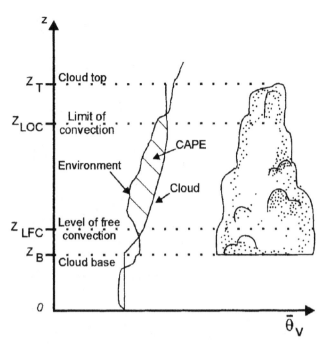

Figure 3.23. Cartoon of an active cloud with four key levels. Profiles of potential temperature for an ascending air parcel and for its environment. Saturation is reached at cloud base (z_{B}). At the level of free convection (z_{LFC}) it becomes warmer than the environment and is positively buoyant. The limit of convection is reached at the level of neutral buoyancy z_{LOC}. Above that level the cloud is overshooting due to the kinetic energy gained during ascent. Cloud top is at z_{T}. The shaded area gives the convective available potential energy CAPE (after Stull, 1988).

continues to ascend as long as it is lighter than the environment. The amount of potential energy available for convective vertical motion is called CAPE and is given by an integral over the area between actual environmental and individual temperature profile in a height–temperature plot (Figure 3.23).

Moist convection exists as shallow and deep convection. Shallow convection is limited to the ABL by the overlying inversion and only non-precipitating cumulus cloud can form. If the overlying inversion can be penetrated, convection may develop deep into the free troposphere. Figure 3.24 shows the related mass flux profiles for shallow and deep convection.

Shallow convection with non-precipitating cumulus clouds

Depending on the geographical location, convective mixing reaches heights up to 4 km above the surface. Evaporation at cloud top increases the water vapor content there and homogenizes the humidity profile but all water substance remains in the zone of convective mixing. Shallow cumuli very

often are randomly distributed with an average spacing of five times the boundary layer depth. One possible explanation of this random distribution is the interaction between gravity waves in the free atmosphere and boundary layer convection (Hauf and Clark, 1989). Shallow cumuli may also be organized in linear streets as a result of boundary-layer rolls (Kuettner, 1971). In a cold-air outbreak over warmer water these streets transform further downstream to open cells and often also to closed cells. Open cells are cloud-free in the interior while closed cells are cloud-covered. Cold-air outbreaks occur mainly over the high-latitude oceans. They cause regional and temporarily substantial heat fluxes into the atmosphere.

Persistent shallow cumuli can be found in the trades. The trades occupy nearly half the surface of the globe and constitute the lower tropospheric Equator-ward flowing branch of the Hadley circulation. They play a central role in atmospheric energetics through the accumulation of latent heat from the ocean and its horizontal export to the tropics. Considerably more than half of the water evaporated into the

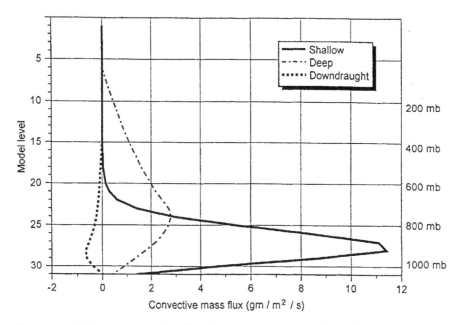

Figure 3.24. Global mean vertical profiles of convective mass fluxes for shallow convection, deep convection and separately for the associated downdrafts. Note the importance of the downdraft-related mass flux. From a GCM simulation (after Miller, 1995).

atmosphere originates from the tropical oceanic boundary layer. In the trades, the boundary layer is characterized by a strong inversion which only occasionally is penetrated. Most clouds evaporate rapidly in the tradewind inversion and cool and moisten the environment, thus counteracting the drying and warming effect of the large-scale subsidence. In that way, shallow moist convection helps to maintain the observed thermal structure of the lower troposphere and has a significant effect on the atmospheric energetics through intensifying the hydrological cycle. Large stratocumulus or stratus decks form polewards of the trades, mainly west of the continents over the cold ocean currents at about 30 degrees north or south of the Equator. Together with the shallow cumuli they exert a cooling effect on the atmosphere.

Deep moist convection with precipitation

To initiate deep convection, either individual air parcels or complete layers have to be lifted to what is called the level of free convection (z_{LFC}, see Figure 3.23). This level is typically found in or above the overlying inversion. The required lifting can be achieved by either sufficiently strong shallow convection or by larger-scale horizontal convergence at low levels and divergence at upper levels. Deep convection is an essential component of the global energy and water cycle through the release of energy, the formation of precipitation, the effect on the magnitude of cloud cover (especially of the widespread anvil ice clouds), the associated radiative processes, and the vertical transports of energy, momentum und moisture. More than half of the global precipitation is caused by deep convection which occurs in a variety of types: mid-level convection, single-cell storms, multi-cell storms, super-cell storms, mesoscale-convective systems or complexes, squall lines, tropical cloud clusters and hurricanes. Their spatial scale ranges between 1 km and several hundred kilometers, while lifetimes of half an hour to several days are found. The net vertical transport effect of all these deep convective systems results from a number of processes interacting in a non-trivial way (see Figure 3.25):

1. A dynamic transport pattern for heat and moisture, and also entropy, with convergence below cloud base, divergence at cloud top and vertical advection in between.
2. Formation of hydrometeors (rain, graupel, hail, ice crystals), their advection with the wind and gravitational settling.
3. Cooling and heating by phase changes.
4. Induced motions in the environment.

As some of these processes counteract one another, such as drying by subsidence and moistening by evaporation of hydrometeors, the net effect is not *a priori* obvious. The basic effect of deep convection on the large-scale environment is usually described in terms of the apparent convective heat source Q_1, being the sum of condensational heating and

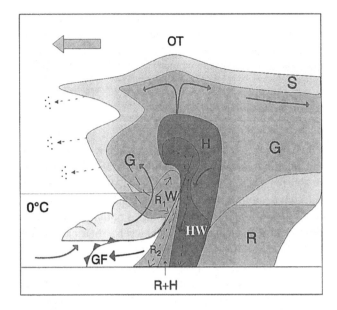

Figure 3.25. Schematic vertical cross-section of a hailstorm illustrating the main dynamic and microphysical features derived from radar observations (after Höller *et al.*, 1994). Storm is moving from right to left. Solid arrows indicate streamlines; dashed arrows represent particle trajectories. Rain (R) is found in region R_1, originating from melting of graupel (G) from the newly grown cell. Some of these raindrops recirculate into the weak echo region (W), where they are carried up to heights well above the freezing level where graupel and snow (S) dominate. Rain in region R_2 can have its origin either from drops passing through R_1 as relatively low trajectories or from melting or shedding hailstones (H) in the main downdraft. The horizontal component of the velocity field causes a size sorting of rain, mixed particles, and wet hailstones (HW) on their way to the ground. Cold air from mid-levels is transported downwards in the outflow, which is separated by a gustfront GF from the ascent of low-level moist air ahead of it. At the cloud edges hydrometeors evaporate and moisten the environment (dashed arrows). Slight overshooting (OT) is indicated directly above the main updraft.

vertical convective transport of heat, and, by the apparent moisture sink Q_2 (Yanai *et al.*, 1973; Emanuel, 1994, Chapter 15). Both Q_1 and Q_2 profiles are functions of the large-scale forcing and vary strongly in time and space. Generalizing, Miller (1995) states that moist atmospheric convection is in approximate statistical equilibrium with its environment. Then temperature and moisture profiles are controlled by convection and the temperature is linked directly to the subcloud layer entropy. Processes involved are (a) the upward transport of subcloud layer entropy, (b) the downward entropy transport in the middle troposphere by compensating environmental subsidence, (c) downdrafts driven by evaporation of precipitation, and (d) upward transport of static

energy, which reduces its value near the surface and increases it aloft.

Of special interest is the storm-related exchange between stratosphere and troposphere. Trace gases are transported within half to one hour directly from the boundary layer to the anvil of a cumulonimbus cloud (Hauf *et al.*, 1995), whose top can penetrate up to two kilometers into the lower stratosphere. Danielsen (1982) proposed a freeze-drying mechanism of lower stratospheric air by which stratospheric water vapor is deposited on highly supersaturated ice crystals which are injected into the stratosphere in the overshooting dome and then sediment back to the troposphere.

So far, we have discussed the convective transports of heat and moisture. But convection transports momentum as well. Clouds and the convective currents act as obstacles to the flow, similar to mountains, causing for example a pressure difference between up- and downstream. Observations during the Global Atmospheric Research Programme (GARP) Atlantic Tropical Experiment (GATE) and in tropical cyclones suggest that convection plays a substantial role in the large-scale momentum budget. The reader is referred to Moncrieff (1992) who developed some archetypal dynamical models to understand and quantify the transport properties of deep convection and to expedite their parameterization in large-scale models. Emanuel (1994, section 15.1.3) lists the effect of convection on atmospheric momentum. Similar to the excitation of waves by shallow convection, deep convection is also a source of stratospheric gravity waves when it penetrates the tropopause.

Slantwise moist convection

Motions on the rotating Earth are subjected to virtual centrifugal forces, such as the Coriolis force. It can be shown that besides the usual convective instability, there is a similar type of instability associated with centrifugal forces. Centrifugal instability is nothing more than convective instability but with inertial accelerations playing the role of gravity. If centrifugal forces are not balanced by the other body forces acting on a volume of air, in particular the pressure gradient force, then lifting occurs, although the atmosphere might be stable with respect to ordinary upright convection. The motion takes the form of thin inclined ascending sheets of cloudy air interspersed with thicker layers of unsaturated, more slowly descending air as was shown theoretically by Emanuel (1994). The centrifugal instability responsible for this *slantwise* convection is referred to as *conditional symmetric instability* (CSI). It is found mainly in baroclinic shear flows such as in extratropical cyclones. It represents a motion which is a combination of vertical convection resulting from gravitational instability and 'horizontal convection' resulting from centrifugal instability.

CSI can be diagnosed on the basis of routine radio soundings. The theory of CSI was successfully applied to explain the banded structure of precipitation in extratropical cyclones (Bennetts and Hoskins, 1979). There is mounting evidence from numerical simulations as well as observations that slantwise convection is not only responsible for the rain bands but also contributes to rapid cyclogenesis. The organization of precipitation, frontogenesis, and explosive marine cyclogenesis depends strongly on the distribution of the stability properties of the atmosphere to both upright and slantwise convection. Reuter and Yau (1993), for instance, showed that the lower-tropospheric air on the warm side of the warm-frontal zone is often stable or neutral with respect to vertical cumulus convection but unstable for slantwise convection. Thus, a complete analysis of convective instability must include the CSI.

Frontal circulations

Extratropical cyclones with their frontal structure reveal a variety of smaller-scale precipitation patterns ranging from 10 to 10^3 km and associated vertical motions from 1 cm s^{-1} to 1 m s^{-1}. Cellular structures, clusters, and the forementioned rain bands are observed. Vertical transport may occur over the entire troposphere. Besides the uplift indicated by the precipitation patterns, downward mixing of upper tropospheric or even stratospheric air into the ABL is also observed. The configuration of the flow within the regions of ascent determines the nature of precipitation patterns (Browning, 1974). Pronounced vertical mixing is associated with cold fronts. See, for instance, Browning and Pardoe (1973) or Figure 168 in Atkinson (1981). In the latter book the subject of circulations in cyclones including tropical cyclones is discussed in detail. Global circulation models do not resolve the small-scale features of extratropical cyclones, especially precipitation patterns. As the numerically calculated rainfall amount broadly matches observed values, the need for parameterization of these patterns is not as urgent as in the case of sub grid-scale convection.

Parameterization of vertical transport in general circulation models

Understanding the global energy and water cycles requires the use of global models. Large-scale models of the atmosphere have a horizontal grid spacing of the order of hundreds of kilometers. Small-scale processes, for example turbulence or convection, cannot be resolved by these models. The task

is thus to express the bulk effects of all these processes in a grid box in terms of resolved variables, a method known as parameterization. For details see the book edited by Emanuel and Raymond (1993).

Parameterization of turbulence

In global atmospheric models the number of vertical levels is of the order of 10 to 30, of which only a few lie within the atmospheric boundary layer. The large-scale model is then not able to represent the evolution of the ABL according to its evolution equations. In practice, the main unknown but required quantities are the vertical fluxes of momentum, heat and moisture at the various model levels. Their calculation has to be based on the grid-averaged meteorological variables and known surface properties. The parameterization of the ABL in a general circulation model can be divided into two parts: the parameterization of the surface fluxes (see section 2.3 in this book) and the determination of the heating and moistening rates within the ABL.

Two different approaches can be distinguished to represent the ABL in a large-scale model. If the ABL is represented by only one or two layers, a bulk approach must be used. The ABL-top is calculated by a prognostic equation and the ABL itself is described by variables which can be understood as vertical averages over the boundary-layer depth. Evolution equations for the bulk properties require the knowledge of the ABL-height, the surface fluxes and the interaction between the ABL and the free atmosphere. Suarez et al. (1983), for instance, use this kind of ABL parameterization in a general circulation model. If the ABL is resolved by three or more layers, the turbulent fluxes at the respective levels can be determined. In a number of general circulation models a flux-gradient relationship is used. This requires the knowledge of a turbulent diffusion coefficient. Louis (1979) has developed such a scheme. It uses empirically determined flux-profile relationships for the surface-layer and the eddy-diffusivity approach above the surface layer. This scheme can be found, for example, in the ECHAM climate model (Roeckner et al., 1992). Brinkop and Roeckner (1995) used what is known as a 1.5 order closure scheme. This improved the performance of a GCM as the ABL was ventilated more efficiently.

Parameterization of convection

Owing to the importance of cumulus clouds for large-scale flow, it is necessary to account for their effects in large-scale models. The time scale of convection is often marginally resolved, with typical time steps ranging from a few minutes to nearly an hour. As strong transports may traverse over many grid points in the vertical direction during a single time step, parameterization of such transports by convection must be non-local. Hauf *et al.* (1995), for example, identified transport paths over the entire depth of the troposphere within half an hour. Thus the representation of convection is different from that of turbulence in the ABL. A parameterization of deep convection or shallow convection aims for proper heating and moistening rates in terms of large-scale atmospheric variables predicted by the model. The possibility of such a functional relationship and its formulation emerged from observational studies in the tropics. Currently used schemes for the representation of cumulus convection in large-scale models of the atmosphere are (a) adjustment schemes, (b) Kuo-type schemes and (c) mass flux schemes.

Parameterization of deep convection

Adjustment schemes are the simplest way to account for deep convective processes in a large-scale model. Cumulus clouds readjust the thermal state towards an equilibrium state whenever it becomes unstable. Moist convective adjustment schemes (Manabe *et al.*, 1965) establish neutral moist stability between two adjacent layers by mass exchange, whenever the two layers are conditionally unstable to vertical ascent.

A further step toward a better physical basis than the simple moist adjustment scheme is the scheme of Betts and Miller (1993). The adjusted profile is chosen to represent thermodynamic structures which are typically observed in convective situations and which resemble a quasi-equilibrium state between large-scale forcing and cumulus convection. The adjustment is only partially done during one time step which makes the scheme a relaxation scheme.

The Kuo-scheme (Kuo, 1974) is based on the observed correlation between large-scale moisture convergence and convection. Convection occurs in deep layers of unstable stratification and is maintained by the moisture supply due to large-scale convergence and evaporation from the surface. It is further assumed that cumulus cloud air mixes into the large-scale environment at the same rate as it is generated by moisture supply. The Kuo-scheme was originally designed for the parameterization of deep convection in tropical hurricanes but is also applied to mid-latitude convection.

The most promising and widely used scheme is the mass flux convection scheme which was first formulated by Ooyama (1971). It is based on the simple picture of convection comprising rapidly rising cloudy air over a small area and slow dry subsidence over a much larger area. A mass flux scheme requires a cloud model as an integral part of it. As many clouds of different type or height are usually present within a GCM grid box, cloud models have to simulate not only the effect of single clouds, but of cloud ensembles. Cloud models vary substantially according to the level of microphysics implemented in the model. Two approaches have been suggested for using cloud models to estimate the quantities required to implement the mass flux theory as a parameterization scheme. The spectral cloud ensemble approach was first introduced by Arakawa and Schubert (1974). They assumed that, within a grid box of a GCM, a spectrum of different clouds existed, where each cloud is characterized by specific values of entrainment rate and cloud top height. Because of its complexity the Arakawa–Schubert scheme is computationally expensive. Thus, several less complex schemes have been developed based on the bulk cloud model approach of Yanai *et al.* (1973) with only one cloud model being applied to a grid box. The mass flux scheme of both Tiedtke (1989) and Gregory and Rowntree (1990) uses this approach. It performs well in both climate and weather forecast models.

The quality of the mass flux parameterization scheme depends crucially on the mass flux at cloud base. This value determines the magnitude of convective heat release while the cloud model distributes the heat in the vertical. No exact theory exists to provide this value and to 'close' the scheme. The various 'closure' assumptions can be broadly classified into 'dynamic' and 'adjustment' types (Gregory, 1995). 'Dynamic' closures relate the cloud base mass flux to the large-scale forcing and thermodynamic fields. Tiedtke (1989)'s closure assumption, for instance, links the mass flux at cloud base to the sub-cloud layer moisture convergence. An example of an 'adjustment' closure is the so-called CAPE-closure. Here the cloud base mass flux is related to the convective available potential energy. This closure assumption is used in the ECMWF forecast model (Nordeng, 1994). For further discussion of the closure problem, see Gregory (1995).

Parameterization of shallow convection

Shallow convection is as important as deep convection for large-scale flow and one has to account for it in large-scale models. In a first attempt, Tiedtke (1989) used the diffusivity approach of a boundary-layer scheme to represent shallow convection by simply using enhanced vertical diffusion coefficients. The performance of the model was remarkably improved. Often the parameterization of shallow convection is done in a similar way as for deep convection but using different specific parameters of the large-scale forcing. The closure problem, however, remains and is as pertinent as for deep convection. Betts and Miller (1993) have developed an adjustment scheme for shallow cumulus convection complementary to their scheme for penetrative cumulus convection. In the mass flux scheme of Tiedke (1989) shallow convection makes use of a slightly modified closure assumption to that used for deep convection and different detrainment rates at cloud top. Gregory and Rowntree (1990) also developed a mass flux scheme for shallow dry convection based on a deep, moist convective one. When applying deep-convective schemes to shallow convection, one should note that the stationarity assumption of deep convection may still be reasonable for the trade-wind cumuli but not for cold-air outbreaks over relatively warm seas.

Parameterization of slantwise convection

So far, slantwise convection is only poorly represented in large-scale models of the atmosphere. Nordeng (1987) has developed a parameterization scheme to represent this process. In his approach the atmosphere is adjusted toward vanishing moist potential vorticity along an absolute momentum surface whenever slantwise convective instability occurs. This scheme considers only the thermodynamical impact of slantwise convection.

Future prospects

Parameterization of clouds and convection in global models is a subject of ongoing research. Forecasts of global warming depend significantly on the quality of cloud representation. Progress will be achieved if the scale gap between the grid size of global models (c. 500 km) and individual clouds (c. 5 km) is bridged. This requires a proper representation of cloud ensembles and their net large-scale effect. Mesoscale and smaller scale numerical model simulations of observed cases serve as a guide for new and more sophisticated but still affordable parameterization schemes (see Section 5.2).

3.4 Precipitating cloud systems

Peter Jonas

Introduction

A major factor in the global water and energy cycles is the efficiency with which water vapor is converted into precipitation. While the importance is clear in the case of the water budget, since precipitation is a major sink for atmospheric water and a source of water at the surface, it must also be stressed that precipitation is crucial for the energy balance due to the ability of precipitation to transport latent, and sensible, heat *relative to the air*. Precipitating cloud systems exist on horizontal scales ranging from a few kilometers to many hundreds of kilometers. The complex processes leading to the formation of precipitation occur on a range of scales much smaller than 1 km.

A variety of cloud systems is responsible for precipitation. In these systems, the microphysical processes involved in precipitation growth are rather different. Figure 3.26 shows the latitudinal variation in average precipitation and indicates the dominant precipitation-bearing cloud systems. In this section, the most important microphysical processes are summarized and descriptions are given of the major types of precipitating cloud systems. More details can be found in several text books including those by Mason (1971), Pruppacher and Klett (1978), Rutledge (1991), Houze and Hobbs (1982) and Houze (1993).

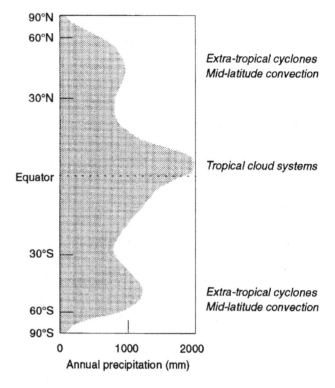

Figure 3.26. Globally averaged annual precipitation showing the types of cloud system associated with the precipitation peaks. (From Sellers, 1965.)

Precipitation microphysics

Clouds are composed of small water drops or ice crystals which grow to form particles sufficiently large to fall from the clouds as precipitation. The basic process by which clouds form is by cooling of the air. Cooling of the air below its dew point, the temperature at which the saturation mixing ratio becomes equal to the actual mixing ratio of the water vapor in the air, will lead to the formation of small droplets or ice crystals, given that there are sufficient solid particles present in the air to act as nuclei. In the atmosphere, cooling may be caused by large-scale ascent of the air, local ascent, mixing between two air masses at different temperatures, radiative cooling and by passage over a cold surface. The amount of water vapor which is removed from the air and converted to liquid or ice as a result of cooling is the difference between the water mixing ratio and the saturated value after cooling, which is strongly temperature dependent, as shown in Figure 3.27. This shows the variation of the saturation mixing ratio over water with temperature at two different pressures, one

representative of conditions in the boundary layer and the other of the middle troposphere. The saturation mixing ratio over ice is also shown.

The small cloud particles which form as the air is cooled may grow to form precipitation. Growth is firstly by deposition from the vapor on to a small nucleus, condensation or sublimation. As the particles become larger, they fall relative to other cloud particles and grow by collision and accretion. The efficiency with which these processes act to convert water made available by cooling of the air into precipitation depends on many factors, not all of which have been fully quantified.

Growth by vapor deposition

In the atmosphere, there are many particles which can act as nuclei for the formation of water drops (see section 3.2). The droplet concentration is determined by the balance between

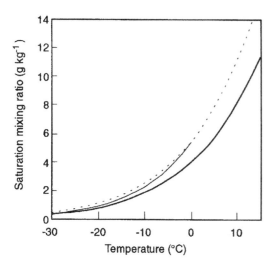

Figure 3.27. Variation with temperature of the saturation mixing ratio over water at 850 hPa (thick solid line), over water at 700 hPa (dashed line) and over ice at 700 hPa (thin solid line).

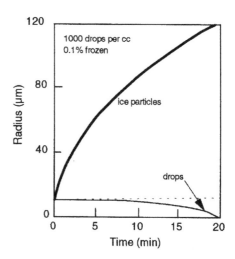

Figure 3.28. The growth of ice spheres and decay of water drops in a mixture subject to a constant updraft of 1 m s^{-1}. The air is assumed to be initially saturated with respect to water. The dashed line shows the growth of the droplets in the absence of any ice particles.

the rate of supply of vapor by cooling and the rate at which it is removed by deposition. Increases in the supply of vapor, for example with increased vertical air velocity in more vigorous clouds, will result in an increase in the number density of particles which are activated to become droplets. The concentration of droplets will be reduced by entrainment of dry air from the environment and therefore, for example, the droplet concentration in stratocumulus clouds is often higher than in cumulus clouds for similar aerosol particle concentrations, despite the higher updrafts in the latter. The droplet concentration is crucial to the rate of growth of the droplets since they compete for the available water. Droplets grow more rapidly by condensation in clouds formed in relatively unpolluted regions than in clouds with high droplet concentrations. The growth rate of droplets by condensation decreases as the radius increases, because the decreasing surface to volume ratio limits the speed with which water can condense. Condensation also tends to produce rather narrow cloud droplet spectra due to the decreasing growth rate as the droplets become larger. Entrainment of dry air into a cloud, which results in partial evaporation of some droplets and the incorporation into the cloud of new nuclei, acts however to broaden these narrow spectra.

The atmosphere normally contains many particles which can act as nuclei for the formation of water droplets. In contrast, there are relatively few particles in the atmosphere which can act as nuclei for the formation of ice crystals, except at low temperatures. Consequently, few clouds become glaciated (composed predominantly of ice crystals rather than water drops), until their temperature is much less

than 0°C, typically around – 15°C. Observations by Moss and Johnson (1994), and earlier work, show wide variations in the fraction of clouds which is glaciated at any given temperature, although the fraction increases as the temperature decreases. There are, however, some observations of convective clouds, in very clean air masses, with significant glaciation even at temperatures only a few degrees below 0°C.

In some clouds, secondary ice nucleation processes may be active; nuclei are produced during droplet freezing (Mossop, 1978). These can increase the ice particle concentrations drastically. Secondary nucleation is sensitive to temperature and to the cloud droplet spectra, broad spectra aiding the formation of secondary ice nuclei. At low temperatures, around – 40°C, nucleation of cloud droplets to form ice crystals will occur even without the presence of a nucleating particle and therefore at such low temperatures very few clouds contain supercooled drops.

The relatively high saturation mixing ratio over water compared with that over ice (see Figure 3.27), combined with lower concentrations of ice crystals than those of water droplets, lead to much more rapid growth of ice crystals by deposition than is the case for droplets. When a cloud of water droplets in air at close to water saturation is cooled, and a small fraction of the droplets freeze, these ice particles grow very rapidly, as shown in Figure 3.28. The rapid growth of the ice crystals at close to water saturation reduces the vapor mixing ratio to a value between that of saturation with respect to ice and water. This leads to the evaporation of the water droplets. Subsequently, growth of the small number of ice particles by deposition from the vapor is rapid. This mechanism, the Bergeron–Findeisen mechanism (see, for example, Mason, 1971, §6.2), is believed to be the dominant mechanism for the growth of precipitation particles in mid-latitude weather systems where precipitating clouds usually extend to temperatures below 0 °C and where melting of precipitation towards the base of the cloud can often be detected using radar by the presence of a 'bright band', or region of enhanced radar reflectivity, just below the 0 °C level.

It is apparent from Figure 3.28 that glaciation may give rise to a rapid change in the number and size of cloud and precipitation particles, and hence in the terms in the water budget of a cloud. Furthermore, the additional release of latent heat associated with cloud glaciation enhances the vigor of clouds, further enhancing the precipitation. Whether or not, and how rapidly, a cloud becomes glaciated has a significant impact on its water budget.

Growth by collision and aggregation

Owing to the very slow rate of growth of droplets by condensation once they have reached about 20 μm in radius, any subsequent growth is predominantly by the collision and aggregation process which is effective only if the initial spectrum is fairly broad. The collection efficiencies of smaller cloud droplets are relatively low, as is the rate at which droplets encounter one another, owing to the small values of the terminal velocities. However, for larger droplets the process is much more effective, and can lead to the rapid growth of precipitation particles. The distance which large growing droplets fall, relative to the smaller cloud droplets, determines the maximum amount of growth that is possible within a cloud of finite depth; however, turbulent air motions increase the relative path of a fraction of the droplets and this increases the possibility of drizzle formation from relatively shallow, but turbulent and long-lived, clouds such as stratocumulus. Growth by aggregation is also enhanced in clouds with high liquid water contents, for example deep clouds or cloud formed at higher temperatures. While the Bergeron–Findeisen process is the mechanism leading to the formation of precipitation in many mid-latitude frontal cloud systems, so-called warm processes involving droplet growth in the absence of ice particles, are important in subtropical convective clouds as well as in low-level layer clouds.

Ice particles in clouds can take many forms as described in Pruppacher and Klett (1978, §2.2): these include snow, hail and graupel. In comparison with the growth of droplets by collision and aggregation, the growth of ice particles is complicated, and a variety of processes are involved in the growth of snowflakes or hailstones. The complexities arise from the different shapes of the ice crystals, a function of the condi-

tions under which they grow by deposition, and the fact that colliding dry crystals often bounce, rather than aggregate to form single particles. Single ice crystals take many forms, including hexagonal plate-like crystals, thin solid or hollow hexagonal columns or complex dendritic crystals. Snowflakes are aggregates of these crystals. Aggregation is assisted by the presence of thin layers of liquid on the surface of the ice crystals. Such layers are formed when the crystals are falling through parts of the cloud containing water drops and when the rate of accretion of droplets is sufficiently high that the latent heat released by the partial freezing of the droplets on impact cannot be removed rapidly enough to allow the surface layers to freeze completely. Consequently, the largest snowflakes form at temperatures in the range 0 °C to − 10 °C (Pruppacher and Klett, 1978, §14.10).

In conditions where ice crystals fall through parts of a cloud containing water droplets, they grow by accretion of water droplets (riming) to form graupel, low density particles consisting of an ice particle on which have frozen many individual droplets. When the vertical air velocity is large and the cloud contains large amounts of liquid water, hailstones are formed. These are higher density particles in which accreted droplets spread out before freezing, due to the limited rate at which latent heat released during droplet freezing can be removed to the air, and are therefore difficult to distinguish visually.

Melting and breakup of precipitation particles

Large raindrops falling at terminal velocity in still air become distorted by the action of aerodynamic forces which are opposed by surface tension. They fall with a flattened lower surface. As the drop becomes larger, the lower surface becomes concave, the distortion increases and the drops become unstable to the extent that drops of diameter larger than about 9 mm break up into a number of smaller drops. This effectively sets an upper limit to the size of rain drops, although few drops of diameter larger than 2 or 3 mm are normally observed.

Drop breakup also results from collisions between rain drops in which the surface tension forces are unable to overcome the forces due to the energy of the colliding drops. Although such collisions are rare, they control the shape of the rain drop size spectrum since they oppose the effects of coalescence growth by creating several smaller drops from the original pair of colliding drops. The spectrum attains an equilibrium shape under the combined effects of collision, coalescence and breakup, as described in 'Characteristics of precipitation' below.

When ice precipitation particles fall through the level at

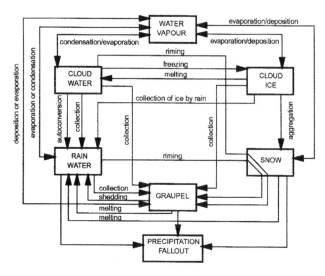

Figure 3.29. Diagram showing many of the processes leading to the formation of solid and liquid precipitation particles and their interactions.

which the wet-bulb temperature of the air is 0 °C into warmer air, they begin to melt. The water is often drawn into the irregularities in the ice particle surface, reducing the size of the mixed ice–water particle and resulting in an increase in the density and fall speed. Due to the strong dependence of fall speed on ice particle density, solid ice particles (hail) will survive a fall during melting much longer than that of low density graupel particles of the same original size. There is no upper limit to the size of ice precipitation particles but large particles composed of ice with a liquid layer on the surface shed water due to the aerodynamic forces on the large particle.

Parameterization of microphysical processes

The complexity of the processes which influence the formation of precipitation is illustrated by Figure 3.29. The interaction between the processes is evident and it is clear that their relative importance will have a major influence on the conversion of water vapor into precipitation within clouds. The range of processes must be included in models of precipitation development. However, they occur on the space and time scales associated with individual clouds (< 5 km and < 1 hr respectively) which are smaller than the grid scale of numerical weather and climate prediction models, and indeed of many cloud models. Consequently, the processes cannot be represented explicitly in such models, but their net effect must be represented in terms of larger-scale model parameters. This representation is referred to as parameterization.

In most models of precipitating cloud systems, the con-

densation and sublimation processes are simply parameterized by sharing the excess water vapor between the cloud and precipitation particles. This parameterization relies on the fact that condensation and sublimation are rapid processes and the supersaturations in clouds are normally small. In the case of mixed-phase clouds containing both water drops and ice particles, for which the liquid and solid water content of the cloud particles are used as model variables, the excess water is simply added to the overall condensed water content, often in some temperature-dependent ratio. However, parameterization of precipitation processes, and of the radiative properties of cloud systems, requires information on the size distribution of the cloud particles, and various schemes have been used to divide the excess water between different size categories and types (for example, snowflakes and graupel). In models with a large grid volume, it is necessary to allow for the variation of specific humidity within the volume which gives rise to condensation in some regions even when the average water content is less than the saturated value. This can be achieved by assuming that condensation occurs at some grid-mean relative humidity less than 100%, although more complex schemes have been developed, making use of observed spatial distributions of specific humidity. A major problem arises when parameterizing cloud microphysical processes because of the difficulty in calculating or specifying the droplet or ice crystal concentration. The former depends on aerosol particle characteristics, and the latter on temperature and the occurrence of secondary ice nucleation processes.

Many models of precipitating cloud systems where warm, non-ice, processes are dominant, employ simple parameterizations of the coalescence growth process based on pioneering work by Kessler (1974). These bulk-microphysical schemes estimate the rate of increase of the water content in precipitation-sized droplets by autoconversion (coagulation of many cloud droplets), collection of cloud droplets by rain drops and condensation onto, and evaporation of, rain drops. Autoconversion is normally parameterized using the cloud water content and some threshold value below which precipitation cannot be formed, and typically is approximated by:

$$\frac{\partial q_r}{\partial t} = \alpha(q_c - q_0) \qquad (3.4.1)$$

where q_r is the precipitation water content, α is a constant and $-q_0$ is the critical value (typically 0.2 to 0.5 g m^{-3}) of the cloud water content, q_c, below which no conversion to precipitation occurs. Modeling of precipitation growth is very sensitive to the values assumed for the constants. Kessler-type schemes are based on the processes within relatively vigorous convective clouds and are not appropriate for conditions in long-lived layer clouds. These are often observed to produce drizzle with water contents much smaller than those necessary before precipitation is observed from convective clouds. Other warm rain processes are similarly parameterized in terms of q_c and q_r.

In frontal cloud systems, and deep convective clouds, numerous empirical relations are needed to parameterize the complex range of ice crystal processes which are indicated in Figure 3.29. Details of a typical scheme are given by Rutledge and Hobbs (1983). In these schemes, which are extensions of the warm-cloud bulk-microphysical parameterization schemes, the model parameters are usually the water content in ice particles of different forms. A major problem arises in parameterizing the nucleation rate since this is a function of aerosol particle concentration which depends on the air trajectory rather than on purely local parameters (Section 3.2).

Precipitation efficiency

Cloud structure and lifetime influence cloud particle growth but the microphysical processes also influence cloud structure and evolution. While this is most evident in the impact of the release and consumption of latent heat on the buoyancy of the air, the lifetime of clouds is also influenced by the rate of conversion of water to precipitation. There is clear evidence from some attempts to modify the weather by artificially enhancing the Bergeron–Findeisen mechanism by introducing artificial ice nuclei, that the production of precipitation can result in the rapid growth and fallout of precipitation particles so that the cloud water content is reduced and the cloud dissipates. In other circumstances however, as will be seen later, the precipitation-generated downdraft sustains a continuing updraft leading to a long-lived cloud system. The efficiency with which water vapor is converted to cloud and precipitation particles is highly variable between different cloud systems. It is also crucial in determining the energy budget of the atmosphere and in determining the magnitude and location of the heat and moisture sources and sinks.

Cloud properties

The above discussion of the microphysical processes leading to the formation of precipitation suggests that the following factors are important when considering whether, and how rapidly, a cloud may produce precipitation. These factors are:

Cloud lifetime, which determines the maximum time available for particle growth, although many cloud particles will have lifetimes much shorter than that of the cloud (for example, within an orographic cloud capping a hill top).

Cloud height and temperature, which determine the amount of water which can be made available in liquid or solid forms.

Cloud depth, which determines the maximum path for growth by accretion and the possibility that part of the cloud will lie in the $-40\,°C$ to $0\,°C$ temperature regime in which the Bergeron–Findeisen process is effective.

Cloud vigor (i.e. updraft strength), which increases the rate of condensation but has a possible opposing influence on the lifetime of individual precipitation elements within a cloud.

The supply of moisture, for example through convergence of the large-scale flow, which determines the availability of water which could be converted into precipitation.

The seeder-feeder mechanism

So far in this section, precipitation growth within a single cloud has been considered. An important mechanism that influences the intensity of precipitation reaching the surface is the seeder-feeder mechanism. When precipitation particles from one cloud (the seeder cloud) fall through a lower cloud (the feeder cloud) they sweep out water from the lower cloud, resulting in an increase in the particle size and hence to an increase in the precipitation rate. Enhancement of the upper-level precipitation may occur even though the lower-level cloud may not be extensive enough to produce significant precipitation itself. The seeder-feeder mechanism has a significant influence on the water budget of several types of cloud systems. As will be seen in 'Characteristics of precipitating cloud systems' below, the process is important both in frontal cloud systems, where it may occur in association with rain bands, and in orographic cloud systems where it may give rise to substantial enhancement of the surface precipitation.

Characteristics of precipitation

Droplets smaller than around 100 µm radius are normally not large enough to survive evaporation while falling through the subsaturated air between cloud base and the surface. Consequently, this size is often used to define the lower limit of precipitation particle size. As a consequence of the growth by coalescence and the breakup of large drops to create smaller drops, the spectra of precipitation droplets larger than 100 µm are often observed to take an equilibrium form in which the droplet concentration decreases exponentially as

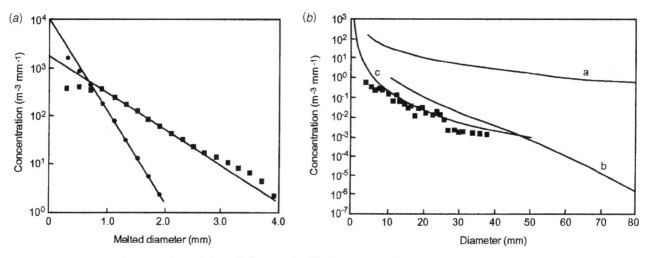

Figure 3.30. (a) Empirical spectra of snowflakes and (b) graupel and hailstones. Snowflake data are for precipitation rates of 0.31 mm hr^{-1} (circles) and 2.5 mm hr^{-1} (squares). The empirical relationships shown for hailstones (a, b and c) relate to observations collected in different conditions; the data points are for observations near Laramie, USA.

the size increases. Although the average rain-drop spectra are of exponential form, spectra measured over short periods may depart significantly from this simple form, while the slope of the average spectra varies between different types of rainfall (e.g. from layer or convective cloud) and with rainfall rate. Empirical relationships have been derived to fit observations of rain-drop spectra. Using surface observations, Best (1950) derived an empirical form of the spectrum:

$$1 - f = \exp\left[-(d_0/a)^{2.25}\right] \tag{3.4.2}$$

where a depends on the rainfall rate R in mm hr^{-1}, d_0 is in mm and f is the fraction of the liquid water in drops with diameters smaller than d_0. The relation between rainfall rate and the slope of the spectrum was given by:

$$a = 1.30\ R^{0.232} \tag{3.4.3}$$

The fall velocity of the droplets varies with size in a complex manner (e.g. Beard, 1976) as a result of the variation of the drag coefficient of a sphere with size and the distortion of the shape of the largest droplets. Combining empirical expressions for the fall speed with the empirical droplet spectrum, estimates can be made of the mass-weighted fall speed of the precipitation as a function of precipitation rate.

The size spectra of solid precipitation particles are more variable than those of rain drops due to the range of possible growth modes. Empirical relations which have been fitted to observed precipitation particles are illustrated in Figure 3.30 although there are even larger variations between individual observations in the case of solid precipitation than for rain drops. While the fall speeds of many ice particles are smaller than those of water drops of similar size because of their

lower average density, dense hailstones have large fall speeds and can only grow in clouds with large updrafts.

Due to the relatively low concentration of precipitation particles compared with those of cloud particles, the amount of water or ice which is present in the form of precipitation is much less than the water content of the cloud drops. While cloud water contents in excess of 1 g m^{-3} are found even in moderate clouds, the precipitation water content in heavy rain of around 50 mm hr^{-1} is only 10^{-3} to 10^{-2} g m^{-3}.

Characteristics of precipitating cloud systems

Precipitation reaching the surface is highly variable in both space and time, even when falling from apparently uniform layer-cloud systems. The variability results from the way in which the cloud systems have developed and the interaction between the microphysical processes of precipitation growth and the structure of the clouds. Generally, the precipitation arising from a particular system may be considered as a combination of a number of intense cells or bands within an area of more widespread and uniform precipitation. As was noted in the first subsection above, the type of cloud system responsible for the largest part of the precipitation reaching the surface is geographically dependent. In mid-latitudes, precipitation from frontal systems associated with the extratropical cyclones dominates in maritime regions, although convective systems dominate in continental regions in summer. The tropical cloud systems are also mostly convective in origin. Several different types of precipitating cloud system are described below, to illustrate the range; other systems are also observed and further details can be found in the comprehensive reviews by Houze and Hobbs (1982), Rutledge (1991) and Houze (1993).

Boundary-layer clouds

In conditions where the atmosphere is stable, precipitating layers of stratocumulus clouds may form. Extensive, semi-permanent, sheets of stratocumulus are found especially off the west coasts of the continents. Although such clouds are seldom deep enough to produce heavy precipitation, the continuous precipitation (often less than 5 mm day^{-1}) may make a significant contribution to the net flux of water to the surface, although measurements of precipitation over the oceans are subject to large uncertainty. The evolution of these clouds is strongly influenced by the balance between radiative and turbulent fluxes of heat. Entrainment instability at cloud top, where entrainment of dry air results in the formation of downdrafts by evaporative cooling, can result in rapid breakup of the cloud layer.

Small-scale convective clouds

Individual cumulus clouds may grow deep enough to produce showers of a few minutes' duration. Typically, precipitation will not form in cumulus less than 1 km deep, but often even considerably deeper clouds fail to produce precipitation. In shallow clouds, 'warm' processes (those not involving the ice phase) result in the formation of precipitation only if the droplet concentration is low, for example in clean maritime air (Section 3.2). The Bergeron–Findeisen process may be important in the production of precipitation by deeper clouds, while in clouds with very low droplet concentrations, ice particles have been observed at temperatures only just below 0 °C. In individual small convective clouds, the duration of the precipitation is often limited by the effects of the drag exerted on the air by the falling precipitation and by evaporation of the precipitation which create a downdraft below the cloud, reducing the supply of water vapor to the cloud, which then dissipates.

Mid-latitude convective systems

In more convectively unstable conditions, the deeper clouds glaciate rapidly, leading to the rapid onset of heavy precipitation by the Bergeron–Findeisen mechanism. Under appropriate conditions of the shear of the horizontal wind with height, the downdrafts associated with precipitation become organized and either sustain the main updraft or create new updraft cells growing at the edge of the decaying cells. This can lead to the development of storm systems with a lifetime of several hours. Depending on the wind profile, the storms may propagate, giving intense localized precipitation along a narrow swath, or they may be almost stationary, giving large accumulations of precipitation over a small area. Storm dynamics are particularly sensitive to the presence of directional wind shear since this influences the relative positions of the downdrafts resulting from precipitation, and the inflow of moisture necessary to sustain the storm development. Several types of convective systems have been identified, depending on the wind and stability profiles.

Multicell storms

Although some single convective cells may develop to produce heavy precipitation, the contribution of such single cell storms to the total precipitation is small (Simpson *et al.*, 1980). Much mid-latitude precipitation over continental regions results from multicell storms persisting for several hours, although the life of the individual cell is less often than one hour. Successive convective cells form along the region of convergence associated with the low-level outflow of air cooled by evaporation of falling precipitation. The cells develop, move through the storm system and decay. While individual cells move with the wind at some mean level (the steering level), the system propagates relative to the wind as new cells form and decay. Depending on the wind shear, new cells may be triggered at almost the same position, and hence the region of intense precipitation may be almost stationary, leading to large accumulations. The new cell formation is often (but not always) on the right forward flank of the storm and thus the multicell storms typically propagate to the right of the mean wind direction.

Supercell storms

Under conditions of strong instability and strong wind shear and veer, supercell storms develop. These are storms which become organized with a persistent updraft–downdraft structure. Supercell storms are often associated with tornadoes. The strong updrafts (up to 50 m s^{-1}) within these storms are favorable to the formation of large hailstones, which become sorted by the organized air flow so that the region of moderate precipitation extends 5–10 km to one side of the track of the storm with a core 1 km or more in diameter, with intense precipitation.

Mesoscale convective systems

Convective storms are commonly grouped within a system which is much larger than the individual storms. These

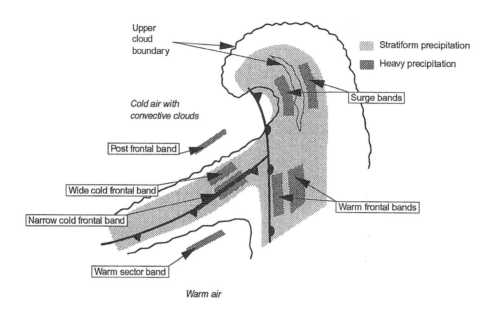

Figure 3.31. Schematic diagram showing the locations in which rain bands may be found in mid-latitude frontal precipitating cloud systems. (From Matejka *et al.*, 1980).

mesoscale convective systems are associated with convergence in the lower troposphere and divergence aloft on scales much larger than the individual storms. Mesoscale convective systems, in addition to containing convective regions, produce large regions (anvil clouds) characterized by lighter and more uniform precipitation. The overall precipitation area often extends for more than 100 km, although the embedded convective cells are narrow.

Mid-latitude squall lines

Another form of mesoscale organization of multicell storms is the mid-latitude squall line. These occur when new convective cells form along the edge of the outflowing air from one or more individual storms, as a line of convective storms. The line may increase in length as it propagates and is characterized by an extensive area of light precipitation falling behind the narrow band of intense precipitation at the leading edge of the system.

Frontal precipitation systems

Although in mid-continental regions such as north America much of the precipitation is from large convective cloud systems, in many mid-latitude regions most precipitation falls from frontal cloud systems. These are associated with cyclonic circulations in which warm air, driven by strong horizontal temperature gradients, rises along a sloping path

over colder air. The ascent is generally slow (*c.* 10 cm s^{-1}) with significant horizontal motion within belts of considerable length, and the shape of these belts has a significant impact on the precipitation pattern (Browning, 1986). Furthermore, precipitation may move horizontally over large distances from the regions where it is generated before reaching the ground. Although the ascent is slow and widespread, the precipitation pattern is often organized and contains a number of regions of intense rainfall within a large region of lighter, more uniform precipitation. The regions of more intense precipitation are arranged in bands of varying width. The range of possible frontal precipitation bands is indicated schematically in Figure 3.31, although all of the possible bands are seldom observed in any one situation.

While some of the rain bands are local enhancements of the widespread, stratiform, precipitation (see for example Houze *et al.*, 1976), other rain bands are clearly convective in nature. The narrow cold frontal rain band is convective and forced by the motion of the cold air as it undercuts the warmer air close to the surface, perhaps as a gravity current. Above the surface cold front, vertical velocities may reach several meters per second, an order of magnitude larger than in other parts of the frontal region. James and Browning (1979) and others have shown that these frontal rain bands are relatively shallow and that the precipitation may be broken up along the front into segments perhaps 20 km long and less than 5 km wide.

In contrast to the narrow cold frontal rain band, wide cold frontal rain bands are formed above the surface and are

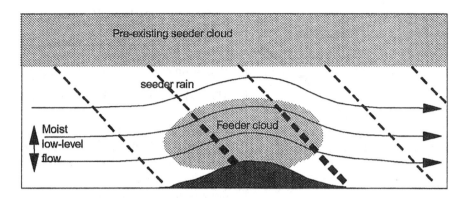

Figure 3.32. The enhancement of precipitation from mid-level stratiform cloud falling through low-level cloud produced by orographic lifting over a range of shallow hills. (From Browning, 1980).

accompanied by shallow convection cells which locally enhance the rain reaching the surface from the stratiform cloud regions through the seeder-feeder mechanism. This type of precipitation band is usually more than 20 km wide. The origins of this band are not certain; Locatelli *et al.* (1992) have suggested that it is the result of local frontogenesis resulting from the evaporation of precipitation falling into the cold dry air below the frontal surface. An alternative explanation is that it results from conditional symmetric instability (Bennetts and Hoskins, 1979) which is released when ascent leads to saturation.

Warm frontal rain bands result from the local enhancement of the low-level precipitation by the seeder-feeder process. Convection at mid-levels results in the formation of the generating cells. The origins of the convection are unclear; possible mechanisms again include conditional symmetric instability, or gravity-wave ducting (Lindzen and Tung, 1976).

Orographic precipitation

Clouds will be formed if air, forced to rise as it passes over a mountain range or an isolated hill, becomes saturated. Clouds may also be formed downwind of a mountain range in the crests of waves triggered by the mountains. Often, but not always, such clouds are of limited extent and the time of passage of air through the clouds (though not the lifetime of the clouds) is sufficiently short that they produce little precipitation. However, it is known that orography has a significant effect on surface precipitation. This arises from several causes.

Over shallow hills, ascent may give rise to the formation of low-level clouds in conditions when the low-level flow is moist. Such situations often occur in the warm sectors of depressions in which precipitation is falling from mid-level clouds. The surface precipitation is then enhanced by the seeder-feeder process as light precipitation from the mid-level clouds falls through the low-level orographic clouds (Browning 1980; Choularton and Perry, 1986), as is shown in Figure 3.32. Water may also be lost from such low-level orographic clouds if they are adjacent to the surface, by direct (or occult) deposition, contributing to the flux of water to the surface.

Ascent forced by hills may also help trigger convection (Browning *et al.*, 1974) which, being the result of stationary forcing, is confined to the area of the forcing and may result in substantial accumulations of precipitation. The orographic triggering of deep convection in some subtropical areas is responsible for very large accumulations of rainfall (Grossman and Durran, 1984). While most of the convection results from uplift on the upwind side of the hills releasing potential instability, it may also result from thermal forcing on the side of the range on which the sun is incident or by low-level convergence in a recirculating flow in the downwind side of the range.

Tropical cloud systems

In tropical regions, most precipitation falls from convective clouds. Although isolated cumulus and cumulonimbus clouds are common, cloud clusters resembling the mid-latitude mesoscale convective systems are responsible for much of the precipitation. The clusters generally form at very low latitudes and, unless they develop into longer-lived tropical storms and move out of the tropics, have lifetimes of less than one day. Although cloud clusters associated with squall lines develop rapidly and propagate at high speed, most of the precipitation results from the more numerous non-squall

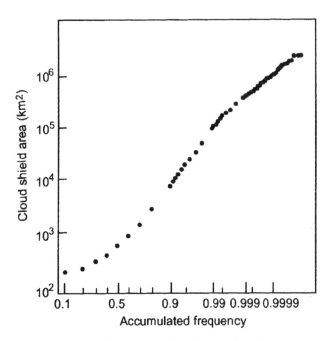

Figure 3.33. Observed cumulative distributions of satellite-derived cloud area for mesoscale convective systems over the western tropical Pacific (5-year average during November–February). (Adapted from Houze, 1993.)

line clusters. Both types of organization result in the occurrence of intense convective precipitation within more widespread stratiform precipitation. Satellite infrared images can be used to determine the frequency of occurrence of mesoscale convective systems as a function of the area of high anvil cloud (Figure 3.33). The observed correlation between cloud area and rainfall suggests that the largest 1% of the cloud systems account for 40% of the precipitation. However, within such systems, a variety of local precipitation patterns is possible. Typically about 40% of the precipitation falls from the stratiform regions. Many studies of the energetics of tropical cloud systems have been made and these suggest that in tropical regions heating due to latent heat release in the middle and upper troposphere may reach 30 K day^{-1}, averaged over 10^5 km^2 and over layers 1 km deep.

Other cloud systems

Other forms of cloud, including fogs resulting from radiative cooling or from cooling and moistening by contact with the surface, and upper-tropospheric wave clouds, seldom produce precipitation although they have a significant impact on the surface radiation budget. The impact of thin clouds is particularly important in the case of thin orographic cirrus, owing to its persistence.

Conclusions

The main factors influencing the development of precipitation within cloud systems, and the variety of those systems, have been described in this section. In attempting to quantify the global water budget, it is necessary first to understand the interactions between the precipitation growth processes and the cloud structure. These interactions are expected to be different in different cloud systems, for example in long-lived layer clouds which are controlled by the balance between turbulent and radiative processes, the rate of conversion of water vapor to precipitation is controlled by cloud depth, lifetime and turbulence. In vigorous convective systems, conversion is dominated by the updrafts within the cells, while in frontal regions, the cloud layer structure and growth by the seeder-feeder process are important in controlling the conversion efficiency. In convective systems also, the wind profile has a strong influence on system lifetime, and hence on the total precipitation.

Over many years, observational studies in many parts of the world have done much to clarify the basic precipitation processes and to identify the factors leading to different types of cloud system organization. In contrast, quantification of the impact of precipitating cloud systems on the larger-scale atmospheric circulation is less advanced. While much has been done to calculate the large-scale heat and moisture budgets in tropical cloud systems, the variety of the precipitating cloud systems at mid-latitudes makes it difficult to generalize from those studies, although progress has been made (see Section 3.3).

The precipitation efficiency is a major area of uncertainty in calculating the spatial distribution of the atmospheric heat budget since it is only by the conversion to precipitation that the sources and sinks of latent heat are separated in space. Similarly, the precipitation efficiency controls the spatial distribution of the atmospheric sink of water. A major problem is therefore to determine the way in which precipitation efficiency depends on the cloud system type and to develop parameterizations which can be used within general circulation models to improve knowledge of the global water budget. This problem can only be tackled by a programme of improved observations of different types of precipitating cloud systems combined with numerical modeling of the processes leading to precipitation formation in the different systems.

References

Ackerman, S. A. and Stephens, G. L. (1987). The absorption of solar radiation by cloud droplets: an application of anomalous diffraction theory. *J. Atmos. Sci.*, **44**, 1574–88.

Albrecht, B. (1989). Aerosols, cloud microphysics, and fractional cloudiness. *Science*, **245**, 1227–30.

Andreae, M. O. (1990). Ocean–atmosphere interactions in the global biogeochemical sulfur cycle. *Mar. Chem.*, **30**, 1–29.

Andreae, M. O. (1995). Climate effects of changing atmospheric aerosol levels. In *World Survey and Climatology*, vol. 16, *Future Climates of the World*, ed. A. Henderson-Sellers, pp. 125–52. Amsterdam: Elsevier Press.

Arakawa, A. and Schubert, W. H. (1974). Interaction of cumulus cloud ensemble with the large-scale environment. Part I. *J. Atmos. Sci.*, **31**, 674–701.

Atkinson, B. W. (1981). *Meso-scale Atmospheric Circulations*. London: Academic Press.

Beard, K. V. (1976). Terminal velocity and shape of cloud and precipitation drops aloft. *J. Atmos. Res.*, **33**, 851–64.

Beard, K. V. (1992). Ice initiation in warm-base convective clouds: an assessment of microphysical mechanisms. *Atmos. Res.*, **28**, 125–52.

Bennetts, D. A. and Hoskins, B. J. (1979). Conditional symmetric instability: a possible explanation for frontal rainbands. *Quart. J. Roy. Meteor. Soc.*, **105**, 945–62.

Best, A. C. (1950). The size distribution of rain drops. *Quart. J. Roy. Meteor. Soc.*, **76**, 16–36.

Betts, A. K. and Miller, M. J. (1993). A new convective adjustment scheme. *Meteorological Monographs*, **24**(46), 107–12.

Boers, R., Jensen J. B., Krummel, P. and Gerber, H. (1996) Microphysical and short-wave radiative structure of winter-time stratocumulus clouds over the Southern Ocean. *Quart. J. Roy. Meteor. Soc.*, **122**, 1307–39.

Boucher, O. and Lohmann, U. (1995). The sulfate-CCN-cloud albedo effect. A sensitivity study with two general circulation models. *Tellus*, **47B**, 281–300.

Brinkop, S. and Roeckner, E. (1995). Sensitivity of a general circulation model to parameterizations of cloud-turbulence interactions in the atmospheric boundary layer. *Tellus*, **47A**, 197–220.

Browning, K. A. (1974). Mesoscale structure of rain systems in the British Isles. *J. Met. Soc., Japan*, Ser II, **52**, 314–27.

Browning, K. A. (1980). Structure, mechanism and prediction of orographically enhanced rain in Britain. In *Orographic Effects in Planetary Flows*, GARP Pub. Series No 23, 85–114.

Browning, K. A. (1986). Conceptual models of precipitating systems. *Wea. Forecasting*, **1**, 23–41.

Browning, K. A., Hill, F. F. and Pardoe, C. W. (1974). Structure and mechanism of precipitation and the effect of orography in a wintertime warm sector. *Quart. J. Roy. Meteor. Soc.*, **100**, 309–30.

Browning, K. A. and Pardoe, C. W. (1973). Structure of low-level jet streams ahead of mid-latitude cold fronts. *Quart. J. Roy. Meteor. Soc.*, **99**, 619–38.

Cahalan, R. F., Ridgeway, W., Wiscomb, W. J., Bell, T. L. and Snider, J. B. (1994). The albedo of fractal stratocumulus clouds. *J. Atmos. Sci.*, **51**, 2434–55.

Cahalan, R. F., Silberstein, D. and Snider, J. B. (1995). Liquid water path and plane-parallel albedo bias during ASTEX. *J. Atmos. Sci.*, **52**, 3002–12.

Cess, R. D., Zhang, M. H., Minnis, P., Corsetti, L., Dutton, E. G., Forgan, B. W., Garber, D. P., Gates, W. L., Hack, J. J., Harrison, E. F., Jing, X., Kiehl, J. T., Long, C. N., Morcrette, J.-J., Potter, G. L., Ramanathan, V., Sabasilar, B., Whitlock, C. H., Young, D. F. and Zhou, Y. (1995). Absorption of solar radiation by clouds: observations versus models. *Science*, **267**, 496–9.

Charlson, R. J. and Heintzenberg, J. (eds.) (1995). *Aerosol Forcing of Climate*. London: John Wiley.

Charlson, R. J., Langner, J. and Rodhe, H. (1990). Sulfate aerosol and climate. *Nature*, **238**, 22.

Charlson, R. J., Langner, J., Rodhe, H., Leovy, C. B. and Warren, S. G. (1991). Perturbation of the northern hemisphere radiative balance by backscattering from anthropogenic sulfate aerosol. *Tellus*, **43AB**, 152–63.

Charlson, R. J., Lovelock, J. E., Andreae, M. O. and Warren, S. G. (1987). Oceanic phytoplankton, atmospheric sulphur, cloud albedo and climate. *Nature*, **326**, 655–61.

Charlson, R. J., Schwartz, S. E., Hales, J. M., Cess, R. D., Coakley, J. A., Hansen, J. E. and Hoffman, D. J. (1992). Climate forcing by anthropogenic aerosols. *Science*, **255**, 423–30.

Choularton, T. W. and Perry, S. J. (1986). A model of the orographic enhancement of snowfall by the seeder-feeder mechanism, *Quart. J. Roy. Meteor. Soc.*, **112**, 335–45.

Clark, T. L., Hauf, T. and Kuettner, J. P. (1986). Convectively forced internal gravity waves: results from two-dimensional numerical experiments. *Quart. J. Roy. Meteor. Soc.*, **112**, 899–925.

Clough, S. A., Tacono, M. J. and Moncet, J. L. (1992). Line-by-line calculations of atmospheric fluxes and cooling rates: application to water vapor. *J. Geophys. Res.*, **97**, 15,761–85.

Coakley, J. A. Jr., Bernstein, R. L. and Durkee, P. A. (1987). Effect of ship-track effluents on cloud reflectivity. *Science*, **237**, 1020–2.

Csanady, G. T. (1980). *Turbulent Diffusion in the Environment*. Dordrecht: Reidel.

Danielsen, E. F. (1982). A dehydration mechanism for the stratosphere. *Geophys. Res. Lett.*, **9**, 605–8.

Driedonks, A. G. M. and Tennekes, H. (1984). Entrainment effects in the well-mixed atmospheric boundary layer. *Boundary Layer Meteor.*, **30**, 75–105.

Dutton, J. A. (1976). *The Ceaseless Wind: An Introduction to the Theory of Atmospheric Motion.* New York: McGraw-Hill.

Eagan, R. C., Hobbs, P. V. and Radke, L. F. (1974). Measurements of cloud condensation nuclei and cloud droplet size distributions in the vicinity of forest fires. *J. Appl. Meteor.*, **13**, 553–7.

Edwards, J. M. and Slingo, A. (1995). Studies with a flexible new radiation code. I: Choosing a configuration for a large-scale model. *Quart. J. Roy. Meteor. Soc.*, **122**, 689–719.

Ellingson, R. G., Ellis, J. and Fels, S. (1991). The intercomparison of radiation codes in climate models: longwave results. *J. Geophys. Res.*, **96**, 8929–53.

Ellingson, R., Yanuk, D., Gruber, A. and Miller, A. J. (1994). Development and application of remote sensing of longwave cooling from the NOAA polar orbiting satellites. *Photogrammetric Engineering and Remote Sensing*, **60**, 307–16.

Emanuel, K. A. (1994). *Atmospheric Convection.* Oxford: Oxford University Press.

Emanuel, K. A. and Raymond, D. J. (eds.) (1993). The representation of cumulus convection in numerical models. *Meteorological Monographs*, **24**(46).

Evans, K. F. (1993). A general solution for stochastic radiative transfer. *Geophys. Res. Lett.*, **20**, 2075–8.

Falkowski, P. G., Kim, Y., Kolber, Z., Wilson, C., Wirick, C. and Cess, R. (1992). Natural versus anthropogenic factors affecting low-level cloud albedo over the North Atlantic. *Science*, **256**, 1311–15.

Foken, T. (1978). The molecular temperature boundary layer of the atmosphere over various surfaces. *Archiv für Meteorologie, Geophysik und Bioklimatologie*, Serie A, **27**, 59–67.

Fouquart, Y., Bonnel, B. and Ramaswamy, V. (1991). Intercomparing shortwave radiation codes for climate studies. *J. Geophys. Res.*, **96**, 8955–68.

Fu, Q. and Liou, K-N. (1993). Parameterization of the radiative properties of cirrus clouds. *J. Atmos. Sci.*, **50**, 2008–25.

Garratt, J. R., Stephens, G. L., O'Brien, D. M., Dix, M.R. and Wild, M. (1996). CO_2 climate forcing and the Earth's radiation budget. *Geophys. Res. Lett.*, (in press).

Geleyn, J. F. and Hollingsworth, A. (1979). An economical analytical method for the computation of the interaction between scattering and line absorption of radiation, *Contrib. Atmos. Sci.*, **52**, 1–16.

Goody, R., Haskins, R. , Abdou, W. and Chen, L. (1995). Detection of climate forcing using emission spectra. *Earth Research from Space*, **5**, 22–3.

Gordon, H. B. and O'Farrell, S. P. (1997). Transient climate change in the CSIRO coupled model with dynamic sea ice. *Mon. Wea. Rev.*, **125**, 875–907.

Greenwald, T. J., Combs, C. L., Jones, A. S., Randel, D. L. and

Vonder Haar, T. H. (1997). Further developments in estimating cloud liquid water over land using microwave and infrared satellite measurements. *J. Appl. Meteor.*, **36**, 389–405.

Greenwald, T. J., Stephens, G. L., Christopher, S. and Vonder Haar, T. H., (1995). Observations of the global and regional characteristics and radiative effects of marine cloud liquid water. *J. Climate*, **8**, 2928–46.

Greenwald, T. J., Stephens, G. L., Vonder Haar, T. H. and Jackson, D. L. (1993). A physical retrieval of cloud liquid water over the global oceans using special sensor microwave/imager (SSM/I) observations. *J. Geophys. Res.*, **98**, 18,471–88.

Gregory, D. (1995). The representation of moist convection in atmospheric models. In *Proceedings of a Seminar on "Parametrization of Sub-grid Scale Physical Processes"*, pp. 77–113. Reading, United Kingdom: ECMWF.

Gregory, D. and Rowntree, P. R. (1990). A mass flux convection scheme with representation of cloud ensemble characteristics and stability dependent closure. *Mon. Wea. Rev.*, **118**, 1483–1506.

Grossman, R. L. and Durran, D. R. (1984). Interaction of low-level flow with the western Ghat Mountains and offshore convection in the summer monsoon. *Mon. Wea. Rev.*, **112**, 652–72.

Han, Q., Rossow, W. B. and Lacis, A. A. (1994). Near-global survey of effective droplet radii in liquid water clouds using ISCCP data. *J. Climate*, **7**, 465–97.

Hansen, J. E., Lacis, A. A., Ruedy, R. and Sato, M. (1992). Potential climate impacts of Mt. Pinatubo eruption. *Geophys. Res. Lett.*, **19,** 215–18.

Harrison, E. F., Minnis, P., Barkstrom, B. R., Ramanathan, V., Cess, R. D. and Gibson, G. G. (1990). Seasonal variation of cloud radiative forcing derived from the Earth Radiation Budget Experiment. *J. Geophys. Res.*, **95**, 18687–703.

Haskins, R. D., Goody, R. M., and Chen, L. (1997). A statistical method for testing a GCM with spectrally-resolved satellite data. *J. Geophys. Res.*, (in press).

Hauf, T. and Clark, T. L. (1989). Three-dimensional numerical experiments on convectively forced internal gravity waves. *Quart. J. Roy. Meteor. Soc.*, **115**, 309–33.

Hauf, T., Schulte, P., Alheit, R. and Schlager, H. (1995). Rapid vertical trace gas transport by an isolated mid-latitude thunderstorm. *J. Geophys Res.*, **100**, 22957–70.

Haywood, J. M. and Shine, K. P. (1995). The effect of anthropogenic sulfate and soot aerosol on the clear sky planetary radiation budget. *Geophys. Res. Lett.*, **22**, 603–6.

Hegg, D. A., Hobbs, P. V., Gassó, S., Nance, J. D. and Rango, A. L. (1996). Aerosol measurements in the Arctic relevant to direct and indirect radiative forcing. *J. Geophys. Res.*, **101**, 23349–63.

Hobbs, P. V. (ed.) (1993). *Aerosol–Cloud–Climate Interactions.* San Diego: Academic Press.

Hobbs, P. V., Radke, L. F. and Shumway, S. E. (1970). Cloud condensation nuclei from industrial sources and their apparent influences on precipitation in Washington State. *J.*

Atmos. Sci., **27**, 81–9.

Hobbs, P. V., Reid, J. S., Kotchenruther, R. A., Ferek, R. J. and Weiss, R. (1997). Direct radiative forcing by smoke from biomass burning. *Science*, **272**, 1776–8.

Höller, H., Bringi, V. N., Hubbert, J., Hagen, M. and Meischner, P. F. (1994). Life cycle and precipitation formation in a hybrid-type hailstorm revealed by polarimetric and Doppler radar measurements. *J. Atmos. Sci.*, **51**, 2500–22.

Houghton, J. T., Meirafilho, L. G., Callander, B. A., Harris, N., Kattenberg, A. and Maskell, K. (eds.) (1996). *Climate Change 1995*, Cambridge: Cambridge Univ. Press.

Houze, R. A. (1993). *Cloud Dynamics*, Academic Press, 573 pp.

Houze, R. A. and Hobbs, P. V. (1982). Organisation and structure of precipitating cloud systems. *Adv. Geophys.*, **24**, 225–315.

Houze, R. A., Locatelli, J. D. and Hobbs, P. V. (1976). Dynamics and cloud microphysics of the rainbands in an occluded frontal system. *J. Atmos. Sci.*, **33**, 1921–36.

Jackson, D. L. and Stephens, G. L. (1995). A study of SSM/I-derived columnar water vapor over the global oceans. *J. Climate*, **8**, 2025–38.

Jaenicke, R. (1978). Über die Dynamik atmosphärischer Aitkenteilchen. *Berichte der Bunsen-Gesellschaft Physik. Chemie*, **82**, 1198–202.

James, P. K. and Browning, K. A. (1979). Mesoscale structure of line convection at surface cold fronts, *Quart. J. Roy. Meteor. Soc.*, **105**, 371–82.

Jones, A., Roberts, D. L. and Slingo, A. (1994). A climate model study of the indirect radiative forcing by anthropogenic sulfate aerosols. *Nature*, **370**, 450–3.

Kaufman, Y. J., Fraser, R. S. and Mahoney, R. L. (1991). Fossil fuel and biomass burning effect on climate-heating or cooling? *J. Climate*, **4**, 578–88.

Kaufman, Y. J. and Nakajima, T. (1993). Effect of Amazon smoke on cloud microphysics and albedo. *J. Appl. Meteor.*, **32**, 729–44.

Kessler, E. (1974). Model of precipitation and vertical air currents. *Tellus*, **26**, 519–42.

Kiehl, J. T. (1995). On the observed near cancellation between longwave and shortwave cloud forcing in tropical regions. *J. Climate*, **7**, 559–65.

Kiehl, J. T. and Briegleb, B. P. (1993). The relative roles of sulfate aerosols and greenhouse gases in climate forcing. *Science*, **260**, 311–14.

Kiehl, J. T. and Rodhe, H. (1995). Modeling geographical and seasonal forcing due to aerosols. In *Proceedings of the Dahlem Workshop on Aerosol Forcing of Climate*, eds. R. J. Charlson and J. Heintzenberg, pp. 281–96. London: John Wiley.

Kiemle, C., Kästner, M. and Ehret, G. (1995). The convective boundary layer structure from lidar and radiosonde measurements during the EFEDA 91 campaign. *J. Atmos Oceanic Tech*, **12**, 771–82.

King, M. D., Radke, L. F. and Hobbs, P. V. (1993). Optical properties of marine stratocumulus clouds modified by ships. *J. Geophys. Res.*, **98**, 2729–39.

Kuettner, J. (1971). Cloud bands in the earth's atmosphere. *Tellus XXIII*, **4–5**, 404–26.

Kuo, H. L. (1974). Further studies of the parameterization of the influence of cumulus convection on large-scale flow. *J. Atmos. Sci.*, **31**, 1232–40.

Langner, J. and Rodhe, H. (1991). A global three-dimensional model of the tropospheric sulfur cycle. *J. Atmos. Chem.*, **13**, 255–63.

Lau, N-C., and Crane, M. W. (1995). A satellite view of the synoptic-scale organization of cloud properties in midlatitude and tropical circulation system. *Mon. Wea. Rev.*, **123**, 1984–2006.

Li, Z., Moreau, L. and Arking, A. (1997). On solar energy disposition: a perspective from observation and modeling. *Bull. Am. Met. Soc.*, **78**, 53–70.

Lindzen, R. S. and Tung, K. K. (1976). Banded convective activity and ducted gravity waves. *Mon. Wea. Rev.*, **104**, 1602–7.

Locatelli, J. D., Martin, J. E. and Hobbs, P. V. (1992). The structure and propagation of a wide cold frontal rainband and their relationship to frontal topography. In *Preprints, Fifth Conf. on Mesoscale Processes, Atlanta*, pp. 192–6. Amer. Meteor. Soc.

Louis, J.-F. (1979). A parametric model of vertical eddy fluxes in the atmosphere. *Boundary Layer Meteorology*, **17**, 187–220.

Manabe, S., Smagorinsky, J. S. and Strickler, R. F. (1965). Simulated climatology of a general circulation model with hydrological cycle. *Mon. Wea. Rev.*, **93**, 769–98.

Manabe, S. and Wetherald, R. T. (1967). Thermal equilibrium of the atmosphere with a given distribution of relative humidity. *J. Atmos. Sci.*, **24**, 241–59.

Mason, B. J. (1971). *The Physics of Clouds*, 2nd Edition, Oxford University Press, 671 pp.

Matejka, T. J., Houze, R. A. and Hobbs, P. V. (1980). Microphysics and dynamics of clouds associated with mesoscale rainbands in tropical cyclones. *Quart. J. Roy. Meteor. Soc.*, **106**, 29–56.

Miller, M. J. (1995). The significance of convection in NWP and climate modelling. In *Proceedings of a Seminar on 'Parametrization of Sub-grid Scale Physical Processes'*, pp. 43–53. Reading, United Kingdom: ECMWF.

Moncrieff, M. W. (1992). Organized convective systems: archetypal dynamical models, mass and momentum flux theory, and parametrization. *Quart. J. Roy. Meteor. Soc.*, **118**, 819–50.

Moss, S. J. and Johnson, D. W. (1994). Aircraft measurements to validate and improve numerical model parametrisations of ice to water ratios in clouds. *Atmos. Res.*, **34**, 1–25.

Mossop, S. C. (1978). The influence of drop size distribution on the production of secondary ice particles during graupel growth. *Quart. J. Roy. Meteor. Soc.*, **104**, 323–30.

National Research Council (1996). *Aerosol Radiative Forcing and Climate Change*. Washington, DC: National Academy Press.

Nguyen, B. C., Bonsang, B. and Gaudry, A. (1983). The role of the ocean in the global atmospheric sulfur cycle. *J. Geophys. Res.*, **88**, 10903–14.

Nordeng, T. E. (1987). The effect of vertical and slantwise convection on the simulation of polar lows. *Tellus*, **39A**, 353–75.

Nordeng, T. E. (1994). *Extended Versions of the Convection Parametrization Scheme at ECMWF and their Impact upon the Mean Climate and Transient Activity of the Model in the Tropics*. Research Department Technical Memorandum no.206. Reading, United Kingdom: ECMWF.

Novakov, T., Hegg, D. A. and Hobbs, P. V. (1997). Airborne measurements of carbonaceous particles on the East Coast of the United States. *J. Geophys. Res.*, **102**, 30023–30.

Ockert-Bell, M. and Hartmann, D. L. (1992). The effect of cloud type on Earth's energy balance: results for selected regions, *J. Geophys. Res.*, **86**, 9739–60.

Oort, A. and Vonder Haar, T. H. (1976). On the observed annual cycle in the ocean-atmosphere heat balance over the Northern Hemisphere, *J. Phys. Ocean.*, **6**, 781–800.

Ooyama, K. (1971). A theory of parameterization of cumulus convection. *J. Meteor. Soc. Japan*, **49**, 744–56.

Penner, J. E. (1995). Carbonaceous aerosols influencing atmospheric radiation: black and organic carbon. In *Proceedings of the Dahlem Workshop on Aerosol Forcing of Climate*, eds. R. J. Charlson and J. Heintzenberg, pp. 91–108. London: John Wiley.

Penner, J. E., Dickinson, R. and O'Neil, C. (1992). Effects of aerosols from biomass burning on the global radiation balance. *Science*, **256**, 1432–4.

Piexoto, J. P. and Oort, A. H. (1991). *Physics of Climate*. New York: American Institute of Physics.

Platnick, S., Durkee, P. A., Nielsen, K., Taylor, J. P., Tsay, S.-C., King, M. D., Ferek, R. J. and Hobbs, P. V. (1997). The role of background microphysics in shiptrack formation. *J. Atmos. Sci.* (accepted).

Poole, L. R. and McCormick, M. P. (1988). Polar stratospheric clouds and the Antarctic ozone hole. *J. Geophys. Res.*, **93**, 8423–30.

Pruppacher, H. R. and Klett, J. D. (1978). *Microphysics of Clouds and Precipitation*. D. Reidel Pub. Co., Dordrecht, 714 pp.

Radke, L. F., Coakley, J. A. Jr. and King, M. D. (1989). Direct and remote sensing observations of the effects of ships on clouds. *Science*, **246**, 1146–9.

Ramanathan, V., Barkstrom, B. R. and Harrison, E. F. (1989). Climate and the Earth's radiation budget. *Physics Today*, **42(5)**, 22–32.

Randel, D. L., Vonder Haar, T. H., Ringerud, M. A., Stephens, G. L., Greenwald, T. J. and Combs, C. L. (1996). A new global water vapor dataset. *Bull. Am. Met. Soc.*, **77**, 1233–46.

Rangno, A. L. and Hobbs, P. V. (1994). Ice particle concentrations and precipitation development in small continental cumulus clouds. *Quart. J. Roy. Meteor. Soc.*, **120**, 573–601.

Raschke, E. and Rossow, W. B. (1995). 12 Years: The International Satellite Cloud Climatology Project (ISCCP) and its regional projects reported. GEWEX Newsletter, **5(4)**, 3.

Read, W. G., Walters, J. W., Flower, D. A., Froidevaux, L., Jarnot, R. F., Hartmann, D. L., Harwood, R. S. and Rood, R. B. (1995). Upper-tropospheric water vapor from UARS MLS. *Bull. Am. Met. Soc.*, **76**, 2381–9.

Reuter, G. W. and Yau, M. K. (1993). Assessment of slantwise convection in ERICA cyclones. *Mon. Wea. Rev.*, 121, 375–86.

Rind, D., Chiou, E.-W., Chu, W., Oltmans, S., Lerner, J., Larsen, J., McCormick, M. P. and McMaster, L. (1993). Overview of the Stratospheric Aerosol and Gas Experiment II water vapor observations: method, validation, and data characteristics. *J. Geophys. Res.*, **98**, 4835–56.

Rodgers, C. D. and Walshaw, C. D. (1966). The computation of infra-red cooling rate in planetary atmospheres, *Quart. J. Roy. Meteor. Soc.*, **92**, 67–92.

Roeckner, E., Arpe, K., Bengtsson, L., Brinkop, S., Dümenil, L., Esch, M., Kirk, E., Lunkeit, F., Ponater, M., Rockel, B., Sausen, R., Schlese, U., Schubert, S. and Windelband, W. (1992). *Simulation of the Present-day Climate with the ECHAM Model: Impact of Model Physics and Resolution*. Max-Planck-Institut für Meteorologie Report no.93. Hamburg: Max-Planck-Institut.

Rossow, W. B. and Garder, L.C. (1993*a*). Validation of ISCCP Cloud Detections. *J. Climate*, **6**, 2370–93.

Rossow, W. B. and Garder, L. C. (1993*b*). Cloud detection using satellite measurements of infrared and visible radiances for ISCCP. *J. Climate*, **6**, 2342–69.

Rossow, W. B. and Kachmar, B. (1988). *International Satellite Cloud Climatology Project (ISCCP): Description of Atmospheric Dataset*. Washington, DC: NOAA NESDIS report.

Rossow, W. B., Walker, A. W. and Garder, L. C. (1993). Comparison of ISCCP and other cloud amounts. *J. Climate*, **6**, 2394–418.

Rossow, W. B. and Zhang, Y.-C. (1995). Calculation of surface and top of atmosphere radiative fluxes from physical quantities based on ISCCP datasets, Part II: Validation and First results, *J. Geophys. Res.*, **100**, 1166.

Rutledge, S. A. (1991). Middle latitude and tropical mesoscale convective systems. *Rev. Geophys.*, **29** (suppl), 88–97.

Rutledge, S. A. and Hobbs, P. V. (1983). The mesoscale and microscale structure and organization of clouds and precipitation in mid latitude cyclones. VIII: A model for the feeder-seeder process in warm frontal rainbands. *J. Atmos. Sci.*, **40**, 1185–206.

Schiffer, R. A. and Rossow, W. B. (1983). The International Satellite Cloud Climatology Project (ISCCP): The first project of the World Climate Research Programme, *Bull. Am. Met. Soc.*, **64**, 779–84.

Schmetz, J., Geijo, C., Menzel, W.P., Strabala, K., Van de Berg, L., Hulmlund, K. and Tjemkes, S. (1995). Satellite observations of upper tropospheric relative humidity, clouds and wind field divergence. *Beitr. Phys. Atmosph.*, **68**, 345–357.

Schmidt, H. and Schumann, U. (1989). Coherent structure of the convective boundary layer derived from large-eddy simulations. *J. Fluid Mechanics*, **200**, 511–62.

Schubert, S. D., Rood, R. B. and Pfaendtner, J. (1993). An sssimilated dataset for earth science application. *Bull. Am. Met. Soc.*, 2331–2342.

Schwartz, S. E. (1988). Are global cloud albedo and climate controlled by marine phytoplankton? *Nature, 336*, 441–5.

Sellers, W. D. (1965). *Physical Climatology.* Univ. of Chicago Press.

Shaw, G. E. (1983). Bio-controlled thermostats involving the sulfur cycle. *Clim. Change, 5*, 297–303.

Simpson, J. S., Wescott, N. E., Clerman, R. J. and Pielke, R. A. (1980). On cumulus mergers. *Arch. Meteorol. Geophys. Bioklimatol.*, A29, 1–40.

Slingo, A. and Webb, M. J. (1992). Simulation of clear-sky outgoing longwave radiation over the oceans using operational analyses. *Quart. J. Roy. Meteor. Soc.*, 118, 1117–44.

Slingo, A. and Webb, M. J. (1997). The spectral signature of global warming. *Quart. J. Roy. Meteor. Soc.*, 123, 293–308.

Soden, B. and Bretherton, F. P. (1993). Upper tropospheric relative humidity from GOES 6.7 micron channel: method and climatology for July 1987. *J. Geophys. Res.*, 98, 16669–88.

Stephens, G. L. (1978). Radiation profile in extended water clouds: II parameterization schemes. *J. Atmos. Sci.*, 35, 2123–32.

Stephens, G. L. (1980). Radiative properties of cirrus clouds in the infrared region. *J. Atmos. Sci.*, 37, 435–46.

Stephens, G. L. (1984). Review: the parameterization of radiation for numerical weather prediction and climate models. *Mon. Wea. Rev.* 112, 826–67.

Stephens, G. L. (1988). Radiative transfer through arbitrarily shaped optical media, I: A general method of solution, *J. Atmos. Sci.*, 45, 1818–36.

Stephens, G. L. (1990). On the relationship between water vapor over oceans and sea surface temperature. *J. Climate*, 6, 634–45.

Stephens, G. L., Jackson, D. L. and Wittmeyer, I. L. (1996). Global observations of upper-tropospheric water vapor derived from radiance data. *J. Climate*, 9, 305–26.

Stephens, G. L., Slingo, A., Webb, M. J., Minnett, P. J., Daum, P. H., Kleiman, L., Wittmeyer, I. and Randall, D. A. (1994). Observations of the earth's radiation budget in relation to atmospheric hydrology. Part IV: atmospheric column radiative cooling over the worlds' oceans. *J. Geophys. Res.*, 99, 18585–604.

Stephens, G. L. and Tsay, S-C. (1990). On the cloud absorption anomaly. *Quart. J. Roy. Meteor. Soc.*, 116, 671–704.

Stephens, G. L. and Webster, P. J. (1984). Cloud decoupling of surface and planetary radiative budgets. *J. Atmos. Sci.*, 41, 681–6.

Stull, R. B. (1988). *An Introduction to Boundary Layer Meteorology.* Dordrecht: Kluwer Academic Publishers.

Suarez, M. J., Arakawa, A. and Randall, D. A. (1983). Parameterization of the planetary boundary layer in the UCLA general circulation model: formulation and results. *Mon. Wea. Rev.*, 111, 2224–43.

Taylor, K. E. and Penner, J. (1994). Response of the climate system to atmospheric aerosols and greenhouse gases. *Nature*, 369, 734–7.

Tiedtke, M. (1989). A comprehensive mass flux scheme for cumulus parameterization in large-scale models. *Mon. Wea. Rev.*, 117, 1779–1800.

Tiedtke, M. (1995). An extension of cloud-radiation parameterization in the ECMWF model: the representation of subgrid scale variation of optical depth. *Mon. Wea. Rev.*, 124, 745–50.

Trenberth, K. E.and Olson, J. G. (1988). *Intercomparison of NMC and ECMWF Global Analyses: 1980–1986.* Boulder, CO: NCAR Technical Note 301.

Tselioudis, G., Rossow, W. B. and Rind, D. (1992). Global patterns of cloud optical thickness variation with temperature. *J. Climate*, 5, 1484–95.

Twomey, S. (1991). Aerosols, clouds and radiation. *Atmos. Environ.*, 25A, 2435–42.

Twomey, S. A. and Bohren, C. F. (1980). Simple approximations for calculations of absorption in clouds. *J. Atmos. Sci.*, 37, 2086–94.

Twomey, S., Peipgrass, M. and Wolff, T. L. (1984). An assessment of the impact of pollution on global cloud albedo. *Tellus*, 36B, 356–6.

Webb, M. J., Slingo, A. and Stephens, G. L. (1993). Seasonal variations of the clear-sky greenhouse effect: The role of changes in atmospheric temperatures and humidities. *Clim. Dynamics*, 9, 117–29.

Whitlock, C. H., Charlock, T. P., Staylor, W. F., Pinker, R. T., Laszlo, I., Ohmura, A., Gilgen, H., Konzelman, T., DiPasquale, R. C., Moats, C. D., Le Croy, S. R. and Ritchey, N. A. (1995). First global WCRP surface radiation budget dataset. *Bull. Amer. Met Soc.*, 76, 905–22.

Wielicki, B. A., Cess, R. D., King, M. D., Randall, D. A. and Harrision, E. F. (1995). Mission to Planet Earth: role of clouds and radiation in climate. *Bull. Am. Met. Soc.*, 76, 2125–54.

Wittmeyer, I. L. and Vonder Haar, T. H. (1994). Analysis of the global ISCCP TOVS water vapor climatology. *J. Climate*, 7, 325–33.

Wyngaard, J. C. and Brost, R. A. (1984). Top-down and bottom-up diffusion of a scalar in the convective boundary layer. *J. Atmos. Sci.*, 41, 102–12.

Yanai, M., Esbensen, S. and Chu, J.-H. (1973). Determination of bulk properties of tropical cloud clusters from large-scale heat and moisture budgets. *J. Atmos. Sci.*, 30, 611–27.

4 Surface and sub-surface processes

4.1 Estimating surface precipitation

D. Rosenfeld and C. G. Collier

The nature of precipitation systems as it affects rainfall measurement

Rainfall measurements at a point are provided by rain gauges. However, rainfall measurements over large areas are practical only by means of remote sensing, using weather radars and satellites. In fact, measurements from satellites are the only means available for large-scale rainfall measurements over the ocean and inaccessible regions, where a major portion of the global rainfall occurs.

A remote sensing method can be either passive or active. Passive methods rely either on thermal radiation that is emitted from the rain clouds, or on solar radiation reflected from them. Active remote sensing instruments transmit the radiation and measure its reflection from the targets.

Quantitative rainfall information is calculated using known interactions of the measured electromagnetic radiation with the clouds, precipitation, atmosphere and the underlying surface from which the radiation is emitted and reflected. The form of interactions is highly dependent on the wavelength of the radiation, the type of precipitation and its distribution. However, for active systems the electromagnetic radiation is increasingly attenuated at wavelengths much shorter than 10 cm.

The main factors that affect the form of interactions between precipitation clouds and the electromagnetic radiation used to measure rainfall are:

1. the thermodynamic phase of the precipitation – ice or water;
2. the particle size distribution of the precipitation;
3. the origin of precipitation particles aloft, i.e. as snow, graupel (low density ice pellets that grow by riming), hail or water drops;
4. the horizontal gradients of the precipitation intensities across the instrument beamwidth;
5. vertical changes in the precipitation intensity due to evaporation or, in contrast, low level enhancement due to orographic uplifting of the air;
6. the cloud depth and temperatures of cloud base and cloud top for passive sensors.

These factors, as they affect ground-based radar measure-ments of precipitation, are summarized in Figure 4.1. The characteristic features of the nature of precipitating cloud systems noted in the figure are determined mainly by the intensity and extent of the vertical air motions (Houze, 1993). Convective rainfall is characterised by localized intense and deep vertical air motions ($w \approx$ 1–10 m s^{-1}), that cause local heavy ($R >$ about 10 mm hr^{-1}) precipitation showers, originating aloft as hail, graupel or water drops.

Stratiform rain is characterized by extensive weak vertical air motions, resulting in extensive weak to moderate precipitation ($R <$ about 10 mm hr^{-1}). Stratiform precipitation particles which form in clouds reaching above the 0°C level typically originate as ice crystals that often aggregate into snowflakes. Such precipitation (as contrasted with graupel and hail in convective precipitation) can exist when vertical air motion satisfies the condition $|w| \ll |V_t|$, where V_t is the terminal fall velocity of snow particles (1–3 m s^{-1}).

Another important distinction is between mid-latitude and tropical rain cloud systems. The mid-latitude systems are often associated with extratropical cyclones and their fronts. Most of this precipitation is stratiform, especially over the cold land in winter. Schematic illustration of precipitation types in extratropical cloud systems is provided in Figure 4.1.

Most rainfall in the tropics, and in the summer subtropics, occurs in convective rain cloud complexes. The building blocks of these complexes are the convective cells, which undergo distinct growth, maturing and dissipation life cycle stages lasting up to an hour. Cells that grow as part of large convective complexes, such as squall lines (see Figure 4.2), lose their identity during the late mature stage and merge into large stratiform precipitation areas which dissipate after several hours. About 20–40% of the precipitation in the tropics is estimated to be of this stratiform type (Houze, 1993).

Measurement requirements from a water cycle perspective

The requirements for precipitation measurements in hydrology are more stringent than for meteorology. For meteorology and climatology the shorter the integration period

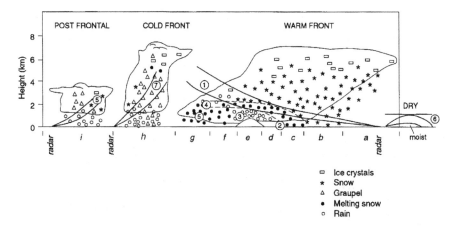

Figure 4.1. Precipitation characteristics that affect their measurement by remote sensing in a schematic extratropical winter cloud system. From right to left are illustrated (*a*) leading edge of a warm front, with snow falling and evaporating from altostratus; (*b*) warm front with snow from nimbostratus with snow; (*c*) rain mixed with melting snow; (*d*) rain produced from melting snow at low altitude; (*e*) low level precipitation enhancement due to the orographic seeder/feeder process; (*f*) rain from melted snow aloft; (*g*) drizzle and clouds without precipitation; (*h*) cold front convective cloud, with graupel that melts into rain, but which occasionally reaches the surface not completely melted; (*i*) post frontal convective clouds, producing small graupel and snow that melts to light rain showers and drizzle. The impact on errors in surface-based radar measurements of precipitation are numbered as follows: 1 radar beam overshooting the shallow precipitation at long range; 2 low-level evaporation beneath the radar beam; 3 orographic enhancement above hills which goes undetected beneath the beam; 4 the bright band; 5 underestimation of the intensity of drizzle because of the absence of large droplets; 6 radar beams bent in the presence of strong hydrolapse; 7 under/overestimation due to updrafts/downdrafts and hail. Areas of strong hydrolapse are those areas where there is strong decreasing moisture with height.

and the smaller the area over which the measurements are made, the more tolerant is the requirement of errors in the data. This is not so for hydrology, probably because there is no linkage provided by an energy cascade through physical systems of large to small scales as there is in the atmosphere, i.e., river flow is not governed by the interaction of physical processes at different scales. Since the requirements for understanding the water cycle demand measurements of precipitation over large areas we consider only the use of ground-based radar and satellite techniques in what follows.

Browning (1990) suggests that, in assessing the science requirements for global distribution of rainfall, equal first priority is given to the acquisition of rainfall statistics, and to the use of rainfall data in the data assimilation stage of numerical weather prediction (NWP) models, including the use of the latter to optimize the generation of climate statistics. Measurements with a resolution of at least 6 hours are needed to provide a useful estimate of the diurnal variation in global precipitation. One application of the global statistics is

the validation of general circulation models (GCMs) whereas another application is that of monitoring changes in the hydrologic cycle that may be due to the greenhouse effect. Observational studies indicate that there are major differences in the latent-heating profiles according to the dynamical organization of the precipitation systems. Areas of convective precipitation give rise to latent-heat sources throughout the depth of the troposphere, with maximum at middle levels. In contrast, latent heat in stratiform precipitation is released in the upper half of the troposphere, above the freezing level. The evaporation and melting of the stratiform precipitation below the 0 °C level causes cooling of the lower troposphere. Due to the large difference in the vertical heating profile between convective and stratiform rainfall, precipitation measurements that are aimed at understanding the hydrologic cycle and global circulation should also measure the convective/stratiform proportions of rainfall. Parts of other chapters in this book address the deposition of latent heat released by the precipitation processes and the way it drives the global circulation.

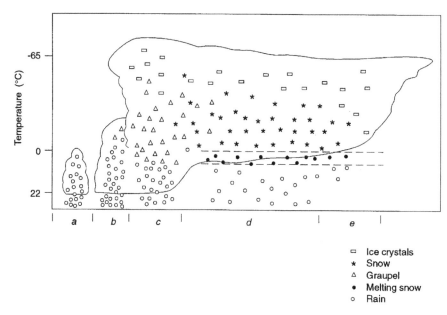

Figure 4.2. Precipitation characteristics that affect their measurement by remote sensing in a tropical cloud system. From left to right are illustrated (a) warm rain showers from all-water clouds; (b) growing convective cells in the leading edge of the squall line, with intense localized showers originating in all-water processes; (c) mature convective cells, producing rainfall mainly as melted graupel; (d) stratiform area, where widespread precipitation is produced as snow melts, causing a change in the rain drop size distribution that reflects the size distribution of the snow flakes, and the evaporation of smaller rain drops; (e) dissipating anvils, with evaporating precipitation aloft.

Principles, capabilities and limitations of ground-based observational methods for estimating precipitation

Ground-based radar: background

As a radar beam rotates about a vertical axis, measurements of the energy back-scattered from precipitation particles in volumes above the ground are made at many ranges out to about 150 km, or more in the tropics and the summer subtropics, and at different azimuths.

The rainfall rate, R, may be expressed as a function of the dropsize distribution $N(D)$, as:

$$R = (\pi/6) \int_0^\infty D^3 N(D) w_t(D) \, \mathrm{d}D \qquad (4.1.1)$$

where w_t is the terminal velocity of the hydrometeors.

If we now assume that $N(D)$ can be described by an exponential function of the form $N(D) = N_0 \exp(-\Lambda D)$, the integral in equation (4.1.1) may be evaluated analytically. A parameter called 'radar reflectivity factor', Z, may be defined as:

$$Z = \int_0^\infty N(D) D^6 \mathrm{d}D \qquad (4.1.2)$$

where $N(D)$ is the dropsize distribution within the resolution cell (Z in $mm^6 \, m^{-3}$, D in mm, $N(D)$ in $mm^{-1}m^{-3}$). Subsequent elimination of Λ produces the following relation between the radar reflectivity, Z and R:

$$Z = AR^B \qquad (4.1.3)$$

where coefficients A and B are empirically determined coefficients.

A great many Z–R power law relationships (4.1.3) have been published; for example, Battan (1973) lists 69 such relationships, with large differences of up to a factor of 5 between rain intensities related to the same reflectivity. Geometrical and timing errors make derivations of Z–R relationships using measurements from rain gauges impractical, except at very close ranges, where general agreement with the Z–R relationships based on dropsize distributions was found, as was first shown by Joss et al. (1970). This is because a radar makes measurements over small volumes above the surface and rain gauges make point measurements at the surface. Therefore, the usual approach is to estimate the rain rate by a power law that was based on a 'representative' dropsize distribution such as $Z = 200 \, R^{1.6}$ (Marshall and Palmer, 1948), and then apply a calibration 'correction' in an effort to minimize the differences between the radar integrated rainfall and that obtained from rain gauges. This has been the basis of most operational rain-measuring radars to date. However, significant effort has been invested in improving algorithms and new radar technologies to minimize, or remove, the need to use data from rain gauges. This work is summarised in the remainder of this section.

Measurement challenges for radars

In making measurements of precipitation using radar we categorize the difficulties as shown in Figure 4.1. In particular we consider here the problems associated with the characteristics of the precipitation. For a more comprehensive review the reader is referred to Atlas et al. (1996).

Variations in dropsize distribution

The different processes leading to the formation of rain drops lead also to differences in the rain dropsize distribution. Convective rain, formed by warm or mixed phase processes, generally produces a dropsize distribution arising from the creation of large drops when drops coalesce, which constitutes the 'convective' Z–R relationship (i.e., region h of Figure 4.1 and regions a–c of Figure 4.2). Warm rain from shallow clouds or low level orographic enhancement (i.e., regions i and 3 of Figure 4.1) lead to many drops or drizzle, which produces relatively low reflectivity for the same rain intensity, as compared with the convective Z–R. The efficient aggregation of snowflakes, especially towards the melting level, causes the dropsize distribution of the snow-melt to be of relatively fewer and larger drops and therefore to have larger reflectivity as compared with that for the convective dropsize distribution of the same rain intensity (see region d of Figures 4.1 and 4.2). A similar effect on the dropsize distribution of stratiform rain is caused by the evaporation of the snow-melted rain in typically subsaturated air below the melting level. The smaller drops evaporate faster than the larger drops, resulting in a decrease of R for a given Z. It should be noted that the same cloud system can undergo changes through these dropsize distribution types during its evolution. Additional smaller scale dropsize distribution variations occur due to sorting of the drops in time and space as they fall through a sheared wind field.

Observational evidence for these changes in the Z–R relationships was first found by Joss et al. (1970), who proposed that the characteristics of the echoes could be used to identify the character of the rain. Collier et al. (1983) and others recognized the need to apply different Z–R relationships derived from rain-gauge data for different rainfall regimes. Automated classification procedures for convective and strat-

iform precipitation have been developed and a procedure that classifies precipitation into types relevant to the Z_e–R variations was introduced by Rosenfeld et al. (1995a,b).

Variations of Z_e–R due to underfilling of the radar beam

A large difference often exists between the true reflectivity field and the radar observed reflectivity, Z_e, field, which can invalidate the use of dropsize distribution-based Z–R relationships alone. This is mainly the result of radar beam spreading, which convolves the radar beam pattern with the true reflectivity field. This causes partial beam filling, i.e., the radar beam is not uniformly filled by precipitation, which, due to the nonlinear conversion of Z to R, causes a relative overestimate of the areal rainfall in storms having large rain rate gradients, especially at the longer ranges.

Variations of Z and Z_e–R with height

Vertical reflectivity and Z_e–R profiles change with height differently for different rain types. Reflectivity decreases very rapidly with height in shallow tropical clouds, which produce rain mainly by warm processes (see Section 3.4). The reflectivity of deep and highly convective clouds remains constant or even increases at heights greater than the freezing level. In stratiform rainfall the reflectivity often does not change much below the freezing level. However, as snowflakes melt they acquire a shell of water, the backscattering cross-section of which is much larger than that from snowflakes or rain drops. The result is enhanced radar echoes from the region of melting, known as a bright-band. The reflectivity decreases rapidly above this region (i.e., region 4 of Figure 4.1). Changes in the vertical reflectivity profile occur also due to evaporation of falling rain or the opposite process, i.e., low level enhancement.

Due to the increasing height of the radar measured volume with range, each rain type has its own distinct dependence of the Z_e–R relationships on the range and on the freezing level. This dependence can be calculated and corrected for in operational systems. Both analytic methods based on using radar data alone and physically based methods using microphysical models are possible.

Z_e–R probability matching method (PMM)

The complex dependence of Z_e–R relationships on the various factors mentioned above and their large variation in time and space make it impossible to derive precise Z_e–R relationships on theoretical considerations alone in most circumstances. Matching the radar measured Z_e to a collection of synchronized rain-gauge measurements of R should provide a Z_e–R relationship that takes into account implicitly all the unknown factors. However, scattergrams of point comparisons between radar and rain gauges are extremely noisy for three main reasons: (a) the large discrepancy between the sample volume of the rain gauge and the radar; (b) timing and navigation mismatches; (c) the large variability of the Z_e–R relationships at small time scales. This made the use of instantaneous point Z_e–R comparisons by regression methods of limited use.

Much of the component of the Z_e–R variability which is related to variations of rain types can be accounted for by the classification of each point in the radar field in terms of the three-dimensional reflectivity structure in its vicinity. The remaining variability after classification is still largely due mainly to sampling, timing and navigation errors. It can be assumed that the measured Z_e and R are derived from the same probability density functions of true R. The probability matching method (PMM) relates R to Z_e by matching R and Z_e pairs that have the same cumulative probability. It can be shown that, under this assumption, the PMM provides the best estimate of the true Z_e–R for the specific rain type. This method has been explored by Rosenfeld *et al.* (1994, 1995*b*).

Polarimetric measurements

Differential reflectivity

The departure of the shapes of precipitation particles from spherical gives rise to different radar reflectivity properties. Seliga and Bringi (1976) related signals in two orthogonal linear polarization planes, horizontal (H) and vertical (V), to dropsize distributions, defining the differential reflectivity,

$$Z_{DR} = 10 \log \left(\frac{Z_H}{Z_V} \right) \qquad (4.1.4)$$

The oblateness of rain drops, when falling at terminal velocity in air, increases with drop volume. Since models for the shape and minor-to-major axis ratio and fall speed data exist, it is possible to relate the dropsize distributions so measured to rainfall rates. Accurate measurements of Z_{DR} can remove much of the measurement errors due to the variability in dropsize distribution. However, the required accuracy of measurements (to within 0.1 dB) require precise matching of H and V channels, and well-matched antenna side lobes, which are not generally available in operational radars.

Unfortunately, further parameters are needed to remove ambiguities due to the effects of the different drop sizes which may be associated with a wide range of rainfall rates. In addition, propagation effects are also detrimental. At best (close ranges, rain only, small scale in time and space), the

Z_{DR} reduces the rain estimation error by a factor of about $\frac{2}{3}$. However, dropsize distribution variability is, in most cases, not the dominant cause of errors in the areal rainfall measurement. Errors arising from the vertical reflectivity profile are usually dominant.

Differential phase shift, K_{DP}

For a Doppler radar additional information is available from the phase difference Φ_{DP} of the returns measured with two polarizations, usually vertical and horizontal. As the radar wave propagates through a region in which the precipitation particles have some degree of horizontal alignment, the horizontally polarized wave progressively lags behind the vertical one, and Φ_{DP} should increase. Sachidananda and Zrnic (1986) have derived the following theoretical relationship between K_{DP} (the gradient of Φ_{DP} along the beam axis) and R:

$$K_{DP} \ (\text{deg km}^{-1}) = 0.03 \ R^{1.15} \ (\text{mm hr}^{-1}) \qquad (4.1.5)$$

K_{DP} has many potential advantages. It is nearly proportional to R and insensitive to dropsize distribution. Because of the near linearity of equation (4.1.5), if changes in Φ_{DP} are measured at two particular range gates it should be possible to estimate the integrated rainfall over the distance between these gates, even when heavy ground clutter completely dominates the reflectivity measurements at the intervening gates. Also, in vigorous convective storms the estimates of rainfall from Z are very error prone because the value of Z may be dominated by the contribution of hail even though the hail contributes negligibly to the integrated precipitation amount. However, K_{DP} should detect the component due to only the rainfall and be insensitive to hailstones provided they tumble as they fall (which may not always be the case).

These methods may well offer the best approach of measuring high rainfall rates, although they are rather insensitive to low rainfall rates, associated with small and nearly spherical drops. It is likely that the best prospects of accurate rainfall measurements over all rainfall rates will result from a combination of several techniques, dynamically adapted to the appropriate conditions.

Principles, capabilities and limitations of space-based observational methods for estimating precipitation

The physical principles of rainfall estimates from space

Rain estimation from space relies on the radiation that is reflected and emitted through cloud tops. Most of the radiation does not penetrate deep into cloud regions containing particles with similar or greater size than the radiation wavelength. Therefore, except for the longest wave bands (shortest frequencies, less than 37 GHz), most of the radiation from precipitation clouds comes from their upper portions, providing information that cannot straightforwardly be directly related to the surface precipitation. This situation leads to the development of a large variety of methods relating these indirect measurements to surface precipitation, using various physical principles. Following are selected methods that represent the major approaches, according to wavebands used and the way the measurements are converted into rain estimates.

Infrared (IR)

The black body temperature of cloud tops is measured routinely at infrared wavelengths by geostationary satellites which provide nearly complete coverage of the globe every half hour. Rain intensities vary with the rate of expansion of the cold ($T < 235$ K) cloud top areas. It is assumed that the expansion of the cloud top is an indicator for the divergence aloft and, hence, to the rate of rising air and precipitation. However, when used over a large area, this method did not show a significant improvement with respect to the simplest possible method which assumes that all clouds with tops colder than a given threshold temperature T precipitate at a fixed rate G mm hr^{-1}. Richards and Arkin (1981) have shown that $T = 235$ K and $G = 3$ mm hr^{-1} is typical for the eastern equatorial Atlantic. This method was developed into the Global Precipitation Index (GPI), which has been extensively used since then. Atlas and Bell (1992) noted that such area methods work well only for a time–space domain that is large enough to include a large number of storms which provide a good representation of the full evolution of convective rain cloud systems (e.g., 2.5° by 2.5° by 12 h). Atlas and Bell suggested that this is so because the area of cold cloud tops is proportional to the amount of air which is convected up and diverges aloft throughout the life cycle of the convective rain cloud complex. The amount of rising air, in turn, is proportional to the integral rain amount. Inherently, the satellite area integral methods are limited to convective rain regimes, performing poorly for clouds associated with extratropical baroclinic disturbances. Classification of clouds into convective and stratiform by the texture of the cloud top temperature showed some improvement for tropical rainfall over land, but failed (along with the rest of the IR methods) in mid-latitude winter systems, because there the 'convective' relation between cold cloud top area and surface rainfall does not apply to largely non-convective cloud systems.

The 'split window'

The atmospheric window around 10 μm is split into two closely spaced wavebands, centered at 10.8 and 12 μm. Clouds have large absorption and emissivity in the longer waveband. Therefore, the 10.8 μm radiation in thin clouds will be contributed from lower and warmer levels as compared with the 12.0 μm waveband, creating a brightness temperature difference between the two channels. Inoue (1987) has shown that cirrus clouds can be distinguished from thicker high clouds by having larger brightness temperature difference. This helps in eliminating thin clouds from consideration as precipitating cloud.

The 'effective radius' of cloud top particles

Very cold cloud top temperature is not always a requirement for precipitation, in which case the IR threshold technique breaks down. An extreme example of this is the plentiful rainfall in parts of Hawaii, which is almost exclusively contributed from clouds with top temperature much warmer than 0°C. The precipitation formation processes require the existence of large cloud droplets and/or ice particles in the cloud, which often spread to the cloud top. These large particles absorb the 1.6 μm and 3.7 μm radiation much more strongly than small cloud droplets. This effect makes it possible to calculate the effective radius (r_{eff} = integral volume divided by integral surface area) of the particles. Rosenfeld and Gutman (1994) have shown that $r_{eff} = 14$ μm can serve to delineate precipitating clouds, regardless of their top temperature.

Microwave (MW)

Microwaves provide the measurements that are physically best related to the actual precipitation, especially in the longest wavebands. The interactions of passive MW with precipitation clouds and the surface are illustrated in Figure 4.3, using two wave bands, shorter (85 GHz) and longer (19 GHz).

(a) Absorption-based measurements
   Water drops have relatively large absorption/emission coefficient, increasing for the higher frequencies. The emission is proportional to the vertically integrated cloud and rain water in the low frequencies, but due to the increased emissivity for the higher frequencies the emission saturates for light rain intensities.

(b) Scattering – based measurements
   Ice particles have relatively small absorption/emission, but they are good scatterers of the MW radiation, especially at the higher frequencies. Therefore, at the high

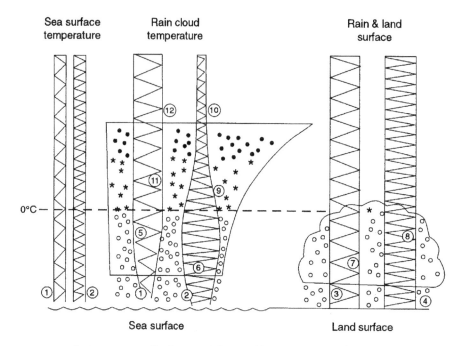

Figure 4.3. The interactions of high (e.g., 85 GHz) and low (e.g., 19 GHz) frequency passive MW with precipitation clouds and the surface. The width of the vertical columns represents the intensity or temperature of the upwelling radiation. In this figure the illustrated features and their demarcations are: (*a*) the small emissivity of sea surface for both low (1) and high (2) frequencies; (*b*) the large emissivity of land surface for both low (3) and high (4) frequencies; (*c*) the emission from cloud and rain drops, which increases with vertically integrated liquid water for the low frequency (5), but saturates quickly for the high frequency (6); (*d*) the signal of the water emissivity at the low frequency is masked by the land surface emissivity (7); (*e*) the saturated high frequency emission from the rain (8) is not distinctly different from the land surface background (4); (*f*) ice precipitation particles aloft backscatter down the high frequency emission (9), causing cold brightness temperatures (10), regardless of surface emission properties; (*g*) the ice lets the low frequency emission upwell unimpeded (11), allowing its detection above cloud top as warm brightness temperature (12).

frequencies (85 GHz) the large scattering from the ice in the upper portions of the clouds makes the ice an effective insulator, because it reflects back down most of the radiation emitted from the surface and from the rain. The remaining radiation that reaches the MW sensor is interpreted as a colder brightness temperature. A major source of uncertainty for the scattering-based retrievals is the lack of a consistent relationship between the frozen hydrometeors aloft and the rainfall reaching the surface.

These physical principles (a and b) have been used to formulate a large number of rain estimation methods. A review and intercomparison is available in Wilheit *et al.* (1994). In general, passive MW rainfall estimates over the ocean were of useful accuracy. However, over the equatorial Pacific, passive

MW does not show significantly improved skill when compared with the simplest infrared method (the global precipitation index, GPI).

Over land the passive MW algorithms can detect rain mainly by the ice scattering mechanism (b). As already noted, this indirect rain estimation method is less accurate. Moreover, rainfall over land from clouds which do not contain significant amounts of ice aloft goes mostly undetected. A passive radiation method that is able to detect such rain is the 'effective radius' method of Rosenfeld and Gutman (1994) mentioned in the previous subsection.

A major limiting factor in the accuracy of passive MW methods is the large footprint, which causes partial beam filling, especially at the higher frequencies. The resolution is greatly improved with the Tropical Rainfall Measurement Mission (TRMM) satellite (Simpson *et al.*, 1988), with a corresponding improvement in the expected accuracy of the MW rain estimates. The TRMM satellite has a radar transmitting at a wavelength of 2.2 cm (active MW) and microwave radiometers (19 to 90 GHz). The resolutions of these instruments range from about 1 km for the visible and infrared radiometer, around 10 km for the microwave radiometers and 250 m for the radar. The radar is expected to provide a quantum jump in the accuracy of instantaneous rain estimates over those previously achieved from space. Since TRRM samples each area between 35° north and south, at best, twice daily, the sampling error is the dominant source of inaccuracy. A combination of the measurements from TRMM-like and geostationary satellites provides the best potential for accurate global precipitation estimates from space.

Sampling errors and validation

All techniques for measuring precipitation suffer from errors arising from sampling. Indeed, these errors can be greater than all the other errors if accumulations are improperly computed, or there is an inappropriate sample in the first place. It is clear that for both radar and satellite techniques the sampling strategy is of prime importance. In tropical regions there can be a significant diurnal cycle in rainfall activity, and the phase and intensity of the cycle may vary from region to region. The low inclination orbit used for TRMM will precess in such a way as to sample a full diurnal range of Equator crossing times over the course of a month. This is not the case for satellites in polar orbit for which the Equator crossing time is always the same. The diurnal cycle may therefore increase the errors due to sampling.

Laughlin (1981) has calculated the resulting sampling error for a range of sample intervals when measurements are averaged over periods of 1 week, 2 weeks and 1 month. For monthly averages over a 280 km square and a sampling interval of 10 hours, appropriate for the TRMM satellite, the sampling error is about 10%. This analysis was carried out using Global Atmospheric Research Programme Atlantic Tropical Experiment (GATE) tropical rain data, but Seed and Austin (1990) have pointed out that for convective systems in other regions, which have shorter decorrelation times than observed for tropical rain, the sampling error is likely to be larger. Therefore, it might be misleading to apply GATE statistics universally.

The validation of satellite algorithms for estimating rainfall accumulations is complex and must be undertaken in ways which ensure that different techniques provide data with similar characteristics, i.e., integration times and coverage.

The effects of data resolution are also important (Morrissey, 1994).

The sampling problem is just as acute for short-period rainfall accumulations derived from radar. Fabry *et al.* (1994) found that the best accumulations are obtained with very high time resolution data. For a given time resolution, there is an optimum spatial resolution that minimizes rainfall accumulation errors. Ogden and Julien (1994) found that the effect of the spatial resolution of radar data depended upon the importance of two processes, namely 'storm smearing' and 'watershed smearing'. Storm smearing occurs when the rainfall data length scale approaches or exceeds the rainfall correlation length (about 2.3 km for convective cells). This tends to decrease rain rates in high intensity regions, and increase rain rates adjacent to low intensity regions. thereby effectively reducing rainfall gradients. It is independent of basin size. Watershed (or measurement area) smearing occurs when the radar grid size approaches the characteristic catchment (or measurement) size (square root of area). In this case, the uncertainty of the location of rainfall within the measurement boundary is increased. It is clearly very important to match the pixel size to the type of measurement required.

Finally, when comparing radar or satellite measurements with rain-gauge measurements it is important to appreciate the sampling errors arising from the point nature of the rain-gauge measurements. Kitchen and Blackall (1992) note that much of the difference between point rain-gauge values and radar measurements is due to gauge sampling problems. This implies that radar adjustment factors derived from rain gauges may actually introduce sampling errors to the radar estimates derived using such factors.

Estimating precipitation from models

Given precipitation estimates from satellite data, procedures are being devised to use the estimates as input to numerical models. However, most numerical models exhibit a characteristic known as 'spin-up' when integrated from an initial state generated from an assimilation of observations and model background field. During the spin-up time the surface boundary layer is established, spurious gravity waves are damped and the model dynamics is established. During this period, up to several days for a global climate model and a few hours for a mesoscale model, large-scale rain produced by the model shows up to about a 40% change.

However, studies of the use of numerical models to aid the estimation of precipitation from satellite data are few. Operational approaches to this problem exist in which the satellite-derived moisture fields are interactively balanced with respect to other model fields even though relative humidity and vertical velocity should be well correlated. Currently rainfall fields derived from models approach the accuracy of observed fields when there is strong synoptic scale or topographic forcing. Numerical models do not yet reproduce convective rainfall with the spatial accuracy necessary for most hydrometeorological studies, although they do reproduce climatology with a useful level of reliability. However, the way in which physical parameterizations are structured has a profound effect upon model output. Also small analysis differences can lead to significant differences in model simulations and predictions (Strensrud and Fritsch, 1994). Variational approaches to model initialization seem a promising approach to gaining further improvements.

Comparison of accuracy

The best measurement of point rainfall overland is that achieved using a rain gauge, but, as pointed out by Bellon and Austin (1986), the use of satellite-based techniques to measure summer afternoon convective rainfall in mid-latitudes becomes better than the use of interpolated rain-gauge measurements when the rain-gauge spacing is greater than about 40 km. However, in most other rainfall types satellite techniques are generally rather poor in mid-latitudes.

The best accuracy for areal rainfall measurements from space is, at present, obtained over the tropical oceans (Wilheit *et al.*, 1994), where the GPI performs as well as passive MW techniques for long period (order month) integrated rainfall. However, errors for individual events may be large because 'warm rain' from shallow clouds is common in some places in the tropics. The passive MW techniques become increasingly advantageous towards higher latitudes where convective rainfall occurs less frequently. Here the best accuracy is achieved by combining passive MW with IR from geostationary satellites. Somewhat lower accuracies of IR techniques are achievable in convective rain over land, due to the large dynamic and microphysical diversity of rain cloud systems. This causes a larger variability between the rainfall and the properties of the upper portions of the clouds. The skill of passive MW techniques is also reduced over land, because its large emissivity reduces greatly the usefulness of frequencies lower than 35 GHz. Nevertheless, results over land at 88.5 GHz reported by Spencer *et al.* (1989) are encouraging.

All techniques suffer to a greater or lesser degree from errors arising from sampling. As already noted, these errors can be greater than all the other errors if accumulations are improperly computed. In tropical regions there can be a significant diurnal cycle in rainfall activity, and the phase and intensity of the cycle may increase the errors due to sampling.

Ground-based radar offers higher accuracy than satellite techniques over the same time–space domain within the quantitative range of the radar (up to about 150 km range). However, when melting snow occurs in the radar beam the performance is significantly degraded. Whilst the probability matching method offers the prospect of reliable adjustment of radar data, the success of this technique may depend upon the level to which physical limitations in the data are minimized before adjustment is undertaken. Rain gauges remain the best method of estimating point rainfall. Finally it is most important that techniques using remotely sensed data are tested against numerical model products in a region with good truth measurements. There is little point in persisting with a method which does not outperform operational models.

Where high spatial and temporal resolution are important for operational management purposes, radar, even though it is more expensive, offers significant advantages. Moreover, radar data can be used to satisfy a variety of other customer needs; e.g., they can be used for aviation severe weather warnings.

Maps of global precipitation

Susskind (1993) reviews the combined use of sounding data from the HIRS (High Resolution Infra-red Radiation Sounder) and the MSU (Microwave Sounding Unit) instrument flying on the TIROS series polar orbiting satellites to prepare estimates of global precipitation distribution. Figure 4.4 shows the mean global precipitation distribution so produced for April and July 1979. Work is on-going to develop the use of the microwave frequencies on the Special Sensor Microwave/ Imager (SSM/I) instrument, which promises further improvements over both land and ocean areas.

However, these data are adversely affected by the limited temporal sampling the satellite orbit provides, soundings only being obtained four times per day. In addition, only a radiatively effective cloud fraction is obtained, and use of sounding data may, in some circumstances, provide estimates which are only a little better, or perhaps even worse, than those derived using a precipitation index derived from geostationary satellite data. Estimates of this kind, obtained for tropical rainfall by Arkin and Janowiak (1993), are also shown in Figure 4.4. Note in this figure that the HIRS/MSU estimates in the tropics are somewhat lower than those made using the GPI technique.

The TRMM satellite (Simpson *et al.*, 1988) aims to improve

(a)

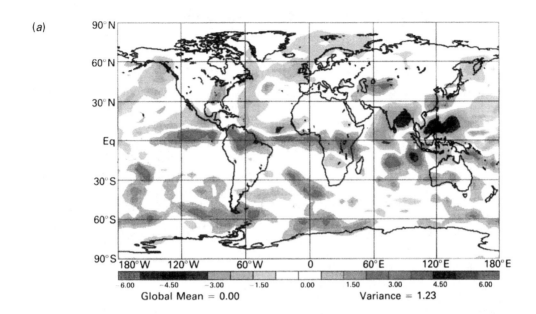

(b)

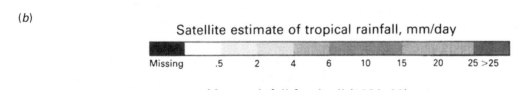

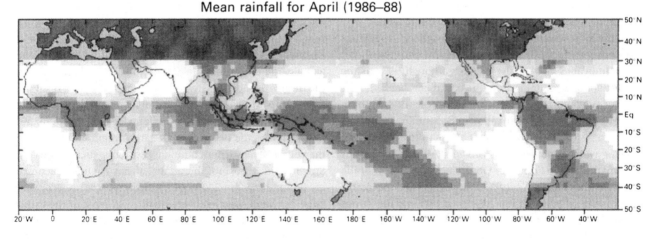

Figure 4.4. Global precipitation estimates (mm day⁻¹) (a) using HIRS/MSU data, April 1979 (from Susskind, 1993); (b) using geostationary infrared Global Precipitation Index data for April 1986–88 (from Arkin and Janowiak, 1993).

estimates of tropical rainfall using complementary data from visible, infrared and passive and active microwave satellite instruments (see above). Work is also under way to assess the feasibility of an active radar system on a polar orbiting satellite.

Whilst satellite estimates of global precipitation are not yet reliable in all circumstances, they do now provide data which can considerably enhance precipitation climatologies over the oceans derived solely from ground-based measurements (Arkin and Ardanuy, 1987) which are often used to assess the performance of global numerical weather prediction models in reproducing climatology.

4.2 Air–sea fluxes and their estimation

Peter K. Taylor and Kristina B. Katsaros

Introduction: Importance of the air–sea fluxes

The ocean absorbs over three-quarters of the shortwave (0.3 to 3 μm wavelength) radiative energy from the sun which reaches the Earth's surface. This absorption takes place in the upper few meters of the ocean (e.g., Paulson and Simpson, 1981). If the wind is less than about 3 m s^{-1}, diurnal sea surface temperature changes of a few kelvin can occur (Price *et al.*, 1986) but normally turbulence distributes the heat down through tens of meters, to the depth of the seasonal thermocline. Thus, in contrast to land surfaces, only a fraction of the absorbed heat is available for immediate, local transfer to the atmosphere either through the exchange of longwave radiation (3 to 50 μm wavelength), or by the sensible and latent heat fluxes (the turbulent transfer of heat, and of water vapor with its latent heat of evaporation). Annual average solar heating is greatest at latitudes less than about 30° and, together, the North Atlantic and Pacific Oceans transport about 2 PW (2×10^{15} W) polewards through latitude 24° N (Bryden, 1993; Trenberth and Solomon, 1994). The transfer of this heat to the atmosphere at higher latitudes is an important component of the Earth's climate system. For those of us who live in northern Europe it has a very tangible effect on our lives!

The mean freshwater flux into the atmosphere is the difference between evaporation and precipitation, and contributes to the buoyancy flux in the ocean. Although 90% of the freshwater that evaporates from the ocean falls back into the ocean as precipitation, the remainder represents about one-third of the terrestrial precipitation (e.g., Schmitt, 1994).

Our present knowledge of these air–sea fluxes is derived from the weather observations provided by merchant ships participating in the Voluntary Observing Ship system of the World Weather Watch. A subset of these ships and the instrumentation used on them was described by Kent and Taylor (1991). For climate studies these data have been organized into the Comprehensive Ocean-Atmosphere Data Set, COADS (Woodruff *et al.*, 1987). We will discuss the accuracy of the ship observations and the formulae used to calculate the fluxes. A major disadvantage of the ship observations is that adequate coverage is obtained only in areas that are well covered by shipping lanes. Sampling studies (e.g., Legler, 1991; Cayan, 1992*c*) suggest that while much of the northern oceans (the North Atlantic, North Pacific, and Mediterranean) is reasonably well sampled, coverage in the tropical oceans is marginal at best, and the Southern Ocean is not adequately sampled by ships.

Satellite-based instruments provide estimates of all the parameters used to calculate the air–sea energy and water fluxes from ship observations except for the near surface air temperature, while some fluxes may be directly estimated from satellite measurements. Alternatively satellite and *in situ* data may be assimilated into numerical weather forecasting models and the model flux values used. Such methods provide a more uniform global coverage, albeit of varying accuracy. We will discuss the potential of these new data sources, but first we discuss the magnitude and characteristics of the ocean surface heat and water flux in order to put their role in the total energy and water cycles of the planet in perspective.

The global distribution of air–sea fluxes

Mean heat flux distributions

The total heat flux through the ocean surface, F_{tot}, is the sum of the sensible and latent heat fluxes, F_H, F_E, and the net shortwave and longwave radiative fluxes F_{SW}, F_{LW}. Estimates of the global distribution of air–sea heat fluxes have been computed by various authors (e.g. Budyko, 1963; Esbensen and Kushnir, 1981; da Silva *et al.*, 1994). Figure 4.5 shows the global variation of the total flux in January and and July. In the tropics the ocean is heated throughout the year, resulting in an annual mean value of several tens of W m^{-2} along the Equator and in the northern Indian Ocean. At middle and high latitudes the ocean is heated in summer and cooled in the winter. Over western boundary currents, such as the Gulf Stream and Kuroshio, the annual mean cooling reaches a maximum of 100 to 200 W m^{-2}; however, these areas of very large flux values are relatively small. Over most of the extra-tropical ocean, annual mean flux is less than 30 W m^{-2} and, in many areas, the sign of this net flux is uncertain. Nevertheless, because of the large areas involved, the total heat transferred by these small net fluxes is potentially important. In the North Atlantic, the line of zero net heat flux extends from Central America to Europe with the large cooling rates to the north extending across the ocean to the Norwegian Sea.

The role of the individual flux components is illustrated in Figure 4.6 which shows the annual net flux and the January and July flux values for three representative sites in the North Atlantic. Site A at about 0° N 20° W is 'Equatorial' in the region of net annual heating. Variations in the fluxes occur as the

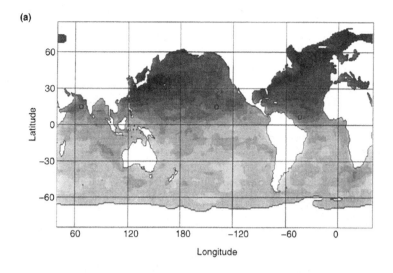

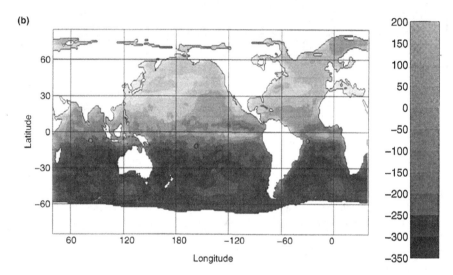

Figure 4.5. Distribution of total heat flux into the ocean (W m⁻²) for (*a*) January, and (*b*) July
(see Josey *et al.*, 1996). The darker values represent ocean cooling and the zero flux isopleth is
marked (contour interval: 50° W m⁻²).

Intertropical Convergence Zone migrates north or south; flux values are somewhat smaller in January compared with July. However, at this site the surface heating, and the partition between the different fluxes, remains substantially the same throughout the year. The 'Mid-latitude' site B (situated near the Azores at 40° N 20° W) is typical of the large region where contrasting winter and summer conditions result in a small net surface flux. In winter the solar heating is small and the cooling by sensible, latent, and longwave fluxes dominates. In summer the insolation is large, being similar to the Equatorial Site (A) and, despite continued surface cooling, the net flux is into the ocean. In fact, similar summer warming of order 100 W m⁻² occurs over almost all the North Atlantic

since further north, where the insolation is reduced, the turbulent fluxes are also less. Finally, the Gulf Stream site C (about 40° N 60° W) is typical of a western boundary current region where large annual net cooling occurs. Even here there is net warming of the ocean in summer, similar in magnitude to that at the Mid-latitude Site (B). The large net annual cooling in this region is the result of exceedingly large turbulent fluxes in winter when cold dry air from North America flows over the warm Gulf Stream waters.

The comparative importance of the various components of the total heat flux is clearly seen from Figure 4.6. The net longwave flux varies between about 30 W m⁻² under cloudy skies to about 80 W m⁻² for clear skies. Except over the Gulf

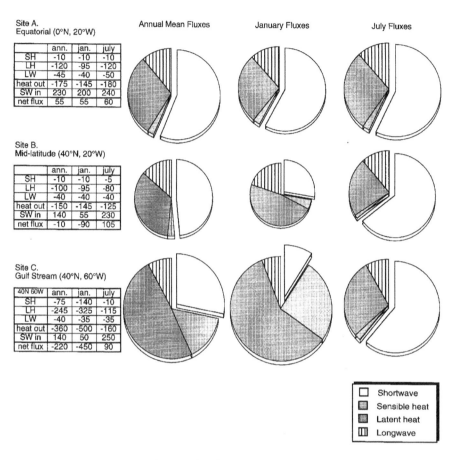

Site A.
Equatorial (0°N, 20°W)

	ann.	jan.	july
SH	-10	-10	-10
LH	-120	-95	-120
LW	-45	-40	-50
heat out	-175	-145	-180
SW in	230	200	240
net flux	55	55	60

Site B.
Mid-latitude (40°N, 20°W)

	ann.	jan.	july
SH	-10	-10	-5
LH	-100	-95	-80
LW	-40	-40	-40
heat out	-150	-145	-125
SW in	140	55	230
net flux	-10	-90	105

Site C.
Gulf Stream (40°N, 60°W)

40N 60W	ann.	jan.	july
SH	-75	-140	-10
LH	-245	-325	-115
LW	-40	-35	-35
heat out	-360	-500	-160
SW in	140	50	250
net flux	-220	-450	90

Annual Mean Fluxes January Fluxes July Fluxes

☐ Shortwave
▨ Sensible heat
▦ Latent heat
▥ Longwave

Figure 4.6. Pie charts showing the annual and January and July mean heat fluxes at three representative sites in the North Atlantic Ocean (data based on Isemer and Hasse, 1987). Site A: Equatorial (0° N, 20° W); Site B: Mid-latitude (40° N, 20° W); Site C: Gulf Stream (40° N, 60° W). The area of the pie chart slices is approximately proportional to the magnitude of the flux components. The surface heating occupies the slice extending clockwise from the top (12 o'clock) position; the surface cooling components are plotted anticlockwise from the top. The magnitude of the net surface heating or cooling is represented by the degree to which the cooling or heating slices extend beyond the 6 o'clock position.

Stream, the latent heat flux is typically two to three times this longwave cooling. The sensible heat flux is typically one-tenth of the latent flux. Again, an exception is the Gulf Steam in winter where the air temperatures are particularly cold. Cold water has a low saturation vapor pressure resulting in weaker forcing of the latent heat flux even when the air is very dry, so that sensible heat flux becomes proportionally more important for cold ocean regions.

To summarize, at most oceanic sites, the net annual flux is the difference between contrasting summer and winter conditions, and the net flux in any month is the residual of individual fluxes which are an order of magnitude greater. To obtain the mean total heat flux between atmosphere and ocean requires accurate and precise estimates of each flux value.

Mean freshwater flux distribution

Determination of the freshwater flux requires knowledge of the precipitation. The precipitation is not measured on ships but is determined from the 'present weather' descriptions reported in the ships' weather observations, a method developed by Tucker (1961). Each of the weather codes which relate to precipitation has an associated formula which describes the precipitation rate in terms of the likely contribution from three continuous rain categories: light, moderate, or heavy. The relationship between these categories, and the rain rate which they represent, was found empirically. Global estimates derived by Baumgartner and Reichel (1975) were recently used by Wijffels *et al.* (1992) in a study of the global freshwater

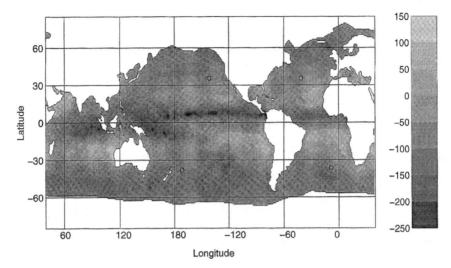

Figure 4.7. A map of evaporation minus precipitation over the global ocean (Josey *et al.*, 1996). The darker values represent excess of precipitation over evaporation and the zero freshwater flux isopleth is marked (contour interval: 50 cm yr^{-1}).

balance for the ocean. Dormand and Bourke (1978) adapted Tucker's scheme for tropical regions (by taking temperature into account) and produced a more detailed precipitation analysis for the Pacific Ocean (Dorman and Bourke, 1979) and for the North Atlantic (Dorman and Bourke, 1981). The latter was used by Schmitt *et al.* (1989) to determine the freshwater transport. Unfortunately, the accuracy of all these 'present-weather' based precipitation estimates is likely to be poor and there is no good standard with which to assess them. A new optical rain gauge for use on buoys was developed for the Tropical Ocean Global Atmosphere (TOGA) experiment (McPhaden, 1993) and this, or similar techniques, may eventually provide a calibration standard.

Allowing for the uncertainty in the precipitation estimates, a global estimate of the net air-sea water flux is shown in Figure 4.7. Precipitation dominates in the tropical regions and at high latitudes. Evaporation dominates in mid latitudes. Large horizontal gradients occur, particularly where evaporation in the trade wind areas creates a moist boundary layer which feeds water vapor into the deep convection of the Intertropical Convergence Zones.

Variability of the fluxes

For many climate studies, knowledge of the interannual variability of the fluxes is required. Because of the large number of ship reports, most recent studies have considered the northern oceans, particularly the North Atlantic Ocean. Michaud and Lin (1992) suggested that interannual variability tended to be masked by the long-term error trends in the

COADS data; for example the changing bias in the wind data. However, Cayan (1992*a,b,c*) described the connection between heat flux anomalies, atmospheric circulation, and sea surface temperature (SST) patterns. Deser and Blackmon (1993) used Empirical Orthogonal Function analysis to identify a quasi-decadal dipole oscillation in SST and surface air temperature between the east coast of Canada and the southeast coast of the USA. Gulev (1995*a,b*) also found a bimodal variation ithe meridional heat transport on decadal scales. Norris and Leovy (1994) described the correlation between SST anomalies and interannual variations in stratiform cloudiness. Studies of other oceans include the moisture variations associated with el Niño in the tropical Pacific Ocean (Weare, 1984), the Mediterranean Sea where Garrett *et al.* (1993) found an interannual variability of 15 W m^{-2} in the total heat flux, and the Indian Ocean for which Jones *et al.* (1993) have recently analyzed a new set of flux estimates.

Determination of the surface fluxes using ship data

The parameterizations for the radiative fluxes

Since it has generally not been considered practicable to equip large numbers of merchant ships with radiometers, radiation estimates are based on the cloud observations reported by ships' officers (for a recent review see Katsaros, 1990). For shortwave insolation the Reed (1977) formulae for the daily mean net shortwave flux F_{sw} (equation 4.2.1) was recommended by the comparative studies of Frouin *et al.* (1988*a*) and Dobson and Smith (1988). The latter found a site-dependant long-term bias of -1 to $+12$ W m^{-2} and a monthly mean rms error of about 8 W m^{-2}. However, using comparisons with data from several air–sea interaction experiments, Katsaros (1990) found this formula to be biased high by about 20 W m^{-2}. The Reed (1977) formula is:

$$F_{SW} = F_{SW_0}(1 - c_n n + 0.0019\phi)(1 - \alpha) \qquad (4.2.1)$$

where F_{SW_0} is the shortwave insolation at the surface under clear skies (Seckel and Beaudry, 1973), $c_n (= 0.62)$ the cloud attenuation factor which is applied to the fractional cloud cover n, and ϕ the noon solar elevation in degrees. Almost all studies use the albedo, α, given by Payne (1972). Equation 4.2.1 should only be used for $0.3 \leq n \leq 1$ and $F_{SW} = F_{SW_0}$ should be used for $n < 0.3$ (Gilman and Garrett, 1994). Seckel and Beaudry (1973) stated that in low latitudes their value for F_{SW_0} is similar to (but not calculated from) the Smithsonian Table (List, 1963) value at the top of the atmosphere, reduced by a transmission factor of 0.7. Da Silva *et al.* (1994) followed the alternative method of Rosati and Miyakoda (1988) which explicitly calculates the direct and diffuse components of F_{SW_0}, assuming a constant transmission factor of 0.7. However, this may not be optimal for all latitudes.

For estimating longwave radiation Katsaros (1990) found similar performance amongst the models compared. Biases were generally less than 10 W m^{-2} and the rms scatter significantly less than the shortwave parameterizations. However, a comparison by Gilman and Garrett (1994) suggested that, while the formulae tested agreed to about ±3 W m^{-2}, all gave a lower value of F_{LW} for the Mediterranean by about 17 W m^{-2} compared with a calculation using the Bunker (1976) formula (with seasonally varying cloud types). The latter appeared to be more compatible with balancing the heat budget (Garrett *et al.*, 1993). Similarly, a comparison by Schiano *et al.* (1993) found that almost all the formulae tested underestimated F_{LW} by 25 to 40 W m^{-2}. Based on these results, Bignami *et al.* (1995) proposed a new formula which was claimed to have negligible bias and an rms error for daily averages of 9 W m^{-2}. However, these Mediterranean results may not be valid in other areas (Josey *et al.*, 1997).

The bulk formulae for sensible and latent heat

The fluxes are calculated from the meteorological observations using the 'bulk formulae'. These can be derived from the relationships between the surface flux of a quantity and its vertical profile, predicted on the basis of dimensional analysis and boundary layer similarity theory (see for example, Geernaert, 1990a). Thus the fluxes of momentum (wind stress, τ), sensible heat (F_H), and latent heat (F_E), are calculated from:

$$\tau = \rho C_D(u - u_0)^2 \qquad (4.2.2a)$$

$$F_H = \rho C_H(u - u_0)(t - \gamma z) - t_0) \qquad (4.2.2b)$$

$$F_E = \rho C_E(u - u_0)(q - q_0) \qquad (4.2.2c)$$

where u, t, and q are the observed wind speed, air temperature, and specific humidity, and u_0, t_0, and q_0 are the corresponding values at the sea surface; ρ is the air density. The air temperature must be corrected for the adiabatic lapse rate, γ, between the surface and the observation height z. The value of the transfer coefficients, C_D, C_H, and C_E will be discussed below; taking humidity as an example, the Dalton number, C_E, may be defined by:

$$C_E = k^2\left(\ln\left(\frac{z_q}{z_{0_q}}\right) - \Psi_q\right)^{-1}\left(\ln\left(\frac{z_u}{z_0}\right) - \Psi_m\right)^{-1} \qquad (4.2.3)$$

where (for humidity and momentum respectively) z_q, z_u are the measurement heights, z_{0_q}, z_0 are roughness lengths, and Ψ_q, Ψ_m are stratification functions which depend on the stability (the ratio of the observation height to the Monin–Obukhov length) and the values assumed for the von Karman constant k (Frenzen and Vogel, 1994). Using marine data, (Edson *et al.*, 1991) present a comparison of different forms for the stratification functions which suggests that stability effects can be adequately accounted for in applying the bulk formulae. Under neutral conditions the stratification functions Ψ_q, $\Psi_m = 0$, and the 10-m neutral value of the transfer coefficient, C_{E10n} may be defined by equation (4.2.3) by setting $z_q, z_u = 10$.

Measured values for the transfer coefficients

The transfer coefficients are determined by air–sea interaction experiments in which the flux is measured by the eddy correlation or inertial dissipation method combined with accurate observations of the mean meteorological variables (e.g. Smith, 1989). Recent reviews are provided by Geernaert (1990b) and Smith (1988).

For open ocean wind stress, the formulae used in many studies were obtained by Large and Pond (1981, 1982) or Smith (1980, 1988). Recently Yelland *et al.* (1997) used a very large data set from the Southern Ocean to confirm the Smith (1980) formula for open-ocean conditions. As the wind decreases below about 2 m s^{-1}, Smith (1988) suggested that C_D would increase due to viscous effects. However, the significantly greater C_D values for winds below about 5 m s^{-1} predicted by Wu (1994) appear to be a better fit to the data of Bradley *et al.* (1991) and Yelland and Taylor (1996).

For latent heat, Smith (1989) reviewed previous studies and suggested a constant Dalton number ($10^3 C_{E10n} = 1.2 \pm 0.1$) for winds between 4 and 14 m s^{-1}. The Humidity EXchange Over the Sea (HEXOS) experiment results (DeCosmo *et al.*, 1996) also suggest a near constant value (any increase being less than 15–20%) with ($10^3 C_{E10n} = 1.12 \pm 0.24$) for winds up to 18 m s^{-1}. Given that the drag coefficient, and hence the roughness length, increases with wind speed, a constant value for the Dalton number implies (equation 4.2.3) that the humidity roughness length decreases with increasing wind speed. In explanation, Liu *et al.* (1979) suggested that, whereas momentum is transferred by pressure differences as well as by molecular forces, heat and water are transferred only by molecular diffusion which becomes less efficient as the roughness increases and sheltering occurs between the wave troughs.

For sensible heat, determination of the Stanton number over the ocean has been hindered by salt contamination of the sensors (e.g. Katsaros *et al.*, 1994). Smith (1988) suggested ($10^3 C_{H10n} = 1.0$) which corresponds with the suggestion of Friehe and Schmitt (1976), based on the ratio of the thermal to species diffusivities (Prandtl/Schmidt numbers), that ($C_E / C_H = 1.16$). However, the HEXOS results (DeCosmo *et al.*, 1996) suggest ($C_E \approx C_H$) to the accuracy of the determination.

The transfer coefficients, C_{E10n}, C_{H10n}, have been found to increase as the wind speed decreases below about 2 m s^{-1} (Bradley *et al.*, 1991) as predicted by Liu *et al.* (1979). At very high wind speeds the heat fluxes are expected to be modified by the evaporation of spray. Andreas (1992) and Andreas *et al.* (1995) suggested that, whereas the turbulent fluxes are proportional to wind speed, spray production might be expected to increase at roughly the third power of the wind speed. Thus they calculate that for wind speeds up to 15 m s^{-1} the spray contributes no more than 10% of the total heat flux; however, at 20 m s^{-1} and above the spray contribution is of similar magnitude to the surface turbulent flux. However, the feedback on the surface evaporation due to changes in the mean surface humidity profiles caused by the evaporating spray was not fully included. This may substantially modify the estimated flux (Katsaros and deLeeuw, 1994; Andreas, 1994). The fluxes of heat and water vapor have not yet been measured in the field for winds over 20 m s^{-1}.

Errors in the observations

If the heat fluxes are to be obtained to 10 W m^{-2} then, from equation (4.2.2), the required order of accuracy for the observations is about $\pm 2\,°C$ for the SST, dry and wet bulb temperatures (or about 0.3 g kg^{-1} for specific humidity) and the winds should be estimated to $\pm 10\%$ or better, say about 0.5 m s^{-1} (see, for example, Taylor, 1984, 1985). Compared with these stringent standards, changes in instrumentation can produce significant but spurious 'climatic trends' in the data. For example, an apparent increase in the mean wind speeds (Ramage, 1987; Wright, 1988) has been attributed to the increasing proportion of wind estimates obtained from anemometer readings rather than visual ('Beaufort') estimates (Cardone *et al.*, 1990), to a varying mix of observer nationalities (Isemer, 1992), or to the changing interpretation of visual estimates (Ward, 1995). The trend does not occur in weather ship wind data (Isemer, 1995) and correction schemes have been suggested (Ward, 1995). There is a continuing effort to accurately calibrate visual wind estimates against anemometer winds (Lindau, 1995; Kent and Taylor, 1997) since the wind scale used can significantly affect heat flux estimates (Isemer and Hasse, 1991; Young *et al.*, 1995) and different wind stress climatologies can produce significantly different estimates of ocean circulation (Boening, 1995). Removing the trends in sea surface temperature records has similarly been the subject of detailed research effort (see for example, Parker and Folland, 1995).

In the VSOP-NA (Voluntary Observing Ship Special Observing Programme – North Atlantic) project (Kent *et al.*, 1993a), a significant effort was made to determine the instrumentation used on each of the participating merchant ships. The results were then analyzed according to instrument type and exposure, ship size and nationality, and other factors. Correction algorithms were devised; for example the day-time error in dry bulb temperature values was found to depend on the solar radiation and relative wind speed (Kent *et al.*, 1993b). The uncorrected data were found to underestimate the net turbulent transfer of heat from sea to air by about 8%. However, comparison with weather ship data suggested that a larger net flux correction of 15% might be needed. These estimated errors were smaller than first expected, partly because some of the errors cancel (Figure 4.8).

Balancing the ocean heat budget

In applying the bulk formulae (equation 4.2.2) to meteorological observations from merchant ships, various approximations are normally made. It is assumed that $u_0 = 0$ which ignores any sea surface current; t_0 is taken to be the sea temperature measured by the ship, and q_0 the specific humidity at 98% saturation (to allow for salinity effects) at that sea temperature, thus ignoring the surface skin effect (see below) and any near surface thermocline. Since the measurement heights are normally not known, a constant value is used (or 10 m implied by default). The lapse rate correction to the air temperature is frequently neglected. In some studies the fluxes have been calculated from monthly mean values of the meteorological variables; the resulting biases (Esbensen and Reynolds, 1981; Simmonds and Dix, 1989) may vary in sign and magnitude both geographically and seasonally, being of the order of 10% for the latent heat flux (Gulev, 1994; Josey *et al.*, 1995).

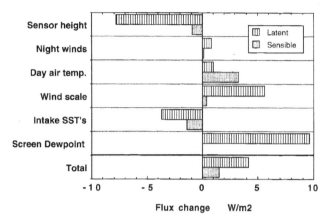

Figure 4.8. Change in the calculated annual mean fluxes for the North Atlantic due to correcting for errors in the ship observations: effect of sensor height; bias in visual winds at night; daytime solar heating of ship; visual wind conversion scale; bias in engine room intake SST data; underestimate of dewpoint depression using thermometer screens. (After Kent and Taylor, 1995.)

To allow for these various biases, and in an attempt to correct for observation errors, the transfer coefficients used to calculate global heat flux fields have usually been increased above the values found from air–sea interaction experiments. Often the magnitude of these increases has been determined by applying a constraint on the area integrated fluxes. For example, Bunker and Worthington (1976) used transfer coefficients increased by 10% from the then accepted experimental values to allow for measurement errors. In recalculating Bunker's fluxes for the North Atlantic, Isemer and Hasse (1987) initially chose lower values for C_E compared with Bunker and Worthington (1976) but their consideration of the implied meridional heat flux (Isemer et al., 1989) led to a revision upwards. Oberhuber (1988) increased his transfer coefficients for similar reasons. Kent and Taylor (1995) compared these various flux calculation methods by applying each to the VSOP–NA data set. Compared with values calculated with the Smith (1988) transfer coefficients and uncorrected ship observations, they showed that the increase in the fluxes ranged from 21% (Oberhuber, 1988), through 31% (Isemer and Hasse, 1987) to 35% (Bunker, 1976), values significantly greater than the 6% to 15% increase justified by the ship measurement errors. Esbensen and Kushnir (1981) did not explicitly increase their flux estimates, although their use of the skin transfer coefficients of Liu et al. (1979) with the bulk sea temperatures was equivalent to a 16% increase.

For the Mediterranean and Red Seas, the advective oceanic heat transport can be estimated, allowing all terms in the oceanic heat budget to be determined. Bunker et al. (1982) found that the use of the Bunker and Worthington (1976) transfer coefficients did not balance the heat budget and suggested that an increase in the calculation of turbulent flux was required. Compared with estimates derived using the Smith (1988) transfer coefficients this increase would be equivalent to about 75% extra. However, Garrett et al. (1993) demonstrated that, considering both heat and freshwater budgets, it was more likely that the insolation was being overestimated. Gilman and Garrett (1994) related this bias to the atmospheric aerosol content and suggested that solar insolation should be reduced by 20 W m^{-2} (10%); they also calculated an increased longwave cooling (by 10 W m^{-2} or about 14%). The residual heat imbalance suggested that, at most, the sensible and latent heat flux estimates were underestimated by about 6% which is compatible with the likely measurement errors.

Da Silva et al. (1994) present global flux calculations, modified by various constraints on the heat and freshwater budgets. The required flux adjustments are almost independent of the constraints applied. Insolation is decreased by about 10%, longwave cooling increased by about 2%, latent heat flux is increased about 14%. However, recent work on the Josey et al. (1996) climatology suggests that a uniform global correction may not be appropriate. Relatively larger corrections to the surface cooling may be required in the poorly sampled Southern Ocean region.

Thus, in summary, a consistent picture appears to be emerging. Biases in the parameterization of radiation (both long and shortwave) may have resulted in overestimation of the heat flux into the ocean. On that assumption, the correction needed to constrain the heat and water fluxes implies that the sensible plus latent heat flux has been underestimated by up to 30 W m^{-2} in the mean. This bias is compatible with recent estimates of the likely errors in the measurements, particularly in poorly sampled regions; it is less than the bias assumed in many previous studies.

Use of satellite data for air–sea flux determination

Introduction

Satellite sensors have the potential for providing consistent observations over the global ocean. Although the sampling density is far superior to the ship observations over most ocean regions, sampling studies (Salby, 1982a,b) suggest that wide swath sensors flown on a pair of polar orbiting satellites are needed to provide routinely the coverage sought (Taylor, 1984). The next subsections will review the present status of satellite measurements; it can be summarized as follows. Satellite data have been in routine use for determining the SST field for a number of years. Measurements of wind and waves from altimeters and scatterometers are being used in climate studies. A shortwave radiation product is now available. Techniques exist for longwave radiation, precipitation, and near surface humidity (and hence the latent heat flux) but improvements are still needed. A method to obtain sensible heat flux estimates remains to be developed.

Sea surface temperature

The most accurate remotely sensed sea surface temperature data are obtained using infrared (IR) radiometers to measure the total IR emission from sea and atmosphere. The radiative temperature of the sea surface is estimated by comparing the measured signal at different wavelengths. McClain *et al.* (1985) found an accuracy (rms scatter) for Advanced Very High Resolution Radiometer (AVHRR) data of about ±0.6°C. Improvements to the retrieval algorithms have since been implemented; for a detailed global comparison of different AVHRR algorithms see Wick *et al.* (1992). However, the review of Barton (1995) suggests that the global accuracy of the AVHRR data remains little changed, with the error budget dominated by the instrument calibration accuracy and atmospheric transmission effects. The most reliable global data set remains a combination of satellite and *in situ* data (e.g., Reynolds and Smith, 1994).

Compared with the AVHRR, the Along-Track Scanning Radiometer (ATSR) has improved sensor stability and built-in calibration systems. It provides views of the sea surface at two different incidence angles and hence the potential to estimate atmospheric transmission effects; Zavody *et al.* (1994) describe the instrument and the data processing procedures used. In comparisons with drifting buoy data, Mutlow *et al.* (1994) found that the least rms difference was ±0.36°C achieved by an algorithm which used the three thermal infrared channels (3.7, 10.8 and 12.0 μm) and both the nadir and 55° forward view. For this algorithm the mean difference (ATSR − buoy) was − 0.03°C. However, this apparent agreement implies a bias; since the ATSR measures the skin temperature and the buoy measurements are at 1 m depth, the difference should be equal to the mean skin effect, about − 0.3°C (Schluessel *et al.*, 1990). A similar warm bias in the ATSR data was found by Barton *et al.* (1995) using comparisons with ship-borne radiometer data. Recently Harris *et al.* (1995) have claimed an rms scatter as small as ±0.25°C for an algorithm that only uses the 3.7 and 11 μm nadir view data. However this and other algorithms using the 3.7 μm channel are limited to night time because of contamination by scattered shortwave radiation during the day. This restriction is potentially serious for ATSR which only has a 500 km swath and therefore provides fewer cloud-

clear views of the sea surface compared with AVHRR (2700 km swath).

Despite the possible bias and limited swath, the accuracy of the ATSR data now appears to be sufficient to determine the relative biases between the UK Meteorological Office analysis (based mainly on ship data) and the drifting buoys (Mutlow *et al.*, 1994). The ships appear to be biased relatively high by about 0.3°C (night) to 0.55°C (day). Indeed, the magnitude and variability of the surface skin effect (typically a few tenths °C) are now critical factors in determining the accuracy of the ATSR SST data. The air–sea heat fluxes at any instant depend on the skin temperature. However, because the heat capacity of the surface skin is very small, there are advantages in considering the skin temperature to be a function of the bulk temperature and the surface fluxes. Different algorithms for predicting the skin effect were tested by Kent *et al.* (1995). The Saunders (1967) model predicted an increase of the skin effect with surface cooling, a decrease with increasing wind speed, and a lack of dependence on solar heating. Compared with an observed skin effect variation of nearly ±0.3°C, the preferred formulae (Saunders, 1967, or modifications of the Grassl, 1976 or Soloviev and Schluessel, 1994) achieved a scatter of about ±0.2°C. Although a significant part of this scatter would have been due to errors in the observations, improvements to the formulae are still needed. Special consideration must also be given to the situation of low wind speed and high insolation when the cool-skin effect is masked by the formation of a near surface thermocline (e.g., Katsaros, 1980; Soloviev and Vershinsky, 1982; Price *et al.*, 1986) whose depth and temperature offset is not readily predicted from meteorological observations alone.

Winds and waves

Wind velocity can be estimated using a satellite scatterometer. This instrument emits microwave radiation and measures the level of backscatter from the sea surface which is related to the surface roughness and thus varies with wind speed. Three or more antennae are used to illuminate sequentially the sea surface from different angles and hence obtain a wind direction estimate. For the European Remote Sensing (ERS) satellite series, significant effort has been invested in improving the C-band (5.3 GHz) scatterometer data products which are calculated off-line (Bentamy *et al.*, 1994). The C-band model has been fitted empirically to buoy wind data (Quilfen, 1994). For wind speed the bias was less than 10 cm s^{-1} and the rms difference was 1.2 m s^{-1} over a range of 2 to 20 m s^{-1}. For wind direction the rms difference was 15°. The scatterometer thus offers great promise for global wind velocity determination. The ERS scatterometer samples a 500 km swath with 50 km spatial resolution. That on the Advanced Earth Observing System (ADEOS) satellite had two swaths, one to either side of the satellite, and provided 25 km resolution.

Estimation of the surface heat fluxes using the bulk formulae requires an estimate of the wind speed but not the direction. Because the natural microwave emissions from the sea surface vary with, for example, the amount of foam cover, wind speed estimates may be obtained from microwave radiometer data. The Special Sensor Microwave/Imager (SSM/I) on satellites in the US Defense Meteorological Satellite Programme samples 1400 km swaths in conical scans across the ground track with 50 km resolution. Currently two satellites carry these sensors operationally. A disadvantage of the sensor is that the wind estimates are contaminated by thick clouds and heavy rain (e.g., Mognard and Katsaros, 1995).

Estimates of wind speed and significant wave height, H_s can be obtained from radar altimeters. These variables are obtained by determining the power of the reflected radar signal and the shape of the reflected pulse; Brown (1990) describes the physics involved. Data are obtained along a narrow swath at nadir with a footprint of about 10 km. Wind speed estimates from the ERS-1 altimeter have been verified by Queffeulou et al. (1994). The ERS-1 values were 5% lower than colocated buoy data over the range 1 to 15 m s^{-1} with a 1.5 m s^{-1} standard deviation. However, large wind speed differences (up to 6 m s^{-1}) sometimes occurred which were not entirely due to wind speed variability, and a correlation with H_s was found.

H_s estimates from satellites are consistent with buoy data, once biases have been removed using linear corrections (Carter et al., 1992; Cotton and Carter, 1994a; Queffeulou et al., 1994). The data are now being used for wave climate studies: for example, for the Southern Ocean (Campbell et al., 1994; Josberger and Mognard, 1996) and the global ocean (Cotton and Carter, 1994b).

Shortwave and longwave radiation

Radiometers are used to measure the radiation intensity at the orbit height of the satellite at a number of different wavelengths. Based on this information, and knowledge of the radiative characteristics of different atmospheric constituents, the emission and absorption of radiation by the atmosphere is calculated and hence the surface radiative fluxes are estimated. The review of Katsaros (1990) concluded that the surface shortwave irradiance can be estimated from geostationary satellite radiance data to about 10% accuracy for daily averages. The WCRP Surface Radiation Budget (SRB) climatology project (Whitlock et al., 1993) has chosen two different shortwave algorithms: one a simplified optical physics model, the other an iterative radiative transfer model.

The difference between the results is considered an indication of their accuracy. Over the ocean the difference in the average net shortwave flux is generally less than 10 W m^{-2} although locally they reach 25 W m^{-2}. Comparison with limited ground truth suggests a bias of less than 10 W m^{-2} (WCRP, 1993). Achieving these accuracies depends on the quality of the satellite calibration which has varied with time (Whitlock et al., 1993).

Surface longwave irradiance can be retrieved from temperature and humidity profiles estimated using the vertical sounders on operational, polar orbiting satellites (Darnell et al., 1983), possibly combined with geostationary visible and infrared radiometer data for the cloud information (Frouin et al., 1988b). Katsaros (1990) suggests that the limitation in these techniques is the accuracy of the TOVS profiles, and discusses other methods of estimating longwave radiation, for example from satellite cloud cover estimates (Frouin et al., 1988a). Progress in longwave estimation for the SRB project is being limited by the uncertainty in the available ground truth measurements (WCRP, 1993).

Precipitation

Estimation of precipitation is considered in section 4.1. Estimation methods available using satellite data have also been reviewed by Arkin and Ardanuy (1989). They include Highly Reflective Cloud (HRC), Outward Longwave Radiation (OLR), and passive microwave techniques. HRC and OLR methods are applicable to tropical convective rainfall. Passive microwave data have been used to produce a global climatology (Chang et al., 1993); however, it is less accurate in tropical regions unless calibrated against infrared methods (Huffman et al., 1993).

Near surface humidity

Passive microwave radiometers can determine the total atmospheric water vapor content by measuring the microwave intensity at two or more chosen frequencies. The determination of near surface atmospheric humidity from estimates of total water vapor was first suggested by Liu and Niiler (1984). This method relies on there being, at least over a monthly mean, a correlation between the total water vapor and the humidity content of the surface boundary layer. It has recently been applied to SSM/I data to calculate evaporation over the global oceans (Esbensen et al., 1993). Biases of the order of 2 g kg^{-1} were found due to the variations in the mean atmospheric humidity profile (see for example, Taylor, 1982, who reviewed earlier attempts to estimate total water vapor from near surface humidity). For calculating global latent heat flux, Jourdan and Gautier (1995) achieved improved accuracy by blending ship and satellite estimates. A promising advance is the development by Schulz et al. (1993) of an algorithm to estimate directly the water content in the lower 500 m of the atmosphere from SSM/I data. Since this layer normally lies within the surface mixed layer over the ocean, the humidity content is well correlated with near surface humidity. An accuracy of 1.2 g kg^{-1} was achieved in comparison with individual radiosonde ascents and averaging would reduce this error significantly.

Surface flux estimates from numerical models

The estimation of surface fluxes from numerical models of the atmosphere is considered in Chapter 2. Models offer the potential of assimilating all available data, using realistic physical constraints, to produce a global surface flux field. However, at present a large effort is being devoted to reducing the significant biases which exist in the model-derived surface fluxes. For example, Foreman et al. (1994) using flux estimates for 1993 found that in the tropics the UK Meteorological Office Model fields heated the ocean by 100 W m^{-2} more than those from the ECMWF model. A similar problem had been identified by Gleckler et al. (1994) in the Atmospheric Model Intercomparison Project, AMIP (Gates, 1992) and attributed to incorrect modeling of the effect of clouds on the radiation budget. Substitution of the Earth Radiation Budget Experiment (ERBE) values for cloud radiative forcing (in place of the model values) resulted in realistic implied ocean heat transports. Recently da Silva and White (1995) showed that the Goddard Earth Observation System assimilation model and the National Centers for Environmental Prediction reanalysis results significantly underestimated fractional cloud cover outside the tropics compared with either ship observations or the satellite-based Surface Radiation Budget project.

We can speculate as to why it should be difficult to obtain accurate surface flux values over the ocean from large-scale models. Over typical open ocean areas there is an inversion capped 'mixed' layer of order 1 km deep with cumulus or stratocumulus clouds beneath the inversion. The air–sea temperature difference is less than 1 °C and the surface aerodynamic drag is less than a tenth of that for a relatively smooth land area. Under these conditions the evaporation and radiative cooling at the cloud top are sources of buoyancy and turbulent kinetic energy of similar magnitude to the surface fluxes. Cloud top entrainment introduces dry air into the 'mixed' layer and thus modifies the surface latent heat flux. However, often these processes occur intermittently, the cloud layer may be decoupled from the surface turbulence, and the 'mixed' layer is only mixed to the extent that the time

mean equivalent potential temperature profile is nearly constant with height. In order to predict the surface fluxes from the conditions in the free atmosphere it will be necessary accurately to represent these subgrid scale cloud top processes in the model either by parameterization or by explicit incorporation of a 'mixed' layer model. Currently a limitation of most large-scale numerical models is an inadequate vertical resolution to parameterize the turbulent fluxes properly.

Summary – the need to combine data sources

In summary, our present knowledge of the surface fluxes has depended on parameterization formulae applied to the weather observations from merchant ships. Recent studies have identified errors in these observations, and biases in the estimated flux values. When these are corrected it appears to be possible to produce balanced heat and water budgets to within a few W m^{-2}. However, the geographical distribution of the ship data is very poor; outside the northern oceans, sampling errors alone are likely to be large and only mean values over several years can be calculated.

Satellite data are used for measuring SST and are starting to become available with a useful accuracy for wind speed, significant wave height, and surface shortwave radiation. Such data sets now span several years. Scatterometer data offer great potential for wind velocity estimates but at present only a restricted data period is available. Satellite estimates of precipitation, longwave radiation, and near surface humidity are possible but require improvement. The way forward would appear to be to use the ship data in areas that are well sampled to verify the model and satellite-based flux estimates. Combined products might be produced in a similar manner to the present production of SST fields. Because the areas of good ship sampling are in the northern hemisphere, where radiosonde sampling is also good, it is likely that the model performance will deteriorate in other areas. However, the quality of the satellite data should remain the same and allow further verification of the model estimates.

The bulk formulae that form the basis for all present air–sea flux estimation schemes require further testing for low (< 4 m s^{-1}) and high (> 20 m s^{-1}) wind speeds. In tropical areas where winds are light, small-scale gustiness may be important in driving the fluxes; such effects are not explicitly resolved in numerical forecast models. In high winds the role of sea spray in the net flux should be evaluated by direct measurements. The lack of surface measurements is most severe for the evaluation of precipitation estimation methods over the sea. New techniques for use on buoys have been tried, but there is certainly room for new and innovative ideas in this area of research.

4.3 The ocean's response to the freshwater cycle

Raymond W. Schmitt

Introduction

The oceans are the largest reservoir of water on Earth, with 97% of the free water, and experience an estimated 86% of the planet's evaporation and 78% of its precipitation. Despite this dominant status, the oceanic freshwater cycle is very poorly understood. As there is now good evidence that the hydrologic cycle over the ocean plays a significant role in long-term climate fluctuations, an improved description of its major elements should have high priority. Here three aspects of the response of the ocean to freshwater fluxes are reviewed:

1. the role of the ocean as the main return path for the atmospheric branch of the water cycle (mass conservation);
2. the barotropic (or depth independent) response of the ocean to surface mass fluxes; and,
3. the baroclinic (or depth dependent) response to buoyancy fluxes induced by concentration or dilution of dissolved salts by evaporation, precipitation and runoff.

The most climatically important baroclinic response is the impact on the large-scale overturning convection cell known as the thermohaline circulation, which is responsible for much of the poleward heat flux. Paleoclimatic data and model results indicate that high latitude freshwater fluxes can directly affect the strength of the thermohaline circulation, especially in the North Atlantic Ocean, where runoff makes a substantial contribution to the basin scale buoyancy flux. Thus, there is a great need better to define the patterns and amplitude of the oceanic water cycle. Small changes in these patterns may have dramatic consequences for the much smaller terrestrial water cycle: a shift of only 1% of Atlantic rainfall to the central United States would be sufficient to double the discharge of the Mississippi river. Some of the techniques discussed elsewhere in this volume will contribute to an improved understanding of the oceanic water cycle. A goal of this section is to motivate such efforts by reviewing recent progress in understanding the oceanic flows resulting from mass conservation and barotropic and baroclinic dynamics. Implications for climate models are also discussed. Finally, some of the oceanic measurements required to describe and monitor its water cycle are reviewed.

Mass conservation

Meridional fluxes

A general pattern of net evaporation in middle latitudes and net precipitation in the tropics and higher latitudes is seen in maps of the global oceanic water flux (Section 4.2). This leads to the necessity for compensating flows in the ocean. That is, the ocean must transport water into the evaporation zones and away from the precipitation regions, in order to maintain sea level over the long term (Schmitt, 1995). Evaporation rates of up to 3 m yr^{-1} and precipitation rates of up to 5 m yr^{-1} lead to the requirement for substantial return flows in the ocean. Integrated over the globe, oceanic evaporation is estimated to be 13.5 Sv (1 Sv = 1 Sverdrup = 10^6 m^3 s^{-1} = 10^9 kg s^{-1}), and oceanic precipitation sums to 12.2 Sv, according to the budget of Baumgartner and Reichel (1975). However, most water does not stay long in the atmosphere (residence time 10–14 days), so the net differences of surface water flux are smaller when integrated over large areas. The magnitude of the north-south return flows in the ocean can be estimated by integrating the net fluxes in meridional bands. For our purposes, the evaporation minus precipitation ($E-P$) climatologies of Schmitt *et al.* (1989) for the Atlantic and Baumgartner and Reichel (1975) for the rest of the ocean are used. As the major gradients are meridional, a zonal integration in 5° latitude bands is presented, which reveals the pattern of meridional transport (Figure 4.9, from Schmitt and Wijffels, 1993). This can be compared with estimates of the water vapor transport in the atmosphere (Peixoto and Oort, 1983). As noted by Wijffels *et al.* (1992), the oceanic and atmospheric transports are roughly equal and opposite; meridional river flows are one or two orders of magnitude smaller. The ocean/atmosphere hydrologic cycle accounts for a significant portion of the poleward heat transport on the planet. For instance, a water transport of 0.6 Sv corresponds to 1.5 petawatts (= 10^{15} W = PW) of latent heat flux. At 25° N, the ocean sensible heat flux is about 2 PW (Bryden, 1993) so the (separate) heat transport associated with the ocean/atmosphere hydrologic cycle (0.8 PW) is quite substantial. As discussed later, ocean data alone can be used to estimate the latent heat component of meridional heat transport, thus permitting the specification of a large part of the total planetary energy budget.

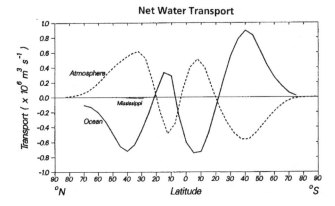

Figure 4.9. The meridional transport of water by the global ocean and atmosphere, in Sverdrups ($= 10^6$ m^3 s^{-1}). The oceanic estimate derives from the integration of the E – P values shown in Section 4.2, plus river discharges into the ocean. The actual meridional transport by rivers alone is small; an estimate for the Mississippi is shown. The atmospheric transport of water vapor comes from Peixoto and Oort (1983).

Interbasin transports

Notable differences exist in the amounts of water gained or lost by each ocean. An excess of precipitation is estimated for the North Pacific, especially in the eastern tropical regions, and a dominance of evaporation is expected in the Atlantic. The Pacific–Atlantic difference is thought to be maintained by water vapor transport across Central America and the lack of any similar transport into the Atlantic from Africa. Schmitt *et al.* (1989) point out that the Atlantic is relatively narrow, and thus under the influence of continental air over a greater fraction of its area. The western tropical Pacific also gains water vapor exported from the Indian Ocean. As a result, the North Pacific is notably fresher than the other oceans, especially the North Atlantic.

The differences in surface water fluxes also mean that there is a necessity for interbasin transport in the ocean. This could be accomplished by a number of different paths, given the multiply-connected nature of the ocean basins. Baumgartner and Reichel (1975) constructed a scheme which assumed zero water transport across the Atlantic Equator, and integrated their surface flux estimates relative to that point. However, Wijffels *et al.* (1992) showed that this arbitrary assumption should be replaced by direct oceanographic transport measurements in Bering Strait. The scheme for ocean water transports they developed differs dramatically from the earlier work (Figure 4.10), which had confused the flux convergence in the Arctic basin with a flux. Indeed, they find that most of the North Pacific water gain (nearly a Sverdrup), is transported through Bering Strait. An assumption of zero

water flux across the Pacific Equator appears to be nearer the truth.

The interbasin water transport is accompanied by a salt transport. Since there is no significant atmospheric pathway for salt, the salt flux through coast to coast sections within a given subsection of the interconnected oceans must be the same. That is, the northward salt flux across 24° N in the Pacific is equal to that through Bering Strait and the southward transport across any zonal section in the Atlantic. This constraint permits calculation of the freshwater flux divergence between sections. Such direct ocean estimates can be used to evaluate integrations of surface fluxes. When complemented by western boundary current measurements, such as the Gulf Stream transport through Florida Straits, and Ekman (wind-driven) transport estimates, the salinity and geostrophic velocity fields from a hydrographic section can be used to calculate the salt flux. As noted by Schmitt (1995), the salt flux can be subtracted from the mass flux to give the net (fresh) water flux divergence between sections in the same basin (or channel). Two examples are given there, one for the freshwater gain between Bering Strait and 24° N in the Atlantic, the other for the gain between 24° N in the Pacific and 24° N in the Atlantic. Estimates of the flow through Bering Strait (Coachman and Aargaard, 1988) and 24° N data analyzed by Hall and Bryden (1982) allow a calculation of the freshwater input between the sections. This gain is estimated to be about 0.1 Sv. In contrast, the climatologies sum to about zero water gain in the Arctic and Atlantic north of 24°. This would indicate the necessity to increase high latitude precipitation or runoff estimates (or decrease evaporation) in the climatologies, though the discrepancy is not substantial compared with the uncertainties which must apply to both types of data (Figure 4.11). Also available for comparison is the recent freshwater flux estimate of Friedrichs and Hall (1993) for 11° N. This latitude is at an extremum in water transport, and also indicates that the climatologies have insufficient water gain to the north. However, because of the uncertainties associated with stronger, and more variable, Ekman and eddy fluxes at 11° N, the error bars are larger than at 24° N.

Similar calculations can be performed with data from 24° N in the Pacific. In combination with the Atlantic data, an estimate of the net oceanic water transport across 24° N is possible. With the reported mean salinities at 24° N of 34.6 for the Pacific (Bryden *et al.*, 1991) and 35.156 for the Atlantic (Hall and Bryden, 1982), a Bering Strait salt transport of 26‰ Sv, and the salt fluxes given by the velocity–salinity correlation across the sections, a net oceanic southward water flux of 0.31 Sv is obtained. This value is insensitive to the precise value of the Bering Strait flux; the interannual variability of

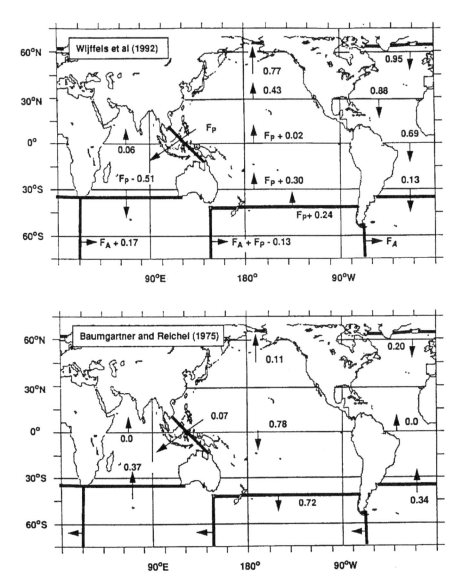

Figure 4.10. The net transport of water by the oceans, according to Wijffels *et al.* (1992) in Sverdrups (= 10^6 m^3 s^{-1}) (top). Contrast this scheme with that of Baumgartner and Reichel (1975) (bottom).

0.1 Sv (out of 0.8) reported by Coachman and Aagaard (1988) contributes an uncertainty of only 0.01 Sv in the estimate of net water gain north of 24° N. The uncertainty in the velocity-salinity correlation in the two sections is a more complex issue; however, it should behave like the heat flux estimates, since salinity and temperature are well correlated within a section. The oceanic heat flux at 24° N in the Atlantic has been found to be consistent for three different occupations, so there is reason to believe that salt flux estimates will be equally stable.

The result of a net southward transport of water by the ocean of 0.31 Sv can be combined with southward meridional river flows of about 10% of this value. The net ocean/river water flow of 0.34 Sv is in reasonable agreement with the northward atmospheric water transport at this latitude (Peixoto and Oort, 1983). This is an interesting result, as there has been speculation that the 'missing petawatt' (Bryden, 1993) resulting from the discrepancies among the meridional heat flux estimates due to the Earth's radiation budget, atmospheric transport and ocean transport, might be found in unresolved latent heat transport in the atmosphere. However, since the return flow would be in the ocean, the data do

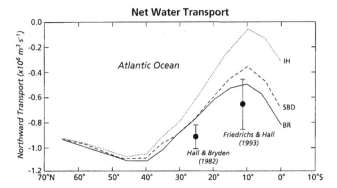

Net Water Transport

Figure 4.11. Meridional transport of water in the Atlantic ocean, according to three different surface flux climatologies. All are summed relative to an estimated Arctic southward export due to the Bering Strait throughflow and the water budget of the Arctic itself. The climatologies are those of Baumgartner and Reichel (1975) (solid line), Schmitt *et al.* (1989) (dashed line) and the combination of Isemer and Hasse (1987) evaporation estimates with Dorman and Bourke (1981) precipitation values (dotted line). Also shown are ocean based estimates at 24° N by Hall and Bryden (1982) and 11° N by Friedrichs and Hall (1993).

not support an increase of more than double the presently estimated meridional transport of water at 24° N, which would be required for the extra petawatt (Schmitt, 1995).

The more complete set of zonal lines collected during the World Ocean Circulation Experiment (WOCE) will permit even more points of comparison between the surface estimates and ocean flux divergences. Indeed, preliminary results indicate that the mid-latitude southern hemisphere oceans carry half the heat flux and twice the water flux of the northern hemisphere oceans. This striking asymmetry suggests that the atmosphere/ocean system has multiple ways of satisfying the global radiation budget (i.e., an enhanced latent heat flux in the atmosphere/ocean system can make up for reduced sensible heat transport in the ocean). It is quite possible that the enhanced water flux over the southern ocean helps to decrease the heat-flux-carrying thermohaline circulation there by freshening the surface waters and limiting deep convection. In a stronger CO_2 greenhouse climate it is expected that the hydrologic cycle will intensify. Could it intensify sufficiently to limit the thermohaline circulation in the North Atlantic?

This question is of obvious first order importance, yet we will have trouble addressing it because our knowledge of precipitation and evaporation over the ocean is so rudimentary. The difficulties of making precipitation measurements at sea, the paucity of data for calculating evaporation, and lingering uncertainty about the bulk formula combine to make the net air–sea–water exchange a very poorly known

quantity (Section 4.2). As can be seen in Figure 4.11, there are significant discrepancies between available climatologies for the North Atlantic which serve to illustrate the problem. When summed over the North Atlantic the Schmitt *et al.* (1989) numbers yield 10^5 m^3 s^{-1} more evaporation than the Baumgartner and Reichel (1975) summation. This is over half the flow of the Amazon. Even greater discrepancies arise when more recent climatologies are considered for the evaporation field. For instance, Isemer and Hasse (1987) have revised the Bunker (1976) estimates for the North Atlantic. While they adjust the exchange coefficients in the bulk formulae downward, their revision of the Beaufort wind scale leads to stronger trade winds and increased evaporation in the tropics and subtropics. Their estimate of 20 to 40 cm yr^{-1} more evaporation south of 40° N suggests that an additional freshwater loss of nearly 4×10^5 m^3 s^{-1} may occur over the North Atlantic. Thus, current estimates of net $E - P$ over the basin differ by 2.5 times the flow of the Amazon (1.9×10^5 m^3 s^{-1}). Since the North Atlantic is a small, relatively well sampled basin, we can infer only that the uncertainties for the other ocean basins are even larger. The use of oceanic section data to constrain estimates of meridional water fluxes will be an important tool in refining our understanding of the global hydrologic cycle.

Dynamical response of the ocean to surface water fluxes

Barotropic response: the Goldsbrough–Stommel circulation

The exchange of water between ocean and atmosphere has consequences for ocean flows beyond the requirement of mass conservation. The simplest (yet largely neglected) response is that governed by the planetary vorticity balance. Goldsbrough (1933) first showed how surface mass fluxes due to evaporation and precipitation drive depth independent (barotropic) flows in the ocean, in what may be viewed as a precursor to the well-known Sverdrup relation. That is, it is readily shown (Huang and Schmitt, 1993) that the steady vorticity balance in the ocean interior is given by:

$$\beta_f V = k \cdot \nabla \times \tau + f(E - P) \qquad (4.3.1)$$

where f is the Coriolis frequency, β_f its meridional gradient, τ is the wind stress and V is the meridional velocity in the ocean. Goldsbrough (who considered only the $(E - P)$ term above) suggested that a subtropical-gyre-like (anticyclonic) circulation would result from a basin interior precipitation regime and an evaporative western boundary. This is of course quite unrealistic; the interior Goldsbrough circulation must be closed by adding western boundary currents with

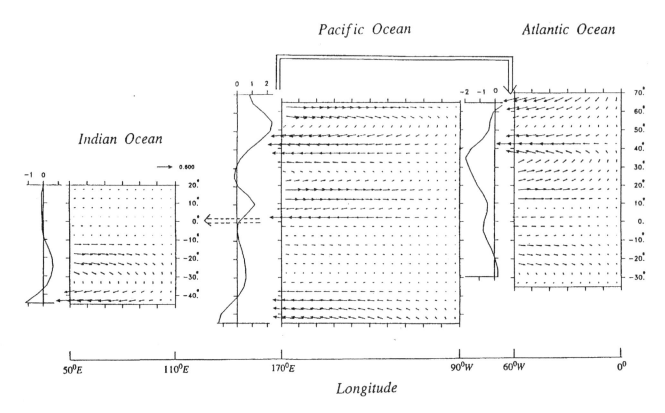

Pacific Ocean　　　*Atlantic Ocean*

Indian Ocean

Longitude

Figure 4.12. The Goldsbrough–Stommel circulation in the global ocean, according to Huang and Schmitt (1993). Each arrow shows the horizontal mass flux integrated over a 5 degree square, in Sverdrups. Along the left edge of each basin, there is a curve indicating the northward transport within the western boundary (in Sv) which is required to close the circulation. The Pacific to Atlantic transport via the Arctic is shown, and is included in the boundary current transports. The Pacific to Indian transport is depicted schematically.

viscous or inertial dynamics. The complete gyres can be referred to as the Goldsbrough–Stommel circulation.

Huang and Schmitt (1993) have presented a preliminary calculation of the Goldsbrough–Stommel circulation for the world ocean (Figure 4.12). While $E-P$ is smaller than Ekman convergence (1 m yr^{-1} vs. 20 m yr^{-1}), the flows generated are generally opposite to the wind driven currents. In the North Atlantic, a western boundary current is required which reaches 2 Sv southward at 35° N, in direct opposition to the Gulf Stream. Other basins have similarly strong boundary currents. These adverse flows are likely to affect the separation latitude of the western boundary currents. For the Gulf Stream, the shift may be as large as 75 km to the south. As there are strong horizontal temperature gradients associated with the western boundary currents, any small changes in separation latitude could have large impact on SST distributions in the western portions of the subtropical gyres. Changes in mid-latitude SST patterns are now believed to have direct effects on the circulation of the atmosphere (Kushnir, 1994).

Buoyancy forcing due to freshwater fluxes

Evaporation or precipitation serves to concentrate or dilute the dissolved salt content of surface sea water, and thus directly affects its density. The haline density flux is given by:

$$\beta F_s = \frac{\beta S \rho_0 (E-P)}{\rho_s(1-S)} \cong \beta S(E-P) \qquad (4.3.2)$$

where (now) $\beta = 1/\rho(\partial\rho/\partial S)$ is the haline contraction coefficient, and ρ_0 and ρ_s are the densities of pure and sea water; F_s being salt flux or transport, and S is salinity. The haline contribution to the density flux is often comparable to that of the thermal density flux ($=\alpha Q$, where Q is the net heat flux and $\alpha(=-1/\rho(\partial\rho/\partial T)$ is the thermal expansion coefficient). Schmitt et al. (1989) compared the annual mean thermal and haline density fluxes in the North Atlantic by contouring the absolute value of the ratio of the two components (Figure 4.13a,b). They find that in the subtropical gyre the haline density flux is often larger than the thermal density flux. Indeed, though the central subtropical gyre is conventionally

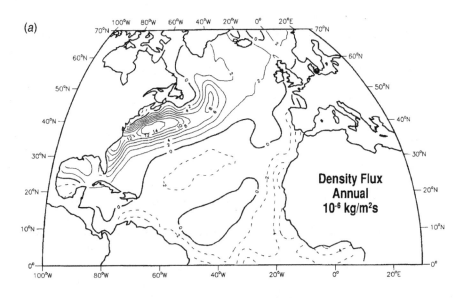

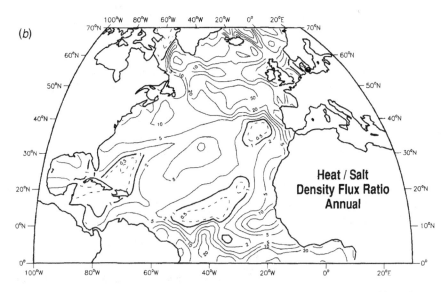

Figure 4.13. (a) The net flux of density into the North Atlantic Ocean as estimated from the separate contributions of heat and water fluxes to sea water density. (From Schmitt *et al.*, 1989.) (b) The absolute value of the ratio of thermal to haline buoyancy fluxes in the open ocean North Atlantic. Coastal runoff and ice processes are neglected. (From Schmitt *et al.*, 1989.)

thought to be an area of density loss (or buoyancy gain), the Schmitt *et al.* maps suggest that substantial portions of it actually experience a density gain due to the evaporation-driven haline density flux. While uncertainties are such that we do not know the sign of the density flux in many areas, a density gain would indicate that the salinity maximum water formed in mid-gyre is actually part of a haline-driven convection cell, rather than simply the result of Ekman convergence of buoyant water.

At other latitudes the general pattern shows the thermal density flux to be greater. In high latitudes the thermal flux dominates in the central subpolar gyre by an order of magnitude or more. Such dominance is somewhat surprising, since the thermal expansion coefficient of sea water becomes small at low temperatures. However, the amplitude of the hydrologic cycle is reduced at low temperatures, due to the small water-vapor carrying capacity of cold air (Clausius-Clapeyron relation). Also, the North Atlantic has anomalously

Table 4.1. *The contribution of continental runoff to the mass budget of the North Atlantic, in 10^4 m^3 for 10 degree latitude bands*

The second column shows the total mass gain from precipitation, runoff, marginal sea exchanges and evaporation. The third column lists the total estimated runoff. The fourth column displays the ratio of runoff to the total water flux. Runoff is a substantial fraction of the water budget between 0 and 10° N and north of 40° N, and nearly negligible between 10 and 40° N.

Latitude range	$(P+R+M-E)$	R	$R/(P+R+M-E)$
60–70	5.4	3.8	0.70
50–60	9.8	4.5	0.46
40–50	3.1	5.6	1.81
30–40	−20.2	1.8	−0.09
20–30	−28.5	2.3	−0.08
1–20	−24.6	3.2	−0.13
0–10	34.3	26.9	0.78

warm waters at high latitudes, which enhances the heat loss to the atmosphere. Finally, the Schmitt *et al.* (1989) flux ratio is based entirely on open ocean air–sea exchanges. Runoff, sea ice formation, transport, and melting, as well as marginal sea exchanges can be a substantial part of the net density flux in many latitude bands.

Role of runoff

To examine this last point, the fractional contribution of the runoff to the net mass flux in each 10 degree latitude band is given in Table 4.1. This fraction varies strongly with latitude. In mid latitudes, where evaporation tends to dominate, the runoff may make a negligible contribution. However, within the high and low latitude precipitation bands, runoff can easily dominate the mass flux.

A more important consequence of runoff involves its contribution to the density flux; however, this is less straightforward to estimate. This is because a salinity representative of the local ocean must be used. In coastal waters this should be specified in conjunction with the length scale of surface salinity variation. Buoyancy-driven coastal currents are a well-known phenomenon. However, for the purposes of judging the impact on the thermohaline circulation we only need consider the larger scale effects of density forcing at the boundary. Using the estimates of the runoff contribution to the total water flux and a zonally averaged surface salinity (in 10 degree latitude bands), the contributions of net heat flux and total water exchange to zonal averaged density flux can be computed. The zonal totals of the heat and salt density

fluxes are shown as a function of latitude for the North Atlantic in Table 4.2. The absolute value of the heat/salt ratio is given (including runoff) for contrast with Figure 4.13(*b*). Not considered are additional (unknown) contributions to the freshwater flux due to the freezing, transport, and melting cycle of sea ice and, more importantly, ice discharge from the Arctic (Aagaard and Carmack, 1989). Poleward of 70° N the atmosphere provides a moisture flux convergence (Serreze *et al.*, 1995) which, when combined with river discharges, should equal the freshwater discharge from the Arctic. Table 4.2 shows that the density flux is not nearly so dominated by heat transfers in the high (and low) latitude North Atlantic as the Schmitt *et al.* map (Figure 4.13(*b*)) implied. The thermal density gain exceeds the haline density loss by only a factor of 3 in the 50–60 degree latitude range, which can be contrasted with the order of magnitude dominance mapped in the interior. In the tropics the total thermal and haline fluxes are quite comparable, and in the 20–30° N range the haline flux dominates as suggested on the map. Thus, the thermohaline circulation in the Atlantic may be closer to shutdown by hydrologic forcing than the open ocean fluxes alone would indicate.

Recent findings suggest that much of the deep water formation occurs on the shallow shelves surrounding the Nordic Seas, rather than in open ocean convective chimneys (Mauritzen, 1996). As the shelf areas are adjacent to the sources of runoff, the local impact on density can be much greater than the zonal average for the basin. Coastal freshwater sources drive along-shore currents (to the right in the northern hemisphere) which can be the most prominent regional flows. Examples are the Norwegian Current and the Labrador Current, though the latter is also driven by the large-scale wind patterns of the subpolar gyre.

The thermohaline circulation and climate models

The role of the haline density flux in large-scale ocean dynamics was first examined by Stommel (1961) with a simple box model. He found multiple stable states of the system, which was forced by both heating/cooling and evaporation/precipitation. Bryan (1986) showed how a general circulation model would display instability of the thermohaline circulation when restoring boundary conditions were replaced by flux boundary conditions for the salinity. Freshening of high latitude surface waters makes them too buoyant to sink and form deep water, thus shutting off the thermohaline circulation for a time. The circulation re-establishes itself (via a sudden 'flushing') once vertical mixing processes warm the deep water sufficiently that the buoyancy difference disappears. Quite a number of models of varying complexity have

Table 4.2. *The estimated thermal and haline density fluxes into the North Atlantic in ten degree latitude bands*

The second column shows the estimated thermal density flux. The third column lists the interior haline density flux due to evaporation and precipitation. The fourth column lists the total haline density flux including continental runoff. The fifth column displays the absolute value of the heat/salt density flux ratio based on the complete water budget. These numbers may be contrasted with the map of interior flux ratios given in Figure 4.13(*b*). In contrast to the map, the total density fluxes suggest that the freshwater contribution is important in nearly all latitude bands.

Latitude range	Thermal density flux (10^6 kg s^{-1})	Interior salt density flux (10^6 kg s^{-1})	Total salt density flux (10^6 kg s^{-1})	Heat/salt flux ratio
50–60	8.5	−1.46	−2.71	3.1
40–50	11.8	0.68	−0.84	14.0
30–40	22.9	4.42	5.41	4.2
20–30	−3.0	8.15	7.63	0.4
10–20	−8.4	7.33	6.48	1.3
0–10	13.2	−1.9	−9.02	1.5

focused on the sensitivity of the thermohaline circulation to freshwater forcing (Stocker and Wright, 1991; Weaver and Sarachik, 1991; Huang *et al.*, 1992; Rahmstorf, 1995). The essential point is that the 'conveyor belt' of the general circulation is very responsive to the freshwater flux at high latitudes. Many of these models display fluctuations on centennial to millennial time scales. However, Delworth *et al.* (1993) have carefully examined the behavior of a realistic circulation model, and found significant variability on 40–60 year time scales. Advection of heat and salt play an important role in the variability, which appears to be governed by the strength of a baroclinic gyre set up in mid-basin. The persistence of sea-surface temperature anomalies is an important element, so the common relaxation boundary conditions, which couple the ocean too tightly to the atmosphere, are not appropriate. Huang (1993) has pointed out the importance of real freshwater flux as a surface boundary condition for models, in part because of the Goldsbrough–Stommel circulation, which is missing when the commonly applied fictitious salt flux is used as a boundary condition.

Also of interest is the modeling study of Shaffer and Bendtsen (1994), who find a great sensitivity of the thermohaline circulation to the transport through Bering Strait. Because this supplies freshwater to the Atlantic, it acts as a negative feedback on the thermohaline conveyor belt. They propose that a more variable climate would be realized with a higher sea level and thus greater Bering Strait transport. Such conditions may have been achieved during the last interglacial period, when the Greenland ice cores indicate that the climate had less stability than at present (Dansgaard *et al.*,

1993). Other paleoclimate records suggest that sudden changes in the North Atlantic thermohaline circulation occurred many times during the last deglaciation (Lehman and Keigwin, 1992), when substantial discharges of meltwater are thought to have occurred at high latitudes. Coupled ocean–atmosphere models such as those of Manabe and Stouffer (1995) also display strong sensitivity of the thermohaline circulation to freshwater pulses, with the associated change of surface air temperature being especially large in the North Atlantic. Further work on the extreme sensitivity of the thermohaline circulation to the hydrologic cycle has been done by Rahmstorf (1995), whose coupled model runs suggest that temperature changes of several degrees in only a few years are possible.

There is also historical evidence that perturbations of the hydrologic cycle do influence the climate system. In particular, fresh salinity anomalies in the subpolar basin have been observed to survive for the order of a decade while being advected by the surface gyre. The Great Salinity Anomaly (GSA) of the 60s–80s was observed for 15 years as it traveled around the entire subpolar North Atlantic gyre (Dickson *et al.*, 1988). There is strong evidence that the GSA had substantially suppressed deep water formation in the Greenland and Norwegian Sea and flux in the western boundary currents. The GSA is thought to have arisen from excess ice discharge from the Arctic through Fram Strait. An increase in the annual export of ice of only 20% (Aagaard and Carmack, 1989) could have provided the necessary freshwater anomaly.

A potentially related process has been identified by Deser and Blackmon (1993). They found a mode of Atlantic surface and air temperature variability with a dipole structure concentrated in the western North Atlantic, by examining 90 years of temperature data. Winter temperatures east of Newfoundland are out-of-phase with those off the US mid-Atlantic states. The mode has an amplitude of 1°C and a 10–12 year period. Interestingly, it appears to be related to the ice extent in the Labrador sea, with Labrador sea ice leading low temperatures off Newfoundland by one or two years. Advection of a low salinity cap formed by the melting ice, which prevents mixed layer deepening and the release of ocean heat, would be consistent with such a relation.

Thus, there is substantial evidence that the high latitude thermohaline circulation has intimate ties to climate on decadal and longer time scales. It is therefore important to define further its sensitivity to perturbations in the freshwater flux. Both improved models and an expanded observational base are required.

It is also important to note that strong haloclines in the tropics, formed by the rainfall under the ITCZ, have a significant impact on upper ocean mixing and heat exchange processes there. Lukas and Lindstrom (1991) found that the surface mixed layer of the western tropical Pacific was limited by salinity stratification; an isothermal layer extended to a greater depth than the mixed layer. The interval with a strong stable salinity gradient has been termed the 'barrier layer' because of its inhibition of vertical mixing. Delcroix and Hénin (1991) describe variations in the surface salinity field of the southwestern tropical Pacific which are correlated with the estimated freshwater flux from 2–3 months prior. The seasonal migration of the rain band associated with the South Pacific Convergence Zone appears to be responsible for the annual cycle in surface salinity in that area, while interannual variations may be related to the El Niño cycle. Recently, Anderson, Weller and Lukas (1996) have reported on observations and modeling of the effects of freshwater input in the western equatorial Pacific as part of the Tropical Ocean Global Atmosphere (TOGA) Coupled Ocean Atmosphere Response Experiment (COARE). New techniques for rainfall measurement were successfully employed. They found that the two-day averaged freshwater and heat flux contributions to the buoyancy flux were negatively correlated, which tended to reduce the variance in the total buoyancy flux and resulted in a cycling between temperature and salinity control of the mixed layer depth. They point out that the atmosphere is quite sensitive to small SST variations in the 'warmpool' region of the tropical Pacific and that SST is very dependent on changes in air–sea heat fluxes because of the shallow mixed layer. As the barrier layer insulates the mixed layer from the cooler waters in the thermocline, adiabatic ocean dynamics can have less effect on the SST. A much stronger local coupling between atmosphere and ocean is suggested when an accurate hydrologic cycle is incorporated into models. Thus, we have incentives from both high and low latitude oceanic processes for making improved estimates of the air–sea water transfers.

The ocean observing problem

Having discussed some of the fundamental aspects of the ocean hydrologic cycle and its importance to ocean dynamics, it is now useful to identify approaches with which we will be able to improve open-ocean estimates of water exchange with the atmosphere. As there is no 'reference standard' against which to calibrate such estimates, we will be best served by a multitude of approaches. Ocean-based assessments of the water cycle will be important complements to the remote sensing and numerical weather prediction schemes described elsewhere in this volume and will be the primary means of validating such techniques over most of the globe.

Goals for accuracy in net freshwater flux over the ocean can be set by comparison with requirements for the heat flux. A common goal for oceanographic programs is an uncertainty in net heat flux of 10 W m^{-2} (e.g. WOCE). During TOGA COARE it is estimated that accuracies were well within this criterion, though in general it is difficult to meet. Two different bases for comparison with the water flux can be used: (a) the mass flux equivalent to a given latent heat flux, and (b) the buoyancy flux equivalent. The impact of the latent heat of vaporization on the heat budget is the more stringent requirement, with only 1 cm per month water loss being equivalent to 10 W m^{-2}. This applies to evaporation (and condensation) only, since the phase change for precipitation occurs high in the atmosphere. Nevertheless, for many purposes (mass budget and barotropic response) it is the total water flux that is required so this represents a formidable accuracy goal. The buoyancy flux that is equivalent to 10 W m^{-2} is a more tractable 5 cm per month or more. This is important for estimating the baroclinic response of the ocean, including climate-related phenomena such as the formation of 'barrier layers' in the tropics and freshwater capping in high latitudes. A goal of 5 cm per month accuracy in the water budget over climate-scale areas of 2 degrees of latitude by 5 degrees of longitude for monthly averages should be achievable over some portions of the oceans with a combination of *in situ* measurements, remote sensing techniques and the data-assimilating models used for numerical weather prediction. Section 4.2 has reviewed a number of measurement techniques for their potential contributions to the problem of observing the water cycle over the ocean. In the following, the potential for expanded salinity measurements to constrain the oceanic water cycle is discussed.

Salinity

Sea surface salinity (SSS) has long been regarded as a good indicator of net surface water exchange. SSS is controlled by $E-P$, advection and mixing with underlying water. The highest SSS values are found in the middle of the subtropical gyres, due to evaporation by the trade winds and convergence by Ekman pumping. Low SSS values are found in both low and high latitudes, where precipitation and runoff exceed evaporation. With a sufficiently well-resolved salinity field, and estimates of surface velocities, horizontal mixing rates and mixed layer depths, one could infer mean $E-P$ from ocean data alone. Salinity has the particular advantage of responding to the integral of the water flux, providing an effective average over the short space/time scales and intermittency of precipitation. A change of salinity of 0.035 psu in a 50 m surface mixed layer would occur with a 5 cm water flux; which is well within the measurement errors of available systems. With data of sufficient spatial resolution, the detailed patterns of the mean water flux should be discernible. However, our present knowledge of the space and time scales of SSS variations is poor. About 30% of 1° squares in the ocean have never had an SSS measurement (Lagerloef *et al.*, 1995). Because of the sparsity of *in situ* measurements, and the lack of a remote sensing technique, features such as the climatically significant Great Salinity Anomaly (Dickson *et al.*, 1988) are at present detected only by serendipity. Recent modeling studies show a predictive value for salinity measurements in decadal time-scale climate variations (Griffies and Bryan, 1997). Thus, an important priority for future ocean monitoring is the improvement of surface and upper-ocean salinity measurements. Such data are necessary for establishing the magnitude of the storage, advection and mixing terms in the freshwater budget. Ultimately, the assimilation of salinity data into ocean models will provide an important technique for constraining the estimates of surface water fluxes from satellites and meteorological models.

Fortunately, the measurement of salinity has become more routine due to the development of compact, stable conductivity sensors. It is now possible to obtain reliable salinity records from shipboard thermosalinographs, conductivity-temperature depth (CTD) packages incorporated into towed vehicles and long term records from moorings (McPhaden *et al.*, 1990) or drifting buoys. In addition, expendable and autonomous profilers are now available, and a remote sensing technique has been proposed by Lagerloef *et al.* (1995).

It is the autonomous profilers which may provide the most cost-effective means of maintaining a global monitoring capability for salinity. R. Davis (Scripps Institution of Oceanography) has deployed hundreds of floats (Autonomous Lagrangian Circulation Explorer, ALACE) that sit at a selected depth for 1–4 weeks, then rise to the surface to report position and profile data via a satellite link. Recent work has demonstrated that these can be fitted with sensors to measure salinity (Schmitt, 1997). The deep cycling is a distinct advantage for salinity measurements, as the frequent returns to a climatically stable water mass allows correction of any long-term sensor drifts. A globe-spanning array of 3000 such floats could be deployed with 300 km spacing at a cost less than that of maintaining a research vessel. Yet the array would provide far more real time temperature and salinity data than any ship possibly could.

Conclusions

This section has attempted to survey the essential elements of the hydrologic cycle over the ocean, provide a sense of its fundamental importance to the ocean circulation and note some new oceanic measurement techniques which promise to reveal much about the water cycle in the years to come. While there remain great uncertainties in our estimates of airsea water fluxes, it is clear that the water cycle drives a number of responses in the circulation of the ocean. Mass conservation requires intrabasin and interbasin flows and vorticity conservation leads to interior and boundary currents which cannot be neglected in oceanic models. In addition, the water flux makes a substantial contribution to the surface buoyancy, often exceeding the influence of heating and cooling, especially when runoff is taken into account. As the buoyancy fluxes control the thermohaline circulation, which has a direct impact on climate, there is a great need to improve our understanding of the oceanic water cycle. Fortunately, a number of emerging ocean measurement techniques promise to provide a means of quantifying the water cycle over the sea. In particular, new tools for monitoring the upper ocean salinity field are becoming available, which will allow it to be used as an integral measure of the net water flux.

Other approaches to assessing the ocean surface water fluxes are discussed elsewhere in this volume. Numerical weather prediction models provide estimates of oceanic water fluxes every 6 hours. However, rather little *in situ* data from the ocean surface enters those models and only direct oceanic measurements can provide a means of validating the developing remote sensing and modeling capabilities over most of the globe. The combination of remote sensing, modeling and new ocean observations will provide a much improved understanding of the largest component of the water cycle. Developing an accurate description of the heat and freshwater fluxes at the ocean surface is essential to the advance of climate modeling on both seasonal to interannual and decadal to centennial time scales. Thus, an investment in research on the oceanic water cycle will pay high returns in the climate research and prediction programs of the next decades.

4.4 Physical and physiological feedback constraining evaporation from land surfaces

J. L. Monteith

Evaporation and surface temperature

Natural surfaces such as soil, vegetation and water are rarely in thermal equilibrium with the air passing over them because of spatial and temporal changes in sources and sinks of heat. The immediate physical consequence of any such change is an increase or decrease of surface temperature in a direction that brings the system closer to thermal equilibrium. In principle therefore, surface temperature plays a dominant role in stabilizing the exchange of both sensible and latent heat between the atmosphere and terrestrial surfaces and their substrates. In practice, the temperature of most natural surfaces is rarely constant in time or uniform in space so that representative measurements are hard to obtain as a step towards estimating fluxes of heat and water vapor.

Penman (1948) discovered how surface temperature could be eliminated from the equations that describe the components of heat balance at the Earth's surface so that rates of evaporation, in the first instance from free water surfaces, could be estimated from standard measurements of climate. Independently, his contemporary Budyko (1948) achieved the same result using an iteration scheme. Almost 50 years later, such methods remain the basis for most practical and theoretical developments in evaporation science, although many extensions and simplifications have appeared in the literature. For example, Priestley and Taylor (1972) expressed the latent heat of evaporation as a fraction α of net available energy H (W m^{-2}) (usually the difference between a radiative gain and the rate of storage in a substrate) multiplied by a temperature-dependent factor $\Delta/(\Delta + \gamma)$ where Δ is the rate of change of saturation vapor pressure with temperature (Pa K^{-1}), and γ is the psychrometer constant (Pa K^{-1}).

In this section, the scope of Penman's equation is extended by incorporating systems of feedback, both physical and physiological, that constrain and stabilize the exchange of water between vegetation and the atmosphere. Vegetation is assumed to be both uniform enough and extensive enough for the development of a convective boundary layer characteristic of the underlying surface.

Interactive extensions of the Penman formula

Physical feedback

The Penman formula is now often used in a form that includes a physical or 'aerodynamic' conductance g_a (m s^{-1}) approximately proportional to windspeed but depending also on surface roughness and atmospheric stability; and a physiological or 'surface' conductance g_s (m s^{-1}) depending mainly on leaf area and on the number, size and behavior of the stomatal pores through which leaves of all land plants lose water vapor to the atmosphere. In this context, a 'conductance' is the reciprocal of a resistance in electrical analogues of heat and vapor transfer where fluxes are equivalent to currents and potentials such as temperature, and vapor pressure are equivalent to voltages. Using these conductances, the conventional Penman–Monteith (PM) equation (Monteith, 1981) for evaporation can be recast to obtain the non-dimensional evaporation rate of Priestley and Taylor (1972) in the form

$$\alpha = \frac{1+C}{1+h/g_s} \qquad (4.4.1)$$

where

$$C = \rho c_p g_a D / \Delta H \qquad (4.4.1a)$$

and

$$h = g_a \gamma / (\Delta + \gamma) \qquad (4.4.1b)$$

D (Pa) is a saturation vapor pressure deficit representative of the surrounding area, usually measured at a reference height of 1 to 2 m; and the thermodynamic properties of air are specified by density ρ (kg m^{-3}) and specific heat c_p (J kg^{-1} K^{-1}).

The dimensionless compound variable C can be regarded as a 'climate number' because it depends on vapor pressure deficit, net radiation, windspeed and air temperature. As discussed later, however, the 'climate' of an extensive site, as specified by C, depends, in part, on the state and behavior of the underlying surface. This dependency can be explored by combining the non-dimensional PM equation with a model of the convective boundary layer (CBL) such as that described by McNaughton and Spriggs (1989). The CBL is capped by an inversion of temperature and its behavior is dominated by exchanges of heat and water vapor at its interface with the surface. During most of the day and over land, the surface usually acts as a source of sensible heat transferred to the CBL by convection. In consequence, the inversion usually rises during the day from a nocturnal height of the order of 100 m to a maximum of several kilometers in the early afternoon.

McNaughton and Spriggs tested their model by comparing estimates of CBL height with measurements at Cabauw in the

155

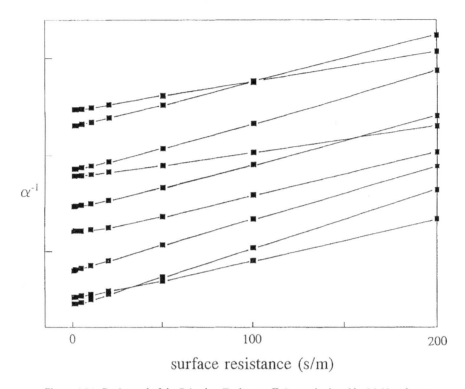

surface resistance (s/m)

Figure 4.14. Reciprocal of the Priestley–Taylor coefficient calculated by McNaughton and Spriggs (1989) from a Convective Boundary Layer model and plotted as daily means for a fixed value of surface resistance. Each set of points represents mean values for one day and the sets are offset vertically to minimise overlapping. The vertical scale is therefore arbitrary and the divisions on the vertical axis are 0.5 apart. Values of α_m range from 1.30 to 1.62 (mean 1.42 ± 0.10) and values of $d\alpha^{-1}/dr_s$ from 0.0013 to 0.0029 (mean 0.0022 ± 0.051).

Netherlands, reported by Driedonks (1981). These estimates were derived from diurnal measurements of (a) net heat supplied to the surface by radiation; and (b) temperature and humidity extending upwards from screen height to the inversion capping the CBL. Having demonstrated that predictions of CBL growth agreed well with measurements on most, though not all the days examined, McNaughton and Spriggs explored the dependence of daily mean values of α on the suface conductance g_s (assumed to be constant in time) and over a physically uniform area with horizontal dimensions large compared with the depth of the CBL. In the analysis that follows, g_s is again assumed constant throughout the day although, in practice, it changes with weather.

For nine Cabauw days, McNaughton and Spriggs plotted α against the logarithm of surface resistance $r_s = 1/g_s$. However, because such plots are sigmoid they do not encourage further analysis. Furthermore, the original PM equation yields similar sigmoid plots of α as a function of $\ln(r_s)$ so this type of presentation contains no visual information about the interaction of the CBL with the surface.

A more useful summary of CBL behavior can be obtained by noting from equation (4.4.1) that the reciprocal of α should increase linearly with r_s, when C is constant so that $[\partial\alpha^{-1}/\partial r_s]$ $(1 + C) = h$. This result suggests that the relation between α^{-1} and r_s may be worth exploring when α is estimated from the McNaughton–Spriggs model. In Figure 4.14, daily mean values of α^{-1} are plotted as a function of r_s for the Cabauw set of 9 days. On all days, the relation is effectively linear, not just from 20 to 200 s m^{-1} as shown but to 10^4 s m^{-1} at least. On several days, linearity also extends down to $r_s = 0$. The estimated response of evaporation rate to a diurnally invariant value of surface resistance can therefore be expressed as

$$\alpha^{-1} = \alpha_m^{-1} + fr_s \qquad (4.4.2)$$

where α assumes a maximum value of α_m when $r_s = 0$ and $f = d\alpha^{-1}/dr_s$. The total differential is appropriate here because other parameters, especially C (through its dependence on vapor pressure deficit, temperature, and aerodynamic resistance), change in response to a change of r_s. The *indirect* impact of the increase of vapor pressure deficit and therefore of C with increasing r_s (numerator of equation 4.4.1) largely offsets the *direct* impact of increasing r_s (denominator). In consequence, most values of $d\alpha^{-1}/dr_s$ given by the slopes of lines in Figure 4.14 fall in the range 0.0013 to 0.0022 m s^{-1} compared with larger values of the order of 0.01 m s^{-1} for $(\partial\alpha^{-1}/\partial r_s)$ from equations (4.4.1) and (4.4.1a) with C held constant. This comparison demonstrates that humidification of the CBL by evaporation at the surface is an effective form of negative feedback that stabilizes rates of transpiration when stomatal conductance changes in response to a physical stimulus such as a decrease in water available to roots or to a biological stimulus such as the expansion of a leaf canopy.

The analysis also draws attention to the error of assuming that the sensitivity to r_s of evaporation from an *extensive and uniform* land surface can be estimated from the PM equation without making a complementary change in the vapor pressure deficit and therefore in the climate number. To explore this change, it is instructive to eliminate r_s from equations (4.4.1) and (4.4.2) so that the Priestley–Taylor coefficient for the *actual* rate of evaporation can be written as

$$\alpha = \alpha_m \frac{X-(1+C)}{X-\alpha_m} \qquad (4.4.3)$$

where $X = h/f$ is the ratio of $[(\partial\alpha^{-1}/\partial r_s)(1+C)]$ (obtained from the PM equation with C constant) to $d\alpha^{-1}/dr_s$, the value obtained with CBL feedback as summarized by equation (4.4.2). The reciprocal of X is therefore a non-dimensional measure of the strength of feedback associated with the humidification of the atmosphere by evaporation.

Equation (4.4.3) implies that provided X exceeds α_m (a condition usually satisfied) the *actual* rate of evaporation *decreases* linearly with increasing C in contrast to equation (4.4.1) where the rate for a constant climate number (including the special case of potential transpiration with $r_s = 0$) *increases* linearly with C. This complementary behavior is shown in Figure 4.15 where α is plotted as a function of C. The upper line with a positive slope corresponds to *potential* evaporation, i.e., $\alpha = 1 + C$ (equation 4.4.1 with g_s infinite) when C has a minimum value of 0.4 so that $\alpha_m = 1.4$ (average value from the Cabauw data set). The three lines with negative slope correspond to *actual* (i.e., sub-potential) rates of evaporation when X assumes values of 2, 3 and 4.

For the case $X = 2\alpha_m$, the value of α obtained from equation (4.4.3) is $2\alpha_m - (1 + C)$, plotted as a dashed line in Figure 4.15. As the corresponding value for potential evaporation is $(1 + C)$, actual and potential lines have slopes of $+1$ and -1 respectively and the sum of potential and actual rates is $2\alpha_m$ for all values of C. This special case was adopted by Bouchet (1963) as a device for estimating actual from potential rates of

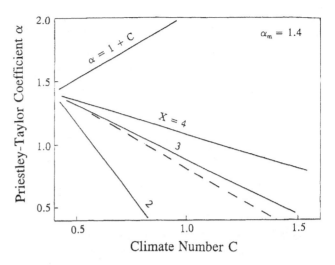

Figure 4.15. Dependence of Priestley–Taylor coefficient α on Climate Number for potential evaporation ($\alpha = 1 + C$) and actual evaporation from equation (4.4.3) with $X = 2,3$ and 4. The dashed line is $X = 2\alpha_m$ equivalent to Bouchet's hypothesis (see text).

evaporation. In general, however, the slope of the relation between α and C for actual evaporation (Figure 4.15) will not complement the slope for potential evaporation so that Bouchet's relation will be invalid. McNaughton and Spriggs (1989) demonstrated this failure for several days in the Cabauw set.

Is the *linear* increase of α^{-1} with increasing surface resistance (a) peculiar to the Cabauw site because it is close to the North Sea; and/or (b) an artefact of the McNaughton–Spriggs CBL model; or (c) a common feature of CBL behavior? This question has not yet been answered but the analysis that follows would still be valid, albeit in a more cumbersome form, if α were some prescribed function of r_s. As a major caveat, however, diurnal changes in surface resistance, characteristic of all types of vegetation, are likely to make the dependence of daily mean values of α on *real* mean values of resistance more complex than equation (4.4.2) suggests. This point will not be explored further here but to indicate how an element of physiological realism can be introduced, the implications of stomatal closure in response to atmospheric 'demand' are now considered.

Physiological feedback

There is extensive evidence both from observation and from theory (e.g., Leuning, 1995) that the stomatal resistance of *individual leaves* on plants with a good water supply responds to changes of photosynthetic rate and therefore to changes of irradiance in the visible spectrum (I), ambient concentration of CO_2 (ϕ), and ambient temperature (T).

Many workers have also reported that stomatal resistance increases when the saturation vapor pressure deficit of ambient air (D) is increased, but the sensor and physiological mechanism responsible for this dependence have never been identified. (It is even less likely that stomata respond to *relative* humidity as suggested by Ball *et al.* 1987.) Jarvis (1976) suggested that the four responses could be treated as independent so that conductance g_s could be expressed as

$$g_s = g_{sm} f_1(I) f_2(\phi) f_3(T) f_4(D) \tag{4.4.4}$$

where f_1, f_2, etc. are functions each with a maximum of 1 when the associated variable is optimal and where g_{sm} is the maximum conductance achieved in principle (but rarely, if ever, in practice) when all four functions are maximal.

Recent laboratory and theoretical work suggests that leaves do *not* respond directly to saturation vapor pressure deficit or to relative humidity *per se* but to the rate (E) at which they lose water by transpiration (Dewar, 1995; Leuning, 1995; Mott and Parkhurst, 1991). Conveniently, the relation between conductance and transpiration rate is often almost linear so that $f_4(D)$ can be replaced by the expression $(1 - E/E_m)$ where the scaling factor E_m is a fictitious maximum rate of evaporation at which the stomatal conductance would be effectively zero (Monteith, 1995). It is then possible to rewrite equation (4.4.4) in the form

$$g_s = g_m (1 - \alpha/\alpha_p) \tag{4.4.5}$$

where $g_m = g_{sm} f_1(I) f_2(\phi) f_3(T)$ (see equation 4.4.4) and α_p is a physiologically determined Priestley–Taylor coefficient corresponding to a maximum evaporation rate of E_m i.e.

$$\alpha_p = \lambda E_m (\gamma + \Delta) / \Delta H \tag{4.4.6}$$

where λ is the latent heat of vaporization of water.

The response of a canopy of leaves to environmental factors represents the complex integration of responses by individual leaves of different age exposed to different microclimates and transpiring at different rates. Kelliher *et al.* (1995) assembled reassuring evidence that canopy conductances for a wide range of species are mainly two to three times the corresponding values for single leaves, a range that is plausible on physical grounds. However, it is difficult to *predict* the equivalent canopy values of functions f_1, f_2, etc. from single leaf values without recourse to a complex model. A common alternative route is to *measure* the rate of evaporation from a canopy under a wide range of environmental conditions and then, as an act of faith, to fit functions appropriate for single leaves using weather measured above the top of the canopy.

Examples of this procedure were reviewed by Shuttleworth (1989) who demonstrated that radiation and humidity functions were similar for several European pine forests and for

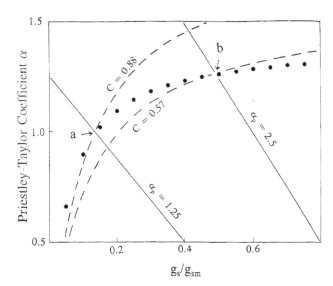

Figure 4.16. Dependence of Priestley–Taylor coefficient α on non-dimensional surface conductance g_s/g_{sm} as given by a CBL model (solid circles), Penman–Monteith equation (dashed lines), and physiological equation (solid lines) (details in text).

Amazonian rainforest. Temperature functions were also similar for two of the pine forests but differed for the rainforest.

Combined physical and physiological feedback

Physical and physiological processes interact when an air mass passes over an extensive area of vegetation. Based on the assumption that vegetation is uniform, Figure 4.16 demonstrates the nature of this interaction for uniform vegetation in terms of three independent relations between daily mean values of the normalized Priestley–Taylor coefficient α/α_m and the normalized stomatal conductance g_s/g_m where $\alpha_m = 1.40$ and $g_m = 0.05$ m s^{-1} were chosen as appropriate scaling factors. The net input of heat to the system was assumed to be 300 W m^{-2} independent of the state of the surface, the aerodynamic conductance was 50 s m^{-1} and the air temperature at screen height was fixed at 20 °C. (In practice, during a long spell of dry weather with constant radiation input, mean air temperature would tend to increase and mean vapor pressure would decrease so that mean vapor pressure deficit would increase. In principle, these secondary effects could be accounted for by iteration but were neglected here.)

The components of Figure 4.16 are as follows.

1. Full points define the dependence of α on g_s as defined by the McNaughton–Spriggs model of the planetary boundary layer reduced to equation (4.4.2) with f set at 0.002. The

locus is not a straight line because increasing the *supply* of water to the atmosphere (by increasing the surface conductance and therefore the evaporation rate) reduces the *demand* as specified by the saturation deficit and therefore by the climate number.

2. Two straight lines represent the dependence of stomatal conductance on transpiration rate (and vice versa) as expressed by equation (4.4.5) with values of $\alpha_p = 1.25$ and 2.5 for the left-hand and right-hand lines respectively.

3. Two curved dashed lines corresponding to estimates of α from the non-dimensional PM equation (equation 4.4.1) for two fixed values of the climate number C, chosen as explained below and with $h = 0.01$ (so that $X = h/f = 5$). These loci, too, are not straight lines because although an increase in surface conductance implies an increase in transpiration rate, the response is partly offset by a decrease in leaf temperature and therefore by a smaller gradient of vapor pressure between sub-stomatal cavities and the ambient air.

For a hypothetical system in a state of equilibrium, all three functions must pass through the same point, thereby defining values of α and g_s. This condition was achieved by setting the climate number at 0.88 for the left-hand curve and at 0.57 for the right-hand curve equivalent to a decrease in saturation deficit of about 53% (assuming no change in net radiation, temperature or aerodynamic resistance). For the triple points of intersection identified by 'a' and 'b', corresponding values were: $\alpha = 1.02$ and 1.27; $g_s/g_m = 0.16$ and 0.50 equivalent to $g_s = 0.008$ and 0.025 m s^{-1} respectively. The second figure of each pair is characteristic of well-watered vegetation (Monteith, 1965; Kelliher, 1995).

By eliminating α from equations (4.4.2) and (4.4.3), the climate number can be expressed as the following function of surface conductance:

$$C = X - 1 - (\alpha_p/\alpha_m)(X - \alpha_m)(1 - g_s/g_m) \qquad (4.4.7)$$

Equation (4.4.7) implies that for a given surface conductance and when the climatologically dependent parameters X, α_m and g_m are held constant, C will increase as water becomes less available to vegetation so that α_p decreases.

In general, as stomata close in response to a reduction in the *supply* of water to roots, the atmosphere becomes drier, thereby increasing the '*demand*' for water (as defined by the climate number). To a limited extent, therefore, the status of the atmosphere is maintained by physical feedback. Conversely, physiological feedback in the form of stomatal closure acts to maintain plant water status as atmospheric demand increases. Acting together, these processes stabilize interactions between vegetation and the atmosphere. When

equations (4.4.2) and (4.4.5) are replaced by more complex functions, or when a boundary-layer model is operated with shorter timesteps (e.g., 1 hour), feedback will continue to operate in the same direction but with differences in scale.

Extension to carbon fluxes

In daylight, the net rate at which vegetation assimilates carbon by photosynthesis is proportional to the product of the stomatal conductance of the canopy and the difference between the effective external and internal CO_2 concentrations of the canopy treated as a big leaf. As the concentration difference is usually much more conservative than the conductance (see, for example, Tanner and Sinclair, 1983), the conductance scale in Figure 4.16 can be treated as an *approximate* scale for photosynthesis rate or even as biomass accumulation rate when, as commonly observed, the daily loss of carbon by respiration is a conservative fraction of the daily gain by photosynthesis. It follows that the carbon gained per unit of water lost (proportional to the quantity often referred to loosely as a 'water use efficiency') is proportional to the ratio of g_s/g_m to α and therefore to the *inverse* slope of lines drawn through the origin of the axes in Figure 4.16.

Using the two intersection points already specified, the water use efficiency is proportional to 0.40 (0.50/1.27) for the right-hand intersection point and 0.13 (0.16/1.02) for the left-hand intersection point. The fractional decrease in efficiency corresponding to a change in α_p from 2.5 to 1.25 is therefore about 67%. The fractional reduction in evaporation rate given by the decrease in α is 27%. This is substantially smaller than the reduction in photosynthesis rate because when stomata close in response to a shortage of water, an increase of foliage temperature increases the vapor pressure gradient across stomata, partly offsetting the impact of a larger stomatal resistance. Depending on other variables, notably irradiance and ambient CO_2 concentration, a small increase of leaf temperature might induce stomata to open slightly, partly offsetting the response to water stress but such compensation would be small.

To summarize, when vegetation is stressed by a shortage of water, growth is likely to be inhibited not simply by stomatal closure but also by a decrease in the amount of biomass produced per unit of water transpired because the atmosphere becomes drier.

Conclusions

Starting from a set of simple algorithms for the relation between the Priestley–Taylor coefficient and surface conductance it is possible to explore qualitatively and even to assess quantitatively how physical and physiological processes interact to determine (a) the daily loss of water from an extensive area of vegetation; (b) the climate number C and therefore the mean daily saturation deficit of air near the surface; and (c) the amount of water lost per unit of assimilated carbon. When more precise daily estimates are needed or for time scales of the order of an hour, the analysis could be made more precise by substituting a full CBL model for equation (4.4.2). The Penman–Monteith equation is virtually independent of time scale but for more precise work should include a term for heat storage in the substrate. The empirical dependence of stomatal conductance on transpiration rate (equation 4.4.5) was obtained from steady-state measurements in the laboratory and has not yet been tested in the field, where the relation is bound to be more complex because of diurnal changes in the amount of water stored within plants and in the accessibility of water to roots.

4.5 Soil water

P. J. Gregory

Introduction

Flow of water in saturated and unsaturated soils obeys laws that are similar to those governing the flow of heat or electricity but the 'transport coefficient' (hydraulic conductivity) depends on the water content (or potential). In unsaturated, non-swelling soils, the 'driving force' for vertical flow is the sum of the gravitational potential and matric potential (a result of adhesive and cohesive forces), and in swelling soils the overburden potential (dependent on the pressure of soil above the point under consideration) is added. Darcy's Law can be combined with the continuity equation to give a partial differential equation governing the flow which can be solved for specific boundary conditions to give the spatial and temporal distributions of soil water. However, the application of this general theory to problems in the field is fraught with problems largely because soils are frequently not homogeneous continua; there is considerable heterogeneity in soil properties and macropores may disrupt the continuity of the matrix.

Principles of vertical water movement

Quantitative description of the flow of fluids through saturated porous media was first made by Darcy (1856) following experimental observations of the flow of water in pipes, open channels and filter beds of sands. Darcy's Law, from which most considerations of water movement in soils have been developed, states that:

$$q = -K \Delta H/L \qquad (4.5.1)$$

where q is the flux density, K is the hydraulic conductivity and $\Delta H/L$ is the gradient of hydraulic head (H being the hydraulic head of water, and L the thickness of the sandbed). Inherent in this approach to describing water flow is the assumption that the medium can be regarded as a continuum so that hydraulic properties are uniform throughout some representative elementary volume (i.e., it ignores the detailed definition of pore geometry and connectivity).

Darcy's Law on its own describes only steady-state flow in which the gradient of hydraulic head and the flux remain constant throughout the conducting system. The extension of this Law to unsteady-state processes, in which the magnitude of the flux and the gradient of hydraulic head may vary in direction and with time, is easily satisfied by the introduction of the law of conservation of mass. The continuity

equation states that the rate of change of flux with distance must equal the rate of change of volumetric water content (θ) with time (t). For flow in one dimension, this gives:

$$\frac{\partial \theta}{\partial t} = \frac{\partial}{\partial x}\left(K \frac{\partial H}{\partial x} \right) \qquad (4.5.2)$$

For vertical flow, the hydraulic head comprises a pressure head (H_p) and a gravitational head (an elevation above a specified reference datum, z). Because $\delta z/\delta z = 1$, equation (4.5.2) can be re-written as:

$$\frac{\partial \theta}{\partial t} = \frac{\partial}{\partial z}\left[K\left(\frac{\partial H_p}{\partial z} + 1 \right) \right] \qquad (4.5.3)$$

In unsaturated soils, there is no positive pressure head (i.e., the water is at sub-atmospheric pressure) and the 'driving force' for water movement is the resultant force arising from the gradients of gravitational, matric (deriving from the interaction of water with the solid soil surfaces and their geometry) and solute potentials, though the latter is only of consequence when there is significant flow via pathways containing differentially permeable surfaces such as water flow into roots and vapor flow through the soil matrix. In most soils, gradients of solute potential are unimportant in the movement of liquid water and the changes in water content can be described using the Richards (1931) equation, which for one-dimensional, horizontal flow is:

$$\frac{\partial \theta}{\partial t} = \frac{\partial}{\partial x}\left[K\left(\Psi_m \right)\frac{\partial \Psi}{\partial x} \right] \qquad (4.5.4)$$

Conductivity is a function of the matric potential (Ψ_m) component of the water potential (Ψ). For vertical flow, Ψ consists of both gravitational (Ψ_g) and matric components so that equation (4.5.4) becomes:

$$\frac{\partial \theta}{\partial t} = \frac{\partial}{\partial z}\left[K\left(\Psi_m \right)\frac{\partial \Psi_m}{\partial z} \right] + \frac{\partial K(\Psi_m)}{\partial z} \qquad (4.5.5)$$

Equations (4.5.4) and (4.5.5) show that in unsaturated soils, hydraulic conductivity is a function of matric potential. However, soil water relations show hysteresis. As a soil dries, air is drawn into pores but because soils comprise pores of non-uniform diameter with variable degrees of connectedness, there are different relations between water content and matric potential depending on whether or not the soil is drying or wetting. This hysteretic behavior means that K is not uniquely related to soil water content.

In defining the gradients of potential, it is usually assumed that conditions are isothermal. Although this is rarely the case in most soil profiles (particularly towards the surface), the assumption introduces only slight errors unless one is concerned with situations where there is significant water vapor movement in dry soils with a pronounced temperature gradient (Philip and de Vries, 1957).

The hydraulic conductivity (K) differs between soils, and for a given soil decreases rapidly with water content typically by 6–9 orders of magnitude between saturation and permanent wilting point (Ψ_m of -1.5 MPa). Typically, K at saturation is 10^{-5} m s^{-1} for a sand and 10^{-7} m s^{-1} for a clay. As water content decreases, the larger pores become air filled and the cross-sectional area of conducting pores decreases rapidly. Poiseuille's Law states that the volumetric flow rate in a pore is proportional to the fourth power of the radius. For example, as the soil dries and the size of water-filled pores changes from a radius of 1 mm to 0.1 mm, the flow rate is reduced by a factor of 10^4 and the average flux density (which is proportional to K) varies by 10^2. The empty pores have to be circumvented so, as drying occurs, tortuosity (a measure of the pathlength) increases. The consequence of this is that K changes very rapidly between saturation and unsaturation and continues to decrease as drying proceeds.

Darcy's Law views the flow of water relative to the solid particles, but what if the particles themselves are moving? In clay soils containing 2:1 clays such as smectites, substantial shrinkage can occur on drying, with swelling on re-wetting. Philip (1969, 1970) set out the essential theory that has now been widely adopted in such soils (Smiles, 1995). Philip introduced an additional potential (the overburden potential, Ψ_o) which defines the work involved in vertically displacing the wet soil as the soil volume changes. The overburden potential is related to the total vertical stress (i.e. the pressure of overlying soil) above the point under consideration and, for a saturated soil, is equal to it. In unsaturated soils, the overburden is shared between the soil water and the solid matrix, and therefore Ψ_o is less than the pressure of the overburden. The physical behavior of swelling soils is different from that of non-swelling soils (Smiles, 1995). For example, gravity will tend to oppose infiltration into a swelling soil because, in a soil swelling normally, the increase in potential energy accompanying swelling is greater than the decrease in potential energy of water entering the soil. Moreover, the effect of gravity is reduced in comparison with the effect of capillary forces so that 'gravity-free' flow persists for longer than in non-swelling systems.

Macropores

Application of the soil-water transport theory described previously is hindered by the fact that soils are rarely homogeneous materials and frequently contain continuous channels such as worm or root holes. Current theory for water flow (such as the Richards' equation; equation 4.5.4) breaks down in such circumstances because the water potential within a soil element is not uniform but might be substantially greater for water within macropores than for water in the immediately adjacent matrix. Indeed, the presence of pores that are continuous over macroscopic distances is a key determinant of soil structure. Macropores is a term that is used loosely to describe continuous channels in soils such as natural cracks and fissures, worm or root holes, and man-made slots. Although generally small in relation to the total volume of pores (0.32% Germann and Beven, 1981; 0.51% Watson and Luxmoore, 1986), they are important because the total water flow through them during and immediately after rain may be considerable even when such channels are few in number and sparsely distributed. For example, Smettem and Collis-George (1985) determined that a single, continuous macropore of 0.3 mm diameter could conduct more water than the rest of a 100 mm diameter soil sample. A feature of macropore flow (sometimes referred to as preferential flow) is its short duration which is typically only two to three times as long as the period of water application to the soil surface.

Macropores have been defined in several arbitrary ways (by size, by volume fraction, by tension at which they are emptied of water, by infiltration rate) but in terms of water movement they are probably best defined as pores of different radii in which the flux density occurring in the minimum pore size, with unit hydraulic gradient, is greater than or equal to the saturated matrix hydraulic conductivity (Chen and Wagenet, 1992). Beven and Germann (1982) and Chen and Wagenet (1992) describe similar conceptual approaches to macropore flow and stress that the occurrence of macropore flow depends on the surface boundary conditions rather than the water content (or water potential) of the bulk matrix. Assuming that macropores are continuous from the surface downwards and that the soil matrix is homogeneous, they outline three situations leading to water flow:

1. Macropore control. With surface water just sufficient for ponding (no positive head), and an application rate greater than the matrix hydraulic conductivity plus saturated hydraulic conductivity of the macropores, water flows in the macropores and in the matrix producing a wetting front. The rapid flow in the macropores also results in some transfer of water into the soil matrix because

Table 4.3. *Potential rates of flow in individual cylindrical macropores, and potential infiltration rates into the soil with fractional macropore areas of 0.01 and 0.05*

Pore diameter (μm)	Tension (kPa)	Flow rate (m³ s⁻¹)	Potential infiltration (mm h⁻¹) Fractional macropore area	
			0.01	0.05
2000	0.0015	3.8×10^{-6}	4.4×10^{4}	2.2×10^{5}
1000	0.003	2.4×10^{-7}	1.1×10^{4}	5.5×10^{4}
200	0.015	3.8×10^{-10}	4.4×10^{2}	2.2×10^{3}
20	0.15	3.8×10^{-14}	4.4×10^{0}	2.2×10^{1}

Note: From White (1985).

of the difference in water potential between the matrix and macropore. If the application of water continues, the macropores will fill with water before the matrix is saturated.

2. Application control. If the application rate of water is greater than the conductivity of the soil into which infiltration is occurring, but less than the sum of the soil matrix conductivity and the macropore saturated conductivity, then the macropores will be partially wetted, giving rise to a partially saturated hydraulic conductivity. An application rate of water greater than the conductivity of the matrix results in macropore flow and, if the application continues for long enough, all macropores will eventually reach saturation.

3. Matrix control. When the application rate is less than or equal to the matrix conductivity, all of the applied water will flow through the matrix. The direct infiltration into the macropores can be neglected because they contribute little to the total surface area.

These descriptions indicate that quantification of flow in macropore systems is not possible simply by extending Darcy's Law because the assumption of homogeneity of hydraulic properties over some representative elemental volume is no longer valid. Theoretically, the relationship between flow rate and size of macropore can be determined by assuming laminar flow induced by gravity in vertical pores. Flow in cylindrical pores is proportional to the fourth power of the radius (Poiseuille's Law) and flow in slits is proportional to the cube of the width (Childs, 1969). Figure 4.17 shows the effect of pore and slit size on the flux of water. White (1985) calculated the potential flow rate in individual pores within the macropore size range, and the potential rate of infiltration when the macropores occupied either 1% or 5% of the surface area assuming that the area between the macropores made no contribution to infiltration (Table

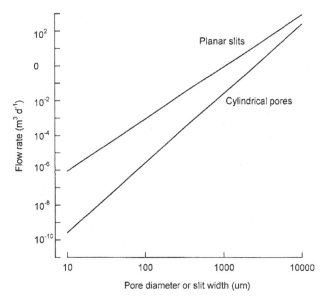

Figure 4.17. The relationship between flow rate and pore size of circular or planar pores of unit length with unit hydraulic gradient.

4.3). As macropore diameter increased from 20 μm, the potential infiltration rate by macropores increased rapidly (Table 4.3).

The potential rates shown in Table 4.3 are rarely achieved in practice because soil pores are irregular in shape and continuity, have rough walls, and contain entrapped air. For example, Ehlers (1975) measured an average flow rate of 3×10^{-8} m³ s⁻¹ for earthworm burrows of 2000 to 5000 μm diameter suggesting that, under natural conditions, even rates of flow in the relatively smooth pores made by worms are 100-fold smaller than the potential values shown in Table 4.3. Much remains to be done to combine potential flow theory with boundary-layer flow theory to develop a theoretical framework allowing the prediction of macropore flow under the range of conditions existing in soils.

Fingering during infiltration

The infiltration of water into layered soils sometimes results in a non-uniform wetting front with the production of distinct and spatially separated flow pathways called fingers. Such unstable wetting fronts are particularly prevalent when a fine-textured soil overlies a coarser textured subsoil. There are many theoretical attempts to explain this behavior but most fail to explain why an initially well-distributed flow field should sometimes break into fingers. Clearly such behavior lies outside the theoretical construct provided by Darcy's Law but many soil profiles comprise layered soils.

Hillel and Baker (1988) offer a relatively simple explanation for such behavior. The basic hypothesis is that a spatially distributed flow field tends to constrict where the flow accelerates. If the field is wide, this constriction may cause the flow field to break into discrete, separated streams. For a soil with two layers where the top layer is less conductive than the sublayer, when the wetting front is only in the top layer it is characterized by a low value of potential. When the front arrives at the interface, the infiltrating water is at too low a potential to enter into the coarser pores of the sublayer. So, the front pauses at the interlayer and more water arriving causes the potential to increase. When the potential at the interlayer plane has increased to a value allowing entry into the smallest pores forming a continuous network in the sublayer, water will begin to enter at various locations. As the potential continues to rise, larger pores start to conduct water until the effective water-entry potential is reached at which point the sublayer is able to conduct the total flux that the top layer is capable of delivering. If the conductivity of the sublayer at the effective water entry potential exceeds the flux through the top layer, then the water velocity in the pores necessarily increases across the interlayer plane and only a fraction of the sublayer conducts the water being delivered by the top layer. The result is fingering (Hillel and Baker, 1988). Experimental results confirm this hypothesis for soils exhibiting both sharp and diffuse wetting fronts and the resultant quantitative model allows predictions of the fraction of wetted soil (Baker and Hillel, 1990). Such considerations are important because they provide another means whereby water and solutes flowing in soils may bypass most of the soil volume.

Heterogeneity and determination of hydraulic conductivity

From the preceding descriptions, it is clear that a knowledge of the soil hydraulic conductivity (K) is a prerequisite for calculating flow. However, in practice it is difficult to obtain a representative elemental volume of soil on which to make measurements. For example, Anderson and Bouma (1973) found that the saturated hydraulic conductivity of the B horizon (subsoil) of a silt loam decreased from an average of $6.5\,\mathrm{m\ day^{-1}}$ to $1.0\,\mathrm{m\ day^{-1}}$ as sample height increased from 5 cm to 17 cm. This difference was explained by considering the vertical patterns of cracks between soil peds (i.e., the individual soil structures). Because hydraulic conductivity of a sample is influenced strongly by the 'necks' in irregular pores, and the probability of a small crack in a vertical flow system increases as the system lengthens, so K decreases with increasing sample length. However, beyond a certain length, a narrow crack always occurs so that K becomes independent of sample size. Bouma (1989) discusses the effects of sample size, pore connectivity, and other sampling issues that give rise to substantial variations in the measured values of K particularly as a consequence of macropores.

Chen *et al.* (1993) proposed a method for estimating matrix and macropore conductivities from field measurements of water content following saturation of a soil profile. The method is based on a two-domain model (Chen and Wagenet, 1992) in which the total saturated hydraulic conductivity K_s is:

$$K_s = K_{sh} + K_{sm} \tag{4.5.6}$$

where K_{sh} is the matrix saturated hydraulic conductivity and K_{sm} is the macropore saturated hydraulic conductivity. The method further assumes that drainage is a three-stage process which is initially dominated by macropores, proceeds to mixed drainage from both macropores and matrix, and finally changes to drainage of the matrix alone. When two-domain behavior exists, a logarithmic plot of soil volumetric water content (θ) against time should give a plot of two straight lines with a discernible breakpoint. Although the breakpoint does not indicate the transition of macropore to matrix flow, such a plot, requiring only measurements of $\theta(z,t)$, can be analysed to provide values of K_{sh} and K_{sm}. The method is simple, although the accuracy is highly dependent on the accuracy of the measurements (Chen *et al.*, 1993).

Because measurements of K can be time-consuming, expensive and technically demanding, there is a continued, and increasing, interest in predicting soil hydraulic properties from more easily measured and available soil properties such as particle size distribution and water retention data. When $\theta \leq \theta_{sh}$, the relationship between matrix hydraulic conductivity (K_h) and the matrix water content (θ) is (Campbell, 1974):

$$K_h = K_{sh}\left(\frac{\theta}{\theta_{sh}}\right)^{2b+3} \tag{4.5.7}$$

where θ_{sh} is the saturated water content of the matrix and b is a constant. This function and other such exponential functions have proved useful for many soils (van Genuchten and Nielsen, 1985) and can be extended to describe more complex soils where macropore flow is evident. For example, Ross and Smettem (1993) have indicated that the hydraulic properties of some soils can be conceived of as resulting from a series of overlapping pore-size distributions and described by summing simple functions.

Scaling from point to catchment

Developments in the numerical simulation of water and solute flow (Belmans *et al.*, 1983; Campbell, 1985; Ross, 1990) have led to the realization that a major constraint to the application of such models is the limited availability of soil hydraulic properties and, in particular, accurate and spatially integrated values of soil hydraulic conductivity (K). The spatial variability of hydraulic properties has significant consequences when attempting to predict the response of a catchment. For example, Sharma and Luxmoore (1979) simulated the components of the water balance for a grassland catchment and compared the integrated response of vertical soil columns of different scales with the results for average scale factors (mean, mode and median). They showed that the water balance simulated by the average factors was different from that simulated by either the normal or log-normal distribution (Table 4.4). The major difference was in the amount of surface runoff predicted, which although a small proportion of the total water balance, varied 15-fold depending on the scale. Many hydrologic problems are regional in nature but the data available for hydrologic processes are often only at a field-plot scale so it is therefore necessary to scale from the local to the regional level. This scaling is usually achieved using one of three approaches: scaling, stochastic analysis or inverse modeling.

Scaling techniques have been widely used to derive areal values from the results of one-dimensional simulations of soil water flow (e.g., Warrick *et al.*, 1977; Hopmans, 1987). Scaling theory, based on the concept of similar media, pro-

Table 4.4. *Annual, simulated water balance (values in mm) of a watershed with an annual precipitation of 1076 mm using different methods for scaling*

	Normally distributed	Log normally distributed	Mean
Evapotranspiration	637.7	646.5	634.8
Drainage	394.1	410.1	451.2
Runoff	44.9	20.7	2.9
Change in soil water storage	−0.3	−0.7	0.2

Note: From Sharma and Luxmoore (1979).

vides a framework for expressing the variability of various soil hydrological properties using a single parameter. Such scaling factors are obtained from easily measurable parameters such as soil particle size analysis or the moisture characteristic curve. However, many soil properties are also spatially dependent so that stochastic approaches (based on generating soil hydraulic properties using Monte-Carlo techniques) have become favored (e.g. Russo and Bresler, 1981).

More recently, rather than attempting to start with soil properties and scale up to the catchment, the catchment has been measured and used to derive the spatially averaged soil hydraulic properties. Using 'inverse modeling', soil hydraulic properties are fitted to areal water movement in combination with the Richards' equation (equation 4.5.4) to describe the hydrologic behavior of the catchment utilizing non-linear parameter optimization techniques (Feddes *et al.*, 1993). This approach has the advantage that it is independent of experimental design and only requires that the appropriate initial and boundary conditions are known. The unknown parameters in the model are determined by matching the model to the available data. The determination of the 'effective' soil hydraulic properties at catchment scale, although without theoretical justification, may be of considerable practical benefit especially where boundary conditions (e.g. rooting depth) and hydraulic properties vary concurrently.

4.6 Surface runoff and subsurface redistribution of water

Eric F. Wood

This section discusses the pathways for water at the surface and the redistribution of water below the surface soil zone. The modeling of the terrestrial water balance, as used in GCMs, is discussed in Section 4.4. The principles of infiltration and vertical movement of moisture in the unsaturated zone is discussed in detail in Section 4.5. In general, the topic of Section 4.6 has often been referred to as 'runoff generation' in the hydrology literature, and there are a number of excellent reviews (for example, Freeze, 1974) and books which focus on the problem (e.g. Kirkby, 1978; Anderson and Burt, 1990). In this section, we will briefly review the various pathways and potential modeling approaches for modeling subsurface water flow and runoff generation.

The problem appears deceptively simple: for any particular rainfall event determine the portion of incident rainfall which reaches the stream to become discharge during or immediately after the rainfall event. This surface runoff is measured as the storm hydrograph. The residual rainwater is (often) assumed to infiltrate into the soil, redistribute either vertically or horizontally (down the hillslope) and either return to the atmosphere through evapotranspiration or contribute to streamflow over an extended period of time (baseflow discharge.) It is important to be able to estimate the 'effective rainfall', which is the rainfall volume equal to the volume of the stream hydrograph, and to know the time transformation between the effective rainfall and the stream hydrograph which is controlled by catchment characteristics that govern its retardation and diffusive effects. For the modeling of the land surface hydrology within land–atmospheric coupled systems, it has been shown from intercomparisons carried out within the Project for the Intercomparison of Land Parameterization Schemes (PILPS) (for example, the results presented in Chen *et al.*, 1997) that modeling runoff volumes correctly is critical to accurate modeling of latent and sensible heat fluxes.

There is a broad spectrum of processes that are involved in the generation of runoff and streamflow and in the redistribution of infiltrated water. Figure 4.18, from Beven (1986), presents some mechanisms for runoff production. As discussed by Beven (1986), Sklash (1990) and Wood *et al.* (1990) among others, the response of any particular catchment may be dominated by a single mechanism, or by a combination, depending on the magnitude of the rainfall event, the antecedent soil moisture conditions of the catchment, and/or the heterogeneity in soil hydraulic properties. Thus, during any particular storm, different mechanisms may generate runoff from different parts of a catchment. Surface runoff from these

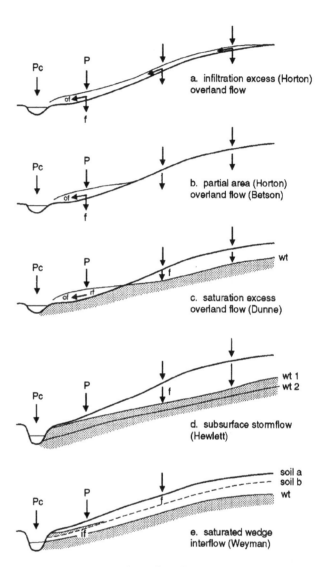

Figure 4.18. Mechanisms of runoff production. (From Beven, 1986.)

(partial) contributing areas may be generated by either the infiltration excess mechanism (*a* and *b* in Figure 4.18) where the rainfall rate exceeds the infiltration capacity of the soil, or from rainfall on areas of soil saturated by a rising water table even in high permeability soils. Such 'saturation excess overland flow', is represented as mechanisms (*c*) to (*e*) in Figure 4.18.

The resulting hydrograph is often divided into a number of components depending on the pathway and time delay from the time the rain drop hits the ground surface to the time it enters the stream. These components are usually divided into the direct surface runoff, subsurface (or interflow) discharge and groundwater baseflow discharge. Direct surface runoff results from rain falling on saturated surfaces (either onto partial-saturated areas due to rain rates greater than the local infiltration rate or onto near-stream saturated areas resulting from emerging water tables). These saturated contributing areas expand and contract during and between rainfall events. Subsurface discharge from the unsaturated soil zone (often referred to as interflow) often arises from either flow through macropores in the upper soil (see below) or through downslope flow through the unsaturated soil matrix, often enhanced due to vertical anisotropic soil properties – especially the occurrence of an impeding soil layer to vertical infiltrating water (McCord et al., 1991)

It is also known as a result of using natural tracers to determine the source of streamflow, that in many catchments a significant portion of the storm hydrograph is derived from subsurface water that is displaced from soil and groundwater by the incoming infiltrated rainwater (e.g. see Sklash et al., 1976; Stewart and McDonnell, 1991; Yoshida et al., 1995. Sklash, 1990, provides a review of such tracer studies) . It is clear that the modeling of runoff needs to deal with the spatial patterns of antecedent moisture conditions, soil hydraulic properties, and rainfall intensities, with the expectations that under many conditions these patterns will be highly variable in space and time.

Observed spatial variability

During the last fifteen years, there has been an increasing awareness of the problems posed by the natural spatial heterogeneity found in catchment characteristics and meteorological inputs on runoff generation, soil moisture and evapotranspiration. Some results for specific models are discussed in Wood et al., 1988; Woolhiser and Goodrich, 1988; Beven et al., 1988; Goodrich, 1990, and Grayson et al., 1992a; reviews include Beven, 1983a and Wood, 1994. Variability caused by precipitation variability (including partial precipitation over a catchment), topography, different soil and vegetation types is readily apparent in spatial and temporal pattern of soil moisture, runoff and evaporation. Even variations within seemingly homogeneous units (soil or vegetation) can also cause significant variations in soil moisture and its redistribution as well as the resulting fluxes of water (runoff and evaporation.)

To appreciate the natural variability in runoff that can occur, consider the field results observed by Hjelmfelt and Burwell (1984). Measurements of the runoff from forty 0.01-ha plots (adjacent to each other) were obtained over one season with storm totals ranging from 6 mm to 96 mm. The coefficient of variation for storm runoff volumes across the

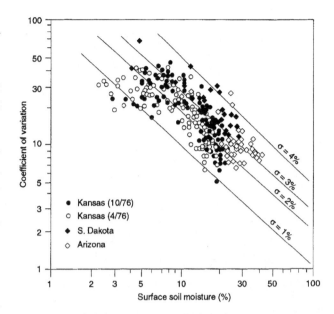

Figure 4.19. Coefficient of variation versus average soil moisture. (After Bell et al., 1980.)

forty plots ranges from 0.071 to 1.09, showing dramatic variations in runoff over a very small area. In addition, there was considerable temporal scatter in the runoff ratios for individual plots. Thus even at the smallest experimental plots, significant variability exists.

Variation in soil moisture

As discussed above and in Section 4.5, surface soil moisture plays a significant role in the infiltration of rainwater and the subsequent subsurface redistribution. There has been a significant number of investigations into how infiltration rates, hydraulic conductivity, porosity, soil moisture contents and other soil characteristics vary in space (e.g. Nielsen et al., 1973; Bell et al., 1980; Viera et al., 1981). Beven (1983a) provides a review of this research and how it influences runoff generation. Figure 4.19, from Bell et al. (1980) shows the relationship between measured surface soil moisture within agricultural fields to its coefficient of variation. The data show that the variance in soil moisture is fairly constant across a wide range of soil moisture, implying that this may be due to the underlying soil properties. It must be noted that this was for homogeneous agricultural fields and larger variability would be expected for natural catchments.

Macropores and redistribution of soil water

Beven and Germann (1982) and Germann (1990) provide extensive reviews of macropores and their important role in

(a)

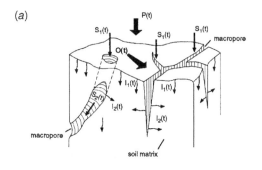

(b)

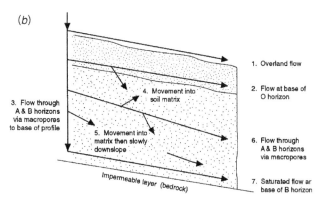

1. Overland flow

2. Flow at base of O horizon

3. Flow through A & B horizons via macropores to base of profile

4. Movement into soil matrix

5. Movement into matrix then slowly downslope

6. Flow through A & B horizons via macropores

7. Saturated flow ar base of B horizon

Figure 4.20. (*a*) Schematic of water flows into soil with macropores: P(t), precipitation; I_1(t), infiltration into the matrix from the surface; I_2(t) infiltration into the matrix from the walls of the macropores at the soil surface; S_1(t), seepage into the macropores at the soil surface; S_2(t), flow within the macropores; O(t), overland flow. (From Beven and Germann, 1982.) (*b*) Flow paths for the movement of water through a shallow soil with macropores. (From Beven and Germann, 1982.)

subsurface water flow. Figure 4.20, taken from Beven and Germann (1982) provides a framework for understanding the role of macropores in infiltration and redistribution of water. Field studies have grouped macropores into four categories: pores formed by (*a*) the soil fauna (earthworms, gophers, moles, etc.), (b) plant roots, (c) soil cracks and fissures, and (d) natural soil pipes. Landuse has a significant role in the formation and changes in soil macropores, and in many catchments it is widely accepted that the magnitude and shape of the storm hydrograph is controlled by subsurface flows due to macropore flow (Beven and Germann, 1982). This suggests that macropores enhance both the vertical infiltration of water, and the lateral downslope redistribution.

Modeling approaches

The terrestrial water balance, which for a control volume, may be written over a time interval d*t* as:

$$< d\theta > = < P > - < E > - < Q > \qquad (4.6.1)$$

where θ represents the moisture in the soil column, E the evaporation from the land surface into the atmosphere, P the precipitation, and Q the net runoff. The spatial average for the control volume is represented by $\langle \cdot \rangle$. Through the parameterization of equation (4.6.1), the water balance model become a 'distributed' or a 'lumped' model, and in particular by the averaging of the variables. By distributed, we refer to a model which accounts for the spatial variability in hydrologic inputs, processes and parameters. This accounting can be done in two ways: (a) deterministically, in which case the actual patterns of variability are represented – examples of such models include the European Hydrological System model, SHE, (Abbott *et al.*, 1986; Bathurst, 1986) and the three-dimensional models of Binley *et al.* (1989), or Paniconi and Wood (1993) or (b) statistically, in which case the patterns of variability are represented statistically – examples being models like TOPMODEL (Beven and Kirkby, 1979) and its variants (see Moore *et al.*, 1988; Grayson *et al.*, 1992*a*; Wood *et al.*, 1992) in which topography, soils and vegetation play an important role in the distribution within a catchment; or models in which variability is implicitly assumed (see for example Liang *et al.*, 1994).

By a lumped model, we mean a model that represents the catchment (or control volume) as being spatially homogeneous with regard to inputs, processes, and parameters – usually through averaging. There is a wide number of hydrologic models of varying complexity that do not consider spatial variability. These range from the well-known unit hydrograph (for the runoff response to effective rainfall) to complex, atmospheric-biospheric Soil-Vegetative-Atmospheric Transfer schemes (SVAT) models. Examples of the latter are the Biosphere-Atmosphere Transfer Scheme (BATS) (Dickinson, 1984) and the Simple Biosphere Model (SiB) (Sellers *et al.*, 1986.) An extensive review of modeling approaches is beyond the scope of this section, as is a review of modeling 'comparison' studies. Loague and Freeze (1985) compared a range of rainfall-runoff models, some calibrated to observations and others using measured parameters such as soil hydraulic conductivity, to attempt to determine performance across a range of models. The ability of models to simulate observations has also been discussed by Grayson *et al.* (1992*b*) which resulted in provocative comments by Smith *et al.* (1994). The reader is encouraged to read these and similar papers on the parameterization of hydrologic models to understand these issues further (for example, Beven, 1989; Jackman and Hornberger, 1993; and Beven, 1995, 1996).

Over the last ten years, the most significant new approach in modeling land surface hydrology is the development of

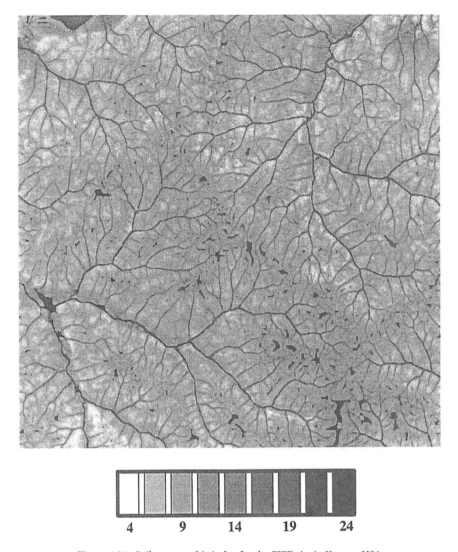

Figure 4.21. Soil topographic index for the FIFE site in Kansas, USA.

coupled water and surface energy balance models. These models have utilized experimental data collected through climate field programs (like the First ISLSCP Field Experiment, FIFE; HAPEX-Sahel; or the Boreal Ecosystem Atmospheric Study, BOREAS) (see Section 5.1) and are forced by incoming radiation and precipitation. The development of these models has been motivated by the need to improve the land surface parameterization in climate models. Section 4.4 presents these models in greater detail than can be done in this section.

The intercomparison of this class of models is taking place under PILPS (Henderson-Sellers and Brown, 1992; Henderson-Sellers *et al.*, 1995.) A significant finding of the comparisons to date is the strong interaction between the water and energy cycles (Chen *et al.*, 1997; Liang *et al.*, 1997; Letten-

maier *et al.*, 1996). Errors in modeled runoff are reflected in compensating errors in evapotranspiration (latent heat) and thus sensible heat. Further studies with a SVAT model have shown the importance of subgrid variability in precipitation (see Liang *et al.*, 1996*b*).

There is still a wide number of models that can not consider variability in their response at scales smaller that the catchment or grid scales. For the remaining portion of this section we will restrict our discussion to model formulations that have a 'distributed' structure in that they represent variability in inputs (radiation and precipitation), topography, soils, and vegetation – so called 'distributed' hydrologic models. Because of the limitations of space, the discussion will not be exhaustive but will attempt to represent the various 'classes' of the distributed models.

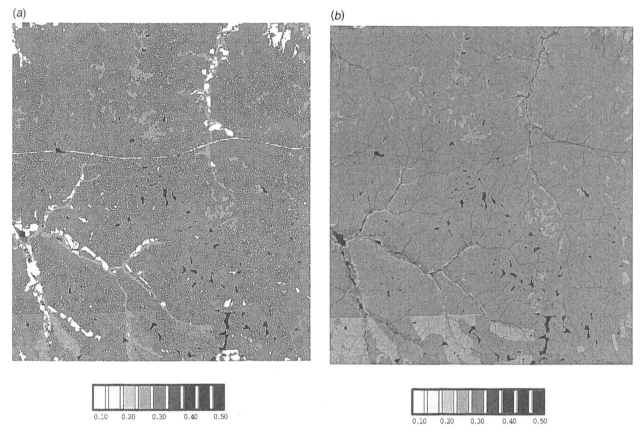

(a)　　　　　　　　　　　　　　　　　　　　(b)

0.10　0.20　0.30　0.40　0.50　　　　　　　0.10　0.20　0.30　0.40　0.50

Figure 4.22. (*a*) Modeled surface zone soil moisture for 1845 GMT, August 17, 1987 for the FIFE site in Kansas, USA. (*b*) Modeled lower zone soil moisture for 1845 GMT, August 17, 1987 for the FIFE site in Kansas, USA. (*c*) Modeled water table depth for 1845 GMT, August 17, 1987 for the FIFE site in Kansas, USA.

Three-dimensional distributed models of subsurface redistribution

Distributed models have the possibility of describing (and predicting) different pathways for the water movement at the surface and through the subsurface, whereas lumped models can not. The three-dimensional Richards equation (Richards, 1931) can be used to represent water movement in both the saturated and unsaturated soil zone and is often regarded as being the most 'physically based' representation of water movement through soil. It can be written, with pressure head ψ as the dependent variable, as:

$$S(\psi)\frac{\partial \psi}{\partial t} = \nabla \cdot [K_s K_r(\psi)\nabla(\psi + z)] \qquad (4.6.2)$$

where $S(\psi)$ is the specific moisture capacity, t is time, z is the vertical coordinate (positive upwards) and the hydraulic conductivity K is expressed as a function of the conductivity at saturation, K_s, and the relative conductivity, K_r. The specific moisture capacity, hydraulic conductivity and thus the vol-

umetric soil moisture, θ, which are nonlinear functions of the pressure head, are usually parameterized using relationships such as those developed in van Genuchten and Nielsen (1985) among others.

Equation (4.6.2) is solved through numerical methods applied to finite difference or finite element approximations. For example, Paniconi and Wood (1993) represented equation (4.6.2) using a finite element Gelerkin discretization in space and a finite difference discretization of the time derivative term. The set of numerical equations are augmented by appropriate boundary conditions (for example, boundary moisture fluxes which for the surface would be infiltrated or evaporated water). These models are usually run at a fine grid resolution (on the order of meters) and high time resolution (on the order of tens of seconds per time step). The theoretical basis of the model and the detailed model domain has not been translated into model predictions that well match observations of moisture fluxes in the unsaturated zone. It is this author's belief that the poor match between model predictions and point observations is related to two causes. The

(c)

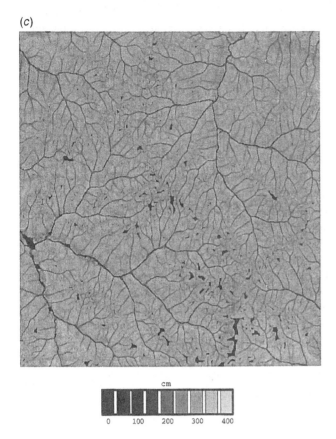

cm

0   100   200   300   400

Figure 4.22 (c)

first cause is the inability to measure and specify soil properties at the model grid resolution where the highly nonlinear (with respect to soil properties) unsaturated flow equations must be solved. The second cause arises from not including important, related processes like flow through soil macropores (see Beven and Germann, 1982 and Germann, 1990 for reviews) and surface run-on, both of which are important at these model grid scales. Since these models rely on local gradients, the small scale variability in soil properties (on the order of one to tens of meters) can have a significant effect on model predictions.

There has been more success in utilizing equation (4.6.2) for saturated flows, and this in fact forms the basis of current groundwater models. For saturated conditions, equation (4.6.2) becomes a linear equation which facilitates its solution. Additionally, for many saturated flow problems, the boundary conditions are specified through pressure (head) along the boundary. While these models have appeared to give good matches for groundwater flows, when coupled to a transport equation (that predicts the movement of a constituent in the groundwater) these predictions tend to be poorer. This seems

to indicate that even for saturated conditions, model-predicted local flow velocities may be in error, but that variables that are more spatially uniform (like pressure fields) are better predicted. In addition, it implies that predicted flows, averaged over a larger domain, may compare well to observations.

Topographically based distributed models of subsurface redistribution

Through a number of field studies during the 1960s, many of which are reviewed in Kirkby (1978) and in Anderson and Burt (1990), there was an increased recognition that subsurface flows play a significant role in the spatial variation of soil moisture within a catchment. During interstorm periods, the subsurface redistribution of soil moisture, with the soil properties, sets up spatially variable initial conditions for infiltration (during rain events) and evaporation (between events). In addition it was shown that for areas of relatively shallow soils, the dominant control on the subsurface flows is topography. While models based on equation (4.6.2) can explicitly incorporate variations in topography, difficulties with this approach led Beven and Kirkby (1979) to formulate a model (TOPMODEL) for transforming rainfall into runoff that utilizes a spatially variable topographic index to predict the extent of saturated areas which generate surface runoff. The model utilizes their hydrologic similarity theory in which locations within a catchment with the same topographic index respond in a similar manner to the same inputs. A similar approach for predicting areas of saturated soils has been derived independently by O'Loughlin (1981, 1986), for more general profiles of saturated hydraulic conductivity, and Grayson (1990) and Grayson *et al.* (1992a) also developed a terrain-based hydrologic model initially for erosion studies. With widely available digital elevation models (DEM) for topography and digital soil data bases, it is relatively straightforward to calculate the topographic index for an area. Figure 4.21 shows the index for an area of Kansas studied as part of the First ISLSCP Field Experiment (FIFE), a 15 × 15 km study area (Sellers *et al.*, 1988).

The original concepts of TOPMODEL have been extended to consider spatial soil properties (Beven, 1983b), infiltration excess mechanisms for runoff production in addition to saturation excess (Sivapalan *et al.*, 1987; Wood *et al.*, 1988), distributed inputs of water and energy (Famiglietti *et al.*, 1992; Famiglietti and Wood, 1994), and an upper thin layer whose rapid soil moisture response is dominated by local soil and vegetation characteristics (Zion, 1995), and a generalized theory of downslope flows that relaxes the exponential relationship and reliance on local topographic slopes (Ambroise *et al.*, 1996). In this way the model has evolved over time from a rainfall-runoff (water balance) model to a land surface model that could potentially be used in a climate model.

For modeling land surface–atmospheric interactions, the inclusion of a thin upper soil layer appears important in the prediction of the land surface water and energy balance (Liang et al., 1996a). This upper layer responds to atmospheric forcings on much shorter time scales, especially during the warm season, while the subsurface redistribution occurs on longer time scales. Figure 4.22, for the FIFE study area, illustrates this. Figure 4.22(a) shows the soil moisture in the upper 5 cm where its value is dominated by atmospheric forcings and soil properties. Figure 4.22(b) shows the average soil moisture for the lower zone (from 5 cm to the water table). Here we see the influence of topography start to exert itself. Figure 4.22(c) shows the predicted water depth, a variable that is predicted from the TOPMODEL theory. These should be compared with Figure 4.21.

Summary

We have briefly presented the general problem of runoff generation within catchments or grid squares and the modeling of such processes. There is long history of modeling runoff from catchments within the hydrology community. Currently there is fierce debate on the degree that recognizable physical processes can be or should be represented explicitly, and the extent that model constants can be calibrated using observed inputs and outputs. The debate includes issues such as: does representing more processes improve model performance, and does it reduce the need for calibration? This debate can be seen in papers such as Beven (1989, 1996), Grayson et al. (1992b), Jackman and Hornberger (1993) and Smith et al. (1994).

It is clear that performance of the models in calculating the correct runoff volumes and timing directly affects the determination of all other water and energy flux and storage terms; i.e. the water and energy cycles are tightly coupled. This conclusion is clear from the PILPS-2c intercomparison (see Lettenmaier et al., 1996). It is less obvious – but it appears to be supported by studies – that models which explicitly consider subgrid variability in inputs and catchment/grid characteristics (soil, vegetation, topography) can better represent water and energy fluxes, especially at small temporal scales (Wood, 1997; Sellers et al., 1995; Liang et al., 1996b). This latter conclusion is still open to considerable discussion in the literature.

References

Aagaard, K. and Carmack, E. C. (1989). The role of sea ice and other freshwater in the Arctic circulation. *J. Geophys. Res.*, **94**, 14,485–98.

Abbott, M. B., Bathurst, J. C., Cunge, J. A., O'Connell, P. E. and Rasmussen, J. (1986). An introduction to the European Hydrological System – Systeme Hydrologique European SHE. *J. Hydrol.*, **87**, 45–77.

Ambroise, B., Beven, K. J. and Freer, J. (1996). Towards a generalisation of the TOPMODEL concepts: topographic indices of hydrological similarity, *Water Resour. Res.*, 2135–45.

Anderson, D. L. and Bouma, J. (1973). Relationships between hydraulic conductivity and morphometric data of an argillic horizon. *Soil Science Society of America Proceedings* **37**, 408–13.

Anderson, M. G. and Burt T. P. (eds.) (1990). *Process Studies in Hillslope Hydrology*. Chichester: John Wiley & Sons, 539 pp.

Anderson, S. P., Weller, R. A. and Lukas, R. B. (1996). Surface buoyancy forcing and the mixed layer of the Western Pacific Warm Pool: observations and 1-D model results. *Journal of Climate*, **9** (12), 3056–85.

Andreas, E. L. (1992). Sea spray and the turbulent air-sea heat fluxes. *J. Geophys. Res.*, **97**, 11429–41.

Andreas, E. L. (1994). Reply. *J. Geophys. Res.*, **99**, 14345–50.

Andreas, E. L., Edson, J. B., Monahan, E. C., Rouault, M. P. and Smith, S. D. (1995). The spray contribution to net evaporation from the sea: a review of recent progress. *Boundary-Layer Meteorol.*, **72**, 3–52.

Arkin, P. A. and Ardanuy, P. E. (1989). Estimating climatic-scale precipitation from space: a review. *J. Climate*, **2**, 1229–38.

Arkin, P. A. and Janowiak, J. E. (1993). Tropical and sub-tropical precipitation in *Atlas of Satellite Observations related to Global change*, ed. R. J. Gurney, J. L. Foster and C. L. Parkinson, pp. 165–80, Cambridge University Press.

Atlas, D. and Bell, T. L. (1992). The relation of radar to cloud area-time integrals and implications for rain measurements from space. *Mon. Wea. Rev.*, **120**, 1997–2008.

Atlas, D., Rosenfeld, D. and Jameson, A. R. (1996). Evaluation of radar rainfall measurement: steps and mis-steps. In *Weather Radar Technology for Water Resources Management*, ed. B. Braga and O. Massambani, pp. 1–60. Montevideo: UNESCO Press.

Baker, R. S. and Hillel, D. (1990). Laboratory tests of a theory of fingering during infiltration into layered soils. *Soil Science Society of America Journal*, **54**, 20–30.

Ball, J. T, Woodrow, I. E., and Berry, J. A. (1987). A model predicting stomatal conductance and its contribution to the control of photosynthesis under different environmental conditions. In *Progress in Photosynthesis Research, Vol.IV*, ed. J. Biggins, pp. 221–34. Dordrecht: Martinus Nijhof Publishers.

Barton, I. J. (1995). Satellite-derived sea surface temperatures: current status. *J. Geophys. Res.*, **100**(C5), 8777–90.

Barton, I. J., Prata, A. J. and Cechet, R. P. (1995). Validation of the ATSR in Australian Waters. *J. Atmos. & Oceanic Tech.*, **12**, 290–300.

Bathurst, J. C. (1986). Physically-based distributed modeling of an upland catchment using the Systeme Hydrologique Européen. *J. Hydrol.*, **87**, 79–102.

Battan, L. J. (1973). *Radar Observation of the Atmosphere*. Chicago: University of Chicago Press, 324 pp.

Baumgartner, A. and Reichel, E. (1975). *The World Water Balance*. New York: Elsevier, 179 pp.

Bell, K. R., Blanchard, B. J., Schmugge, T. J. and Witczak, M. W. (1980). Analysis of surface moisture variations within large-field sites. *Water Resour. Res.*, **16**(4), 796–810.

Bellon, A. and Austin, G. L. (1986). On the relative accuracy of satellite and raingauge rainfall measurements over middle latitudes during daylight hours. *J. Climate App. Met.*, **25**, 1712–24.

Belmans, C., Wesseling, J. G. and Feddes, R. A. (1983). Simulation model of the water balance of a cropped soil: SWATRE. *J. Hydrol.*, **63**, 271–86.

Bentamy, A., Quilfen, Y., Queffeulou, P. and Cavanie, A. (1994). *Calibration of the ERS-1 Scatterometer C-Band Model*. Technical Report DRO/OS-94-01, Brest: IFREMER, 72 pp.

Beven, K. J. (1983a). Surface water hydrology: runoff generation and basin structure. In *Contributions in Hydrology*, U.S. National Report 1979–1982 to IUGG, *Rev. of Geophy. and Sp. Phy.*, **21**(3), 721–9.

Beven, K. J. (1983b). *Introducing Spatial Variability into TOPMODEL: Theory and Preliminary Results*. Technical Report, Dept. Environ. Sc., Univ. of Va., Charlottsville.

Beven, K. J. (1986). Runoff production and flood frequency in catchments of order *n*: an alternative approach. In *Scale Problems in Hydrology*, ed. V. K. Gupta, I. Rodriguez-Iturbe and E. F. Wood, pp. 107–31. Dordrecht: Reidel.

Beven, K. J. (1989). Changing ideas in hydrology: the case of physically-based models, *J. Hydrol.*, **105**, 157–72.

Beven, K. J. (1995). Linking parameters across scales: subgrid parameterizations and scale dependent hydrological models. In *Scale Issues in Hydrological Modelling*, eds. J. D. Kalma and M. Sivapalan, pp. 263–82. Chichester: John Wiley.

Beven, K. J. (1996). The limits of splitting: hydrology, *The Sci. of the Total Environ.*,**183**, 89–97.

Beven, K. and Germann, P. (1982a). Water flow in macropores. 11. A combined flow model. *J. Soil Sci.* **32**, 15–29.

Beven, K. J. and Germann, P. (1982b). Macropores and water flow in soils, *Water Resour. Res.*, **18**, (5), 1311–25.

Beven, K. J. and Kirkby, M. J. (1979). A physically-based variable contributing area model of basin hydrology. *Hydrol. Sci. J.*, **24**(1), 43–69.

Beven, K. J., Wood, E. F. and Sivapalan, M. (1988). On hydrological heterogeneity: catchment morphology and catchment response. *J. Hydrol.*, **100**, 353–75.

Bignami, F., Marullo, S., Santoleri, R. and Schiano, M. E. (1995). Longwave radiation budget in the Mediterranean Sea. *J. Geophys. Res.*, **100**(C2), 2501–14.

Binley, A. M., Elgy, J. and Beven, K.J. (1989). A physically based model of heterogeneous hillslopes: 1. Runoff production. *Water Resour. Res.*, **25**(6), 1219–26.

Boening, C. W. (1995). Effects of different wind stress climatologies on the north Atlantic circulation: model results. In *Internat. COADS Winds Workshop*, Kiel, Germany, 31 May–2 June 1994, pp. 171–4. NOAA/ERL.

Bouchet, R. J. (1963). Evapotranspiration réele et potentielle. *Gen. Assembly Berkley, Publ.*, **62**, Int. Assoc. Sci. Hydrol., 134–42.

Bouma, J. (1989). Using soil survey data for quantitative land evaluation. *Advances in Soil Science*, **9**, 177–213.

Bradley, E. F., Coppin, P. A. and Godfrey, J. S. (1991). Measurements of sensible and latent heat flux in the western equatorial Pacific Ocean. *J. Geophys. Res.*, **96** (Supplement), 3375–89.

Brown, G. S. (1990). Quasi-specular scattering from the air-sea interface. In *Surface Waves and Fluxes*, Volume 2: *Remote Sensing*, ed. G. L. Geernaert and W. J. Plant, pp. 1–39. Dordrecht: Kluwer.

Browning, K. A. (1990). Rain, rain clouds and climate. *Quart. J. Roy. Meteor. Soc.*, **116**, 1025–51.

Bryan, F. (1986). High-latitude salinity effects and interhemispheric thermohaline circulations. *Nature*, **323**, 301–4.

Bryden, H. L. (1993). Ocean heat transport across 24° N latitude. In: *Interactions Between Global Climate Subsystems*, ed. G. A. McBean and M. Hantel, pp. 65–75. Geophysical Monograph 75, American Geophysical Union.

Bryden, H. L., Roemmich, D. H. and Church, J. A. (1991). Ocean heat transport across 24° N in the Pacific. *Deep-Sea Research*, **38**(3), 297–324.

Budyko, M.I. (1948). *Evaporation under Natural Conditions.* GIMIZ, Leningrad, English trans. by Israel Prog. Sci. Trans., Jerusalem (1963).

Budyko, M. I. (1963). *Atlas of Heat Balance of the World.* Glabnaia Geofiz. Observ.

Bunker, A. F. (1976). Computations of surface energy flux and annual air-sea interaction cycles of the North Atlantic Ocean. *Mon. Wea. Rev.*, **104**, 1122–40.

Bunker, A. F., Charnock, H. and Goldsmith, R. A. (1982). A note on the heat balance of the Mediterranean and Red Seas. *J. Mar. Res.*, **40** Supplement, 73–84.

Bunker, A. F. and Worthington, L. V. (1976). Energy exchange charts of the North Atlantic Ocean. *Bull. Amer. Meteor. Soc.*, **57**(6), 670–8.

Campbell, G. S. (1974). A simple method for determining unsaturated hydraulic conductivity from moisture retention data. *Soil Science*, **117**, 311–15.

Campbell, G. S. (1985). *Soil Physics with BASIC.* Amsterdam: Elsevier.

Campbell, W. J., Josberger, E. G. and Mognard, N. M. (1994). Southern Ocean wave fields during the Austral winters 1985–1988, by Geosat Radar Altimeter. In *The Polar Oceans and their Role in Shaping the Global Environment*, pp. 421–34. *Geophysical Monograph* **85**, American Geophysical Union.

Cardone, V. J., Greenwood, J. G. and Cane, M. A. (1990). On trends in historical marine wind data. *J. Climate*, **3**, 113–27.

Carter, D. J. T., Challenor, P. G. and Srokosz, M. A. (1992). An assessment of Geosat wave height and wind speed measurements. *J. Geophys. Res.*, **97**, 11383–92.

Cayan, D. (1992*a*). Latent and sensible heat flux anomalies over the Northern Oceans: driving the sea surface temperature. *J. Phys. Oceanogr.*, **22**, 859–81.

Cayan, D. (1992*b*). Latent and sensible heat flux anomalies over the Northern Oceans: the connection to monthly atmospheric circulation. *J. Climate*, **5**, 354–69.

Cayan, D. (1992*c*). Variability of latent and sensible heat fluxes estimated using bulk formulae. *Atmosphere-Ocean*, **30**(1), 1–42.

Chang, A. T. C., Chiu, L. S. and Wilheit, T. T. (1993). Oceanic monthly rainfall derived from SSM/I. *EOS, Transactions of the American Geophysical Union*, **74**(44), 505–13.

Chen, C., Thomas, D. M., Green, R. E. and Wagenet, R. J. (1993). Two-domain estimation of hydraulic properties in macropore soils. *Soil Science Society of America Journal*, **57**, 680–6.

Chen, C. and Wagenet, R. J. (1992). Simulation of water movement and chemicals in macropore soils. 1. Representation of the equivalent macropore influence and its effect on soil water flow. *J. Hydrology*, **130**, 105–26.

Chen, T. H. *et al.* (1997) Cabauw experimental results from the Project for Intercomparison of Land-Surface Parameterization Schemes, *J. Climate*, **10**(6), 1194–215.

Childs, E. C. (1969). *An Introduction to the Physical Basis of Soil Water Phenomena.* New York: John Wiley & Sons.

Coachman, L. K., and Aagaard, K. (1988). Transports through Bering Strait: annual and interannual variability. *J. Geophys. Res.*, **93**, 15,535–9.

Collier, C. G., Larke, P. R. and May, B. R. (1983). A weather radar correction procedure for real-time estimation of surface rainfall. *Quart. J. Roy. Meteor. Soc.*, **104**, 589–608.

Cotton, P. D. and Carter, D. J. T. (1994*a*). Cross calibration of TOPEX, ERS-1 and Geosat wave heights. *J. Geophys. Res.*, **99**(C12), 25025–33.

Cotton, P. D. and Carter, D. J. T. (1994*b*). Inter-annual variability in global wave climate from satellite data. *SPIE* 2319, 174–80.

Dansgaard, W., Johnsen, S. J., Clausen, H. B., Dahl-Jensen, D., Gundestrup, N. S., Hammer, C. U., Hvidberg, C. S., Steffensen, J. P., Sveinbjornsdottir, A. E., Jouzel, J. and Bond, G. (1993). Evidence for general instability of past climate from 250 KYR ice core record. *Nature*, **364**, 218–20.

Darcy, H. (1856). *Les Fontaines Publiques de la Ville de Dijon.* Paris: Dalmont.

Darnell, W. L., Gupta, S. K. and Staylor, W. F. (1983). Downward longwave radiation at the surface from satellite measurements. *J. Clim. Appl. Met.*, **26**, 7–87.

da Silva, A. M. and White, G. (1995). A comparison of surface marine fluxes from NMC and GEOS-1 reanalyses. *Workshop on results from the GEOS-1 five-year assimilation*, Goddard Space Flight Center, Greenbelt, Maryland, 6–8 March 1995, (unpublished).

da Silva, A. M., Young, C. C. and Levitus, S. (1994). *Atlas of Surface Marine Data 1994.* NOAA Atlas NESDIS 7, U.S. Dept. of Commerce, National Oceanic and Atmospheric Administration, Washington D.C.

DeCosmo, J., Katsaros, K. B., Smith, S. D., Anderson, R. J., Oost, W., Bumke, K. and M. Grant, A. L. (1996). Air-sea exchange of sensible heat and water vapour over whitecap sea states. *J. Geophys. Res.* **101**(C5), 12001–16.

Delcroix, T. and Hénin, C. (1991). Seasonal and interannual variations of sea surface salinity in the tropical Pacific ocean. *J. Geophys. Res.*, **96**(C12), 22,135–50.

Delworth, T., Manabe, S. and Stouffer, R. J. (1993). Interdecadal variations of the thermohaline circulation in a coupled ocean-atmosphere model. *J. Climate*, **6**, 1991–2011.

Deser, C. and Blackmon, M. L. (1993). Surface climate variations over the North Atlantic Ocean during winter: 1900–1989. *J. Climate*, **6**, 1743–53.

Dewar, R. C. (1995) Interpretation of an empirical model for stomatal conductance in terms of guard cell function. *Plant, Cell and Env.*, **18**, 365–72.

Dickinson, R. E. (1984). Modeling evapotranspiration for three-dimensional global climate models. In *Climate Processes and Climate Sensitivity*, eds. J. E. Hanson & T. Takahashi, Geophysical Monograph **29**, Maurice Ewing, 5, pp. 58–72.

Dickson, R. R., Meincke, J., Malmberg, S. A. and Lee, A. J. (1988). The 'Great Salinity Anomaly' in the northern North Atlantic 1968–1982. *Prog. Oceanogr.*, **20**, 103–51.

Dobson, F. W. and Smith, S. D. (1988). Bulk models of solar radiation at sea. *Quart. J. Roy. Meteor. Soc.*, **114**, 165–82.

Dorman, C. E. and Bourke, R. H. (1978). A temperature correction for Tucker's ocean rainfall estimates. *Quart. J. Roy. Meteor. Soc.*, **104**, 765–73.

Dorman, C. E. and Bourke, R. H. (1979). Precipitation over the Pacific Ocean, 30° S to 60° N. *Mon. Wea Rev.*, **107**, 896–910.

Dorman, C. E., and. Bourke, R. H. (1981). Precipitation over the Atlantic Ocean, 30° S to 60° N. *Mon. Wea. Rev.*, **109**, 554–63.

Driedonks, A. G. M. (1981). *Dynamics of the Well-Mixed Boundary Layer.* Sci. Rep. WR 81–82., KNMI, deBilt, Netherlands.

Edson, J. B., Fairall, C. W., Mestayer, P. G. and Larsen, S. E. (1991). A study of the inertial-dissipation method for computing air-sea fluxes. *J. Geophys. Res.*, **96**(C6), 10689–711.

Ehlers, W. (1975). Observations of earthworm channels and infiltration on tilled and untilled loess soil. *Soil Science*, **119**, 242–9.

Esbensen, S. K., Chelton, D. B., Vickers, D. and Sun, J. (1993). An analysis of errors in Special Sensor Microwave Imager evaporation estimates over the global oceans. *J. Geophys. Res.*, **98**(C4), 7081–101.

Esbensen, S. K. and Kushnir, Y. (1981). The heat budget of the global ocean: an atlas based on estimates from surface marine observations. Climate Research Institute, Oregon State University.

Esbensen, S. K. and Reynolds, R. W. (1981). Estimating monthly averaged air-sea transfers of heat and momentum using the bulk aerodynamic method. *J. Phys. Oceanogr.*, **11**, 457–65.

Fabry, F., Bellon, A., Duncan, M. R. and Austin, G. L. (1994). High resolution rainfall measurements by radar for very small basins: the sampling problem re-examined. *J. Hydrol.*, **161**, 415–28.

Famiglietti, J. S. and Wood, E. F. (1994). Multi-scale modeling of spatially-variable water and energy balance processes. *Water Resour. Res.*, **30**(11), 3061–78.

Famiglietti, J. S., Wood, E. F., Sivapalan, M. and Thongs, D. J. (1992). A catchment scale water balance model for FIFE. *J. Geophys. Res.*, **97**(D17), 18997–9008.

Feddes, R. A., de Rooij, G. H., van Dam, J. C., Kabat, P., Droogers, P. and Stricker, J. N. M. (1993). Estimation of regional effective soil hydraulic parameters by inverse modelling. In *Water Flow and Solute Transport in Soils* ed. D. Russo and G. Dagan, pp. 211–31. Berlin: Springer-Verlag.

Foreman, S. J., Alves, J. O. S. and Brooks, N. P. J. (1994). *Assessment of Surface Fluxes from Numerical Prediction Systems.* Forecasting Research Division Technical Report 104, Meteorological Office, Bracknell, 9 pp. + figs.

Freeze, R. A. (1974). Streamflow generation. *Rev. Geophys.*, **12**(4), 627–47.

Frenzen, P. and Vogel, C. A. (1994). On the sensitivity of the phim function to k: a corrected illustration for the turbulent kinetic energy budget in the ASL. *Boundary-Layer Meteorol.*, **68**, 439–42.

Friedrichs, M. A. M. and Hall, M. M. (1993). Deep circulation in the tropical North Atlantic. *J. Mar. Res.*, **51**(4), 697–736.

Friehe, C. A. and Schmitt, K. F. (1976). Parameterisation of air-sea interface fluxes of sensible heat and moisture by the bulk aerodynamic formulae. *J. Phys. Oceanogr.*, **6**, 801–9.

Frouin, R., Gautier, C., Katsaros, K. B. and Lind, R. J. (1988a). A comparison of satellite and empirical formula techniques for estimating insolation over the oceans. *J. Clim. Appl. Met.*, **97**, 1016–23.

Frouin, R., Gautier, C. and Morcrette, J.-J. (1988*b*). Downward longwave irradiance at the ocean surface from satellite data: methology and in situ validation. *J. Geophys. Res.*, 93(C1), 597–619.

Garrett, C., Outerbridge, R. and Thompson, K. (1993). Inter-annual variability in Mediterranean heat and water fluxes. *J. Climate*, **6**, 900–10.

Gates, W. L. (1992). AMIP: the atmospheric model intercomparison project. *Bull. Am. Meteorol. Soc.*, **73**(12), 1962–70.

Geernaert, G. (1990*a*). The theory and modeling of wind stress with applications to air-sea interaction and remote sensing. *RAS*, **2**, 125–49.

Geernaert, G. L. (1990*b*). Bulk parameterisations for the wind stress and heat fluxes. In *Surface Waves and Fluxes. Volume 1 – Current Theory*, ed. G. L. Geernaert and W. J. Plant, pp. 91–172. Dordrecht: Kluwer.

Germann, P. F. (1990). Macropores and hydrologic hillslope processes. In *Process Studies in Hillslope Hydrology*, ed. M. G. Anderson and T. P. Burt, pp. 327–64. Chichester: John Wiley.

Germann, P. and Beven, K. (1981). Water flow in soil macropores. 1. An experimental approach. *J. Soil Sci.*, **32**, 1–14.

Gilman, C. and Garrett, C. (1994). Heat flux parameterizations for the Mediterranean Sea: the role of atmospheric aerosols and constraints from the water budget. *J. Geophys. Res.*, **99**(C3), 5119–34.

Gleckler, P. J., Randall, D. A., Boer, G., Colman, R., Dix, M., Galin, V., Helfand, M., Kiehl, J., Kitoh, A., Lau, W., Liang, X.-Z., Lykossov, V., McAvaney, B., Miyakoda, K. and Planton, S. (1994). *Cloud-radiative Effects on Implied Oceanic Energy Transports as Simulated by Atmospheric General Circulation Models*. PCMDI Report 15, Lawrence Livermore National Laboratory, Livermore, CA., 13 pp.

Goldsbrough, G. (1933). Ocean currents produced by evaporation and precipitation. *Proc. Roy. Soc. London*, **A141**, 512–17.

Goodrich, D. C. (1990). Geometric simplification of a distributed rainfall runoff model over a range of basin scales Ph.D. dissertation, Department of Hydrology and Water Resources, University of Arizona, Tucson, AZ, 361 pp.

Grassl, H. (1976). The dependence of the measured cool skin of the ocean on wind stress and total heat flux. *Boundary-Layer Meteorol.*, **10**, 465–74.

Grayson, R. B. (1990) A terrain-based hydrologic model for erosion studies. Ph.D. thesis, Univ. of Melbourne, Parkville, Victoria, Australia, 375 pp.

Grayson, R. B., Moore, I. D. and McMahon, T. A. (1992*a*). Physically based hydrologic modeling, 1, a terrain based model for investigative purposes, *Water Resour. Res.*, **28**(10), 2639–58.

Grayson, R. B., Moore, I. D. and McMahon, T. A. (1992*b*). Physically based hydrologic modeling, 2, is the concept realistic?, *Water Resour. Res.*, **28**(10), 2659–66.

Griffies, S. M. and Bryan, K. (1997). Predictability of North At-

lantic multidecadal climate variability. *Science*, **275** (5297), 181–3.

Gulev, S. K. (1994). Influence of space-time averaging on the ocean-atmosphere exchange estimates in the North Atlantic mid latitudes. *J. Phys. Oceanogr.*, **24**, 1236–55.

Gulev, S. K. (1995*a*). Inter-annual and annual variability of statistical characteristics of synoptic ocean-atmosphere interaction processes in mid-latitudes. *Atmosphere-Ocean*, (submitted).

Gulev, S. K. (1995*b*). Long term variability of sea–air heat transfer in the North Atlantic Ocean. *Int. J. Climatol.*, **15**, 825–52.

Hall, M. M. and Bryden, H. L. (1982). Direct estimates and mechanisms of ocean heat transport. *Deep-Sea Res.*, **29**, 339–59.

Harris, A. R., Saunders, M. A., Foot, J. S., Smith, K. F. and Mutlow, C. T. (1995). Improved sea surface temperature measurements from space. *Geophys. Res. Letters*, **22**(16), 2159–62.

Henderson-Sellers, A. and Brown, V. B. (1992). Project for the intercomparison of land surface parameterization schemes (PILPS): first science plan. *GEWEX Tech. Note*, IGPO Publ. Series No. 5, 53 pp.

Henderson-Sellers, A., Pitman, A. J., Love, P. K., Irannejad P. and Chen, T. H. (1995). The project for the intercomparison of land surface parameterization schemes (PILPS): phases 2&3. *Bull. Amer. Met. Soc.*, **76**, 489–503.

Hillel, D. and Baker, R. S. (1988). A descriptive theory of fingering during infiltration into layered soils. *Soil Science*, **146**, 51–6.

Hjelmfelt, A. T. and Burwell, R. E. (1984). Spatial variability of runoff. *J. Irrig. Drain. Eng.*, **110**(1), 46–54.

Hopmans, J. W. (1987). A comparison of various methods to scale soil hydraulic properties. *J. Hydrol.*, **93**, 241–56.

Houze, R. A. Jr. (1993). *Cloud Dynamics*. London: Academic Press.

Huang, R. X. (1993). Real freshwater flux as a natural boundary condition for salinity balance and thermohaline circulation forced by evaporation and precipitation. *J. Phys. Oceanogr.*, **23**, 2428–46.

Huang, R. X., Luyten, J. R. and Stommel, H. M. (1992). Multiple equilibrium states in combined thermal and saline circulation. *J. Phys. Oceanogr.*, **22**, 231–46.

Huang, R. X., and Schmitt, R. W. (1993). The Goldsbrough-Stommel Circulation of the World Oceans. *J. Phys. Oceanogr.*, **23**(6), 1277–84.

Huffman, G. J., Adler, R. F., Keehn, P. R. and Negri, A. J. (1993). Global precipitation estimates using microwave and IR data. In *Analysis Methods of Precipitation on a Global scale – Global Precipitation Climatology Centre GEWEX Workshop*, Koblenz, Germany, 14–17 September 1992, pp. A49–A53. Geneva: WCRP, WMO.

Inoue, T. (1987). An instantaneous delineation of convective rainfall areas using split window data of NOAA-7 AVHRR. *J. Met. Soc. Japan*, **65**, 469–81.

Isemer, H.-J. (1992). Comparison of estimated and measured

marine surface wind speed. In *International COADS Work-shop*, Boulder, Colorado, 13–15 January 1992, pp. 143–58. NOAA Environmental Research Labs.

Isemer, H.-J. (1995). Trends in marine surface wind speed: ocean weather stations versus voluntary observing ships. In *Internat. COADS Winds Workshop*, Kiel, Germany, pp. 68–84. NOAA/ERL.

Isemer, H.-J. and Hasse, L., (1987). The Bunker Climate Atlas of the North Atlantic Ocean. Vol.2: *Air–Sea Interactions*. Berlin: Springer-Verlag, 252 pp.

Isemer, H.-J. and Hasse, L. (1991). The scientific Beaufort equivalent scale: effects on wind statistics and climatological air–sea flux estimates in the North Atlantic Ocean. *J. Climatol.*, **4**(8), 819–36.

Isemer, H.-J., Willebrand, J. and. Hasse, L. (1989). Fine adjustment of large scale air–sea energy flux parameterizations by direct estimates of ocean heat transport. *J. Climate*, **2**(10), 1173–84.

Jackman, M. B. and Hornberger, G. M. (1993). How much complexity is warranted in a rainfall-runoff model? *Water Resour. Res.*, **2**, 37–47.

Jarvis, P. J. (1976). The interpretation of the variations in leaf water potential and stomatal conductance found in canopies in the field. *Phil. Trans. Roy. Soc. Lond. B*, **273**, 593–602.

Jones, C. S., Legler, D. and. O'Brien, J. J. (1993). Variability of surface fluxes over the Indian Ocean: 1960–1989. *The Global Atmosphere-Ocean System*, **3**(2–3), 249–72.

Josberger, E. G. and N. M. Mognard (1996). Southern Ocean monthly wave fields for austral winters 1985–1988, by Geosat Radar Altimeter. *J. Geophys. Res.*, **101**(C3), 6689–96.

Josey, S. A., Kent, E. C., Oakley, D. and Taylor, P. K. (1996). A new global air-sea heat and momentum flux climatology. *International WOCE Newsletter*, **24**, 3–5 + figs.

Josey, S. A., Kent, E. C. and Taylor, P. K. (1995). Seasonal variations between sampling and classical mean turbulent heat flux estimates in the North Atlantic. *Annales Geophysicae*, **13**, 1054–64.

Josey, S. A., Oakley, D. and Pascal, R. W. (1997). On estimating the atmospheric longwave flux at the ocean surface from ship meteorological reports. *J. Geophys. Res.*, (accepted).

Joss, J., Schram, K., Thams, J. C. and Waldvogel, A. (1970). On the Quantitative Dermination of Precipitation by Radar. Scientific Communication No. 63. Research Dept. of Federal Commission on the Study of Hail Formation and Hail Suppression, Ticinese Observatory of the Swiss Central Meteorological Institute, Federal Institute of Technology, Zurich, 38 pp.

Jourdan, D. and C. Gautier. (1995). Comparison between global latent heat flux computed from multisensor (SSM/I and AVHRR) and from in situ data. *J. Atmos. & Oceanic Tech.*, **12**, 46–72.

Katsaros, K. (1980). The aqueous thermal boundary layer. *Boundary-Layer Meteorol.*, **18**, 107.

Katsaros, K. B. (1990). Parameterization schemes and models for estimating the surface radiation budget. In *Surface Waves and Fluxes, Vol. 2, Remote Sensing*, ed. G. L. Geernaert and W. J. Plant, pp. 339–368. Dordrecht: Kluwer.

Katsaros, K. B., DeCosmo, J., Lind, R. J., Anderson, R. J., Smith, S. D., Kraan, C., Oost, W., Uhil, K., Merstayer, P. G., Larsen, S. E., Smith, M. J. and de Leeuw, G. (1994). Measurements of humidity and temperature in the marine environment during the HEXOS main experiment. *J. Atmos. and Oceanic Tech.*, **11**, 964–81.

Katsaros, K. B. and deLeeuw, G. (1994). Comment on 'Sea spray and the turbulent heat fluxes' by E. L.Andreas. *J. Geophys. Res.*, **99**, 14339–43.

Kelliher, F. M., Leuning, R., Raupach, M. R. and Schulze, E. D. (1995). Maximum conductances for evaporation from global vegetation types. *Agric. and Forest Met.*, **73**, 1–16.

Kent, E. C., Forrester, T. N. and Taylor, P. K. (1995). A comparison of oceanic skin effect parameterisations using shipborne radiometer data. *J. Geophys. Res.*, **101**(C7), 16649–66.

Kent, E. C. and Taylor, P. K. (1991). *Ships Observing Marine Climate: a Catalogue of the Voluntary Observing Ships Participating in the VSOP-NA*. Marine Meteorology and Related Oceanographic Activities 25. Geneva: World Meteorological Organisation. 123 pp.

Kent, E. C. and Taylor, P. K. (1995). A comparison of sensible and latent heat flux estimates for the North Atlantic Ocean. *J. Phys. Oceanogr.*, **25**(6), 1530–49.

Kent, E. C. and Taylor, P. K. (1997). Choice of a Beaufort Equivalent Scale. *J. Atmos. & Oceanic Tech.*, **14**(2), 228–42.

Kent, E. C., Taylor, P. K., Truscott, B. S. and Hopkins, J. A. (1993*a*). The accuracy of voluntary observing ships' meteorological observations. *J. Atmos. & Oceanic Tech.*, **10**(4), 591–608.

Kent, E. C., Tiddy, R. J. and Taylor, P. K. (1993*b*). Correction of marine daytime air temperature observations for radiation effects. *J. Atmos. & Oceanic Tech.*, **10**(6), 900–6.

Kirkby, M. J. (ed.) (1978). *Hillslope Hydrology*. Chichester: John Wiley.

Kitchen, M. and Blackall, R. M. (1992). Representative errors in comparisons between radar and gauge measurements of rainfall. *J. Hydrol.*, **134**, 13–33.

Kushnir, Y. (1994). Interdecadal variations in North Atlantic sea surface temperature and associated atmospheric conditions. *J. Climate*, **7**, 141–57.

Lagerloef, G. S. E., Swift, C. T. and Le Vine, D. M. (1995). Sea surface salinity: the next remote sensing challenge. *Oceanography*, **8**(2), 44–50.

Large, W. G. and Pond, S. (1981). Open ocean momentum flux measurements in moderate to strong winds. *J. Phys. Oceanogr.*, **11**, 324–36.

Large, W. G. and Pond, S. (1982). Sensible and latent heat flux measurements over the ocean. *J. Phys. Oceanogr.*, **12**, 464–82.

Laughlin, C. R. (1981). On the effect of temporal sampling on the observation of mean rainfall. In *Precipitation Measurements from Space*, ed. D. Atlas and O. Thiele, D59–D66, Workshop Reports, NASA.

Legler, D. (1991). Errors in 5-day mean surface wind and temperature conditions due to inadequate sampling. *J. Atmos. & Oceanic Tech.*, **8**(5), 705–12.

Lehman, S. J. and Keigwin, L. D. (1992). Sudden changes in North Atlantic circulation during the last deglaciation. *Nature*, **356**, 757–62.

Lettenmaier, D. P., Lohmann, D. Wood, E. F. and Liang, X. (1996). PILPS-2c draft workshop report: Report of a Workshop Held at Princeton University, October 28–31, Internet http://earth.princeton.edu.

Leuning, R. (1995). A critical appraisal of a combined stomatal-photosynthesis model for C3 plants. *Plant, Cell and Env.*, **18,** 339–55.

Liang, X., Lettenmaier, D.P. and Burges, S.J. (1994). A simple hydrologically based model of land surface water and energy fluxes for general circulation models. *J. Geophys. Res.*, **99**(D7), 14415–28.

Liang, X., Wood, E. F. and Lettenmaier, D. P. (1996*a*). Surface soil moisture parameterization of the VIC-2L model: evaluation and modification. *Global and Planetary Change*, **13**, 195–206.

Liang, X., Lettenmaier, D. P. and Wood, E. F. (1996*b*). A one-dimensional statistical-dynamic representation of subgrid spatial variability of precipitation in the two-layer VIC model. *J. Geophys. Res.*, **101**(D16), 21403–22.

Liang, X., Wood, E. F. and Lettenmaier, D. P. (1997). Initial results of PILPS(2C) land surface scheme intercomparisons using large scale data sets. *77th AMA Meeting*, **13**(9.3).

Lindau, R. (1995). A new Beaufort equivalent scale. In *Internat. COADS Winds Workshop*, Kiel, Germany, 31 May–2 June 1994, 232–52. NOAA/ERL.

List, R. J. (1963). *Smithsonian Meteorological Tables (6ᵗʰ revised edition)*. Washington, D.C.: Smithsonian Institution.

Liu, W. T., Katsaros, K. B. and Businger, J. A. (1979). Bulk parameterization of air–sea exchanges of heat and water vapour including the molecular constraints at the interface. *J. Atmos. Sci.*, **36**, 1722–35.

Liu, W. T. and. Niiler, P. P (1984). Determination of monthly mean humidity in the atmospheric surface layer over oceans from satellite data. *J. Phys. Oceanogr.*, **14**, 1451–7.

Loague K. M. and Freeze, R. A. (1985). A comparison of rainfall-runoff modeling techniques on small upland catchments, *Water Resour. Res.*, **21**(2), 229–48.

Lukas, R. and Lindstrom, E. (1991). The mixed layer of the Western Equatorial Pacific Ocean. *J. Geophys. Res.*, **96** (Suppl.), 3343–57.

Manabe, S., and Stouffer, R. J. (1995). Simulation of abrupt climate change induced by freshwater input to the North Atlantic Ocean. *Nature*, **378**, 165–67.

Marshall, J. S. and Palmer, W. (1948). The distribution of rain-drops with size. *J. Met.*, **5**, 165–6.

Mauritzen, C. (1996). Large scale circulation and formation of dense overflow waters in the Nordic Seas and the Arctic Ocean, Part 1. *Deep-Sea Research*, **43** (6), 769–806.

McClain, E. P., Pichel, W. G. and Walton, C. C. (1985). Comparative performance of AVHRR-based multichannel sea surface temperatures. *J. Geophys. Res.*, **90**, 11587–601.

McCord, J. T., Stephens, D. B. and Wilson, J. L. (1991). Hysteresis and state-dependent anistropy in modeling unsaturated hillslope hydrologic processes. *Water Resour. Res.*, **2,** 1501–18.

McNaughton, K. G. and Spriggs, T. W. (1989). A mixed-layer model for regional evaporation. *Boundary-Layer Met.*, **34**, 243–62.

McPhaden, M. J. (1993). TOGA-COARE optical rain gauge measurements: workshop report. *TOGA Notes*, **13**, 18–19.

McPhaden, M. J., Frietag, H. P. and Shephard, A. J. (1990). Moored salinity time series measurements at 0°, 140° W. *J. Atmos. & Oceanic. Tech.*, **7**, 569–75.

Michaud, R. and Lin, C. A. (1992). Monthly summaries of merchant ship surface marine observations and implications for climate variability studies. *Clim. Dyn*, **7**(1), 45–55.

Mognard, N. and Katsaros, K. B. (1995). Statistical comparison of the Special Sensor Microwave Imager and the Geosat altimeter wind speed measurements over the ocean. *Global Atmos. Ocean System*, **2**(4), 291–9.

Monteith, J. L. (1981). Evaporation and surface temperature. *Quart. J. Roy. Met. Soc.*, **107**, 1–27.

Monteith, J. L. (1995). A reinterpretation of stomatal responses to humidity. *Plant, Cell and Env.*, **18**, 357–64.

Moore, I. D., O'Loughlin, E. M. and Burch, G. J. (1988). A contour-based topographic model for hydrological and ecological applications. *Earth Surf. Proc. Landforms*, **1,** 305–20.

Morrissey, M. L. (1994). The effect of data resolution on the area threshold methods. *J. App. Met.*, **33**, 1263–70.

Mott, K. A. and Parkhurst, D. F. (1991). Stomatal response to humidity in air and helox. *Plant, Cell and Env.* **14**, 509–15.

Mutlow, C. T., Zavody, A. M., Barton, I. J. and Llewellyn-Jones, D. T. (1994). Sea surface temperature measurements by the Along Track Scanning Radiometer on the ERS-1 satellite: early results. *J. Geophys. Res.*, **99**(C11), 22575–88.

Nielsen, D. R., Biggar, J. and Erh, K. (1973). Spatial variability of field measured soil-water properties, *Hilgardia*, **42**, 215–60.

Norris, J. R. and Leovy, C. B. (1994). Inter-annual variability in stratiform cloudiness and sea surface temperature. *J. Climate*, **7**, 1915–25.

Oberhuber, J. M. (1988). An Atlas Based on the COADS Data Set: the budgets of heat, buoyancy and turbulent kinetic energy at the surface of the global ocean. *Report 15*, Max-Planck-Institut fur Meteorologie, 20 pp.

Ogden, F. L. and Julien, P. Y. (1994). Runoff model sensitivity to radar rainfall resolution. *J. Hydrol.*, **158**, 1–18.

O'Loughlin, E. M. (1981). Saturation regions in catchments and their relations to soil and topographic properties. *J. Hydrol.*, 229–46.

O'Loughlin, E. M. (1986). Prediction of surface saturation zones in natural catchments by topographic analysis. *Water Resour. Res.*, **22**, 794–804.

Paniconi, C. and Wood, E. F. (1993). A detailed model for simulation of catchment scale subsurface hydrologic processes. *Water Resour. Res.*, **29**(6), 1601–20.

Parker, D. E. and Folland, C. K. (1995). Correction of instrumental biases in historical sea surface temperature data. *Quart. J. Roy. Meteor. Soc.*, **121**(522), 319–67.

Paulson, C. A. and Simpson, J. J. (1981). The temperature difference across the cool skin of the ocean. *J. Geophys. Res.*, **86**, 11044–54.

Payne, R. E. (1972). Albedo of the sea surface. *J. Atmos. Sci.*, **29**, 959–70.

Peixoto, J. P. and Oort, A. H. (1983). The atmospheric branch of the hydrological cycle and climate. In *Variations in the Global Water Budget*, ed. A. Street-Perrott *et al.*, pp. 5–65. Boston: Reidel.

Penman, H. L. (1948). Natural evaporation from open water, bare soil, and grass. *Proc. Roy. Soc. Lond.*, **A193**, 120–45.

Philip, J. R. (1969). Moisture equilibrium in the vertical in swelling soils. 1. Basic theory. *Australian Journal of Soil Research* **7**, 99–120.

Philip, J. R. (1970). Reply to note by E. G. Youngs and G. D. Towner on hydrostatistics and hydrodynamics in swelling soils. *Water Resour. Res.*, **6**, 1248–51.

Philip, J. R. and de Vries, D. A. (1957). Moisture movement in porous materials under temperature gradients. *Trans. Amer. Geophys. Union*, **38**, 222–32.

Price, J. F., Weller, R. A. and Pinkel, R. (1986). Diurnal cycling: observations and models of the upper ocean response to diurnal heating, cooling, and wind mixing. *J. Geophys. Res.*, **91**, 8411–27.

Priestley, C. H. B. and Taylor, R. J. (1972). On the assessment of surface heat flux and evaporation using large-scale parameters. *Mon. Wea. Rev.*, **100**, 81–92.

Queffeulou, P., Bentamy, A., Quilfen, Y. and Tournadre, J. (1994). *Cross validation of ERS-1 and TOPEX POSEIDON wind and wave measurements*. Technical Report DRO-OS 94-08, Brest: IFREMER.

Quilfen, Y. (1994). *ERS-1 Wind Scatterometer Products and Applications*. CERSAT New Letter. Brest: IFREMER.

Rahmstorf, S. (1995). Bifurcations of the Atlantic thermohaline circulation in response to changes in the hydrological cycle. *Nature*, **378**, 145–9.

Ramage, C. S. (1987). Secular change in reported surface wind speeds over the ocean. *J. Climate and Appl. Meteor.*, **26**, 525–8.

Reed, R. K. (1977). On estimating insolation over the ocean. *J. Phys. Oceanogr.*, **7**, 482–5.

Reynolds, R. W. and Smith, T. M. (1994). Improved global sea surface temperature analyses using optimum interpolation. *J. Climatol.*, **7**, 929–48.

Richards, F. and Arkin, P. A. (1981). On the relationship between satellite observed cloud cover and precipitation. *Mon. Wea. Rev.*, **109**, 1081–93.

Richards, L. A. (1931). Capillary conduction of liquids through porous mediums. *Physics*, **1**, 318–33.

Rosati, A. and Miyakoda, K. (1988). A general circulation model for upper ocean simulation. *J. Phys. Oceanogr.*, **18**, 1601–26.

Rosenfeld, D., Wolff, D. B. and Amitai, E. (1994). The window probability matching method for rainfall measurements with radar. *J. App. Met.*, **33**, 683–93.

Rosenfeld, D., Amitai, E. and Wolff, D. B. (1995a). Classification of rain regimes by the 3-dimensional properties of reflectivity fields. *J. App. Met.*, **34**, 198–211.

Rosenfeld, D., Amitai, E. and Wolff, D. B. (1995b). Improved accuracy of radar WPMM estimated rainfall upon application of objective classification criteria. *J. App. Met.*, **34**, 212–23.

Rosenfeld, D. and Gutman, G. (1994). Retrieving microphysical properties near the tops of potential rain clouds by multispectral analysis of AVHRR data. *J. Atmos. Res.*, **34**, 259–83.

Ross, P. J. (1990). *SWIM – a Simulation Model for Soil Water Infiltration and Movement*. Adelaide: CSIRO Division of Soils.

Ross, P. J. and Smettem, K. R. J. (1993). Describing soil hydraulic properties with sums of simple functions. *Soil Sci. Soc. America J.*, **57**, 26–9.

Russo, D. and Bresler, E. (1981). Soil hydraulic properties as stochastic processes. 1. An analysis of field spatial variability. *Soil Sc. Soc. America J.*, **45**, 682–7.

Sachidananda, M. and Zrnic, D. S. (1986). Differential propagation phase shift and rainfall rate estimation. *Radio Science*, **21**, 235–47.

Salby, M. L. (1982a). Sampling theory for asynoptic satellite observations, I. Space-time spectra, resolution, and aliasing. *J. Atmos. Sci.*, **39**, 2577–600.

Salby, M. L. (1982b). Sampling theory for asynoptic satellite observations, II. Fast fourier synoptic mapping. *J. Atmos. Sci.*, **39**, 2601–14.

Saunders, P. M. (1967). The temperature at the ocean-air interface. *J. Atmos. Sci.*, **24**, 269–73.

Schiano, M. E., Santoleri, R., Bignami, F., Leonardi, R. M., Marullo, S. and Boehm, E. (1993). Air-sea interaction measurements in the west Mediterranean Sea during the Tyrrhenian Eddy Multi-Platform Observations Experiment. *J. Geophys. Res.*, **98**(C2), 2461–74.

Schluessel, P., Emery, W. J., Grassl, H. and Mammen, T. (1990). On the bulk-skin temperature difference and its impact on satellite remote sensing of the sea surface temperature. *J. Geophys. Res.*, **95**(C8), 13341–56.

Schmitt, R. W. (1994). *The Ocean Freshwater Cycle*. OOSDP Background Paper 4, Dept. of Oceanography, Texas A & M Univ, College Station, Texas, 40 pp.

Schmitt, R. W. (1995). The ocean component of the global water cycle. *Reviews of Geophysics*, **33**, (Supplement, Pt. 2), 1395–1409.

Schmitt, R. W. (1997). New approaches to the salinity problem. *CLIVAR Exchanges*, **2**(1), 4–6.

Schmitt, R. W., Bogden, P. S. and Dorman, C. E. (1989). Evaporation minus precipitation and density fluxes for the North Atlantic. *J. Phys. Oceanogr.*, **19**, 1208–21.

Schmitt, R. W. and Wijffels, S. E. (1993). The role of the ocean in the global water cycle. In *Interactions Between Global Climate Subsystems*, ed. G. A. McBean and M. Hantel, pp. 77–84. Geophysical Monograph 75. American Geophysical Union.

Schulz, J., Schlussel, P. and Grassl, H. (1993). Water vapour in the atmospheric boundary layer over oceans from SSM/I measurements. *Internat. J. Remote Sensing*, **14**(15), 2773–89.

Seckel, G. R. and Beaudry, F. H. (1973). The radiation from sun and sky over the Pacific Ocean (Abstract). *Trans. Am. Geophys. Union*, **54**, 1114.

Seed, A. W. and Austin, G. L. (1990). Variability of summer Florida rainfall and its significance for estimation of rainfall by gauges, radar and satellite. *J. Geophys. Res.*, **95**, D3, 2207–15.

Seliga, T. A. and Bringi, V. N. (1976). Potential use of radar differential reflectivity measurements at orthogonal polarisation for measuring precipitation. *J. App. Met.*, **15**, 69–75.

Sellers P. J. *et al.* (1995). Effect of spatial variability in topography, vegetation cover and soil moisture on area-averaged surface fluxes: a case study using FIFE 1989 data, *J. Geophys. Res.*, **100**(D21), 25607–29.

Sellers, P. J., Hall, F. G., Asrar, G., Strebel, D. E. and Murphy, R. E. (1988). The first ISLSCP field experiment (FIFE). *Bull. Am. Met. Soc.*, **69**, 22–7.

Sellers, P. J., Mintz, Y., Sud, Y. C. and Dalcher, A. (1986). A simple biosphere model (SiB) for use within general circulation models. *J. Atmos. Sci.*, **43**, 505–31.

Serreze, M. C., Barry, R. G. and Walsh, J. E. (1995). Atmospheric water vapor characteristics at 70° N. *J. Climate*, **8**(4), 719–31.

Shaffer, G. and Bendtsen, J. (1994). Role of the Bering Strait in controlling North Atlantic ocean circulation and climate. *Nature*, **357**, 354–7.

Sharma, M. L. and Luxmoore, R. J. (1979). Soil spatial variability and its consequences on simulated soil water balance. *Water Resour. Res.* **15**, 1567–73.

Simmonds, I. and Dix, M. (1989). The use of mean atmospheric parameters in the calculation of modelled mean surface heat fluxes over the world's oceans. *J. Phys. Oceanogr.*, **19**, 205–15.

Simpson, J., Adler, R. F. and North, G. R. (1988). A proposed Tropical Rainfall Measuring Mission (TRMM) Satellite. *Bull. Amer. Met. Soc.*, **69**, 278–95.

Sivapalan, M., Beven, K. J. and Wood, E. F. (1987). On hydrological similarity, 2, A scaled model of storm runoff production. *Water Resour. Res.*, **23**, 2266–78.

Sklash, M. G. (1990). Environmental isotope studies of storm and snowmelt runoff generation. In *Process Studies in Hillslope Hydrology*, ed. M. G. Anderson and T. P. Burt, pp. 401–35. Chichester: John Wiley.

Sklash, M. G., Favolden R. N. and Fritz, P. (1976). A conceptual model of watershed response to rainfall, developed through the use of oxygen-18 as a natural tracer. *Can. J. Earth Sci.*, **13**, 271–83.

Smettem, K. R. J. and Collis-George, N. (1985). The influence of cylindrical macropores on steady-state infiltration in a soil under pasture. *J. Hydrol.* **52**, 107–14.

Smiles, D. E. (1995). Liquid flow in swelling soils. *Soil Sci. Soc. Amer. J.*, **59**, 313–18.

Smith, R. E., Goodrich, D. C., Woolhiser D. A. and Simanton, J. R. (1994). Comment on 'Physically based hydrologic modeling, 2, is the concept realistic?', *Water Resour. Res.*, **30**(3), 851–4.

Smith, S. D. (1980). Wind stress and heat flux over the ocean in gale force winds. *J. Phys. Oceanogr.*, **10**, 709–26.

Smith, S. D. (1988). Coefficients for sea surface wind stress, heat flux and wind profiles as a function of wind speed and temperature. *J. Geophys. Res.*, **93**, 15467–74.

Smith, S. D. (1989). Water vapour flux at the sea surface. *Boundary Layer Meteor.*, **47**, 277–93.

Soloviev, A. V. and Schluessel, P. (1994). Parameterisation of the cool skin of the ocean and of the air-ocean gas transfer on the basis of modelling surface renewal. *J. Phys. Oceanogr.*, **24**, 1339–46.

Soloviev, A. V. and Vershinsky, V. N. (1982). The vertical structure of the thin surface layer of the ocean under conditions of low wind speed. *Deep-Sea Res.*, **29**, 1437.

Spencer, R. W., Goodman, H. M. and Hood, R. E. (1989). Precipitation retrieval over land and ocean with the SSM/I: Identification and characteristics of the scattering signal. *J. Atm. Ocean. Tech.*, **6**, 254–73.

Stewart M. K. and McDonnell, J. J. (1991). Modeling base flow soil water residence times from deuterium concentrations. *Water Resour. Res.*, **27**(10), 2681–95.

Stocker, T. F., and D. G. Wright, (1991). Rapid transitions of the ocean's deep circulation induced by changes in surface water flux. *Nature*, **351**, 729–32.

Stommel, H. (1961). Thermohaline convection with two stable regimes of flow. *Tellus*, **13**(2), 224–30.

Strensrud, D. J. and Fritsch, J. M. (1994). Mesoscale convective systems in weakly forced large-scale environments. Part III Numerical simulations and inspection for operational forecasting. *Mon. Wea. Rev.*, **122**, 2084–404.

Susskind, J. (1993). Water vapour and temperature. In *Atlas of Satellite Observations related to Global Change*, ed. R. J. Gurney, J. L. Foster and C. L. Parkinson, pp. 89–128. Cambridge: Cambridge University Press.

Tanner, C. B. and Sinclair, T. R. (1983). Efficient water use in crop production. In *Limitations to Efficient Water Use in Crop Production*, ed. H. M. Taylor, W. R. Jordan and T. R. Sinclair, pp. 1–27. Madison, Wis.: American Society of Agronomy.

Taylor, P. K. (1982). Remote sensing of atmospheric water content and sea surface latent heat flux. In *Annual Technical Conference*, Liverpool, December, 1982, pp. 265–72. Reading: Remote Sensing Society.

Taylor, P. K. (1984). The determination of surface fluxes of heat and water by satellite microwave radiometry and in situ measurements. In *NATO ASI, Series C: Mathematical and Physical Sciences: Large-scale oceanographic experiments and satellites*, **128**, ed. C. Gautier and M. Fieux, pp. 223–46. Boston: Reidel.

Taylor, P. K. (1985). TOGA surface fluxes of sensible and latent heat by in situ measurement and microwave radiometry. *Third session of the JSC/CCCO TOGA Scientific Steering Group*, Scripps Institution of Oceanography, La Jolla, Ca. Geneva: WMO. 30 pp.

Trenberth, K. E. and Solomon, A. (1994). The global heat balance: heat transports in the atmosphere and ocean. *Clim. Dyn.*, **10**, 107–34.

Tucker, G. B. (1961). Precipitation over the North Atlantic Ocean. *Quart. J. Roy. Meteor. Soc.*, **87**, 147–58.

van Genuchten, M. Th. and Nielsen, D. R. (1985). On describing and predicting the hydraulic properties on unsaturated soils. *Annals of Geophys.*, 3, 615–28.

Viera, S. R., Nielsen, D. R. and Biggar, J. (1981). Spatial variability of field measured infiltration rates. *Soil Sci. Soc. Amer. J.*, **45**, 1040–8.

Ward, M. N. (1995). Corrections for wind trends 1949–1988. In *Internat. COADS Winds Workshop*, Kiel, Germany, 31 May–2 June 1994, pp. 102–19. NOAA/ERL.

Warrick, A. W., Mullen, G. J. and Nielsen, D. R. (1977). Scaling field-measured soil hydraulic properties using a similar media concept. *Water Resour. Res.* **13**, 355–62.

Watson, K. W. and Luxmoore, R. J. (1986). Estimating macroporosity in a forest watershed by use of a tension infiltrometer. *Soil Sci. Soc. Amer. J.*, **50**, 578–82.

WCRP (1993). Fifth session of the working group on radiative fluxes: summary of main discussions and recommendations, San Diego, California, 8–12 Feb. 1993. Geneva: WMO.

Weare, B. C. (1984). Inter-annual moisture variations near the surface of the tropical Pacific Ocean. *Quart. J. Roy. Meteor. Soc.*, **110**(464), 489–504.

Weaver, A. J., Sarachik, E. S. (1991). The role of mixed boundary conditions in numerical models of the ocean's climate. *J. Phys. Oceanogr.*, **21**, 1470–93.

White, R. E. (1985). The influence of macropores on the transport of dissolved and suspended matter through soil. *Advances in Soil Sci.*, **3**, 95–120.

Whitlock, C. H., Charlock, T. P., Staylor, W. F., Pinker, R. T., Laszlo, I., DiPasquale, R. C. and Ritchey, N. A. (1993). In

WCRP Surface Radiation Budget Shortwave Data Product Description – Version 1.1. NASA Tech. Memorandum 107747. Hampton, Virginia: Langley Research Center, 28 pp.

Wick, G. A., Emery, W. J. and Schluessel, P. (1992). A comprehensive comparison between satellite measured skin and multichannel sea surface temperature. *J. Geophys. Res.*, **97**(C4), 5569–95.

Wijffels, S. E., Schmitt, R. W., Bryden, H. L. and Stigebrandt, A. (1992). Transport of freshwater by the oceans. *J. Phys. Oceanogr.*, **22**(2), 155–62.

Wilheit, R., Adler, R., Avery, S., Barrett, E., Bauer, P., Berg, W., Chang, A., Ferriday, J., Grody, N., Goodman, S., Kidd, C., Kniveton, D., Kummerow, C., Mugnai, A., Olsen, W., Getty, G., Shibata, A. and Smith, E. (1994). Algorithms for the retrieval of rainfall from passive microwave measurements. *Remote Sensing Reviews*, **11**, 163–94, Harwood Academic Publishers.

Wood, E. F. (1994). Scaling, soil moisture and evaporation in runoff models. *Advances in Water Resources*, **17**, 25–34.

Wood, E. F. (1997). Effects of soil moisture aggregation on surface evaporative fluxes. *J. Hydrol.*, **190**, 397–412.

Wood, E. F., Lettenmaier, D. P. and Zartarian, V. (1992). A land surface hydrology parameterization with sub-grid variability for general circulation models. *J. Geophys. Res.*, **97** (Atmosphere), 2717–28.

Wood, E. F., Sivapalan, M. and Beven, K. (1990). Similarity and scale in catchment storm response. *Reviews of Geophysics*, **28**(1), 1–18.

Wood, E. F., Sivapalan, M., Beven, K., and Band, L. (1988). Effects of spatial variability and scale with implications to hydrologic modeling. *J. Hydrol.*, **102**, 29–47.

Woodruff, S. D., Slutz, R. J., Jenne, R. L. and Steurer, P. M. (1987). A comprehensive ocean-atmosphere data set. *Bull. Am. Meteor. Soc.*, **68**(10), 1239–50.

Woolhiser, D. A. and Goodrich, D. C. (1988). Effect of storm rainfall intensity patterns on surface runoff. *J. Hydrol.*, **102**, 335–54.

Wright, P. B. (1988). On the reality of climate changes in wind over the Pacific. *J. Climatology*, **8**, 521–7.

Wu, J. (1994). The sea surface is aerodynamically rough even under light winds. *Boundary Layer Meteor.*, **69**, 149–58.

Yelland, M. J., Moat, B. I., Taylor, P. K., Pascal, R. W., Hutchings, J. and Cornell, V. C. (1997). Measurements of the open ocean drag coefficient corrected for air flow disturbance by the ship. *J. Phys. Oceanogr.*, (submitted).

Yelland, M. J. and Taylor, P. K. (1996). Wind stress measurements from the open ocean. *J. Phys. Oceanogr.*, **26**, 4541–58.

Yoshida, H., Hashino, M., Tamura T. and Muraoka, K. (1995). Formation process of streamwater chemistry in small forested mountain basin, *J. Hydrosci. and Hydraulic Engineering*, **13**(2), 83–97.

Young, C. C., da Silva, A. M. and Levitus, S. (1995). The effect of a revised Beaufort Equivalent Scale on momentum and heat fluxes over the global oceans. In *Internat. COADS Winds*

Workshop, Kiel, Germany, 31 May–2 June 1994, pp. 287–301. NOAA/ERL.

Zavody, A. M., Gorman, M. R., Lee, D. J., Eccles, D., Mutlow, C. T. and Llewellyn-Jones, D. T. (1994). The ATSR data processing scheme developed for the EODC. *Int. J. Remote Sensing*, **15**(4), 827–43.

Zion, M. S. (1995). Use of operational satellite and radiosonde data to estimate radiation forcings for hydrologic models. Master of Science Thesis, Dept. Civil Eng and Oper. Res., Princeton University, Princeton, 125 pp.

5 Use of small-scale models and observational data to investigate coupled processes

5.1 Mesoscale field experiments and models

Jean-Claude André, Joël Noilhan and Jean-Paul Goutorbe

Introduction

Over a period of almost ten years we have seen a number of major land-surface experiments taking place in different parts of the world. HAPEX, FIFE, EFEDA, and others, are names and acronyms that are now known to a large number of meteorologists, hydrologists and climate modelers. This chapter addresses the rationale behind them and explains why and how they were decided, planned and organized.

We have to go back to the early 1980s, when climate modellers first gathered and asked for improved description of land-surface processes, and especially of interactions between the atmosphere, vegetation, soil and hydrology (WCRP, 1981). It took scientists two years to develop the HAPEX ('Hydrological-Atmospheric Pilot Experiment') concept, under the auspices of the 'World Climate Research Programme' (WCRP, 1983). At that time the main emphasis was on the upscaling problem, i.e. recognizing the large gap between scales at which land-surface processes would have to be parameterized in general circulation models of the atmosphere (i.e. of the order of a few hundred kilometers) and the scale at which physical knowledge on water and energy vertical transfers was available (i.e. at the patch scale, of the order of a few hundred meters). Parameterization was from the very beginning the driving issue behind these mesoscale experiments.

In response to this parameterization problem, WCRP developed together with the 'International Satellite Land-Surface Climatology Programme' (ISLSCP) the concept of mesoscale experiments. HAPEX-MOBILHY (André et al., 1986) and FIFE (Sellers et al., 1988) were the two forerunners in a series of experiments, with spatial extent of either 15 km (FIFE) or 100 km (HAPEX-MOBILHY). The emphasis was then on coupling between the atmosphere and hydrology for HAPEX-MOBILHY, and on estimating area-averaged land-surface parameters from airborne and satellite remote sensing for FIFE. At the beginning it was quite reasonably decided to experiment above flat continental surfaces, and to select areas where the land-surface cover would be fairly simple to characterize (contrast between agricultural land and conifer forest for HAPEX-MOBILHY, and natural grassland for FIFE). These experiments took place at mid-latitudes, very close to

the laboratories and agencies which were involved in organizing them.

Only later on was it clearly realized that upscaling and remote sensing were not the only two issues of interest. The HAPEX concept has also been used to understand feedback mechanisms between the surface and the atmosphere. In the Sahel, for instance, a large increase in albedo due to drought could strengthen air subsidence, leading to a further decrease in rainfall (Charney, 1975). HAPEX-Sahel was accordingly designed with these radiative feedbacks in mind. More generally it became progressively clear that more attention should be given to the interaction with the vegetation and the ecosystem distribution, both from the atmospheric and hydrologic point of views. It also was recognized that the coupling between the geosphere and the biosphere was of particular importance over land surfaces: not only should water and energy exchanges be under experimental investigation, but also carbon transfer would have to be studied in detail at the same time. These issues were addressed by the 'International Geosphere-Biosphere Programme' (IGBP), especially through its core project called 'Biospheric Aspects of the Hydrological Cycle' (BAHC) (BAHC, 1993).

As a result of this progressive widening of their aims and objectives, and of a concomitant and progressive improvement in their design, the mesoscale land-surface experiments can now be characterized by a few crucial features:

1. They encompass an area large enough to be significant with respect to the scales used in climate models (i.e. from a few tens of kilometers to one or two hundred kilometers).
2. They encompass an area with enough heterogeneities in land-surface cover that they make upscaling studies possible, and lead to non-trivial aggregate values of land-surface parameters and fluxes. Of particular importance is the fact that they must produce measurements at various scales, so that the water budget can be closed at these intermediate scales.
3. They effectively address the coupling between the atmosphere, vegetation, soil and hydrology. This means that all these processes have to be simultaneously measured, and consequently that the very large number of point

measurements, necessarily scattered over the whole area, have to be accomplished by a large variety of measuring techniques.

4. They are intimately coupled to numerical mesoscale modeling, both in the atmosphere and for the hydrology. This is necessary, as will be shown below, not only to give a better interpretation framework to the various measurements, but also to provide a crucial supplementary tool for upscaling studies and for deriving 'effective' values of parameters and fluxes at the scale of the total domain.

5. They are organized in such a way that they provide the ground truth for developing, calibrating and validating remote sensing of land-surface parameters and fluxes, as satellite observations are the only feasible way to derive these parameters for the entire globe.

Most, if not all, of the above features have indeed been considered when planning the mesoscale field experiments to be reported below, and their experimental design took the corresponding constraints into account. There is one constraint which was not initially part of the design, and which turned out later to be of crucial importance: the duration of the field phase. A short-duration mesoscale field experiment may indeed be adequate to document and study a number of physical processes, as well as the upscaling methodology; however, it is now clear that this is not sufficient to test the long-term validity of the various models to be developed and calibrated from these experiments. This is particularly important for the models which describe the upper-soil and surface-water budgets, for which annual imbalance may be traced either to inaccuracies in the models themselves, or to physical effects associated with a particularly dry or wet year. It is then easily understood why programs such as PILPS (Project for Intercomparison of Land-surface Parameterization Schemes; see Henderson-Sellers *et al.*, 1995) have called for much longer duration, ideally multi-year, measurements, even though they apply only for a few individual sites and not the entire mesoscale domain. This new idea of longer duration experiments has progressively been added into the actual design of mesoscale field programs.

Multiscale measurement strategy in land-surface experiments

The main thrust of large-scale experiments, as already mentioned, is toward upscaling from local to regional scales. A multiscale measurement strategy has been developed and used in several large-scale experiments such as HAPEX-MOBILHY, FIFE , EFEDA or HAPEX-Sahel. To study the interactions between the continental surface and the atmosphere it

is necessary to choose a site which is large enough to allow observations of the exchanges within the planetary boundary layer, so as to define the contributions of the mesoscale atmospheric circulation, and to obtain a reasonable sampling of radiometric observations from space. On the other hand, the strategy must allow comparisons of optical and remote-sensing measurements with surface observations used for validation. The observational techniques and experimental devices which must be used at the surface therefore require small-scale deployment.

The spatial scale of the experiment reflects these two objectives: a large domain (of the order of $100 \times 100 \, \text{km}^2$) is defined to permit the collection of measurements on the larger scale (meteorology and satellite remote sensing). Inside this region a number of so-called 'super sites', each approximately $20 \times 20 \, \text{km}^2$, are equipped to concentrate experimental efforts on representative areas (see Figure 5.1). Within each supersite three or more subsites are selected in order to sample correctly the various land uses (for instance vine, bare soil, irrigated crops and natural vegetation for EFEDA). The basic set of measurements performed at the subsites are the components of the radiation budget, surface fluxes, biometric parameters, soil parameters, soil moisture and ground-truth for remote-sensing studies. The link between these local measurements is provided by airborne measurements of turbulent fluxes and radiometric properties in the optical, thermal and microwave domains. The airborne flux measurements provide a direct measurement of fluxes of sensible and latent heat and of momentum at a scale of 5 to 20 km, compatible with the supersite scale. The flux aircraft flies long legs close to the surface (at an altitude of about 50 to 100 m above ground level) and also close to the top of the boundary layer. The flight plans are designed both for sampling fluxes and estimations of mean gradients.

The airborne remote-sensing programme is also designed to interpolate local surface measurements. In particular, local flights at supersite-scale are designed and phased with ground measurements. For instance, for HAPEX-Sahel, this included optical measurements of the bidirectional reflectance for calibration of the optical sensors and surface soil-moisture measurements for the calibration of microwave sensors. A ground-truth programme for thermal infrared was also carried out. Airborne remote-sensing imagery was acquired as much as possible at the time of satellite overpasses.

The boundary-layer development has to be monitored on special days using sounding stations at the supersites. The operational soundings are also used for the large-scale meteorological analysis. Rainfall and surface meteorology can be monitored with a network of automatic weather stations. HAPEX-Sahel took advantage of the EPSAT (Estimation des

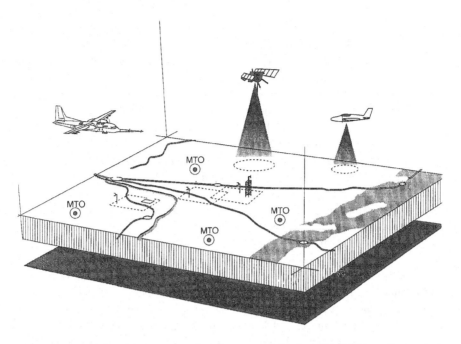

Figure 5.1. Schematic representation of a mesoscale land-surface field experiment.

Pluies par SATtellite, Lebel *et al.*, 1992) network of over 100 recording rain gauges.

Each of these large-scale programs has a hydrologic component to check independently areal evaporation on the long time scale. However, the hydrologic program relies heavily on existing operational networks and is very dependent on the location. For instance it was possible to implement a distributed hydrologic model in the case of HAPEX-MOBILHY, while the same investigation was not possible for HAPEX-Sahel owing to the lack of drainage network. In the time domain, hydrology, soil moisture and, for some programs, vegetation studies, span several years. Soil moisture was for instance monitored for 2 years at 12 sites during HAPEX-MOBILHY and for 3 years at 6 sites during HAPEX-Sahel.

In order to facilitate the handling and interpretation of the data, a central data base has to be established. One of the examples was the FIS (FIFE Information System; Strebel *et al.*, 1990) developed for the FIFE experiment and later adapted for HAPEX-Sahel.

Mesoscale atmospheric and hydrologic models as aggregation tools

It has been known for a long time that, given sufficient horizontal advection, relatively homogeneous turbulent fluxes of heat and moisture can be found at some height above the surface, even when the land-surface exhibits a fairly high degree of heterogeneity. This same idea is behind the strategy of using detailed atmospheric modeling, in this case mesoscale modeling, to study the way upscaling is taking place.

It has also been known for just as long that hydrologic catchments are physical 'aggregators' for the water budget over complicated surfaces, e.g. those with significant topography. This idea is behind the alternative strategy of using distributed hydrologic modeling to derive effective values of water fluxes, and especially of evapo(transpi)ration fluxes.

We develop these two strategies below.

Atmospheric mesoscale modeling

Depending on the scale at which one wants to start the upscaling process, it is necessary to use one of the two kinds of atmospheric models: hydrostatic or non-hydrostatic.

The hydrostatic assumption is valid in the atmosphere as long as one does not address too small horizontal scales or phenomena with too intense vertical velocities. It can be shown (e.g. Holton, 1972) that the vertical hydrostatic equilibrium is indeed satisfied at horizontal scales of a few kilometers or more. One should then use non-hydrostatic models for studying upscaling from the very small scales, as when addressing the influence of horizontal advection between adjacent patches with significantly different properties (differences in either surface cover, or physical soil properties, or soil-moisture content, etc.), or the influence of small-scale circulations triggered by strong horizontal contrasts. One should alternatively turn to hydrostatic models to integrate upscale, from a few kilometers up to the size of the experimental domain or to the size of the grid box of the climate model. Hydrostatic modeling is indeed used for aggregating over patchy landscapes with relatively slow variations of surface properties in space. Hydrostatic models are also, at least for the time being, more easily 'operationally' implemented, as they are simpler to feed with realistic large-scale surface conditions (as derived, for example, from satellite remote sensing; see below) and atmospheric forcings.

Non-hydrostatic atmospheric models are mostly used to study upscaling in idealized cases. Very often they are implemented in a two-dimensional version, where different surface conditions are given for the upstream and downstream regions, and where the larger-scale atmospheric forcing is taken from a single upwind radiosounding. This methodology then gives access to very important information such as the height and downwind distance at which fluxes reach homogeneity, and consequently how one must proceed to estimate aggregate values at these scales. One should, however, consider with some caution such idealized modeling studies when they deal with too abrupt, and sometime unrealistic, changes in surface properties. In such cases one would indeed tend to overestimate the intensity of mesoscale circulations developing inside the boundary layer. Furthermore, though such circulations are said to modify the atmospheric structure and hence the water demand and the surface evapo(transpi)ration, there has been up to now no experimental *in situ* proof, at least to our knowledge, that it is indeed so. Thus there is a pressing need for the development of appropriate mesoscale field experiments to observe and quantify these processes. Further discussion of these points is given below.

The importance of mesoscale circulations generated by landscape heterogeneities has been addressed from numerical experiments with mesoscale non-hydrostatic models. Avissar and Chen (1993) and Chen and Avissar (1994) showed for instance that vertical heat fluxes associated with mesoscale circulations generated by land-surface wetness discontinuities are often larger than turbulent fluxes, especially in the upper part of the boundary layer. It should, however, be noted that, although a broad range of large-scale atmospheric conditions and landscapes (soil wetness, albedo, roughness) were considered, moist processes that develop at the mesoscale were not investigated. Based on the previous works, Lynn *et al.* (1995) nevertheless proposed a parameterization of the subgrid-scale mesoscale heat fluxes in GCMs, using the subgrid-scale spatial variability of surface sensible heat flux, and the characteristic structure of the landscape (characteristic length-scale of land discontinuities, distribution of surface wetness). Given the strong sensitivity of me-

soscale fluxes to soil wetness, such a parameterization scheme should also take into account the spatial variability of precipitation, as well as the time evolution of soil moisture and surface fluxes at the subgrid scale.

Turning now to atmospheric hydrostatic mesoscale modeling, the proposal to use it to help derive aggregation methods, and the strategy to do it, goes back to when HAPEX experiments were designed (André and Bougeault, 1988). In this approach the three-dimensional hydrostatic mesoscale model is implemented in a careful manner (see, for example, Noilhan *et al.*, 1991). One has first to ensure that it is physically and meteorologically driven in a correct way. This is usually done by using, for a parameterization scheme at the grid-size scale (typically 3 to 10 km), a one-dimensional transfer scheme of the SVAT-type (Soil-Vegetation-Atmosphere Transfer Scheme), which has been previously validated against a number of independent point measurements. The larger-scale forcing is then directly prescribed from meteorological analysis (or could alternatively be taken from coupling with a global model). Surface and subsurface conditions have to be prescribed with care and realism. Soil properties (soil type, conductivities, etc.) are usually taken from digitized information (see, for example, Mascart *et al.*, 1988). Surface parameters relating to land cover are inferred from satellite remote sensing. The determination from space of the NDVI (Normalized Difference Vegetation Index, i.e. normalized difference betwen radiances in visible and thermal infrared channels), both for 'instantaneous' values and seasonal-to-annual evolution, allows for the classification of vegetative cover into main types. It is then more or less straightforward to deduce from these vegetation types what are the surface parameters (albedo, leaf-area index, fractional vegetation cover, minimum stomatal resistance, etc.), or to deduce some of these parameters from more sophisticated remote-sensing techniques using different wavelengths (as is the case with microwaves, or multipolarization techniques, etc.). Finally, the total domain of the mesoscale model is chosen so that it encompasses an area five to ten times larger than the experimental domain (to avoid possible perturbation by lateral boundary conditions), with a grid size of a few kilometers.

It should be emphasized here that remote sensing, and especially satellite remote sensing, is more than just a complementary tool for providing boundary conditions to numerical models. It also has its own potential for giving insight into aggregation and upscaling. Smaller scale radiative surface properties are indeed being averaged out at the satellite pixel size. The pixel sizes range from fairly small, 10 m or so in the case of the SPOT and Landsat, to medium, 1 to 5 km in the case of operational meteorological satellites, either polar-or-

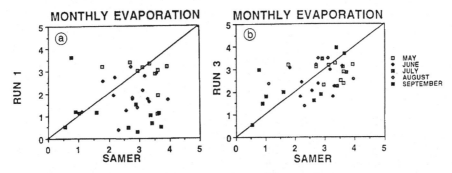

Figure 5.2. Scatter diagram between measured and hydrologically modelled values of monthly accumulated evapo(transpi)ration for the HAPEX-MOBILHY experiment, for two versions (*a*) and (*b*) of the hydrologic model. (See Ottlé and Vidal-Madjar, 1994.)

biting or geostationary, and to even larger, 10 km or more in the case of microwave or radar observations. Careful consideration of the relations between these various pixel values does of course bring insight into the way individual heterogeneous elements average. Upscaling techniques can then be constructed, which are based mainly upon remote sensing. We shall not, however, develop this approach here, but shall only describe some particular examples when discussing the FIFE experiment (see below). Readers interested in more details are referred to review articles such as those by Schmugge and Becker (1991) and Hall *et al.* (1992).

Hydrologic mesoscale distributed modeling

The mesoscale approach can also be used for hydrologic catchment modeling. In this case the hydrologic model has to be a distributed model, i.e. include a quite detailed spatial description of water routing, both at the surface and possibly in the deeper layers. The grid size is then chosen according to the topographic features of the area, of the order of 1 to 10 km depending upon the particular location within the domain. The size of the model domain must correspond to a catchment encompassing the experimental area. Such hydrologic models are usually calibrated against a long time series (10 years or so) of measured precipitation inputs and water outflows. Calibration consists of determining the various coefficients describing timing and amount of water transfer between the various reservoirs (run off and storage, surface and subsurface), so that the simulated outflow agrees with measurements.

Once calibration is achieved, the model can be applied to simulate the particular field experiment. Input is then the daily measured precipitation and potential evaporation fields, and the outputs are the river outflow, the level of the water table, the soil-water storage and the evapo(transpi)-ration. Improved SVAT schemes can also be coupled to me-

soscale catchment models in order to give a more detailed simulation of evapo(transpi)ration. This approach is for example described in Ottlé and Vidal-Madjar (1994), where the production function of the hydrologic model was changed for a two-layer surface scheme with the objective to improve the surface evapotranspiration (see Figure 5.2*a*). The inputs of the modified hydrologic model were kept unchanged (precipitation and potential evaporation) and the system was operated with a time step of one day.

The pioneering work of Ottlé and Vidal-Madjar gave rise to an ambitious project of coupling between a mesoscale atmospheric model and the same distributed hydrologic model (Habets *et al.*, 1995). In order to ensure a full interaction between the atmosphere and the hydrology, it is necessary to bridge the gap in time scales by including an explicit representation of the daily cycles of the surface energy and water budgets. Such a method is summarized in Figure 5.3. The main idea is to achieve the coupling between the two regional models through a common interface. This is done by using a surface parameterization scheme which can be implemented over a time scale of several years. The surface scheme, embedded within the hydrologic model, can be used either in a forcing mode by imposing the observed atmospheric quantities at screen level or with a full interaction with the atmosphere when the mesoscale meteorological model is used. The ouputs of the surface scheme are the four components of the surface energy balance as well as the soil water content and the outflow composed by the runoff and the gravitational drainage. This latter quantity, accumulated over one day, is the input for the surface layer in the hydrological model. Then, the water is subdivided into storage, lateral flow and infiltration to the underground layer where the model calculates the horizontal transfers to and within the water table. In this coupling, the time step for the surface scheme is several minutes, while the time step for the hydrologic component is kept to one day. This improved

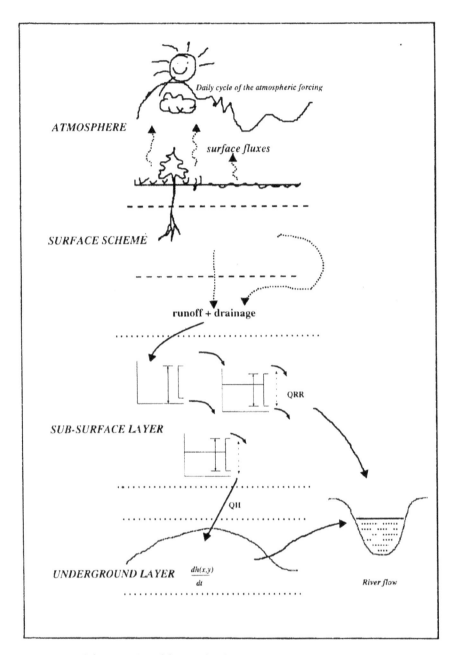

Figure 5.3. Schematization of the coupling between the atmosphere and the hydrology.

methodology is now under development for the HAPEX-MO-BILHY area, taking full advantage of the existing data base and of previous significant works concerning the modeling of the hydrology (see the next subsection for some earlier hydrological results obtained for the HAPEX-MOBILHY program) and of the atmosphere at mesoscale, as well as of the excellent knowledge of the spatial variability of soil and vegetation in that region.

It should be emphasized here that such an approach has a useful built-in constraint, i.e. conservation of total water. The integrated evapo(transpi)ration flux at the catchment scale is consequently very trustworthy. The use of distributed modeling furthermore leads to an estimation of the spatial distribution of evaporation and soil-moisture storage, as exemplified below.

Examples and results of aggregation from land-surface experiments

FIFE 1987/89

The FIFE experiment mostly addressed the aggregation problem from the remote-sensing perspective, which is best summarized in Sellers and Hall (1992). Even in the case where aggregation is achieved through numerical modeling, it is always necessary to provide 'effective' surface parameters at various spatial scales for the surface forcing of these models. This makes it a necessary step to derive such aggregate surface parameters, something most easily achieved from satellite remotely sensed data.

FIFE was organized having in mind a stratification strategy for experimental sites, in such a way that data collected during the field phase would be best suited for providing adequate and complete ground-truth on the one hand, and for reconstructing the spatially averaged surface values on the other hand. As discussed in Sellers et al. (1992), the individual experimental sites were chosen so as to be representative of topography (valley bottoms, slopes of different aspect and grade, and plateau tops) and management treatment (burned or unburned). This called for a number of 50 or so measurement stations, distributed so as to sample fully the $15 \times 15 \text{ km}^2$ total area of the experiment. This also called for the use of other integrating devices like eddy-correlation measurements from airplanes and remote-sensing instruments on helicopters, aircraft and satellites, with resolution ranging from very high (SPOT and Landsat, 10 m) to medium (NOAA, 1 km) and coarse (GOES, a few kilometers).

We shall discuss here only the results obtained for the latent (LE) and sensible (H) turbulent surface heat fluxes, as these results are quite different and are representative of most contrasting behaviors. Spatial upscaling can indeed be either relatively easy or much more cumbersome, depending on whether or not the processes at play are almost linear or highly nonlinear.

In the case of the turbulent surface flux of evapo(transpi)ration E, processes at work exhibit a simple behavior, and upscaling follows relatively easily. This is due to the fact that (some) spectral vegetation indices appear to be near-linearly related to the fraction of absorbed (green) photosynthetically active radiation (i.e. the so-called FPAR), which in turn relates almost linearly to the derivative of the unstressed surface conductance with respect to incident photosynthetically active radiation (i.e. the so-called PAR). By combining such near-linear relations, one deduces that area estimates of unstressed surface conductance derived from fields of spectral vegetation indices should be scale invariant. Other essential

factors which determine the surface flux of latent heat are the atmospheric demand and the available energy at the surface. As these latter quantities do not usually exhibit a large spatial variability, it can be safely assumed that methods for estimating surface latent heat flux from spectral vegetation indices are more or less scale independent, at least for the various cases and relatively narrow scale range documented during the FIFE experiment. This is best seen in Figure 5.4, from Sellers et al. (1992), where the total latent heat flux estimated directly from area-averaged parameters compares quite favorably to the aggregate value reconstructed from a number of much smaller-scale pixel estimates.

On the other hand, the turbulent surface flux of sensible heat H appears to be much more difficult to deal with. Hall et al. (1992) indeed showed, on the one hand, that uncertainties in estimating the atmospheric demand from the temperature difference $(T_a - T_s)$ between the atmosphere (T_a) and the surface (T_s) were as large as this difference itself. This unfavourable behavior is even worse when going upscale, as the relation between the surface aerodynamic resistance and both the temperature difference driving the atmospheric demand and the sensible heat flux itself is nonlinear. This prevents simple aggregation rules for upscaling sensible heat flux, a problem which will be looked at in the following subsections.

HAPEX-MOBILHY 1986

Once a numerical atmospheric mesoscale model has been implemented and validated for the particular geographical location of a field experiment, it can be used for upscaling purposes. The particular example to be described here follows the strategy proposed by André and Bougeault (1988).

In the case of the HAPEX-MOBILHY experiment, the so-called 'PERIDOT' atmospheric model was first further validated against independent experimental data. Airborne measurements of boundary-layer eddy fluxes were taken at different heights and times during the experiment. These data were not used for calibrating or validating the one-dimensional vertical turbulent transfer scheme of Noilhan and Planton (1989) used in the PERIDOT model, so that they could be used for independent validation. Simulated values of sensible and latent heat fluxes were compared with airborne values, as shown in Figure 5.5 (taken from Noilhan and Lacarrère, 1995). It can be seen that the accuracy of the mesoscale model was sufficiently good that it can quite safely be used for interpolation and upscaling. The way one proceeds is described below.

The full three-dimensional simulation was used for generating the horizontal distribution of surface turbulent fluxes of sensible (H) and latent (LE) heat. By summing the

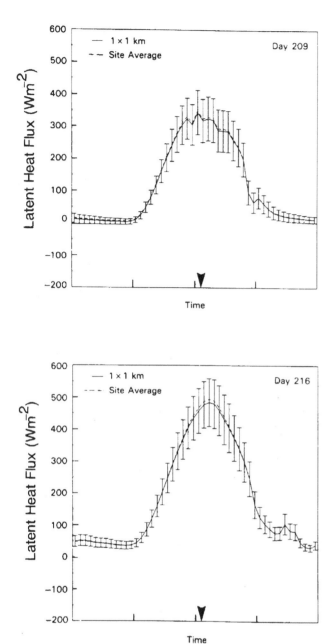

Figure 5.4. Comparison between the site-averaged value of surface latent heat flux obtained using single-site-averaged spatial mean values of driving parameters (dashed line) and its site mean value obtained by aggregating separate calculations on a 1 × 1 km² grid over the site (solid line, with error bars for the standard deviation of individual estimates), for the FIFE experiment. (See Sellers *et al.*, 1992.)

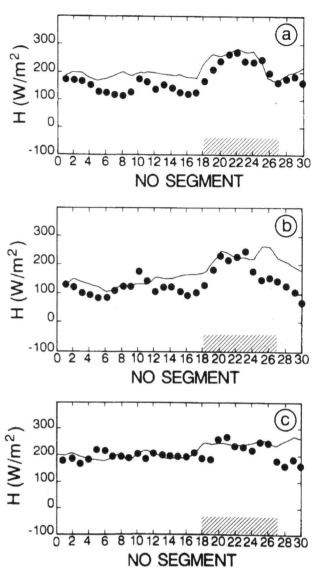

Figure 5.5. Comparison betwen sensible heat fluxes as estimated from eddy-correlation airborne measurements (full circles) and as modeled using a three-dimensional atmospheric mesoscale model (solid lines) for the HAPEX-MOBILHY experiment. (See Noilhan and Lacarrère, 1995.)

individual values over all the model grid points, one has access to the likely value of the area-averaged surface fluxes, called respectively H_a and LE_a. By having access at the same time to the various grid point values, it is fairly easy to test any upscaling methodology. The first test, which corresponds to what is usually done in GCMs, was to determine which is the most frequent vegetation type inside the entire domain, and to compute the surface fluxes for the entire domain by using the one-dimensional transfer scheme fed by parameters corresponding to this dominant vegetation type. In the HAPEX-

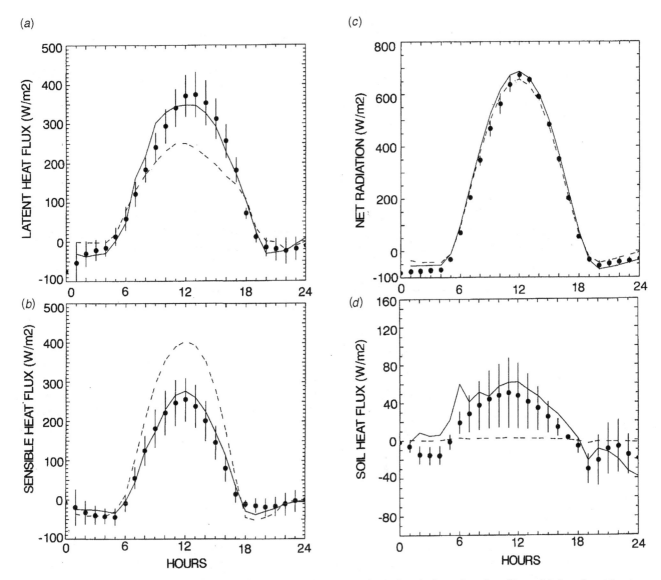

Figure 5.6. Comparison between area-averaged values of diurnally varying values of (*a*) the latent heat flux, (*b*) sensible heat flux, (*c*) net radiation and (*d*) heat flux into the ground as 'measured' from experimental field data interpolated through atmospheric mesoscale modeling (points and vertical bars, showing subgrid-scale variability), and as estimated from a one-dimensional transfer scheme using either dominant (dashed line) or 'effective' (solid line) surface properties for the HAPEX-MOBILHY experiment. (See Noilhan and Lacarrère, 1995.)

MOBILHY case, the dominant vegetation is coniferous forest (by something like 40%), while the rest of the domain is covered by a variety of different shorter agricultural crops and prairies. Values computed this way, denoted by H_d and LE_d, clearly underestimate the averaged values for the tran-spiration and latent heat fluxes in this particular case, as shown in Figure 5.6. This is not a surprise, as subgrid-scale variability is not taken into account in this approach, which, although it is oversimplified, is still used in most numerical weather prediction and atmospheric climate models.

An alternative upscaling method, proposed by Noilhan and Lacarrère (1995) and developed by Li and Avissar (1994), is to estimate 'effective' surface parameters at the domain scale by averaging each of them differently, in such a way that conserved quantities are indeed conserved. One would accordingly average linearly albedos, leaf-area indices and vegetation-cover values, as this is the way these parameters influence the various physical processes. One would, on the other hand, average the roughness lengths logarithmically, as the friction velocity, and hence surface stress, depends

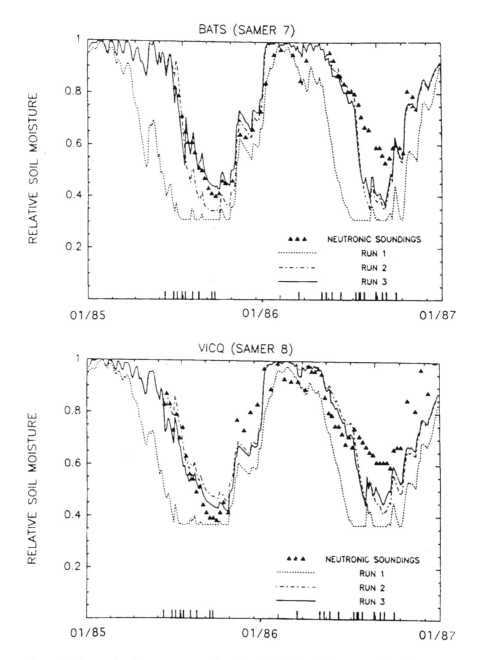

Figure 5.7. Comparison between measured and modeled values of the upper-soil moisture content at two sites during the HAPEX-MOBILHY experiment. (See Ottlé and Vidal-Madjar, 1994.)

logarithmically upon the roughness length, while one would inversely average stomatal resistances, so that conductances, as usual, do add. 'Effective' surface parameters at the domain scale being constructed this way can be used in the one-dimensional transfer scheme to give the so-called effective surface fluxes H_e and LE_e. These values compare well with their average values, as can be assessed in Figure 5.6,

confirming that subgrid-scale variability of land-surface parameters has a significant influence upon area-averaged surface fluxes.

An alternative method for upscaling land-surface processes has been tried in the case of the HAPEX-MOBILHY experiment. The Ottlé and Vidal-Madjar (1994) hydrological model, as updated by including an improved scheme for the evapo(transpi)ration, is run with observed precipitation data. Its monthly averaged grid-point values are then compared with the observed values at surface stations. Such a comparison is shown in Figure 5.2(*b*), where it appears that the individual values still exhibit scatter. Hydrologic modeling is, however, of extreme value for time-averaging processes related to water transfer between the various reservoirs, and particularly the storage of soil moisture, a variable not easily inferred from other methods. Figure 5.7 indeed shows, for the same runs of this same model, the comparison between simulated and measured moisture storage in the upper meter or so of the soil. The hydrologic model displays excellent skill in simulating the annual cycle of soil moisture, due largely to the fact that the model's parameters are usually adjusted in the light of stream flow observed over periods of ten or more years. This indicates that the catchment scale processes have characteristic time scales that can be represented adequately in the models. As described above, present work involves the full coupling between the atmospheric and hydrologic mesoscale models.

EFEDA 1991

Numerical models can be used not only for upscaling purposes, as shown above, but also to interpret results from field experiments. There are in fact cases for which the experimental results viewed alone are ambiguous and difficult to understand, while the simultaneous consideration of output from the corresponding numerical model greatly improves the understanding of the phenomena under scrutiny. Such an example is given below, taken from the EFEDA-1991 program (see Bolle *et al.*, 1993), as modeled by Noilhan *et al.* (1997).

If one considers the latent heat flux map derived from measurements taken by an instrumented aircraft (Jochum *et al.*, 1993) as shown in Figure 5.8(*a*), one is faced with a difficulty in understanding why there are such large differences between latent heat fluxes at moderate height (i.e. 400 m agl) over distances as short as that separating the Barrax and Tomelloso sites (i.e. less than 100 km). It is puzzling to experience such differences, especially if one remembers that the surface latent heat flux over this whole region does in fact vary in the opposite way, with larger evaporation fluxes in the eastern part, due to irrigation, and significantly reduced latent heat fluxes in the western part, due to more continental influence. The explanation can be found only through numerical modeling and the explicit description of land–sea breeze effects. The three-dimensional numerical mesoscale model implemented over the EFEDA region (Noilhan *et al.*, 1997) unambiguously shows the development of a fairly in-

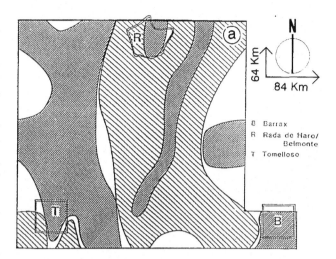

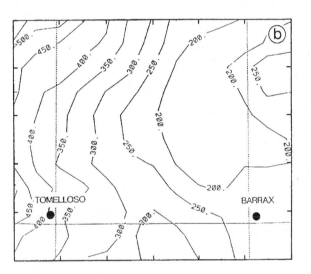

Figure 5.8. Comparison between (*a*) observed (Falcon aircraft, see Jochum *et al.*, 1993) and (*b*) simulated (mesoscale model, see Noilhan *et al.*, 1997) latent heat flux around noon at 400 m above ground for the EFEDA experiment. The grey areas in (a) correspond to regions where the latent heat flux is large, while hatched areas are regions where it is small. Units in (b) are W m⁻².

tense land–sea breeze circulation, leading to moistened air aloft in the eastern part, closer to the Mediterranean, and drier air inland, as shown in Figure 5.8(*b*).

This results in totally different heat-flux vertical profiles over these two locations, as shown in Figure 5.9: the latent heat flux is indeed decreasing with height in the eastern Barrax area, because entrainment of dry air at the top of the atmospheric planetary boundary layer (PBL) is reduced by the advection of moist air. On the other hand, the latent heat flux over the western Tomelloso area is increasing with

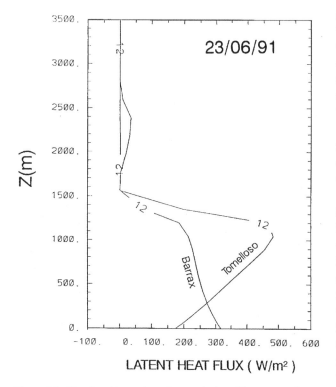

Figure 5.9. Simulated latent heat flux vertical profiles at noon for gridboxes representative of the sites of Tomelloso and Barrax in the EFEDA experiment. (See Noilhan *et al.*, 1997.)

height, due to the strong entrainment of dry air at the top of the PBL. The strong value of entrainment is associated with the rapidly growing PBL and the large discontinuity of specific humidity at the PBL top. Such complicated and interacting processes can be clearly identified only through explicit numerical three-dimensional modeling, as no experimental program will ever be able to gather such a wealth of information.

HAPEX-SAHEL 1992

Aggregation studies in the Sahel cannot be a simple copy of what is done in the temperate areas. For one thing, catchment hydrology cannot play a role as an integrator at large scale since most streams are ephemeral. However, the smaller scale unconnected drainage units, such as ponds, can be used for that purpose. The emphasis given to atmospheric modeling was less, due to the scarcity of upper-air observations in the area, which made an accurate prescription of initial and boundary conditions impossible. For these reasons the modeling methods are of more limited use than for the other large-scale programs reviewed above. Another specific feature is that surface conditions are extremely hetero-

geneous, even at plot scale, because of the patchiness of vegetation. The vegetation is usually composed of a stratum of herbs and a stratum of bushes, the exposed ground being crusted or not. An aggregation problem exists even at the smaller scales. Moving upscale, one is faced with the extreme spatial variability of rainfall, which can vary from 500 mm to 780 mm (annual value for 1992) over a distance of only 5 km.

Besides the general description of the HAPEX-Sahel experiment and results given in Goutorbe *et al.* (1994) and in Prince *et al.* (1995), some more specific considerations are developed below.

At the larger scale of the domain considered for HAPEX-Sahel (100 × 100 km²), the variability in surface conditions has been studied by D'Herbès and Valentin (1997) from SPOT data and ground transects. They came up with a classification in 16 classes based on landform, vegetation, soil crust type and land use. Based on this land-unit classification and on representative *in situ* measurements for each component of the landscape, the theoretical problem of aggregation was investigated by Passerat de Silans *et al.* (1994) at the scale of a 5 × 5 km² catchment using a simple Deardorff-type SVAT model (Deardorff, 1978). The result is that a regional conductance can be defined as a weighted average of the conductances of the various units. The weights are not strictly proportional to the relative surface, but also depend on the moisture availability. This sets a basis for an empirical aggregation scheme if the moisture status of the components can be estimated (e.g., from antecedent rainfall or microwave observations). Furthermore, runoff capabilities being assigned to each class, the map of these can also be used to forecast the aggregated flow of water in each endoreic unit.

A detailed piezometric survey of the whole 100 × 100 km² area has been carried out. Results are summarized in Leduc *et al.* (1997). In spite of large differences between wells, a general picture emerges. The recharge is local and essentially due to infiltration from the temporary drainage system (streams and pools). At most places, where old records are available, the water level is rising. This is interpreted as a recovery after dry years in the 1970s and 1980s, and perhaps also to a change in land use. The yearly recharge over the area has been tentatively estimated as 50 to 60 mm, about 10% of the rainfall amount.

The boundary layer development over the central part the HAPEX-Sahel domain has been simulated by Goutorbe *et al.* (1997). The selection of cases where surface heating is the driving mechanism, with limited large-scale forcing, is the most delicate part of the study. Out of the two months of the intensive period, four diurnal cycles were ultimately selected. The so-called ISBA surface scheme by Noilhan and Planton (1989) has been calibrated for the dominant land cover of the

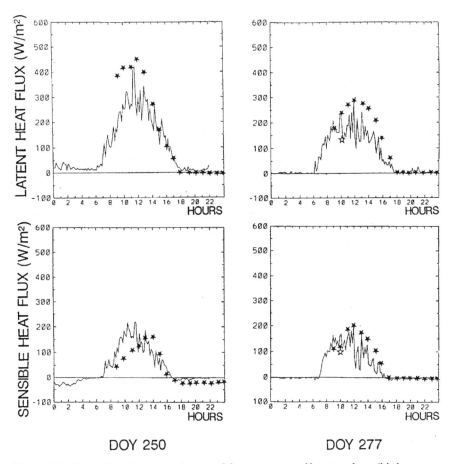

Figure 5.10. Comparison between estimates of the area-averaged latent and sensible heat fluxes from one-dimensional modeled planetary boundary-layer budgets for heat and moisture (full stars) and *in situ* measurements either at the surface (solid line) or by instrumented aircraft (open stars), for two days of the HAPEX-Sahel experiment. (See Goutorbe *et al.* (1997.)

area, namely fallow savanna. The surface scheme is coupled to a one-dimensional version of the ARPEGE model (i.e. the operational weather forecasting model at Météo-France). In order to fit the observations, direct heating of the boundary layer by aerosol absorption of shortwave radiation was introduced. With minor modifications of the local calibration parameters, boundary-layer profiles of temperature and also mixing ratio were reproduced with good accuracy. For this limited number of cases the one-dimensional model has been used to obtain a best fit of the surface fluxes from boundary-layer observations (Figure 5.10). This result is backed up by the observations of Gash *et al.* (1997), that for the nine weeks of the intensive period the evaporative fraction was surprisingly conservative. However, the generality of this result has not yet been demonstrated.

Atmospheric mesoscale modeling activities have been so far quite limited for HAPEX-Sahel, and only one case study has yet been reported (Taylor *et al.*, 1997). The simulation domain covers a 4° × 4° area centered on the HAPEX-Sahel square, with a 10 km grid size. There are difficulties in providing boundary conditions for such a large domain. Initial conditions for soil moisture are particularly critical and largely determine the partition of energy at the surface. They are determined using a simple soil moisture accounting model, which generates moisture patterns from daily rainfall estimates. The model is to some extent able to reproduce the north–south gradient for mean parameters and the sensible heat flux, as measured from the aircraft, but shows less skill for the latent heat flux (Figure 5.11).

Smaller-scale and other experiments

Although the purpose of this paper is to address those experiments aimed at studying land-surface processes at the mesoscale, it is of interest to review briefly results obtained from other experiments, organized at a smaller scale. These experi-

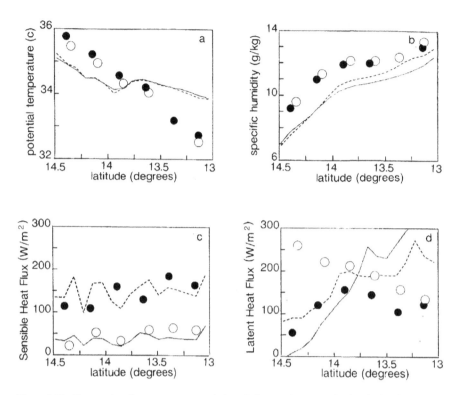

Figure 5.11. Comparison between measured aircraft data and mesoscale simulation for (a) potential temperature, (b) specific humidity, (c) sensible heat flux, and (d) latent heat flux, for the HAPEX-Sahel; experiment. (See Taylor *et al.*, 1997.)

ments were, and still are, usually organized to address a more specific issue, such as the influence of steep topography, the oasis effect, the small-scale transition between different patches of land-surface, etc. A good review of such experiments can be found in BAHC (1993).

The LOTREX experiment, organized in Germany in 1988–89, was aimed mostly at studying the atmospheric processes and upscaling of remotely sensed properties. It also helped significantly the development and improvement of SVAT models. The KUREX experiment, which took place in Russia in 1988–91, was also relevant to mid-latitude agricultural landscape, as is also the REKLIP 1991–99 experiment, which is still running along the Rhine valley. They document spatial variability of surface fluxes and corresponding boundary-layer development, with a particular emphasis upon small-scale non-hydrostatic processes and modeling, at least for the case of REKLIP.

The HEIFE and OASIS experiments, which were organized respectively in China (1990–94) and Australia (1993–96), are related to the study of advection (either at the mesoscale or at the microscale) above underlying surfaces with very contrasting soil moisture. Although their final results are not yet known, they have shed light about the way mesoscale atmos-

pheric structure is responding to such discontinuous surface forcing. The 1991 CODE experiment (Mahrt *et al.*, 1994) was also specifically aimed at documenting the influence of a mesoscale (i.e., 12 km across) irrigated area upon the turbulent fluxes of heat. These fluxes are significantly modified by the internal boundary-layer circulation (in this particular case an inland breeze), so that the fluxes measured from aircraft, even as low as a few tens of meters above ground level, are not representative of the surface fluxes. This indicates the need for care when performing airborne eddy-flux measurements above land surfaces with sharp transitions.

Finally, the ABRACOS experiment in the Amazon forest (1990–94) and the NOPEX experiment in the Swedish forest (1994–96) can be seen as either forerunners or companion experiments of larger-scale programs organized within these two types of biomes. The larger-scale experiments are, respectively, LBA for the tropical forest and BOREAS for the boreal forest (see below). As an interesting observational fact, it is worth mentioning that Cutrim *et al.* (1995) have observed systematic cloud formations over deforested regions in Amazonia, likely to be induced by forest burnings, although triggering by surface roughness discontinuities between the forest and adjacent deforested land could also be of importance.

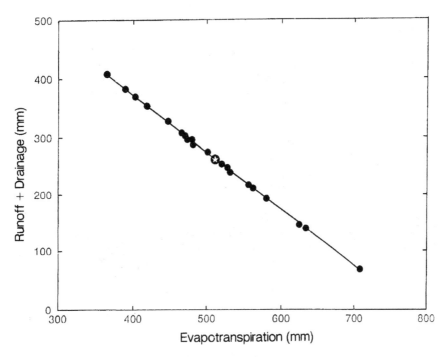

Figure 5.12. Scatter plot of annual runoff and drainage versus evapotranspiration simulated by 23 surface schemes (full circles) using the Cabauw data set (Chen *et al.*, 1997). The *in situ* observation is indicated by the circled star.

Toward longer duration experiments

The study of the continental cycle of the water budget requires consideration of all relevant time scales, from the diurnal to the annual scales. The HAPEX-type experiments focus mainly on the daily cycle. For instance, aggregation studies are conducted for a given soil moisture and plant status. The seasonal time scale has received comparatively little attention. Annual variation of soil moisture is a very important component of a land-surface scheme, as it is closely related to evaporation and runoff, and thus to the partitioning between sensible and latent heat fluxes. Accurate evaluation of the various components of the water budget on the annual basis is crucial for the hydrology (intensity of runoff and gravitational drainage) and for ecological models (i.e. role of the soil-water content in the root zone). The performance of a large variety of land-surface schemes to simulate the soil-moisture budget are presently being examined in the frame of the PILPS project (Henderson-Sellers *et al.*, 1995).

The first problem in the PILPS project is to find relevant data sets with a complete documentation of the atmospheric quantities (forcing variables and surface fluxes), of the soil water budget (soil moisture, runoff and drainage) and of the plants and soil properties. Two data sets have been provided

so far, representative of middle latitude grassland (the Cabauw data, see Beljaars and Bosveld, 1997) and crop (HAPEX- MOBILHY data, see Mahfouf and Noilhan, 1996). Hydrologic characteristics of Cabauw are very simple, as the deep soil is assumed to be permanently saturated throughout the year. Although all schemes used identical forcing data and the land-surface properties were specified with great care, large differences are found in the partitioning of precipitation into evapotranspiration, change in soil-water storage, and runoff. Annual totals of evapotranspiration and runoff simulated by 23 schemes show a range of variation of about 30% for precipitation (Figure 5.12). As a result of large differences in the soil-water partitioning, there is a large scatter in the simulations of the Bowen ratio. Only a limited number of surface schemes have monthly simulation errors of sensible and latent heat fluxes within the upper bounds of observation errors.

Because weekly observations of soil moisture were available over a full year, the HAPEX-MOBILHY data set was used to assess and understand the differences in the soil-moisture simulation (Figure 5.13, taken from Shao *et al.*, 1994). Again, large differences exist in the simulation of all the terms of the energy and water budgets, even with atmospheric forcing data of rather high quality and carefully chosen parameters. In these conditions, the interpretation of simulations of soil

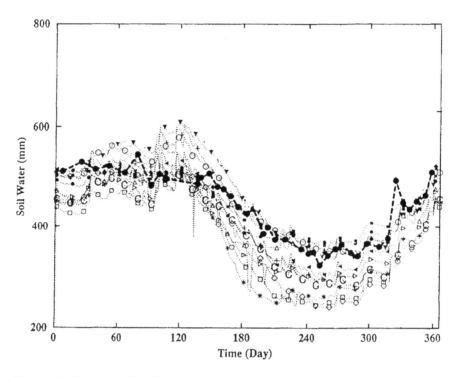

Figure 5.13. Comparison for soil moisture (mm) in the total soil column (1.6 m depth) between the observed values (full circles) and the values from 14 numerical simulations using different surface schemes (other symbols), in the case of the HAPEX-MOBILHY experiment. (See Shao *et al.*, 1994.)

moisture in climate models is difficult, since the atmospheric forcing is less accurate and the specification of land-surface parameters rather crude. A detailed analysis reveals that differences among a wide range of schemes were due to differences both in the model structure (for instance the number of soil layers to represent the root zone) and in the parameterization of individual processes. The analysis shows marked differences in the formulations of gravitational drainage, transpiration and surface evaporation. Finally, the two studies underline the key role played by runoff and drainage in the link between the energy and water budgets.

The studies by Chen *et al.* (1997) and Shao *et al.* (1994) proved how informative can be the use of data sets relevant to the long term to assess our ability to reproduce the annual cycle of surface exchanges. However, part of the discrepancy between simulations and observations may be due to uncertainties in important soil and vegetation parameters, in the atmospheric forcing and in surface fluxes as well as in soil-moisture observations. For instance, for the Cabauw data set, the lack of soil-moisture measurements is a serious shortcoming, especially in view of the spread of simulations of runoff and water drainage. Another weakness of the considered data sets concerns the estimation of evaporation. In

both studies, the latent heat flux was derived as the residual of the energy balance, and it has not been possible to investigate from the experimental point of view the accuracy with which the surface water budget balances over the whole year. For instance, the HAPEX-MOBILHY data allow some evaluation of the consistency of the energy and water budgets for only a limited period of the year. The same comment applies to the ARME (Shuttleworth *et al.*, 1984), FIFE 1987, EFEDA 1991 and HAPEX-Sahel 1992 data sets as well.

There is therefore a critical lack of adequate observational data sets for improving the currently used land-surface schemes. Ideally, one would need many years of measurements for soil moisture, and atmospheric forcing, with accurate estimation of precipitation and radiative fluxes, surface fluxes, including direct observations of evaporation, as well as details on the soil characteristics and on the vegetation, with particular attention to the growing-period phenology. The site of observations should be representative of a large area with respect to the land use, and should be located within an instrumented catchment basin, allowing some estimation of the runoff and drainage flows on a monthly basis. Such an arrangement will likely open the way to investigations where the atmospheric, hydrologic, and biological com-

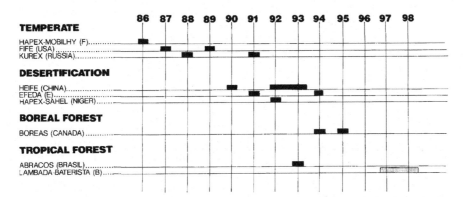

Figure 5.14. The series of major mesoscale land-surface experiments as recommended by WCRP/GEWEX and IGBP/BAHC.

Conclusion and perspectives

We have now reached a kind of plateau in the study of land-surface processes and of their aggregation and parameterization for use within larger-scale numerical climate and weather prediction models. The first series of mesoscale land-surface experiments, and associated numerical models, have delivered a number of methodological results and have allowed the definition of a precise strategy for constructing upscaled, area-averaged effective parameterization schemes. The task of applying this strategy to the real and world-wide distribution of land-surface properties remains, however, to be done effectively. Systematic account for subgrid-scale heterogeneity within a grid-size description of the global land-surface properties has still to be undertaken, a project presently considered by IGBP/BAHC.

The first series of land-surface experiments has not yet covered the full variety of biomes to be accounted for within climate models. Among the biomes for which systematic large-scale experimental investigation is still presently lacking, the tropical rainforest and the high-latitude tundra are probably the most important. There are plans to conduct a major experimental program in the Amazonian basin in 1996–97 (the so-called LBA program; e.g. ISLSCP and IGPO, 1993). No such plans have yet been proposed for the tundra, although this biome is playing a key role in storing carbon, and possibly releasing it under global warming.

Past large-scale experiments have also not properly addressed the interaction with topography. All of these experiments took place over relatively flat terrain, the most non-flat conditions studied up-to-now being gently rolling hills, as in the case of HAPEX-MOBILHY and FIFE. It thus remains to document the various ways topography influences land-surface properties, through hydrology and routing and storing of water at the surface or immediately below it, on the one hand, or through radiative exposition and vertically varying conditions, on the other hand. Such an approach is presently being pursued by the BAHC (1993) program.

Besides all the above issues related to documentation of biomes and surface conditions during experiments of relatively short duration (i.e. lasting for a few months at the most), it has become progressively evident that land-surface processes were exhibiting not only a large spatial heterogeneity, but that they were also of crucial importance in the long range. Firstly the hydrological budget at a particular site has to be measured over a period of a few annual cycles, in order to ensure that a transfer scheme or a multidimensional model which is to be calibrated against these data is properly validated. This is what PILPS is recommending (Henderson-Sellers *et al.*, 1995). Secondly, most ecosystems interact with the land-surface at the multi-annual time scale. The next series of land-surface experiments will consequently have to include longer-term components, with, for example, local sites to be maintained and operated over periods of five and more years (see, for example, WCRP, 1990). A complementary, sometimes alternative, way to address these longer-term effects is to look at transect-type studies, where one can document the various processes along climatological and/or ecological gradients. Such an approach has been pioneered with the so-called BOREAS experiment (Sellers *et al.*, 1995), where two sites with clearly different meteorological and vegetation conditions are being studied over a full annual cycle. It is to be expected that more of this type of experimental program will take place over the next years, after the first series of large-scale land-surface experiments have been completed (Figure 5.14).

5.2 Cloud-resolving models

M. W. Moncrieff and W.-K. Tao

Introduction

Numerical models of the global atmosphere used to predict weather for up to about two weeks ahead (medium range forecasting) can resolve horizontal scales of motion down to about 100 km. In general circulation models (GCMs) used in climate research and prediction, the resolution is typically many hundreds of kilometers. It is well known, however, that processes associated with motion scales too small to be resolved (subgrid-scale processes) are vitally important in the climate system. For example, the energy that drives the weather systems and the general circulation of the atmosphere is ultimately supplied by small-scale processes. It follows that subgrid-scale processes must be approximated and linked to the resolved scales of motion, a procedure called parameterization.

The key processes that need to be parameterized in large-scale models are convection, radiation, fluxes at the land and ocean surfaces, and atmospheric turbulence. Convective parameterization is faced with the challenging problem of providing simple representations of complex processes occurring within equally complex motion systems – a quest that may not be generally attainable. The concept of parameterization dates back to the early days of numerical weather prediction and presently requires a great deal of empiricism. Clouds in GCMs are classified in a simple way; namely, shallow (non-precipitating) convection, deep (precipitating) convection and large-scale (non-convective) condensation. As more realism is sought in both weather prediction and climate research, it is imperative that our understanding of how processes interact and how clouds relate to the larger scales of motion be vastly improved.

When GCMs are coupled to the ocean and land surfaces (coupled climate models) small uncertainties due to the misrepresentation of clouds, for example, can systematically accumulate and cause unacceptably large errors. This demonstrates a pervasive *non-local* issue; that is, effects of clouds at large scales are often manifested through interactions among physical processes rather than in a direct way. This underscores the need for an accurate representation of the collective effects of clouds, rather than details of cloud elements *per se*. It also raises the concept of *coupled processes*, by which we mean the collective effect of microphysical, radiative, and turbulent boundary layer and surface processes coupled to the large-scale environment through cloud-scale dynamics.

Precipitating convective cloud systems are the best example of this physical coupling. They also exhibit organization or coherence on scales much larger than convective elements (e.g., cellular structures in polar air outbreaks over the oceans, rain bands in extratropical cyclones and hurricanes, and various kinds of cloud clusters). Organization is an intrinsic property of convecting fluids. Although this process has been recognized for a long time (Ludlam 1980), its role in the large-scale behavior of the atmosphere is poorly understood.

The climatic role of precipitating clouds was surveyed by Browning (1990) and the role of the hydrologic cycle by Chahine (1992). Tropical cloud systems are especially important to the hydrologic cycle because they produce about two-thirds of the global precipitation, while the accompanying latent heat is a major source of energy for atmospheric motion.

Cloud-resolving models

Numerical models have been applied extensively to the study of cloud-scale and mesoscale processes during the past three decades. The distinctive aspect of these models is their ability to treat explicitly (resolve) cloud-scale dynamics. This requires that the models are formulated from the non-hydrostatic equations of motion that explicitly include the vertical acceleration terms. In contrast, global models are hydrostatic; that is, they can neglect the vertical acceleration because the horizontal/vertical (aspect) ratio is sufficiently large.

Cloud-resolving models (CRMs) resolve the dynamical circulations that not only couple convective cloud systems to the large-scale environment but also interlink the physical processes. This is distinct from GCMs where coupling (if it occurs) is on unrealistically large scales and where convective transport must be parameterized. Rates of change among the three phases of water (microphysics) and cloud interactions with both shortwave and longwave radiation have to be parameterized in both CRMs and GCMs because the intrinsic scales are too small to be resolved even in CRMs.

Since the dynamics of precipitating clouds can be resolved, there is a good prospect that CRMs can be used to improve parameterizations. For this reason, they are the approach of choice of the GEWEX (Global Energy and Water-cycle Experiment) Cloud System Study (GCSS) in its quest to improve the parameterization of cloud systems in global models (GEWEX

Cloud System Science Team, 1993). The principal cloud system types studied are boundary-layer clouds, layer clouds in mid-latitude cyclones, cirrus clouds and precipitating convective cloud systems.

A CRM with a 1-km grid can provide realizations of precipitating convective cloud systems in computational domains considerably larger than a GCM grid volume; this means that quantities derived from these realizations are statistically meaningful. Moreover, the large-scale effects of cloud systems derived from CRMs can be used to improve single-column models which are the building blocks of parameterization schemes for GCMs. This is being implemented through the GCSS working group on precipitating convective cloud systems (Moncrieff *et al.*, 1997)

The second type of non-hydrostatic numerical model, the large-eddy simulation (LES) model, is distinguished from a CRM more in terms of the problem it addresses rather than by fundamental differences in numerical approach. LES models are used to study processes on scales of motion ranging from a few hundred meters to a few kilometers, typically with a grid of about 10 m. In LES models and CRMs alike, the larger eddies which are assumed to contain most of the kinetic energy are calculated explicitly while those smaller than the grid resolution are parameterized. In LES there is an assumption that small scales approximate an inertial subrange where the energy spectrum is in statistical equilibrium and there is an energy cascade from the resolved scales to the dissipation scales. Compared with other uncertainties, this assumption is reasonably satisfied for boundary layer clouds. Because a typical LES (unlike a CRM) model has a domain much smaller than a typical GCM grid volume, it is more appropriate for detailed studies of processes (e.g., cloud-top entrainment) rather than a statistical measure of cloud-system effects on the environment.

Large-eddy simulation models have been widely applied to marine stratocumulus. The fractional area of cloud (cloud fraction) is an important parameter in climate models because it affects the radiation budget; for example, the extensive stratocumulus over regions of ocean upwelling and/or persistent anticyclonic regions. In particular, the cloud fraction is sensitive to the entrainment rate of free tropospheric air into the stratocumulus layer. LES models have recently been used in a model intercomparison study of cloud-top entrainment (Moeng *et al.*, 1995).

In summary, the value of CRMs is their explicit representation of quantities that are difficult to measure from observations. Due to the greater homogeneity of cloud processes on the 1-km scale it should be easier to derive more realistic parameterizations at these scales. For example, comprehensive data of comparable temporal resolution (ten seconds)

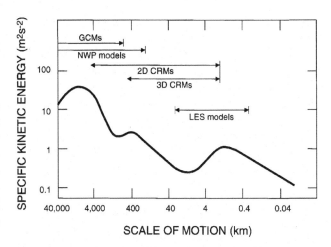

Figure 5.15. Schematic diagram showing the hierarchy of scales represented by various types of models. Note that the CRM, NWP, and GCM domains overlap so CRM realizations can be directly used for parameterization development.

and spatial resolution (one kilometer) are impossible to obtain from field experiments, especially on large spatial and temporal scales.

Use of cloud-resolving model results to derive parameterizations for GCMs

Cloud-resolving models span the gap between cloud-scale processes and the large-scale effects of clouds: this is a key point in their application to parameterization. The overlap between CRM computational domains and the large scales of motion in numerical weather prediction (NWP) and GCMs, shown in Figure 5.15, is very significant. For example, CRMs can have large horizontal domains (thousands of km in two-dimensional models and many hundreds of km² in three dimensions) and therefore span many GCM grid volumes. When used for parameterization studies they should preferably contain microphysics, cloud-interactive radiation and surface flux parameterizations.

Some CRMs have an inner domain of higher resolution which interacts in a two-way sense with an outer domain of lower resolution (e.g., Clark *et al.*, 1996). In fact, numerous subdomains having progressively higher resolution can be treated in these (interactively nested) models. However, there is a limit to this telescoping procedure – computational accuracy requires that the nesting ratio should not exceed about 3:1. In principle, the innermost domain can have comparable resolution to a LES model. Cloud-resolving models with 1-km grids are well suited to studies of precipitating cloud systems which are controlled by large-scale variables such as advection and vertical shear.

The parameterization of precipitating cloud systems is typically a three-step procedure that involves representing (a) the initiation of convection, (b) the sources, sinks and transport by clouds which are functions of the resolved-scale variables and (c) the interaction of cloud systems with the environment. Note that all of these are basic research problems. Nevertheless, parameterization schemes have to be computationally efficient; therefore judicious choices have to be made of what are the *minimum acceptable* levels of sophistication.

Precipitating convection: time scales of hours

During the past two decades, observational analysis, numerical studies and dynamical investigations have established that organized convection is strongly affected by vertical shear (which controls the dynamical structure) and large-scale ascent that generates convective available potential energy. Cloud-resolving models have been widely used to study the dynamics of precipitating convection on short time scales, notably squall lines and mesoscale convective systems. Three-dimensional simulations of these systems using 1-km grids are well within the capability of modern computers. Detailed simulations are a practicable way to study problems such as the initiation of convection, and the effect of various microphysical parameterizations on cloud system structure.

These model-generated results should be evaluated against observations, although this is not always done. An example is Lafore *et al.* (1988) who compared three-dimensional CRM results with dual-Doppler radar observations of a tropical cloud system observed during a field program in west Africa. The vertical flux of horizontal momentum, determined from the radar analysis and the simulation, are compared in Figure 5.16. Considering the uncertainties inherent in both observations and modeling, especially the comparatively poor resolution of the boundary layer, the agreement is encouraging. In most CRMs, the liquid and ice phases are represented in simple bulk terms yet the models validate reasonably well against observations. Note that since squall-line cloud systems are to a significant degree dynamically controlled, the microphysical influences have a smaller effect than when convection occurs in weakly forced conditions.

The physical interpretation of the (synthetic) CRM results is almost as challenging as observational analysis except that model-produced fields are internally consistent. Idealizations help determine key governing principles and set the scene for new physically based parameterization methods. For example, by representing the Lagrangian generation of momentum by the pressure gradient and assuring general

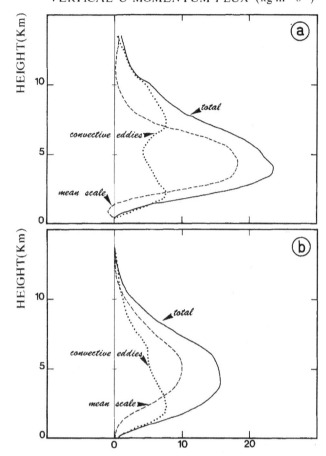

Figure 5.16. Line-normal vertical flux of u-momentum from a west-African squall line study diagnosed from (*a*) radar observations and (*b*) simulation. (From Lafore *et al.*, 1988.)

conservation properties and integral constraints are satisfied, Moncrieff (1992) produced a dynamical theory for momentum transport by squall lines. Figure 5.17 shows the momentum flux obtained from this theory tested against cloud-resolving model results. This theory was also evaluated against observational data by LeMone and Moncrieff (1994). The next step is to test the verified model in a dynamically based parameterization scheme for momentum (e.g., the dynamical model could replace an entraining plume in highly sheared conditions). The overall approach could be followed for some other kinds of cloud systems.

Precipitating convection: nonlinear interaction among processes on long time scales

The parameterization problem is put into perspective by noting that even if an *exact* quantification of cloud-scale

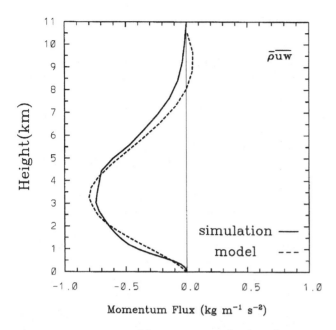

Figure 5.17. Comparison of the u-momentum flux from the archetypal model (broken line) and the result from a cloud-resolving model (solid line). (From Wu and Moncrieff 1996a.)

physics existed this would not be sufficient for the parameterization of cloud systems in large-scale models. This fundamental point echoes the distinction between the interaction between individual particles and the particle-field concepts that are at the root of basic fluid properties (e.g., temperature and velocity) as well as the Navier–Stokes equations controlling atmospheric motion. Such a *field theory* of cloud systems is arguably the ultimate basis of the role of clouds in climate: cloud-resolving models can help address this problem.

An evaluation of the cloud models on time scales of a few hours is compatible with existing *in situ* observational techniques. In contrast, the procurement of *in situ* observational data to evaluate long integrations and large-scale properties is both logistically difficult and prohibitively expensive. Only rarely, as part of large internationally coordinated programs, are comprehensive observations of the effect of clouds on the environment conducted. There are notable exceptions to this situation, namely:

1. The Global Atmospheric Research Programme (GARP) Atlantic Tropical Experiment (GATE) was conducted in the tropical eastern Atlantic, July–September 1974 (Houze and Betts, 1981).
2. The Tropical Ocean Global Atmosphere-Coupled Ocean Atmosphere Response Experiment (TOGA COARE) was

conducted in the tropical western Pacific, November 1992–February 1993 (Webster and Lukas, 1992).
3. An ongoing effort is the Atmospheric Radiation Measurement (ARM) program at sites in Kansas/Oklahoma, over the Tropical Western Pacific and on the North Slope of Alaska (see ARM Science Plan 1996).

A basic question regarding the parameterization of convection in large-scale models is: given the resolved (grid-averaged) profiles of temperature, moisture and velocity, together with large-scale forcing, can the convective response be determined with sufficient realism? A necessary requirement is that a CRM should be able to reproduce the life cycle and bulk effects of cloud systems as the large-scale fields evolve. A numerical study by Grabowski *et al.* (1996) provided an affirmative answer to this question for strongly forced precipitating cloud systems over the tropical oceans. This was a week-long, two-dimensional, numerical study of cloud systems observed during GATE. The evolving large-scale temperature and moisture fields from objectively analysed GATE observations were used to force the model and the horizontally averaged wind field was relaxed to the observed winds. The model contained ice physics, cloud-interactive radiation and a bulk type of surface flux parameterization. The vertical shear and large-scale ascent (during the passage of an easterly wave) organized the convection into distinct regimes (e.g., non-squall clusters, highly organized squall lines, and disorganized or scattered convection). Figure 5.18 is a 3-D model realization of various cloud systems in a situation of evolving large-scale forcing and shear, and is an extension of the two-dimensional Grabowski *et al.* (1996) study. These transitions verify well against those observed by radar.

Two-dimensional cloud-resolving modeling of tropical cloud systems was also performed for a 39-day period (December 5, 1992 through January 12, 1993) during a major westerly wind burst in TOGA COARE. Figure 5.19 shows the mass flux, which is a key quantity needed in most convective parameterization schemes but impossible to obtain directly from observations. It is readily obtained from the cloud-resolving model. This work has also been extended to three spatial dimensions (Wu and Moncrieff, 1996b).

In summary, it has been shown that when evolving large-scale conditions are introduced in a simple way (i.e., as domain-averaged profiles) to force a CRM, realistic representations of the environmental effects of convection are obtained. Convective parameterization schemes in GCMs can be derived from these results; for example, similar to the way Tiedtke (1989) used GATE observations and single-column models (one-dimensional versions of GCMs) to develop convective parameterizations.

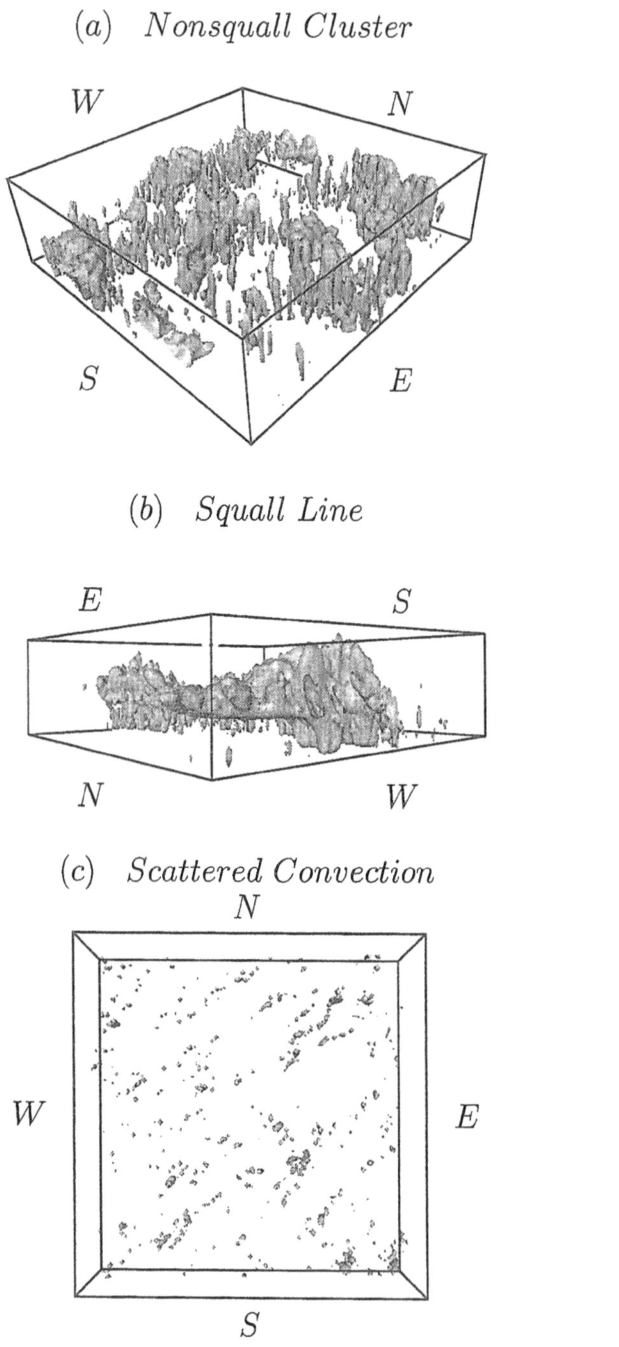

(a) *Nonsquall Cluster*

(b) *Squall Line*

(c) *Scattered Convection*

Figure 5.18. Snapshots of the total condensate field (isosurface 0.1 g kg^{-1}) in a three-dimensional domain (400 km × 400 km × 26 km) for (a) a nonsquall cloud cluster (southeast view, 2 September, 1974), (b) a squall line (northwest view, 4 September, 1974), and (c) scattered convection (top view, 7 September, 1974).

Convection in mid-latitude cyclones

Few cloud-resolving simulations of mid-latitude cyclones have been conducted. Most previous studies used idealized diabatic heating or were dry paradigms of the dynamics of fronts. Dudhia (1993) performed a three-dimensional nested simulation of an Atlantic cyclone and cold front. Various types of bands associated with frontal precipitation were produced. Comparison of parts (a) and (b) of Figure 5.20 shows the evolution of the vertical velocity over a two-hour period. Line elements broadly similar to those observed are evident. Shallow convection can be seen behind the cold front (Figure 5.20(c)) and cumulonimbus lines ahead (Figure 5.20(d)).

Cloud-resolving modeling of frontal cloud systems has the potential to yield quantitative information on the role of moist processes in frontal regions (e.g., the role of diabatic processes on frontal evolution). It will also guide the parameterization of physical processes that cannot be resolved in GCMs. Accurate modeling of shallow, multilayered cloud decks, common in mid-latitude cyclones is particularly challenging.

Starr and Cox (1985) used a two-dimensional CRM to study thin cirrus and altostratus. The role of various physical processes in the formation and maintenance of mid-latitude cases was investigated. This kind of study demonstrates the interplay among dynamics, microphysics, and radiation that strongly modulates the properties of the simulated clouds. Although imposed large-scale ascent had a significant effect, the role of ice habit and size distribution was important, especially through radiative interactions in thin cirrus. Heckman and Cotton (1993), using a non-hydrostatic, nested mesoscale model, have also simulated cirrus clouds in mid-latitudes. They obtained broad agreement between the observed and model-predicted dynamical fields and layered cloud, demonstrating that key properties can be derived without very high resolution.

Issues in convective cloud system modeling

Model design

While CRMs have much promise, they also have limitations. These are anticipated to be alleviated gradually through improvements in model design. Grabowski *et al.* (1996) quantified some key uncertainties. First, periodic lateral boundary conditions are physically restrictive; for example, moisture transported from the ocean to the atmosphere cannot escape from a periodic domain. In long integrations this can feed back to affect the model 'climate' through cloud-radiation

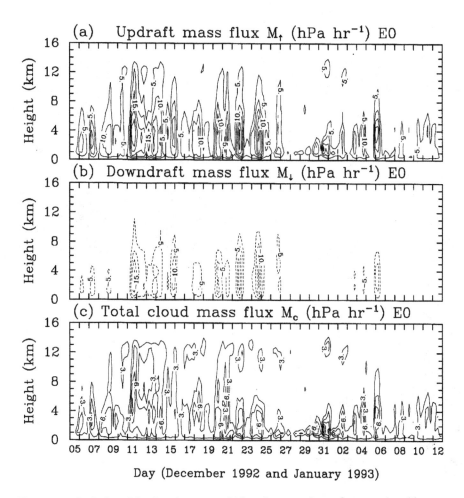

Figure 5.19. Evolution of the domain-averaged 6-hourly mean of mass fluxes produced from a CRM: (*a*) updraft, (*b*) downdraft, and (*c*) total cloud mass fluxes. Contour interval is 5 hPa hr^{-1} for (*a*) and (*b*), and 3 hPa hr^{-1} for (*c*).

interactions (see below). Second, the two-dimensional assumption may be restrictive, although comparisons with three-dimensional integrations suggest (the somewhat surprising result) that this is not the primary issue. Third, the effect of resolution needs to be investigated. While a 1-km grid resolves the mesoscale airflow and, to a lesser degree, the deep convection within it, shallow convection and boundary layer clouds are poorly resolved. Finally, the effects of convection are a function of the environmental state, noting that forcing and shear are largely responsible for organizing precipitating convection.

Mesoscale organization

The concept of mesoscale organization of convection is not only of intrinsic scientific interest but also has implications

for parameterization, especially when large-scale models attain higher resolution (Moncrieff, 1981). This issue has long been expressed but has seldom been addressed in terms useful to parameterization. The mesoscale circulation in the stratiform region of squall lines is often generated by the baroclinic gradients established by small-scale convection (Lafore and Moncrieff, 1989) on a horizontal scale of about 100 km, which is *larger* than the grid scale of some NWP models. However, the statistical basis of parameterization, specifically the Reynolds-averaging concept, requires that cloud area be much *smaller* than the grid area (scale separation).

Scale separation is not a valid assumption for convection organized on a mesoscale, especially in operational regional models where the grid length is rapidly approaching the hydrostatic limit (10 km). Yet, existing parameterization

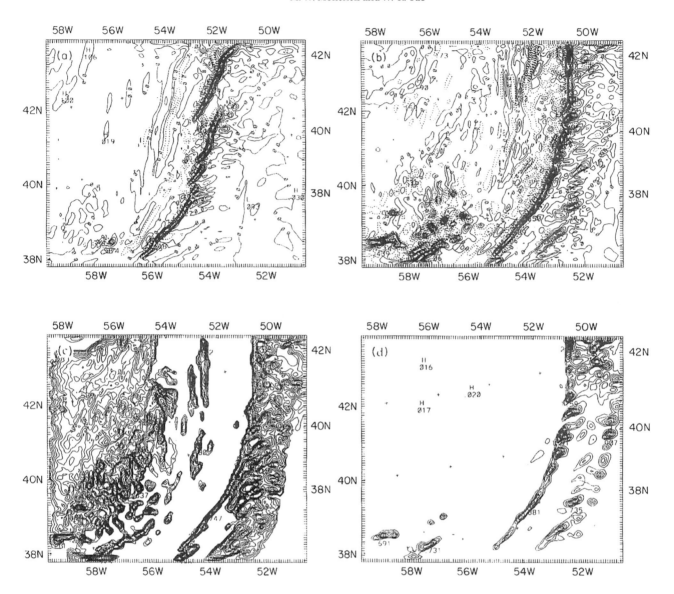

Figure 5.20. Plots from a cloud-resolving simulation of a cold frontal region in an Atlantic cyclone (6.6 km grid). (*a*) Vertical velocity at approximately 1 km (contour interval 0.2 m s⁻¹) at 4 hr with negative values shown by dashed contour; (*b*) same fields at 6 hr; (*c*) cloud water (interval 0.1 g kg⁻¹); and (*d*) rain (interval 0.1 g kg⁻¹). (From Dudhia, 1993.)

methods are generally used without much critical assessment; this is surprising considering that this resolution will soon be achieved by global NWP models. Two issues are worthy of mention. First, organized convection is sometimes associated with up-gradient and up-scale momentum transport (Moncrieff, 1981; LeMone, 1983) which is conceptually distinct from diffusive mixing. This process is presently ignored in convective parameterizations. Second, upper-tropospheric tropical cirrus which strongly interacts with radiative processes is to a large degree produced by organized convection.

Microphysics

There are two main approaches for representing mixed-phase microphysics within clouds. First, water substance is sorted into discrete categories and the evolution of all elements is calculated within each category. This is impractical for the long-term modeling of cloud systems. Second, in the more commonly used bulk microphysics approach (e.g., Kessler 1969, for the liquid phase), it is assumed that the various water categories can be approximated by continuous size distributions. This method is efficient in computational terms but its validity depends on the modeling assumptions. Ice process parameterizations have been progressively introduced into CRMs over recent years. An example of an advanced bulk microphysics scheme is that of Ferrier *et al.* (1995). Bulk ice schemes appear to be reasonably successful but more evaluation is necessary. The pacing issue is again the *minimum* acceptable level of sophistication. This is likely to be dependent on the cloud system type and thus, indirectly, on environmental conditions.

Radiation

Like microphysical processes, the direct calculation of cloud-interactive radiative transfer is computationally prohibitive, so bulk methods are again used. Emission and absorption by water vapor and cloud droplets are included in the two-stream longwave radiation methods used in cloud models. Broadband methods for longwave radiation combine the effects of reflection, emission, and transmission by cloud droplets and air molecules. The treatment of shortwave radiation in CRMs is also based on broadband approximations. An issue is how to parameterize cloud optical thickness, especially in the presence of the ice phase, considering the important impact of radiation heating and cooling profiles within clouds. A major impediment is the dearth of observations available both to improve the parameterization of ice processes and to evaluate the CRM results. The interaction between convective clouds and radiation continues to be studied (e.g., Xu and Randall, 1995).

Extensive cirrus is generated by organized mesoscale cloud clusters, especially in a sheared environment. The area of cirrus resulting from a single system can be enormous and can persist for days. It is not surprising that convectively produced tropical cirrus is a key uncertainty in climate system models.

Surface fluxes

Surface flux parameterizations are reviewed in Viterbo (1994). Bulk schemes similar to those developed in the large-scale modeling community are also used in CRMs. However, the cloud-induced winds in the surface flux schemes can be resolved in CRMs as, for example, in the cited numerical simulations of convection over a tropical ocean. The coupling of a CRM to land-surface schemes has not been much developed, although the effect of land-surface heterogeneities has been introduced into mesoscale models. The impacts of surface flux schemes are likely the most critical for convec-

tion occurring in weakly forced and weakly sheared environments.

Coupling among physical processes

It has been argued that the key issue controlling the physics of cloud systems is the coupling among the various processes (radiation, microphysics, turbulence and surface fluxes) rather than any individual process *per se* (Moncrieff, 1995). This thesis was given credence by the aforementioned work of Grabowski *et al.* (1996). Of course, uncertainties remain but will be alleviated through improving the design of CRMs, and by advancements in computer technology. Difficulties arise due to the inherently nonlinear nature of cloud systems; for example, the transition among cloud system regimes (dynamical determinism), the input of energy (boundary layer and shallow clouds not adequately resolved) and cloud–radiation interaction (microphysical processes). These aspects are not properly treated in climate models. Because cloud-scale dynamics is the principal way this coupling is manifested, it is anticipated that CRMs will yield fundamental advances in the large-scale role of cloud systems.

Use of cloud-resolving model results to derive remote-sensing algorithms

The global measurement of precipitation requires space-based methodology, especially over the ocean. Rainfall over tropical regions is now being measured by the space-based Tropical Rainfall Measuring Mission (TRMM) as described in Simpson *et al.* (1988). CRMs provide synthetic data sets (e.g. ice–water distribution) for developing the remote sensing algorithms.

Retrieval algorithms for latent heating profiles

Understanding of the global energy and water cycle requires quantification of the three-dimensional evolution of rainfall and latent heating in the tropics (Simpson *et al.*, 1988). Since latent heating is a consequence of phase changes of water (vapor, liquid and solid), it can, in principle, be determined from the hydrometeor profiles. However, few observed hydrometeor profiles are yet available and the associated latent heating profiles are not directly measurable. CRMs have been used to develop latent heating retrieval algorithms, as shown by Tao *et al.* (1993). One of these algorithms has been tested on cloud systems in various geographical locations to demonstrate their performance (Figure 5.21).

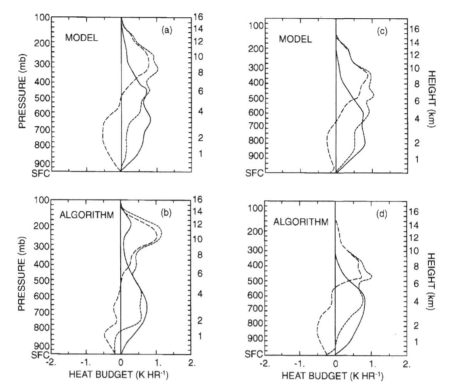

Figure 5.21. Heat budget for a mid-latitude continental squall system during its mature stage. (*a*) Cloud-resolving model estimate, (*b*) profiles derived from the retrieval algorithm. Parts (*c*) and (*d*) are the corresponding results for a tropical oceanic squall line.

Retrieval algorithms for surface rain: CRMs and passive radiative transfer models

There are two types of passive microwave rainfall retrieval, called histogram algorithms and profiling algorithms. The first uses the emission properties of the 10, 19 and 37 GHz channels to obtain monthly surface rainfall over oceanic areas (Wilheit *et al.*, 1991). The second uses the fact that weighting functions for various frequencies peak at different heights within a precipitating atmosphere in order to determine the vertical profile of hydrometeors. The information needing to be retrieved far exceeds the number of independent observations, so CRM data are used to provide the first-guess profiles (Simpson *et al.*, 1988). For example, groups of cumuli with varying fractions of stratiform and convective rain are produced that agree with properties of the cumulus clusters observed over many regions of the tropics (Simpson and Tao, 1993). The simulated cloud structures (vertical distribution of liquid and solid phases of water) are then supplied to a passive microwave radiative model to determine upwelling radiance at the top of the atmosphere. This has led to the derivation of a relationship between rain rate and brightness temperature, which can be used to re-

trieve surface rain rates from brightness temperatures sensed by airborne or spaceborne microwave instruments (e.g., Special Sensor Microwave/Imager (SSM/I) and TRMM). In addition, using a brightness-temperature/rainfall relationship derived from the CRM, and the passive microwave radiative-transfer model, a regression formula has been established for rain retrievals using infrared satellite data in combination with multifrequency microwave measurements.

Another application of CRMs is to determine how cloud hydrometeors interact with radiative processes. For example, Adler *et al.* (1991) found that brightness temperatures at 19, 37 and 86 GHz were significantly affected by ice in the CRM, whereas 10 GHz results showed very little effect. They also found that non-precipitating cloud water, associated with very low or very high rain rates, as well as with significant ice, affected the brightness temperatures.

Retrieval algorithms for surface rain: CRMs and spaceborne radar models

There are two types of algorithms for spaceborne radar. One is a statistical probability method and the other is a deterministic method. The deterministic algorithms must deal with two distinct problems, namely the determination of the correct relationship between reflectivity and rainfall rate and the attenuation of the radar signal. For example, a 14 GHz radar attenuates so seriously that unless corrections are made, heavy surface rainfall ($R \geq 20$ mm hr^{-1}) is severely underestimated.

A CRM has been coupled to a spaceborne radar model for addressing this problem (Yeh *et al.*, 1994). By using synthetic four-dimensional data sets they found significant attenuation in the convective region of tropical cloud systems but much less attenuation in the anvil region. Applying these findings they constructed a hybrid rainfall retrieval algorithm whose performance is quite good and will be pursued in the TRMM rain retrievals over the tropical oceans.

Future developments

Observational studies of cloud processes have traditionally relied on surface-based and aircraft-based measurements. However, even when coordinated with surface-based remote sensing (e.g. acoustic sounders, lidar and radar) the scales of motion that can be intensively measured extend to only a few hundred km^2 (somewhat greater if airborne platforms are used). Moreover, these data sets have inadequate temporal and spatial resolution even to describe a single cloud process. Thus, although measurements can be obtained of the basic cloud microscale processes and their interactions on scales of a few kilometers, this body of knowledge falls short of what is needed for improving climate models.

Cloud-resolving models have advanced greatly since they became of age about two decades ago to study cloud processes. Their application to the challenging problem of understanding the *large-scale effects* of cloud systems is facilitated by the power of modern computers (as in many scientific disciplines). Among others, CRMs can be used to address the following aspects:

1. derive physically based parameterizations for numerical weather prediction models and climate models;
2. test single-column representations of physical processes;
3. complement large field experiments that would otherwise be subcritical in terms of cloud-scale measurements;
4. add value to data sets in situations where standard soundings are the only measurement available;
5. improve the physical basis of ocean–atmosphere interaction in coupled climate models; and
6. help design of space-based and earth-based remote-sensing and the interpretation of the data sets.

Quantification of the subtle and highly non-linear effects of cloud systems in the hydrologic cycle, a primary objective of GEWEX, is arguably the most challenging task faced by future cloud system studies and cloud-resolving models in particular.

References

Adler, R. F., Yeh, H.-Y., Prasad, N., Tao, W.-K. and Simpson, J. (1991). Microwave rainfall simulations of a tropical convective system with a three-dimensional cloud model. *J. Appl. Meteor.*, **30**, 924–53.

André, J. C. and Bougeault, Ph. (1988). On the use of the HAPEX-MOBILHY data for the validation and development of parameterization schemes of surface fluxes. *WCP*, **126**, B1–B18, Geneva.

André, J. C., Goutorbe, J. P. and Perrier, A. (1986). HAPEX-MOBILHY, a hydrological-atmospheric experiment for the study of water budget and evaporation flux at the climatic scale. *Bull. Amer. Meteor. Soc.*, **67**, 138–44.

ARM Science Plan (1996). *Science Plan for the Atmospheric Radiation Measurement Program (ARM)*. U.S Dept. Energy, Office of Scientific and Technical Information, P.O. Box 62, Oak Ridge, TN 37831.

Avissar, R. and Chen, F. (1993). Development and analysis of prognostic equations for mesoscale kinetic energy and mesoscale fluxes for large-scale atmospheric models. *J. Atmos. Sci.*, **50**, 3751–74.

BAHC (Biospheric Aspects of the Hydrological Cycle) (1993). The operational plan. *IGBP Report*, **27**, Stockholm.

Beljaars, A. C. M. and Bosveld, F. C. (1997). Cabauw data for the validation of land surface parameterization schemes. *J. Climate*, **10**, 1172–93.

Bolle, H. J. and 36 co-authors (1993). EFEDA: European field experiment in a desertification-threatened area. *Annales Geophysicae*, **11**, 173–89.

Browning, K. A. (1990). Rain, rainclouds and climate. *Quart. J. Roy. Meteor. Soc.*, **116**, 1025–51.

Chahine, M. T. (1992). The hydrological cycle and its influence on climate. *Nature*, **359**, 373–80.

Charney, J. G. (1975). The dynamics of deserts and droughts. *Quart. J. Roy. Meteor. Soc.*, **101**, 193–202.

Chen, F. and Avissar, R. (1994). The impact of land-surface wetness heterogeneity on mesoscale heat fluxes. *J. Applied Meteor.*, **33**, 1323–40.

Chen, T. H. and 42 collaborators (1997). Cabauw experimental results from the Project for Intercomparison of Land-surface Parameterization Schemes (PILPS). *J. Climate*, **10**, 1194–215.

Clark, T. L., Hall, W. and Coen, J. L. (1996). Source code documentation for the Clark-Hall cloud-scale model: Code version G3CH01. *NCAR Tech. Note*, NCAR/TN-426+STR, 137 pp.

Cutrim, E., Martin, D. W. and Rabin, R. (1995). Enhancement of cumulus clouds over deforested lands in Amazonia. *Bull. Amer. Meteor. Soc.*, **76**, 1801–5.

Deardorff, J. W. (1978). Efficient prediction of ground surface temperature and moisture with inclusion of a layer of vegetation. *J. Geophys. Res.*, **20**, 1889–903.

D'Herbès, J. M. and Valentin, C. (1997). Land surface conditions of the Niamey region: ecological and hydrological implications. *J. Hydrol*, **189**, 18–42.

Dudhia, J. (1993). A nonhydrostatic version of the Penn. State/NCAR mesoscale model: validation tests and simulation of an Atlantic cyclone and cold front. *Mon. Wea. Rev.*, **121**, 1493–513.

Ferrier, B. S., Tao, W-K. and Simpson, J. (1995). A double-moment multiple phase four-class bulk ice parameterization. Part II: Simulations of convective systems in different large-scale environments and comparisons with other bulk parameterizations. *J. Atmos. Sci.*, **51**, 1001–37.

Gash, J. H. C., Kabat, P., Amadou, M., Bessemoulin, P., Billing, H., Blyth, E. M., DeBruin, H. A. R., Elbers, J. A. T., Harrison, G., Holwill, C. J., Lloyd, C. R., Lhomme, J. P., Moncrieff, J. B., Monteny, B. A., Puech, D., Soegaard, H., Tuzet A. and Verhoef, A. (1997). The variability of evaporation during the HAPEX-Sahel intensive observation period. *J. Hydrol.*, **189**, 385–99.

GEWEX Cloud System Science Team (1993). The GEWEX Cloud System Study (GCSS). *Bull. Amer. Meteor. Soc.*, **74**, 387–400.

Goutorbe, J. P., Lebel, T., Tinga, A., Bessemoulin, P., Brouwer, J., Dolman, A. J., Engman, E. T., Gash, J. H. C., Hoepffner, M., Kabat, P., Kerr, Y., Monteny, B., Prince, S., Saïd, F., Sellers, P. J. and Wallace, J. S. (1994). HAPEX-Sahel: A large-scale study of land-surface interactions in the semi-arid tropics. *Annales Geophysicae*, **12**, 53–64.

Goutorbe, J. P., Noilhan, J., Lacarrère, P. and Braud, I. (1997). Modelling of the atmospheric column over the central sites during HAPEX-Sahel. *J. Hydrol.*, **189**, 1017–39.

Grabowski, W. W., Wu, X. and Moncrieff, M. W. (1996). Cloud resolving modelling of tropical cloud systems during Phase III of GATE. Part I: Two-dimensional experiment. *J. Atmos. Sci.*, **53**, 3684–709.

Habets, F., Lacarrère, P., Noilhan, J., Péris, P., Ledoux, E., Ottlé, C. and Vidal-Madjar, D. (1995). Eléments de couplage d'un modèle atmosphérique à mésoechelle avec un modèle hydrologique: application sur le bassin versant de l'Adour dans le cadre de l'expérience HAPEX-MOBILHY 86. *CNRM Internal Report* (available upon request at Météo-France, CNRM, 42 Avenue G. Coriolis, 31057 Toulouse, France).

Hall, F. G., Huemmrich, K. F., Goetz, S. J., Sellers, P. J. and Nickeson, J. E. (1992). Satellite remote-sensing of surface energy balance: success, failures and unresolved issues in FIFE. *J. Geophys. Res.*, **97**, 19061–89.

Heckman, S. T. and Cotton, W. R. (1993). Mesoscale numerical simulation of cirrus clouds – FIRE case study and sensitivity analysis. *Mon. Wea. Rev.*, **121**, 1164–284.

Henderson-Sellers, A., Pitman, A. J., Love, P. K., Irannejad P. and Chen, T. H. (1995). The project for intercomparison of land-surface parameterization schemes (PILPS), Phases 2 and 3. *Bull. Amer. Meteor. Soc.*, **76**, 489–503.

Holton, J. R. (1972). *An Introduction to Dynamic Meteorology.* Academic Press, 509 pp.

Houze, R. A., Jr. and Betts, A. K. (1981). Convection in GATE. *Rev. of Geophys. and Space Phys.*, **19**, 541–76.

ISLSCP (International Satellite Land-Surface Climatology Project) and IGPO (International GEWEX Planning Office) (1993). *A Preliminary Science Plan for a Large-scale Biosphere-atmosphere Field Experiment in the Amazon Basin.* IGPO and ISLSCP, 95 pp.

Jochum, A. M., Mitchels, B. I. and Entstrasser, N. (1993). Regional and local variations of evaporation fluxes during EFEDA. In *Conference on Hydroclimatology.* Anaheim: Amer. Meteor. Soc.

Kessler, E. (1969). On the distribution and continuity of water substance in atmospheric circulations. *Meteorological Monographs*, **32**, 84 pp.

Lafore, J.-P. and Moncrieff, M. W. (1989). A numerical investigation of the organization and interaction of the convective and stratiform regions of tropical squall lines. *J. Atmos. Sci.*, **46**, 521–44.

Lafore, J.-P., Redelsperger, J.-L. and Jaubert, G. (1988). Comparison between a three-dimensional simulation and Doppler radar data of a tropical squall line: transports of mass, momentum, heat and moisture. *J. Atmos. Sci.*, **45**, 3483–500.

Lebel, T., Sauvageot, H., Hoepffner, M., Desbois, M., Guillot, B. and Hubert, P. (1992). Rainfall estimation in the Sahel, the EPSAT-Niger experiment. *Hydrol. Sci. J.*, **37**, 201–15.

Leduc, C., Bromley, J. and Schroeter, P. (1997). Water table fluctuation and recharge in a semi-arid climate: some results of the HAPEX-Sahel hydrodynamic survey (Niger). *J. Hydrology*, **189**, 123–38.

LeMone, M. A. (1983). Momentum transport by a line of cumulonimbus. *J. Atmos. Sci.*, **40**, 1815–34.

LeMone, M. A. and Moncrieff, M. W. (1994). Momentum transport by convective bands: comparisons of highly idealized dynamical models to observations. *J. Atmos. Sci.*, **51**, 281–305.

Li, L. and Avissar, R. (1994). The impact of spatial variability of land-surface characteristics on land-surface heat fluxes. *J. Climate*, **7**, 527–37.

Ludlam, F. H. (1980). *Clouds and Storms: The Behavior and Effect of Water in the Atmosphere.* Pennsylvania State University Press, 405 pp.

Lynn, B., Abramopoulos, F. and Avissar, R. (1995). Using similarity theory to parameterize mesoscale heat fluxes generated by subgrid-scale landscape discontinuities in GCMs. *J. Climate*, **8**, 932–51.

Mahfouf, J. F. and Noilhan, J. (1996). Inclusion of gravitational drainage in a land surface scheme based on the force restore method. *J. Appl. Meteor.*, **35**, 987–92.

Mahrt, L., Sun, J., Veakers, D., McPherson, J. I., Pederson, J. R. and Desjardins, R. L. (1994). Observations of fluxes and inland breezes over a heterogeneous surface. *J. Atmos. Sci.*, **51**, 2484–99.

Mascart, P., Gelpe, J. and Pinty, J. P. (1988). Etude des caractéristiques texturales des sols dans la zone HAPEX-MOBILHY 86. *Note OPGC*, **95**, 37 pp. (available upon request at Observatoire de Physique du Globe de Clermont, 12 Avenue des Landais, 63000 Clermont-Ferrand, France).

Moeng, C.-H., Cotton, W. R., Bretherton, C., Chlond, A., Khairoutdinov, M., Krueger, S., Lewellen, W. S., MacVean, M. K., Pasquier, L., Rand, H. A., Siebesma, A. P., Sykes, R. I., and Stevens, B. (1995). Simulation of a stratocumulus-topped PBL: intercomparison among different numerical codes. *Bull. Amer. Meteor. Soc.*, **77**, 261–78.

Moncrieff, M. W. (1981). A theory of organized steady convection and its transport properties. *Quart. J. Roy. Meteor. Soc.*, **107**, 29–50.

Moncrieff, M. W. (1992). Organized convective systems: archetypal dynamical models, mass and momentum flux theory, and parameterization. *Quart. J. Roy. Meteor. Soc.*, **118**, 819–50.

Moncrieff, M. W. (1995). Mesoscale convection from a large-scale perspective. *J. Atmos. Res.*, **35**, 87–112.

Moncrieff, M. W., Kreuger, S. K., Gregory, D., Redelsperger, J.-L., and Tao, W.-K. (1997). GEWEX Cloud System Study (GCSS) Working Group 4: Precipitating Convective Cloud Systems. *Bull. Amer. Meteor. Soc.*, **78**, 831–45.

Noilhan, J., André, J. C., Bougeault, Ph., Goutorbe, J. P. and Lacarrère, P. (1991). Some aspects of the HAPEX-MOBILHY programme, the data base and the modelling strategy. *Survey in Geophysics*, **12**, 31–61.

Noilhan, J. and Lacarrère, P. (1995). GCM grid-scale evaporation from mesoscale modelling. *J. Climate*, **8**, 206–23.

Noilhan, J., Lacarrère, P., Dolman, A. J. and Blyth, E. M. (1997). Defining area-average parameters in meteorological models for land surfaces with mesoscale heterogeneity. *J. Hydrol.*, **190**, 302–16.

Noilhan, J. and Planton, S. (1989). A simple parameterization of land-surface processes for meteorological models. *Mon. Wea. Rev.*, **117**, 536–49.

Ottlé, C. and Vidal-Madjar, D. (1994). Assimilation of soil moisture inferred from infra-red remote-sensing in a hydrological model of the HAPEX-MOBILHY region. *J. Hydrol.*, **158**, 241–64.

Passerat de Silans, A., Monteny, B. and Lhomme, J. P. (1994). Tentative de spatialisation des paramètres d'un modèle SVAT. Application au bassin de Banizoumbou. In *Proc. Xèmes Journées Hydrologiques*, pp. 411–28 (available at ORSTOM, 911 Avenue Agropolis, BP 5045, 34032 Montpellier, France).

Prince, S. D., Kerr, Y. H., Goutorbe, J. P., Lebel, T., Tinga, A., Bessemoulin, P., Brouwer, J., Dolman, A. J., Engman, E. T., Gash, J. C. H., Hoepffner, M., Kabat, P., Monteny, B., Saïd, F.,

Sellers, P. J. and Wallace, J. S. (1995). Geographical, biological and remote-sensing aspects of the Hydrologic Atmospheric Pilot Experiment in the Sahel (HAPEX Sahel*). Remote-Sensing of the Environment*, **51**, 215–34.

Schmugge, T. J. and Becker, F. (1991). Remote-sensing observations for the monitoring of land-surface fluxes and water budgets. In *Land-surface Evaporation, Measurement and Parameterization*, ed. T. J. Schmugge and J. C. André, pp. 337–47. New York: Springer.

Sellers, P. J. and Hall, F. G. (1992). FIFE in 1992: Results, scientific gains, and future research directions. *J. Geophys. Res.*, **97**, 19091–109.

Sellers, P. J., Hall, F. G., Asrar, G., Strebel, D. E. and Murphy, R. E. (1988). The first ISLSCP field experiment (FIFE). *Bull. Amer. Meteor. Soc.*, **69**, 22–7.

Sellers, P. J., Hall, F. G., Margolis, H., Kelly, B., Baldocchi, D., den Hartog, J., Cilhar, J., Ryan, M., Goodison, B., Crill, P., Ranson, J., Lettenmaier, D. and Wickland, D. (1995). The boreal ecosystem-atmosphere study (BOREAS), an overview and early results from the 1994 field year. *Bull. Amer. Meteor. Soc.*, **76**, 1549–77.

Sellers, P. J., Heiser, M. D. and Hall, F. G. (1992). Relations between surface conductance and spectral vegetation indices at intermediate (100 m² to 15 km²) length scales. *J. Geophys. Res.*, **97**, 19033–59.

Shao, Y., Anne, R. D., Henderson-Sellers, A., Irannejad, P., Thornton, P., Liang, X., Chen , T. H., Ciret, C. Desborough, C., Balachova, O., Haxeltine, A. and Ducharne, A. (1994). Soil moisture simulation. *A Report of the RICE and PILPS Workshop. IGPO Publication Series*, 14, 179 pp.

Shuttleworth, W. J., Gash, J. C. H., Lloyd, C. R., Moore, C. J., Roberts, J. M., de O. Marques, A., Frisch, G., de P. Silva, V., Ribeiro, M. N. G., Molion, L. C. B., de Sa, L. D. A., Nobre, J. C., Cabral, O. M. R., Patel, S. R. and de Moraes, J. C. (1984). Eddy correlation measurements of energy partition for Amazonian forest. *Quart. J. Roy. Meteor. Soc.*, **110**, 1143–62.

Simpson, J., Adler, R. F., and North, G. R. (1988). A proposed satellite tropical rainfall measuring mission (TRMM). *Bull. Amer. Meteor. Soc.*, **69**, 278–95.

Simpson, J. and Tao, W.-K. (1993). The Goddard Cumulus Ensemble Model. Part II: Applications for studying cloud precipitating processes and for NASA TRMM. *Terrestrial, Atmospheric, and Oceanic Sciences*, **4**, 55–96.

Starr, D. O'C. and Cox, S. K. (1985). Cirrus clouds, Part II: Numerical experiments on the formation and maintenance of cirrus. *J. Atmos. Sci.*, **42**, 2682–94.

Strebel, D. E., Newcomer, J. A., Ormsby, J. P., Hall, F. G. and Sellers, P. J. (1990). The FIFE information system. *IEEE Trans. Geosci. Remote Sens.*, **28**, 703–10.

Tao, W.-K., Lang, S., Simpson, J. and Adler, R. F. (1993). Retrieval algorithms for estimating the vertical profiles of latent heat release: their applications for TRMM. *J. Meteor. Soc. Japan*, **71**, 685–700.

Taylor, C. M, Harding, R. J., Thorpe, A. J. and Bessemoulin, P. (1997). A mesoscale simulation of land surface heterogeneity from HAPEX-Sahel. *J. Hydrol.*, **189**, 965–97.

Tiedke, M. (1989). A comprehensive mass flux scheme for cumulus parameterization in large-scale models. *Mon. Wea. Rev.*, **117**, 1779–800.

Viterbo, P. (1994). A review of parameterization schemes for land–surface processes. In *Proceedings: Seminar on Parameterization of Sub-grid Scale Physical Processes*, 5–9 September, 1994. ECMWF.

WCRP (World Climate Research Programme) (1981). *Report of the JSC Study Conference on Land-surface Processes in Atmospheric General Circulation Models*. WCP, **46**, 89 pp. Geneva.

WCRP (World Climate Research Programme) (1983). *Report of the Meeting of Experts on the Design of a Pilot Atmospheric Hydrological Experiment for the WCRP*. WCP, **76**, 70 pp. Geneva.

WCRP (World Climate Research Programme) (1990). *Report of the First Session of the WCRP-GEWEX/IGBP-CP3 Joint Working Group on Land-surface Experiments*. WCRP, **38**, WMO/TD, **370**, 44 pp. Geneva.

Webb, M. J., Slingo, A. and Stephens, G. L. (1993). Seasonal variations of the clear-sky greenhouse effect: the role of changes in atmospheric temperatures and humidities, *Clim. Dynamics*, **9**, 117–29.

Webster, P. J. and Lukas, R. (1992). TOGA COARE: The Coupled Ocean Atmosphere Response Experiment. *Bull. Amer. Meteor. Soc.*, **73**, 1377–416.

Wilheit, T. T., Chang, A. T. C. and Chiu, L. S. (1991). Retrieval of monthly rainfall indices from microwave radiometric measurements using probability density functions. *J. Atmos. Oceanic Technol.*, **8**, 118–36.

Wu, X. and Moncrieff, M. W. (1996*a*). Collective effects of organized convection and their approximation in General Circulation Models. *J. Atmos. Sci.*, **53**, 1477–95.

Wu, X. and Moncrieff, M. W. (1996*b*). Recent progress on cloud-resolving modeling of TOGA COARE and GATE cloud systems. In *Proceedings Workshop on New Insights and Approaches to Convective Parameterization*, November 4–7, 1996. Reading, UK: ECMWF.

Xu, K.-M and Randall, D. A. (1995). Impact of interactive radiative transfer on the macroscopic behavior of cumulus ensembles. Part II: mechanisms for cloud-radiation interactions. *J. Atmos. Sci.*, **52**, 800–17.

Yeh, H.-Y. M., Prasad, N., Meneghini, R., Tao, W.-K. and Adler, R. F. (1994). Model-based simulation of TRMM spaceborne radar observations. *J. Appl. Meteor.*, **34**, 175–97.

6 Examples of the use of GCMs to investigate effects of coupled processes

6.1 The interaction of convective and turbulent fluxes in general circulation models

David Gregory, Andrew Bushell and Andrew Brown

Introduction

Vertical transport by motions with a variety of height scales plays a crucial role in the energetic balance of the atmosphere. In the lowest two kilometers (the so-called atmospheric boundary layer) turbulent motion redistributes heat and moisture from the surface. Deep convection, extending throughout the depth of the troposphere, ventilates the boundary layer, distributing heat and moisture throughout the lowest 10–15 km of the atmosphere and balancing cooling due to radiative processes and upward motion on horizontal scales an order of magnitude larger than that of the convection itself.

In order to simulate the global circulation of the Earth's lower atmosphere accurately, such processes must be included into general circulation models (GCMs) even though the horizontal length scales of such motions are much smaller than the resolution of the models themselves. Over the past 10 years advances have been made in the 'parameterization' of such subgrid-scale processes and this has resulted in improved simulations of present-day climate. However, many problems are still unresolved, especially in the area of the interaction of different processes. This chapter is concerned with one such interface, namely that between the convection and boundary layer parameterizations.

In reality the transition between vertical motions of different scale is a continuous one. However, in models different processes are represented by different physical models which may not be well matched at their interface. This may lead to poor interactions of the methods and, as described in this section, dependencies upon factors such as vertical resolution. Changes in one method used to represent one process may have consequences for the behavior of the representation of other processes even though these are themselves unchanged.

Convection and boundary layer fluxes are parameterized by different techniques which reflect the distinct nature of the scales and character of the vertical motion associated with each. Boundary layer turbulence is usually represented in GCMs by so-called K-theory. In this the vertical flux of heat, moisture and momentum at a level of the atmosphere is taken to be proportional to the vertical gradient of the quantity concerned, with the constant of proportionality being related to the stability of the atmosphere to vertical motion. Such fluxes are local in nature, affecting only the layer just below and just above the level where the flux is estimated. However, recently it has been recognized that a wide range of vertical scales exists within boundary layer turbulence and some recent parameterization schemes aim to represent this wider spectrum of vertical motion. Stull (1994) provides a discussion of the basis of these schemes.

Although the motion within convective clouds can also be viewed as turbulent (especially when the depth of the cloud is shallow), the transport properties of convective clouds are not solely dependent upon the local structure of the atmosphere. In deep convection mass is transported from the lower troposphere to near the tropopause, i.e. through many model levels. Similarly, shallow convection within the boundary layer ventilates the surface layer, transporting mass, energy and moisture into the inversion at the top of the boundary layer. In GCMs these 'non-local' processes are now widely represented by the so-called mass flux approach (see Gregory, 1997, for a review of these parameterization schemes). Simple cloud models are used to estimate the vertical flux of mass within a grid-box of a GCM associated with an ensemble of convective clouds. From this the net effect of convection upon the large-scale atmosphere can be estimated through the assumption that motion within the cloud (driven by latent heat release) is compensated by unsaturated downward motion within the grid box. Such motion in the cloud-free part of the grid box leads to adiabatic warming as air descends. Thus the release of latent heat within the clouds, which occurs on a small scale, is distributed over a larger area. Other processes such as the evaporation of falling rain and snow are also accounted for by these models.

In this chapter some of the known problems associated with the interaction of boundary layer and convective parameterization schemes will be illustrated using results from the Hadley Centre Climate Model, a version of the UK Met. Office Unified Model (Cullen, 1993) with a horizontal resolution of $2.5 \times 3.75°$ lat-long. The model employs a mass flux convection scheme described by Gregory and Rowntree (1990) and a moist boundary layer scheme (Smith, 1990)

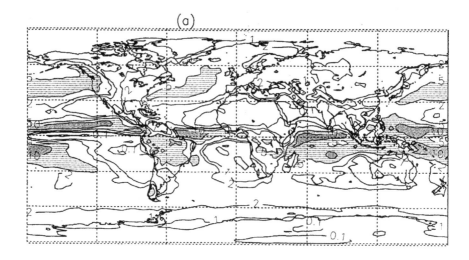

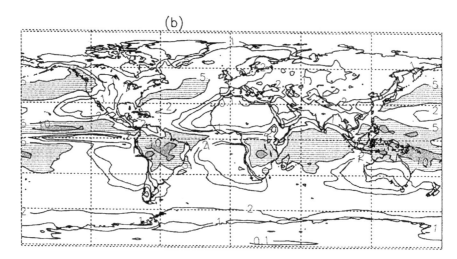

Figure 6.1. Impact of convective downdrafts upon precipitation and surface fields for December, January and February (3 year means). (*a*) Simulated precipitation without downdrafts in the convection scheme, (*b*) simulated precipitation with downdrafts in convection scheme, (*c*) change in mixing ratio of lowest model layer and (*d*) change in surface evaporation on introduction of downdrafts.

which has been extended to include non-local mixing (Smith, 1993). The boundary layer scheme estimates the stability of the atmosphere using a moist Richardson number and so is able to represent mixing in both dry and cloudy boundary layers. These parameterization schemes are used sequentially within the model, with the boundary layer transports being calculated first and the temperature and moisture profiles being updated before convective heating and fluxes are estimated. Generally 19 levels are employed in the vertical with 5 in the boundary layer (below a pressure of 850 hPa). Although the details of results presented here will be sensitive to the particular parameterizations used in the model, the issues they illustrate will be true for most GCMs that use similar methods to represent convection and boundary layer turbulence.

The interaction of convection and boundary layer processes

The influence of convection upon surface and boundary layer processes

Deep convective processes play a large role in determining the structure of the boundary layer with upward motions, triggered from within the boundary layer, removing heat and moisture from the lowest few kilometers of the atmosphere.

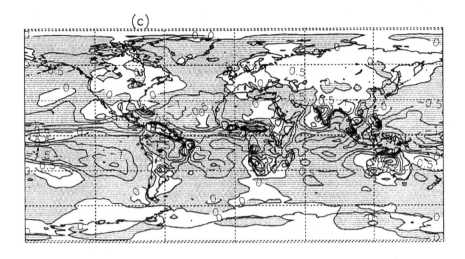

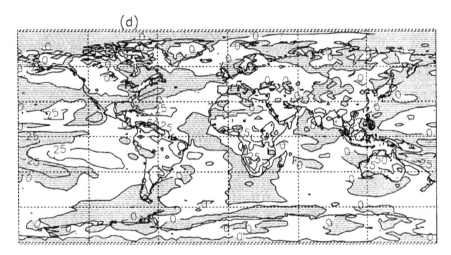

Figure 6.1. (*cont.*)

However, downward motions (downdrafts), initiating in the mid-troposphere through the evaporation of falling precipitation and also downward drag of precipitation particles upon air, transport (relatively) dry, cold air into the boundary layer. The effect of such downdraft processes upon the surface layer is illustrated by Figure 6.1(*c*) and (*d*) which show respectively the difference in mixing ratio at about 1000 hPa and surface evaporation from two simulations of the Hadley Centre Climate Model. One simulation uses a version of the convection scheme with a representation of convective downdrafts; the other simulation considers the effects of convective updrafts only. The results presented are averaged over three December/January/February periods. Global precipitation patterns are also shown from these two simulations (Figures 6.1(*a*), no downdraft in convection scheme; Figure 6.1(*b*), with downdraft in convection scheme). Inclusion of convective downdrafts dries the surface layer, especially in

regions where convection is active. This results in an increase in the amount of evaporation from the surface over much of the tropical oceans. Overall, the introduction of downdrafts is beneficial to the simulation of global climate, increasing precipitation over south America and Africa (Figure 6.1(*a*),(*b*)), in better agreement with observations. Monsoon circulations over Australia during the northern hemisphere winter and over Africa and India during the northern hemisphere summer (not shown) are also enhanced, bringing the model simulation into better agreement with observations.

The influence of boundary layer processes upon convection

Changes in the formulation of the boundary layer scheme can also affect the performance of the convection scheme. Smith (1993) describes a modified version of the boundary layer scheme used in the Unified Model which includes a

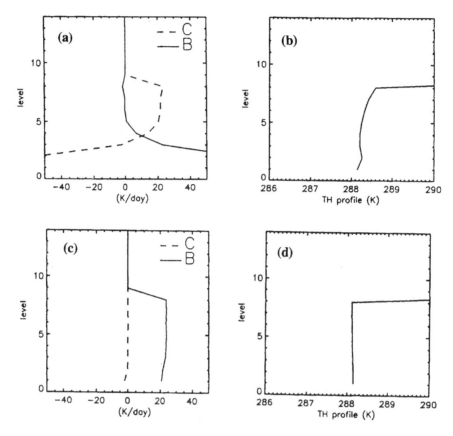

Figure 6.2. (a) and (c) Profiles of heating due to convective (dashed) and boundary layer (solid) parameterizations; (b) and (d) potential temperature profiles for single column experiments using two different versions of the boundary layer scheme; (a) and (b) local mixing only, (c) and (d) non-local mixing terms included.

representation of non-local mixing in unstable conditions. Together with terms representing local turbulence, this revised scheme also includes transports due to eddies which may penetrate to the top of the boundary layer from the surface. Smith (1993) reports that this has beneficial effects on the climatology of the model, increasing the amount of low level cloud and reducing the model's cold bias.

The introduction of the non-local terms upon the activity of the convection scheme is illustrated in Figure 6.2 which shows profiles of the heating rates due to boundary layer and convective processes derived from a single column version of the Unified Model. Such a single column model represents a single grid point of a general circulation model and uses the same physical parameterizations as the larger model. It provides a useful test-bed in which to understand the behavior and interaction of physical parameterizations, allowing comparison with detailed model and observational data. In the simulations here the model is initialized with a profile in which potential temperature was held constant with height

(at 288 K) up to 850 m, being topped by a strong inversion. The model is dry and surface fluxes were held constant throughout the four days of the simulation through the maintenance of a constant temperature difference between the surface and lowest model layer. The thermodynamic structure is highly idealized but is similar to that which might occur over a dry land surface overlaid by a high pressure region. It can also be viewed as a dry proxy of the boundary layer found in the subtropical oceans. The time step used was 60 seconds and the vertical resolution varies between 6 mb at the surface to 35 hPa near 850 hPa, with 13 levels within the boundary layer of the model (as in the high resolution, 31 level, case below; see Table 6.1 for definition of the pressure of model level boundaries). An equilibrium between convective and boundary layer mixing was found to be established after one day – the profiles in Figure 6.2 being those due to the convection and boundary layer schemes once equilibrium has been reached. Further details of the model configuration are provided by Brown (1995).

Table 6.1. *Pressure of model level boundaries (assuming a surface pressure of 1000 hPa) for simulations with 5 and 13 levels in the boundary layer.*

19 level model (5 in boundary layer)	31 level model (13 in boundary layer)
750	745
	780
	810
835	835
	858
	880
905	901
	921
	940
956	957
	972
	985
994	994
1000	1000

With only local mixing being included in the boundary layer scheme, most of the surface flux remains in the lowest two model layers (Figure 6.2(a)), which being thin leads to large temperature increments. Convection is triggered from this surface layer, which exhibits large instability due to the action of the boundary layer scheme, transporting heat upwards, cooling the surface layer while warming the upper part of the boundary layer. The net result of the action of convection and boundary layer mixing is to produce a slight stabilization of the boundary layer (Figure 6.2(b)) compared with the initial temperature which was neutral to dry ascent. With the introduction of the non-local mixing terms in the boundary layer scheme, the surface flux is well distributed through the depth of the boundary layer (Figure 6.2(c)). This maintains the initial neutrality of the initial conditions. Without the large temperature increments in the lowest layers, convection is inhibited.

These simulations raise the question as to which scheme (boundary layer or convection) should be representing buoyant motion within the planetary boundary layer. Inclusion of the non-local effects within the boundary layer scheme reduces the intensity of shallow convection represented by the convection scheme. However, the net effect of the two schemes is little different when only local terms are included in the boundary layer scheme. In the latter case the activity of the boundary layer scheme is reduced while that of the convection scheme is increased. This is not a surprising result as the non-local terms are intended to represent buoyant eddies or plumes which extend through more than one layer

of the model, exactly the type of motion which the convection scheme aims to represent!

Whether such non-local processes should be represented by the convection or boundary layer scheme is still a matter of debate. In reality of course, the atmosphere knows no such distinction, but in models the answer may depend upon the horizontal and vertical scale of the eddies. For example, the mass flux method of parameterizing convective processes relies on the assumption that the cloud area is small compared with the size of the grid box. This may not be the case in some cloudy boundary layers, such as those topped by stratocumulus beneath an inversion, in which the upward motions may have a similar horizontal scale to downward motions. Also the convection scheme of the Unified Model can only represent convective plumes which are at least three model layers deep, i.e., a minimum depth of approximately 300 m in this case, whereas the boundary layer scheme can represent non-local processes with smaller vertical scales. A solution to the problem may be the development of more sophisticated switching algorithms between the convection and boundary layer schemes in differing situations.

The simulation of stratocumulus: an example of the impact of vertical resolution upon boundary layer and convective mixing

Stratocumulus clouds form beneath low-level inversions by a complex interaction between boundary layer turbulence, radiation, convection and cloud microphysical processes. As such they present a significant challenge to the current generation of GCMs. Stratocumulus clouds play an important role in the radiation balance of the atmosphere, especially in subtropical regions where they form under low-level inversions caused by the descent of air forced by radiative cooling. In general, climate models fail to capture the horizontal extent of these clouds. This leads to underestimation of the top of atmosphere albedo and excessive shortwave radiative fluxes at the ocean surface. In coupled ocean-atmosphere models this leads to excessive sea surface temperatures and the need for correction of surface fluxes, the so-called 'flux correction technique', in order to maintain sea surface temperatures at a realistic level.

Figure 6.3 shows a cross-section through the stratocumulus sheet in the south Atlantic along 15° S, with the African coast being situated at 12° E, as simulated by the Hadley Centre Climate Model during northern hemisphere summer (June/July/August) using 19 levels in the vertical, and five within the boundary layer of the model (essentially the lowest 250 hPa of the model). Layer cloud amount is

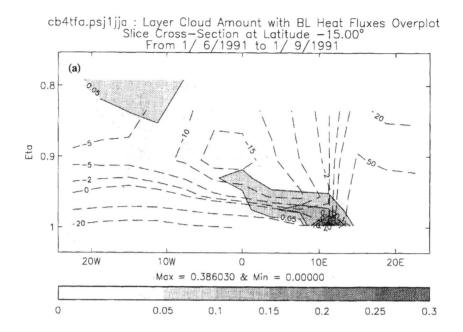

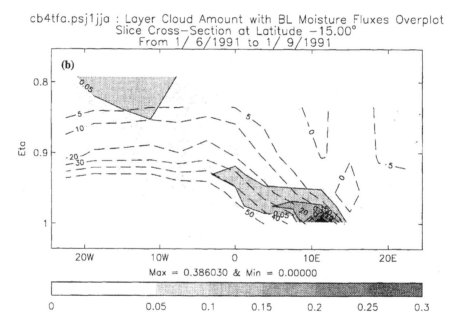

Figure 6.3. Height–longitude cross-sections along 15° S through the south Atlantic stratocumulus sheet showing layer cloud amount. Also shown is (a) boundary layer heat flux and (b) boundary layer moisture flux (Wm⁻²) meaned over a single June, July and August with 5 levels in the model boundary layer. The vertical coordinate is approximately p/p^*, where p^* is the surface pressure.

shown, together with fluxes of heat and moisture due to boundary layer turbulence. The model's representation of boundary layer turbulence incorporates the non-local terms discussed above. Maximum cloudiness is found next to the coast in the lowest model layer. The depth of the cloud increases away from the coast as might be expected as the boundary layer deepens, although discrete jumps in cloudiness are seen due to the coarse vertical resolution. Cloud

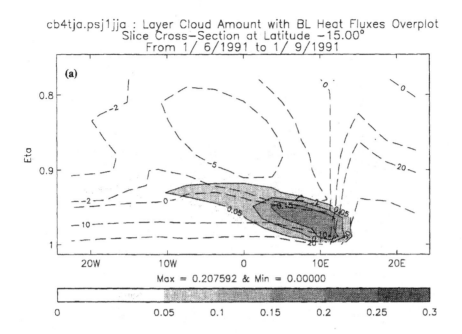

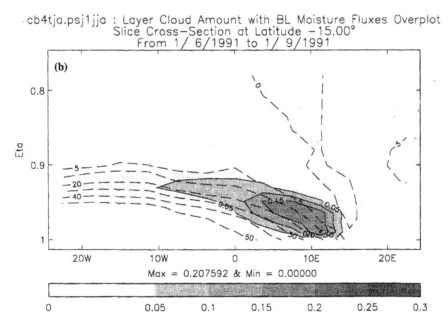

Figure 6.4. As Figure 6.3 but with 13 levels in the model boundary layer.

amounts are low and the top of atmosphere albedo is underestimated by up to 15%.

A contributing factor to the failure of GCMs to simulate stratocumulus adequately is the coarse vertical resolution used in the boundary layer. With 5 levels in the lowest 250 hPa of the atmosphere, the vertical resolution varies between 6 hPa next to the surface and 85 hPa at the top of the boundary layer (see Table 6.1). Increasing the number of model levels in the boundary layer to 13 (with a vertical resolution of 35 hPa at the top of the boundary layer) brings about an increase in cloud cover in the vicinity of the subtropical stratocumulus sheets (Figure 6.4). Maximum cloud amounts are lifted up from the lowest model layer and cloud fractions greater than 5% extend further to the west. The structure of boundary layer heat and moisture fluxes is also seen to be modified by the use of increased vertical resolution. The magnitude of the down-

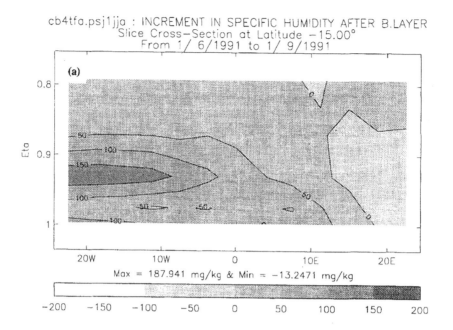

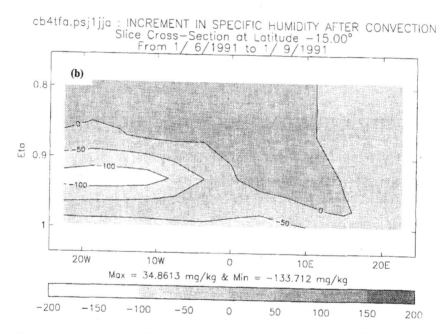

Figure 6.5. Cross-section as Figure 6.3 showing moisture increments ($\times 10^{-3}$ g kg^{-1}) due to (a) boundary layer and (b) convection schemes per model time step (30 minutes) meaned over a single June, July and August with 5 levels in the model boundary layer. The vertical coordinate is approximately p/p^*, where p^* is the surface pressure.

ward heat flux near 900 hPa into the top of the cloud layer is reduced on the meridian. The vertical gradient of the moisture flux across the cloud layer is increased. This increases the amount of water available for the formation of clouds.

As well as leading to changes in the structure of boundary layer fluxes, increased vertical resolution also leads to an increase in the activity of the model's convection scheme within the stratocumulus region. Figures 6.5 and 6.6 show a cross-section through the South Atlantic stratocumulus sheet (as in Figures 6.3 and 6.4) showing the half-hourly moisture

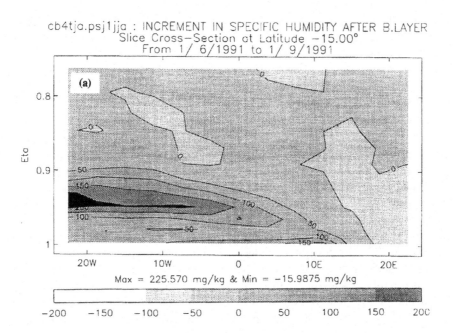

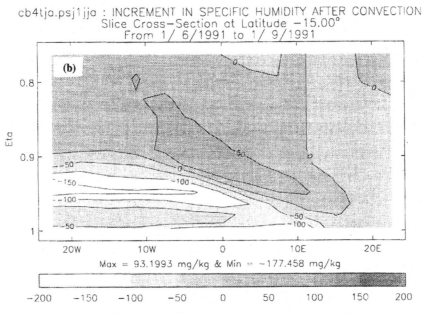

Figure 6.6. As Figure 6.5 but with 13 levels in the model boundary layer.

increment due to the boundary layer and convection schemes for the simulations using 5 and 13 levels in the boundary layer, respectively (see Table 6.1 for pressure of model layers). With coarse vertical resolution, convective increments are small in the vicinity of the stratocumulus sheets. The activity of the convection scheme increases to the west of the meridian where the depth of the boundary layer increases and cloud amounts decrease below 5%. In this region although boundary layer processes increase the moisture content of the boundary layer, convection dries the boundary layer by transporting moisture through the inversion. A similar picture is seen west of the meridian when the vertical resolution of the boundary layer is increased; however, convection is much more active in the stratocumulus region in this simulation. Whereas boundary layer turbulence moistens the subcloud and cloud layer, convection acts to dry the subcloud and lower part of the stratocumulus sheet while moistening the upper part of the stratocumulus. Although decreasing the availability of moisture for cloud formation near cloud base, this increases the moisture supply near the top of the stratocumulus sheet and so tends to increase cloud cover there.

The increase in convective activity that results from increasing the model resolution may be due to several factors. In the scheme described by Gregory and Rowntree (1990) the magnitude of the convection is determined by the buoyancy of a parcel after being lifted one model layer. For convection initiated from the lowest model layer, with 13 levels in the boundary layer, this distance is only a third that when 5 model levels are used. This may result in larger parcel buoyancy being used to determine the initial convective mass flux. Also, as mentioned above, the convection scheme can only represent clouds whose vertical depth is three model layers or greater. With only 5 model layers in the boundary layer this represents a depth of approximately 1000 m, greater than the typical boundary layer depth simulated by the model in the stratocumulus region. This minimum cloud depth is reduced to only 300 m when 13 levels are used. Hence the convection scheme is able to see instability on a much shorter vertical scale and therefore play a more active role in vertical transports within the boundary layer. The increase in convective activity within the stratocumulus region resulting from increased vertical resolution may be realistic. Results from the Atlantic Stratocumulus Transition Experiment (ASTEX) suggest that moisture detrained from cumulus clouds into the stratocumulus layer enhances the water content of the stratocumulus (Miller and Albrecht,

1995). This is consistent with the response of the simulated stratocumulus described above when boundary layer vertical resolution is increased.

Discussion

This section has illustrated some of the interactions between, and current problems associated with the interaction of, convection and boundary layer parameterization schemes. Convection has been shown to modify the structure of the boundary layer and so the flux of energy from the surface to the atmosphere. The activity of the convection scheme has been shown to be sensitive to the nature of physical processes represented by the boundary layer scheme. The nature of the balance between these schemes has also been shown to be dependent upon the vertical resolution of models. It may be difficult for such sensitivity to be totally removed from these models. However, what is important is that boundary layer and convection schemes work together to produce realistic vertical thermodynamic (and wind) structures. A large eddy simulation with high horizontal resolution (such as in Siebesma and Cuijpers, 1995) can explicitly represent the coupling of different processes. The use of such a model to provide a reference against which the performance of parameterizations can be measured should provide useful insights into the problem of the next few years. It is to be hoped that the result of such work will lead to more integrated schemes which bring improvements to the performance of future generations of GCMs.

6.2 Soil moisture–precipitation interaction: experience with two land surface schemes in the ECMWF model

Anton C. M. Beljaars and Pedro Viterbo

Introduction

Soil moisture and precipitation interact strongly. The impact of precipitation on soil moisture is a very direct one: precipitation refills the soil moisture store. The other coupling is through the enhancing effect of soil moisture on evaporation and precipitation (e.g., Rowntree and Bolton, 1983; Mintz, 1984; Garratt, 1993). Although the precipitation response to evaporation is marked, it is not very clear how this response works in detail. Rowntree and Bolton (1983) present the general notion that evaporation increases the moisture content of the troposphere and brings the air closer to saturation and therefore facilitates precipitation. However, given a certain incoming net radiation, soil moisture mainly affects the Bowen ratio (the ratio of sensible and latent heat flux at the surface). In the approximation of the boundary layer as a reservoir for heat and moisture, the equivalent potential temperature (θ_e, determining the condition for convective precipitation) in the boundary layer is not affected by the Bowen ratio at the surface. The approximation is crude because it does not account for the change in condensation level due to changes in the Bowen ratio. Another complication is the entrainment at the top of the boundary layer. Betts *et al.* (1996) and Betts and Ball (1995) demonstrate with observational data and a simple boundary layer model that reduced evaporation, and therefore increased heating at the surface, increases entrainment at the top of the boundary layer and thus increases entrainment of low θ_e air from above the boundary layer. This leads to lower θ_e in the boundary layer, a more stable troposphere and might therefore lead to less convective precipitation. Benjamin and Carlson (1986) and Lanicci *et al.* (1987) discuss the role of differential advection, with a warm/dry southwesterly flow from the Mexican plateau overlying a moist southerly flow from the Gulf of Mexico, giving a capping inversion which allows the buildup of large conditional instability. As a consequence, when the soil is more moist over the Mexican Plateau the heating is smaller, the stabilizing inversion is weaker, and convective precipitation is not inhibited. We will illustrate this mechanism in more detail below.

The two-way coupling between soil moisture/evaporation and precipitation involves a positive feedback leading to amplification of anomalies. It may also lead to a prolonged persistence of dry or wet spells, which was indeed found from a data study by Namias (1958) for the Midwest of the USA between spring and summer and by Oglesby (1991) from model simulations.

It is of course important to have the correct coupling between processes in general circulation models. The positive feedbacks are particularly crucial and potentially dangerous as they may amplify other model deficiencies. The evaporation–precipitation feedback is such an example and is believed to create considerable biases in the precipitation climate of many general circulation models. Arpe *et al.* (1994) show for a number of mid-latitude continental areas with an observed summer maximum in precipitation, that the annual cycle of the ECHAM3 model (ECMWF Hamburg climate model version 3) has a minimum. Also the ECMWF model, running in climate mode, has a pronounced dry bias over continents in summer (Viterbo and Beljaars, 1995, hereafter VB95). Excessive solar radiation at the surface, a characteristic of many general circulation models (Garratt *et al.*, 1993), may act as a catalyst for the model dry bias. The evaporation–precipitation feedback is likely to play an important role in maintaining and enhancing this dry bias. Possible causes are: (*a*) the soil may dry out too quickly because the soil moisture reservoir is too small, and (*b*) the coupling between precipitation and evaporation may be too strong because of the closure in the convection scheme.

It is important to realize that, whatever the origin of the problem, the evaporation–precipitation feedback amplifies the deficiency. This makes the diagnostics of such model problems extremely complicated. It is therefore necessary to have observational material that allows for validation on the process level. To illustrate this in this section we use the recent experience at ECMWF (European Centre for Medium-range Weather Forecasts) with the implementation of the new land surface parameterization scheme. The scheme was extensively tested in stand-alone mode with long time series of observations, and in the global model with five year long integrations at low resolution (T63, triangular truncation with 63 wavenumbers and about 200 km resolution in grid-point space). The final test before operational implementation was a parallel run with data assimilation and 10 day forecasts at operational resolution (T213; about 60 km resolution in grid point space) in July 1993, which happened to be a month with exceptional flooding of the Mississippi river. The precipitation forecasts for the central USA with the new CY48 turned out to be much better than with the old CY47. We will use this example to illustrate the mechanism by

Table 6.2. *Main features of land surface schemes of CY47 and CY48*

	Model cycle 47	Model cycle 48
Number of soil layers	2 prognostic layers 1 climate layer	4 prognostic layers
Layer depths	7 and 42, and 42 cm	7, 21, 72 and 189 cm
Deep boundary conditions	monthly climate	no heat flux, free drainage
Soil properties	constant	dependent on soil moisture
Atmospheric temperature boundary condition	temperature of layer 1	skin layer
Interception reservoir	yes	yes
Root depth	2 layers (49 cm)	3 layers (100 cm)
Aerodynamic resistance in surface layer	Louis *et al.* (1982) with equal roughness lengths for heat and momentum	Monin Obukhov scheme with a smaller roughness length for heat than for momentum (Beljaars and Holtslag, 1991)

which precipitation responds to evaporation over the USA in summer.

The new ECMWF land surface scheme

In July 1993 a new land surface scheme was implemented in the operational ECMWF model. The main difference between the old model cycle 47 (CY47) and the new model cycle 48 (CY48) was in the treatment of the soil hydrology (see Table 6.2 for a summary). CY47 used two prognostic soil moisture and soil temperature layers with depths of 7 cm and 42 cm respectively (see Blondin, 1991 for a more complete description). A third layer served as a boundary condition which was specified according to the climatology by Mintz and Serafini (1992).

A diagnostic study with CY47 based on observations of the FIFE experiment (First ISLSCP Field Experiment, where ISLSCP is the acronym for International Satellite Land Surface Climatology Project) by Betts *et al.* (1993) revealed a number of model deficiencies of which we mention: (*a*) the evaporation in CY47 was heavily constrained by the deep soil climatology, and (*b*) a considerable fraction of precipitation was lost into surface runoff (also in dry soil conditions), because the infiltration rate was limited by a fixed soil moisture diffusion coefficient. The result was a much lower evaporation rate than observed in late summer. Because of the climatological boundary condition, CY47 could not handle soil moisture anomalies.

These model deficiencies inspired the development of a new scheme for the land hydrology (see VB95 for details). The

scheme has four prognostic layers with depths of 7, 21, 72, and 189 cm respectively and roots in the top three layers, no heat flux and free drainage at the bottom and diffusivities and conductivities that depend in a highly nonlinear way on soil moisture according to Clapp and Hornberger (1978). The last point is particularly important because the efficient diffusion of soil water in wet conditions allows a rapid infiltration rate during precipitation which prevents surface runoff to take place. The main source of runoff in CY48 is from drainage from the deep layer.

Furthermore, the new scheme has a skin layer for temperature, which allows for a quick response of the skin temperature to changes in the radiative forcing at the surface. Also the formulation of the aerodynamic resistance was changed. The old scheme used the Louis (1979) formulation to compute the transfer between the lowest model level and the surface together with a surface roughness length that is the same for momentum, heat and moisture. The new scheme has a Monin Obukhov formulation (Beljaars and Holtslag, 1991) with smaller roughness lengths for heat and moisture than for momentum. Together with the land surface changes described above, a few additional model changes were introduced in CY48 (e.g. boundary layer top entrainment and modified air–sea interaction), but they are believed to be less relevant to the discussion in this section.

The new and old schemes have been tested in stand-alone mode by VB95 with the help of three data sets from climatologically different locations: (*a*) FIFE (Kansas, USA) for May to September 1987; (*b*) ARME (Amazonian Rainforest Me-

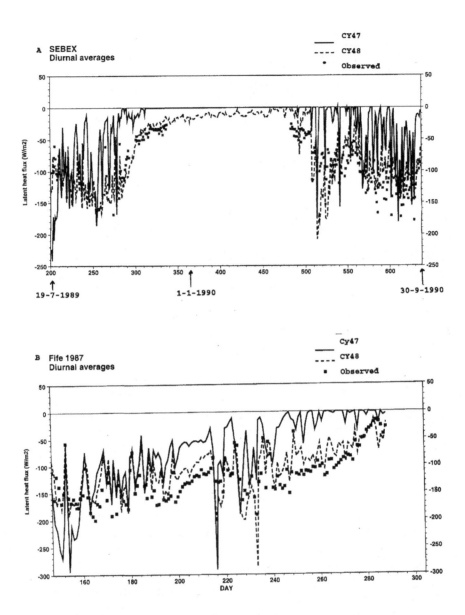

Figure 6.7. Time series of diurnal averages of latent heat flux for (A) SEBEX in the Sahel and (B) FIFE in the centre of the USA. These are results from stand-alone simulations with the land surface scheme and imposed meteorology (observations of wind, temperature, specific humidity, downward radiation and precipitation). CY47 (solid line) and CY48 (dashed line) are compared with observations (solid squares). (Simon Allen of the UK Institute of Hydrology (IH) kindly provided the SEBEX data. SEBEX is a collaborative project between IH and the International Crops Research Institute for the Semi-Arid Tropics Sahelian Center, funded by the UK Overseas Development Administration and the UK Natural Environment Research Council.)

teorological Experiment) for September 1983 to September 1985; and (c) Cabauw (The Netherlands) for the year of 1987. As an example we reproduce here five months of the time series of latent heat flux for FIFE and recently obtained results for SEBEX (the Sahelian Energy Balance EXperiment

from July 1989 to September 1990, see Wallace *et al.*, 1991). The FIFE experimental domain consists of tallgrass prairie on a gently undulating topography. The SEBEX area consists of fallow bushland and the terrain is flat.

From the SEBEX case we see very clearly that the new CY48

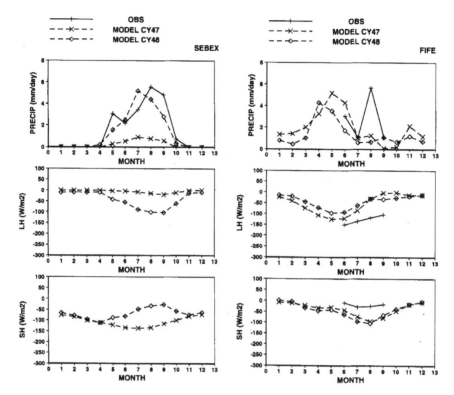

Figure 6.8. Monthly averages of precipitation and latent heat flux from 5-year long climate runs at T63 with the global ECMWF models CY47 and CY48 for the SEBEX and FIFE locations in comparison with observations (monthly averages for FIFE are only available from June until September). Upward fluxes are negative by model convention.

responds much better to precipitation and that the drying-out phase is represented much better than with the old CY47. The dry season starts around day 280 in Figure 6.7(*a*), after which the CY47 evaporation drops to virtually zero within a few days. After a very long dry season, precipitation starts again at about day 500 in Figure 6.7(*a*). CY47 responds to the precipitation events in a very intermittent way with short extremes in the evaporation. However, evaporation cannot be sustained after the precipitation events because most of the precipitation has been lost to surface runoff. The new model CY48 responds much better to precipitation because it has the capacity to recharge the deep-soil layers very rapidly during precipitation and to store enough water for evaporation during subsequent dry spells.

A similar conclusion can be drawn on the basis of FIFE data (see Figure 6.7(*b*)) for the period from 27 May (day 147) to 16 October (day 289) 1987. The dry spells between day 200 and 216 and after day 260 show a clear drying out in the observations, although evaporation during the latter period also decreases because of the decrease of available solar energy. Both model versions underestimate the latent heat flux during the dry spells, but CY48 is much closer to the observed values than

CY47. Model CY47 has a strong tendency to drift towards an evaporation rate that is determined by the deep-soil moisture climatology. After day 250 the climatological soil moisture is so low that CY47 can not sustain any evaporation.

Long global integrations with CY47 and CY48

To investigate possible model drift on seasonal time scales, 5-year integrations were done at T63 resolution with CY47 and CY48 and a specified climatological annual cycle for the sea surface temperature. VB95 present the mean annual cycle of the surface energy balance from these multiyear runs for three locations (FIFE, Cabauw and ARME). Comparison of CY47 and CY48 indicates mixed results: some parameters are improved and others show deterioration. Figure 6.8 reproduces the averaged annual cycles of precipitation, sensible and latent heat fluxes with CY47 and CY48 for FIFE and SEBEX. Looking at model interannual variability for these locations, we believe that the averaged annual cycle of the model is rather robust. This is not the case for the monthly averages of observations particularly for precipitation, which can have large interannual variability. Still, we believe that systematic patterns can be seen from the model results.

Precipitation for SEBEX is much better with CY48, mainly because the tropical convergence zone moves farther north with this model version. CY47 has very low precipitation for this location and evaporation is also very low. The higher precipitation with CY48 leads to higher evaporation, but the maximum evaporation shows a time lag with respect to precipitation. This is to be expected because the soil layers in CY48 have the capability to store precipitation and to make it available for evaporation later in the season. Unfortunately, sensible and latent heat fluxes cannot be compared with observations, because the gaps in the data do not allow the computation of monthly averages. However, the time series of daily latent heat flux values in Figure 6.7 provide a similar message. In the stand-alone simulation with the same prescribed atmospheric forcing, CY48 maintains evaporation much longer into the dry season than CY47. CY47 has the unrealistic characteristic of responding to precipitation with a very short-lived maximum in evaporation.

For FIFE both model versions have rather low precipitation compared with observations in August. In June, CY47 has more than twice the precipitation of CY48 with the observation of 1987 in between. The latent heat flux is underestimated and the sensible heat flux is overestimated. Both model versions have the tendency of drying in summer and CY48 is worse than CY47 in spite of the improved scheme, as could be demonstrated with the stand-alone simulations. In fact, the climatological soil moisture boundary condition of CY47 keeps the evaporation going in June and to some extent in July. All the interactions in CY48 and the lack of a deep-soil moisture constraint allow CY48 to dry out very early in the summer season.

An important systematic signal in all the results is the positive correlation between evaporation and precipitation. When evaporation in CY47 and CY48 is different, precipitation shows a difference with the same sign. In principle, such a correlation gives no information about cause/effect relationships. However, evaporation in CY47 is very much driven by the soil moisture climatological boundary condition, which suggests that precipitation responds to evaporation.

The conclusion we draw here is that precipitation and evaporation are coupled (see Figure 6.8) and that an improved land surface scheme does not necessarily improve the model climate. Deficiencies in other parameterizations (e.g., of model clouds) and complicated interactions between processes may lead to results that are difficult to interpret. It is therefore important to test parameterizations in stand-alone mode, because it is the only way to gain confidence in that particular parameterization.

The precipitation response to evaporation over the USA

July 1993 showed anomalously high precipitation over the central USA with exceptional flooding of the Mississippi (Kunkel *et al.*, 1994). The observed precipitation amounts and anomalies are reproduced in Figure 6.9. During this month the new model version CY48 and the operational model CY47 ran in parallel at full resolution (T213) including data assimilation. This gives us the opportunity to look at averages of short-range forecasts rather than at individual days. The rationale for using averages is that the flow pattern over the USA in July is very persistent. It is often referred to as the USA summer monsoon. Moisture is transported from the Gulf of Mexico in a northward boundary layer flow over Mexico, Texas, and Oklahoma, curving gradually eastward over the plains (Rasmusson, 1971). Severe storms, triggered by upper air disturbances coming from the west, create much of the summer precipitation.

In Table 6.3 we show precipitation and evaporation amounts from 0 to 24, 24 to 48 and 48 to 72 hour forecasts from consecutive days (from the 12 UTC analysis) averaged from 9 to 25 July according to verifying date. The area is between 35–45° N and 90–100° W, centered around the maximum precipitation. The two models produce rather similar precipitation amounts during the first 24 hours of the forecast, but from 24 hours onwards, CY47 dries out very quickly whereas CY48 manages to sustain a reasonable precipitation rate.

The spatial distribution of the 48- to 72-hour precipitation is shown in Figure 6.10 together with observations from SYNOP stations averaged from 9 to 25 July. The difference is remarkable: the maximum precipitation as observed (see also Figure 6.9) is captured much better with model CY48. From Table 6.3 we see that in the area with excessive precipitation, evaporation does not spin down in the same way as precipitation. This suggests that the effects of evaporation are not local.

The geographical distribution of the evaporation difference between CY48 and CY47 is given in Figure 6.11. The differences are large over the western part of the USA and over Mexico, mainly because the parallel run with model CY48 was started on 2 July 1993 with soil moisture at field capacity and the climatology used in CY47 is rather dry in these areas. The decision to use moist soil initial conditions was inspired by the fact that it had been wet over much of the USA and Europe in June. However, it is realized that the wet initial condition is not necessarily realistic throughout the USA.

Comparing Figures 6.10 and 6.11 we see that the area with maximum precipitation difference is northeast of the area with the evaporation difference. To trace the development of the differences, three-day backward trajectories were computed from the averaged flow fields (Figure 6.12). The same

Table 6.3. *Precipitation and evaporation from the July parallel run at different forecast ranges with CY47 and CY48 at T213L31 resolution averaged over the area between 35–45° N and 90–100° W and over all forecasts verifying between 9 and 25 July 1993.*

	Forecast range (hours)	Precipitation (P) (mm/day)	Evaporation (E) (mm/day)	$P - E$ (mm/day)
CY48	0–24	6.62	4.45	2.17
	24–48	7.71	4.67	2.44
	48–72	5.56	4.50	1.06
	96–120	5.08	4.62	0.46
CY47	0–24	6.42	4.41	2.01
	24–48	4.79	4.60	0.19
	48–72	2.60	4.60	−2.00
	96–120	1.92	4.36	−2.44

has been done with CY47 with very similar results, indicating that the difference in precipitation and evaporation has only minor influence on the flow pattern. The boundary layer air at 40° N 95° W originates over the Gulf of Mexico two days earlier and follows a slightly subsiding motion. There is, however, considerable wind shear, with the 850 hPa air following a more westerly trajectory than the near-surface air and therefore experiencing the influence of the land surface over a much longer distance. This differential advection plays a dominant role in the formation of the inversion in the area with maximum precipitation (Benjamin and Carlson, 1986).

The direction of the flow, together with the patterns of precipitation and evaporation difference, suggest that the evaporation influences precipitation in a downstream area. The origin of the difference of precipitation between CY47 and CY48 for the USA summer monsoon regime is illustrated more clearly with the help of the averaged thermodynamic structure of the 78-hour forecasts (local noon) at 40° N, 95° W (Figure 6.13). CY47 develops a pronounced inversion with very warm and dry air above the boundary layer, whereas CY48 has a structure which is much more in agreement with the analysis. The unrealistic inversion structure inhibits convection and develops gradually by differential advection. The excessively warm and dry air at the 850 hPa level in CY47 is the result of excessive heating upstream. CY48 has much less heating because more energy is used at the surface for evaporation and therefore less energy is available to warm the atmosphere. Whether the increased evaporation is more realistic is difficult to say without flux observations. We know from SYNOP verification that the temperature errors are smaller with CY48 but the increased evaporation may compensate for other model biases (e.g., radiation).

CY47 and CY48 differ in a number of aspects of the parameterization of the land surface (see Table 6.2) and therefore it is difficult to say which model changes improve the summer rain over the USA. For instance Koster and Suarez (1994) show that the short-term variability introduced by the interception reservoir affects the precipitation without changing the mean evaporation. However, given the sensitivity of CY48 to the initial condition of soil moisture as demonstrated by Beljaars *et al.* (1996), we believe that the difference in soil water availability between CY47 and CY48 is the dominant factor. More extensive diagnostics on this particular case is given by Beljaars *et al.* (1996).

Two-way precipitation/evaporation coupling

Qualitatively, the two-way coupling between precipitation and evaporation is well established now. The first branch of the coupling is rather direct: precipitation refills the soil moisture reservoir, making water available for evaporation. The second branch is through the influence of evaporation on precipitation. The latter is also well established, but the mechanism is not always clear and it is not obvious to what extent this coupling is local or non-local and to what extent it is different for different geographical areas.

The first branch of the coupling is determined by the surface processes and is parameterized in models as part of the land surface scheme. Many of these schemes exist and most of them have been compared with data sets of short length. Although many of these schemes have the same general characteristics, they may differ considerably in their implementation as is becoming increasingly clear from the Project for Intercomparison of Land surface Parameterization

a)

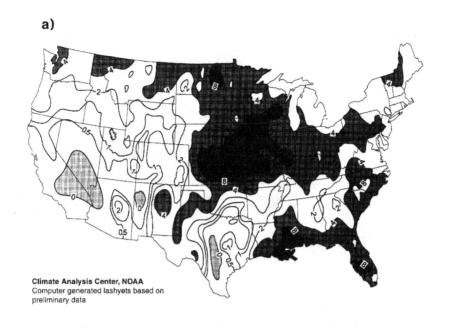

Climate Analysis Center, NOAA
Computer generated lashyets based on
preliminary data

b)

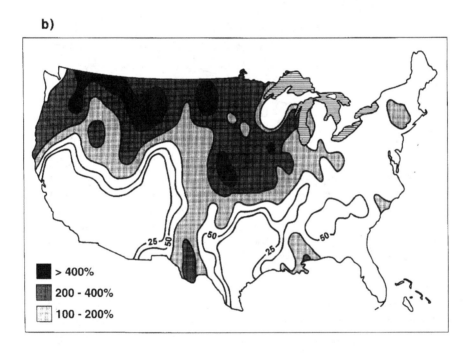

> 400%

200 - 400%

100 - 200%

Figure 6.9. Total observed precipitation over the USA for (*a*) July 1993 and (*b*) the percentage of normal precipitation (*b*) as published by the Weekly Weather and Crop Bulletin (August 3, 1993). The contours in (*a*) are at 0.5, 1, 2, 4, and 8 inches with light and heavy shading above 4 and 8 inches respectively; (*b*) has contours at 25, 50, 100, 200, and 400%; with shading above 100%.

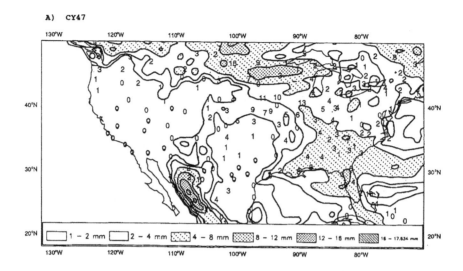

A) CY47

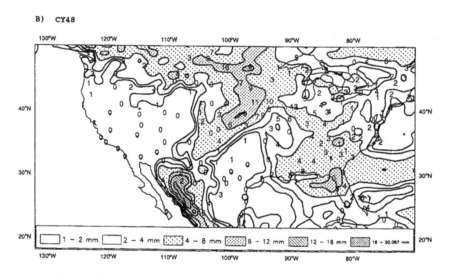

B) CY48

Figure 6.10. Mean forecast precipitation of all 48 to 72 hour forecasts verifying between 9 and 25 July, with (A) CY47 and (B) with CY48. The contours are at 1, 2, 4, 8 mm day^{-1} with shading above 4 mm day^{-1}. The printed numbers are station observations in mm day^{-1}.

Schemes (PILPS, Henderson Sellers *et al.*, 1993). Some key issues are: (*a*) how is precipitation partitioned between runoff and water to refill the soil water reservoir(s), (*b*) how big is the soil water reservoir (Milly and Dunne, 1994) and (*c*) how does soil water in the root zone control stomatal resistance and evaporation. These issues are discussed more extensively in sections 4.4 to 4.6.

In this section we used the model development at ECMWF as an example. We have shown that stand-alone testing of a land surface scheme is extremely powerful because all the atmospheric feedbacks are switched off. It is also felt that long time series are needed to assess the behavior of land surface schemes on the seasonal time scale. A few long data

sets exist (ARME, FIFE, Cabauw, SEBEX) and they have been used successfully, but more high quality data sets are needed to cover a larger variety of climatological regimes.

An improved parameterization does not necessarily improve a model in the short term. Feedbacks and compensating errors may be the reason that an improved scheme leads to larger model errors. It is therefore essential to have data to test parameterizations at the process level (e.g. the land surface scheme). Without data on the process level it will be impossible to disentangle compensating errors. This is very clear in the ECMWF model. The new CY48 land surface scheme is known to be better than the old CY47 from the stand-alone testing, but the model still has a pronounced

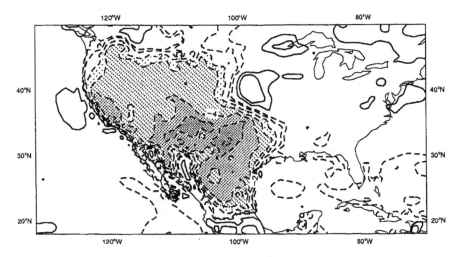

Figure 6.11. Difference between CY48 and CY47 of the 48 to 72 hour evaporation, averaged over verifying dates between 9 and 25 July 1993. Dashed contours indicate negative values, which imply an increase of evaporation (by model convention all downward fluxes are positive). The contours are at 0.5, 1, 2, 3 ... with shading above 2 mm day^{-1}.

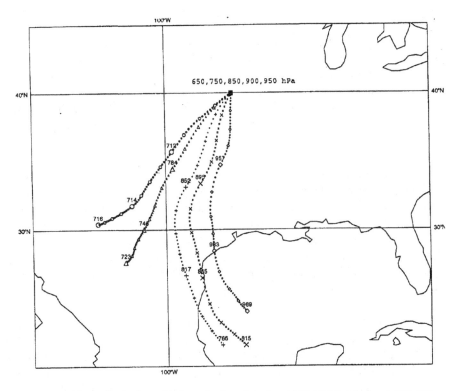

Figure 6.12. Three-day backward trajectories from location 40° N, 95° W at the levels 950, 900, 850, 750 and 650 hPa. The fields at successive forecast ranges 72, 66, 60, ... 12, 6, 0 hours, averaged between verifying dates 9 to 25 July, have been used for the computation of these trajectories. The printed numbers along the trajectories indicate the pressure height at one day intervals. This figure is for CY48, but the results with CY47 are very similar.

231

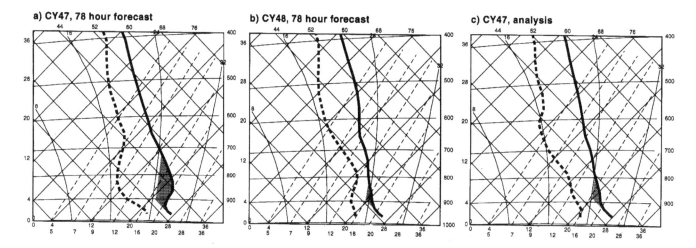

Figure 6.13. Profiles of temperature (solid line) and dewpoint (dashed line) with CY47 (*a*) and CY48 (*b*) of the averages verifying from 9 to 25 July at the forecast ranges of 78 hours. The model location is 40° N, 95° W. (*c*) shows the averaged verifying analysis. The diagram has a number of different coordinate systems: curved horizontal lines are lines of constant pressure, lines at an angle of about + 45° are lines of constant temperature, lines at an angle of about – 45° are lines of constant potential temperature (dry adiabats), the dashed lines are lines of constant specific humidity and the curved nearly vertical lines are moist adiabats. The shading indicates the area where a parcel lifted from the surface has a lower temperature than the surrounding air, i.e. the shaded surface area is a measure for the stability a parcel has to overcome before convection can occur.

dry bias in summer over continental areas. The exact origin of this dry bias is not clear yet. The systematic underestimation of cloud cover is almost certainly one aspect of the problem, but also the the convection parameterization may play a role.

This brings us to the the mechanism by which evaporation influences precipitation (the second branch of the coupling). A number of mechanisms have been mentioned above (see Betts *et al.*, 1996 for a review), but for the USA summer monsoon, the evaporation and heating appear to influence the inversion structure and therefore the convective precipitation one day downstream. Although the USA summer circulation is rather specific to that area, the downstream propagation of anomalies by recirculation of water may be a more general characteristic of the evaporation–precipitation feedback. This positive feedback can introduce long time scales in the atmosphere. If the difference between evaporation and precipitation is small, it takes a long time to empty the soil water reservoir in summer. Therefore anomalies can persist and recent experiments by Beljaars *et al.* (1996) even suggest that some predictive skill exists on the monthly time scale related to the soil moisture reservoir. Such predictive skill can be exploited only when the coupling between processes is modeled realistically. At this stage it is not clear whether models reproduce this feedback in a quantitatively correct way. It is even not clear what kind of data would be

necessary to test this feedback in models. The convection parameterization is obviously an important component, but is very difficult to test in stand-alone mode for conditions over land with a strong diurnal cycle.

The positive feedback loop involving precipitation and evaporation does not only cause drift in climate models but can also lead to dramatic drift in a numerical weather prediction system as shown by Viterbo (1996, chapter 7). Such systems consist of a forecast model and an analysis scheme. The purpose of the latter is to push the atmospheric fields towards the observations and to provide a suitable initial condition for the subsequent forecast. The analysis is not independent of the forecast; short range forecasts (6 hours) are used as first guess fields for the analysis, and for areas and/or parameters for which no data exist, the first guess is kept. In the first implementation of the scheme by Viterbo and Beljaars (1995), no attempt was made to analyze soil moisture; i.e. soil moisture in the analysis evolved as the result of precipitation and evaporation during successive 6-hour forecasts. The result was a pronounced drift to dry soil conditions in summer, very similar to the behavior of many climate models. Soil moisture observations are not available on an operational basis, and currently the only practical way to adjust soil moisture is by using temperature and specific humidity errors from short range forecasts (Bouttier *et al.*, 1993*a,b*). For example, too dry soil leads to a warm and dry drift in the boundary layer in short-range forecasts, and when the relation between drift and soil moisture is known, soil moisture can be corrected. A realistic model is a prerequisite for such a scheme because temperature and specific humidity errors can also be caused by other model biases (e.g., errors in boundary layer diffusion or radiation).

References

Arpe, K., Bengtsson, L., Dümenil, L. and Roeckner, E. (1994). The hydrological cycle in the ECHAM3 simulations of the atmospheric circulation, *NATO ASI Series*, I **26**, 361–77.

Beljaars, A. C. M. and Holtslag, A. M. M. (1991). Flux parametrization over land surfaces for atmospheric models, *J. Appl. Meteor.*, **30**, 327–41.

Beljaars, A. C. M., Viterbo, P., Miller, M. J. and Betts, A. K. (1996). The anomalous rainfall over the USA during July 1993: sensitivity to land surface parametrization and soil moisture anomalies. *Mon. Wea. Rev.*, **124**, 362–83.

Benjamin, S. G., and Carlson, T. N. (1986). Some effects of surface heating and topography on the regional severe storm environment. Part I: Three-dimensional simulations. *Mon. Wea. Rev.*, **114**, 307–29.

Betts, A. K., and Ball, J. H. (1995). The FIFE surface diurnal cycle climate. *J. Geoph. Res.*, **100**, 25679–93.

Betts, A. K., Ball, J. H., and Beljaars, A. C. M. (1993). Comparison between the land surface response of the European Centre model and the FIFE-1987 data. *Quart. J. Roy. Meteor. Soc.*, **119**, 975–1001.

Betts, A. K., Ball, J. H., Beljaars, A. C. M., Miller, M. J. and Viterbo, P. (1996). The land-surface-atmosphere interaction, *J. Geoph. Res.*, **101**, 7209–25.

Blondin, C. (1991). Parameterization of land-surface processes in numerical weather prediction. In *Land Surface Evaporation: Measurement and Parameterization*, ed. T. J. Schmugge and J. C. André, pp. 31–54. New York: Springer-Verlag.

Bouttier, F., Mahfouf, J.-F. and Noilhan, J. (1993a). Sequential assimilation of soil moisture from atmospheric low-level parameters. Part I: Sensitivity and calibration studies, *J. Appl. Meteor.*, **32**, 1335–51.

Bouttier, F., Mahfouf, J.-F. and Noilhan, J. (1993b). Sequential assimilation of soil moisture from atmospheric low-level parameters. Part II: Implementation in a mesoscale model, *J. Appl. Meteor.*, **32**, 1352–64.

Brown, A. R. (1995). *An Evaluation of Single Column Unified Model Performance in the Dry Unstable Boundary Layer*. Met O (APR) Turbulence and Diffusion Note No. 219, Available from Atmospheric Processes Research Division, Meteorological Office, London Road, Bracknell, Berks RG12 2SY United Kingdom.

Clapp, R. B. and Hornberger, G. M. (1978). Empirical equations for some soil hydraulic properties. *Water Resour. Res.*, **14**, 601–4.

Cullen, M. J. P. (1993). The Unified Forecast/Climate Model. *Meteorol. Mag.*, **122**, 81–94.

Garratt, J. R. (1993). Sensitivity of climate simulations to land-surface and atmospheric boundary-layer treatments: a review. *J. Climate*, **6**, 419–49.

Garratt, J. R., Krummel, P. B. and Kowalczyk, E. A. (1993). The surface energy balance at local and regional scales: a comparison of general circulation model results with observations. *J. Clim.*, **6**, 1090–109.

Gregory, D. (1997). The mass flux approach to the parameterization of deep convection. In *The Physics and Parameterization of Moist Atmospheric Convection*, ed. R. K. Smith, NATO ASI Series C-Vol 505, Kluwer Academic Publishers.

Gregory, D. and Rowntree, P. R. (1990). A mass flux convection scheme with representation of cloud ensemble characteristics and stability dependent closure. *Mon. Wea. Rev.*, **118**, 1483–506.

Henderson-Sellers, A., Yang, Z.-L. and Dickinson, R. E. (1993). The project for intercomparison of land-surface parameterization schemes. *Bull. Am. Meteor. Soc.*, **74**, 1335–1349.

Koster, R. D. and Suarez, M. J. (1994). The components of a SVAT scheme and their effects on a GCM's hydrological cycle. *Adv. Water Res.*, **17**, 61–78.

Kunkel, K. E., Changnon, S. A. and Angel, J. R. (1994). Climate aspects of the Mississippi river basin flood. *Bull. Am. Meteor. Soc.*, **75**, 811–22.

Lanicci, J. M., Carlson, T. N. and Warner, T. T. (1987). Sensitivity of the great plains severe-storm environment to soil-moisture distribution. *Mon. Wea. Rev.*, **115**, 2660–73.

Louis, J. F. (1979). A parametric model of vertical eddy fluxes in the atmosphere. *Bound.-Layer Meteor.*, **17**, 187–202.

Louis, J. F., Tiedtke, M. and Geleyn, J. F. (1982). A short history of the operational PBL-parameterization at ECMWF. In *Workshop on Boundary Layer Parameterization*, November 1981. Reading, UK: ECMWF.

Miller, M. A. and Albrecht, B. A. (1995). Surface based observations of mesoscale cumulus-stratocumulus interaction during ASTEX. *J. Atmos. Sci.*, **52**, 2809–26.

Milly, P. C. D. and Dunne, K. A. (1994). Sensitivity of the global water cycle to the water-holding capacity of land. *J. Clim.*, **7**, 506–26.

Mintz, Y. (1984). The sensitivity of numerically simulated climates to land-surface boundary conditions. In *The Global Climate*, ed. J. Houghton, pp. 79–105. Cambridge: Cambridge University Press.

Mintz, Y. and Serafini, Y. V. (1992). A global monthly climatology of soil moisture and water balance. *Clim. Dyn.*, **8**, 13–27.

Namias, J. (1958). Persistence of mid-tropospheric circulations between adjacent months and seasons. In *The Atmosphere and Sea in Motion* (Rossby memorial volume), ed. B. Bolin, pp. 240–8. Rockfeller Institute Press.

Oglesby, R. J. (1991). Springtime soil moisture variability, and North American drought as simulated by the NCAR Community Climate Model I. *J. Climate*, **4**, 890–7.

Rasmusson, E. M. (1971). A study of the hydrology of Eastern North America using atmospheric vapor flux data. *Mon. Wea. Rev.*, **99**, 119–35.

Rowntree, P. R. and Bolton, J. A. (1983). Simulation of the atmospheric response to soil moisture anomalies over Europe. *Quart. J. Roy. Meteor. Soc.*, **109**, 501–26.

Siebesma, A. P. and Cuijpers, A. A. M. (1995). Evaluation of parametric assumptions for shallow cumulus convection. *J. Atmos. Sci.*, **52**, 650–66.

Smith, R. N. B. (1990). A scheme for predicting layer clouds and their water contents in a general circulation model. *Quart. J. Roy. Meteor. Soc.*, **116**, 435–60.

Smith, R. N. B. (1993). Experience and developments with the layer cloud and boundary layer mixing schemes in the UK Meteorological Office Unified Model. In *Workshop on the 'Parameterization of the cloud topped boundary layer'*, 8–11

June. ECMWF, Shinfield Park, Reading RG2 9AX United Kingdom.

Stull, R. B. (1995). A review of parameterization schemes for turbulent boundary layer processes. In *Seminar proceedings on the 'Parameterization of sub-grid scale physical processes'*, 5–9th September 1994, ECMWF, Shinfield Park, Reading RG2 9AX United Kingdom.

Viterbo, P. (1996). *The Representation of Surface Processes in General Circulation Models*. ECMWF report (also PhD thesis, University of Lisbon) ECMWF, Reading, England.

Viterbo, P. and Beljaars, A. C. M. (1995). An improved land surface parameterization scheme in the ECMWF model and its validation. *J. Clim.*, **8**, 2716–48.

Wallace, J. S., Wright, I. R., Stewart, J. B. and Holwill, C. J. (1991). The Sahelian Energy Balance Experiment (SEBEX): ground based measurements and their potential for spatial extrapolation using satellite data. *Advances in Space Res.*, **11**, 131–41.

7 Continental-scale water budgets

7.1 Estimating evaporation-minus-precipitation
as a residual of the atmospheric water budget

Kevin E. Trenberth and Christian J. Guillemot

Introduction

Given the obvious importance of water to humans, it is surprising how poorly the global hydrologic cycle is known. In this section the focus is on the atmospheric branch of the hydrologic cycle, evaporation and precipitation, and the transport of water vapor. Reasons for the scanty knowledge of both moisture in the atmosphere and precipitation stem from the lack of observations over the oceans and the nature of the quantities. Rainfall and clouds often occur on small time and space scales so that a single moisture or precipitation observation is not representative of an area more than a few kilometers in size or for more than a small fraction of a day. Moreover, measurements of precipitation occur only where humans live, in relatively widely spaced locations, and the buckets used to accumulate precipitation may not catch it all, especially under snowy and windy conditions. Satellite data on moisture have been made available to the global analyses from TOVS (TIROS Operational Vertical Sounder), although with mixed results (Liu *et al.*, 1992; Wittemeyer and Vonder Haar, 1994). After July 1987 fields of precipitable water and other quantities from the Special Sensor Microwave Imager (SSM/I) have become available but not in time for the operational analyses at the global weather forecast centres.

One means of obtaining global fields is from the assimilation of observed moisture data into models as part of four-dimensional data assimilation (4DDA) (see Section 2.4), but this can occur reliably only if the moist physics of the model are realistic; otherwise convective adjustments and cumulus parameterizations are apt to be activated so as to change the moisture field to one that the model is comfortable with. Together with the representiveness issues for observations, these factors have led to less attention being given to moisture fields in the global analyses. The consequence has been that the fields of both relative and specific humidity have changed abruptly every time there is a big change in the 4DDA system (Trenberth and Olson, 1988b; Trenberth, 1992). At the European Centre for Medium-range Weather Forecasts (ECMWF), spurious changes in zonal mean relative humidities of 20% are common (Trenberth, 1992). Moisture transport in the tropics is very dependent on the vertical air motions, the so-called divergent part of the wind field, which

has also undergone major changes with changes in the 4DDA system.

This section evaluates the state-of-the-art in computing evaporation E minus precipitation P as a residual from the large-scale atmospheric transports using global analyses from ECMWF and US National Meteorological Center (NMC; now known as the National Centers for Environmental Prediction, NCEP). Results from the retrospective reanalysis of past data using 4DDA (commonly referred to as 'reanalysis') by NASA/Goddard (Schubert *et al.*, 1993) are also included. The technique for deducing $E-P$ as a residual has a long history, although it usually makes use of rawinsonde data directly rather than global analyses via 4DDA. The results of many of these studies are encapsulated by the review of Peixoto and Oort (1983); see also Savijärvi (1988). Most previous studies have not properly accounted for variations in surface pressure p_s (Trenberth 1991) which is important in vertically integrating moisture amounts and transports, and so the lowest fraction of the atmosphere has generally not been correctly accounted for. Recent examples using global analyses are by Roads *et al.* (1992, 1994) for NMC data and Oki *et al.* (1993) for ECMWF data.

Data sets

Global analyses are produced using 4DDA in which multivariate observed data are combined with the 'first guess' using a statistically optimum scheme. The first guess is the best estimate of the current state of the atmosphere from previous analyses produced using a numerical weather prediction (NWP) model. The operational analyses are performed under time constraints for weather forecasting purposes and not for climate purposes. Changes in the NWP model, data handling techniques, initialization, and so on, which are implemented to improve the weather forecasts, disrupt the continuity of the analyses (Trenberth and Olson, 1988b; Trenberth, 1992) and impact the moisture fields.

The ECMWF data used consist of uninitialized analyses four-times daily at T106 resolution. In this notation the 'T' refers to triangular truncation at 106 wavenumbers in the east–west direction using a spherical harmonic representa-

236

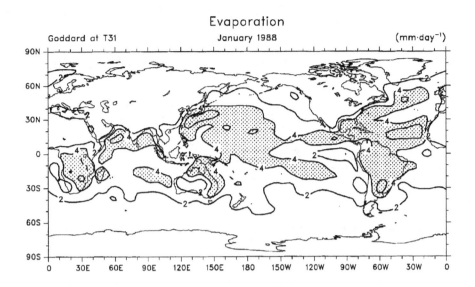

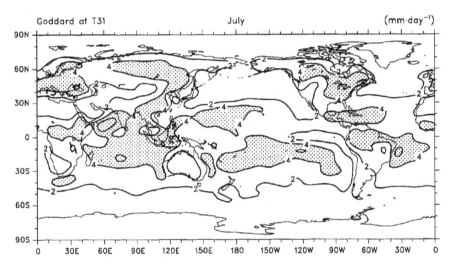

Figure 7.1. For January and July 1988 at T31 resolution, E from the NASA/Goddard reanalysis. The contour interval is 2 mm day⁻¹ and values exceeding 4 mm day⁻¹ are stippled.

tion. At a spectral truncation of T106 resolution, the Gaussian grid is 320 by 160 (51 200) points 1.125° of latitude and longitude apart. The analyses are at 14 or 15 levels (1000, 925, 850, 700, 500, 400, 300, 250, 200, 150, 100, 70, 50, 30, 10 hPa). The 925 hPa level became available only from January 1992. All fields are truncated to T42 for processing (Trenberth and Solomon, 1993). A description and evaluation of the ECMWF analyses is given by Trenberth (1992).

For NMC, the analyses are twice daily at 12 levels (1000, 850, 700, 500, 400, 300, 250, 200, 150, 100, 70, 50 hPa) on a 2.5° grid. The temperature archived by NMC is virtual temperature, but temperatures at each time have been computed from the available fields. NMC sets relative humidity to zero

above 300 hPa; therefore the temperature at those levels is unchanged. Missing data have been replaced with ECMWF analyses (in 1987 missing data occurred 35 times, and since then it has averaged 9 times per year). All data were interpolated to a T42 grid for processing. An evaluation of earlier NMC data and a comparison with ECMWF data were described by Trenberth and Olson (1988a, 1988b).

Fields of p_s were derived for all data sets using real mean topography at T42 resolution using methods given in Trenberth (1992) and as described in Trenberth and Guillemot (1995). This field is not spectrally truncated in order to preserve an accurate depiction of the surface and avoid spurious ripples over the ocean. The ECMWF archive has a 'surface'

Precipitation

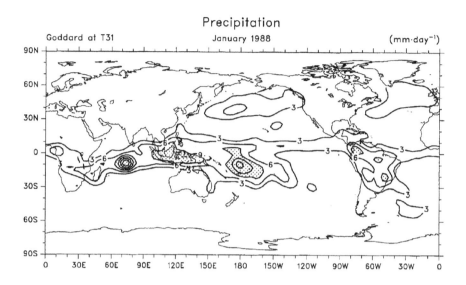

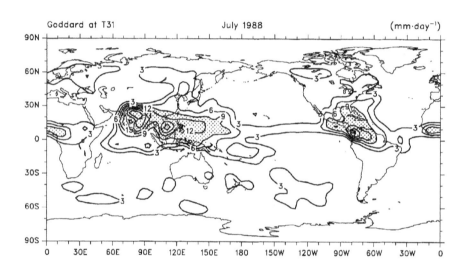

Figure 7.2. For January and July 1988 at T31 resolution, *P* from the NASA/Goddard reanalysis. The contour interval is 3 mm day⁻¹ and values exceeding 9 mm day⁻¹ are stippled.

pressure field included, but it is the surface pressure of the model topography, which includes an enhancement as so-called 'envelope' orography (approximating the envelope of the highest peaks and ignoring the valleys). Prior to 6 March 1991 NMC also had an enhanced 'silhouette' orography (which is similar to envelope orography but uses a silhouette to ensure that the highest peaks are included), but mean orography has been used since. However, as there is no archive of the p_s from NMC available, calculated values are used throughout for the moisture budget.

Monthly mean fields of several quantities are also examined from the recent reanalysis by NASA/Goddard (Schubert *et al.*, 1993) from 1987 to 1989. Evaporation and precipitation are saved from accumulations from the NWP model over 3-hour increments.

For precipitation, new global fields are becoming available from the Global Precipitation Climatology Project (GPCP) (Arkin and Xie, 1994) and consist of a combination of rain-gauge data over land and several satellite algorithms over the ocean. It is not the purpose here to explore the accuracy of these fields, but they are regarded as independent measures of *P* that are useful for evaluating the $E-P$ fields from the other sources, because the dominant spatial and temporal variability originates in *P*, not *E*, as will be discussed.

E – P

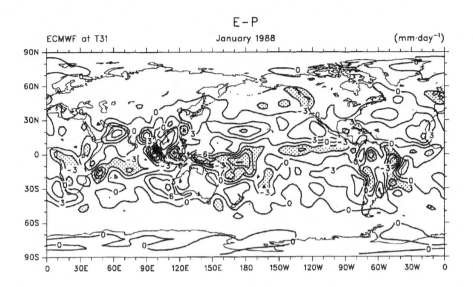

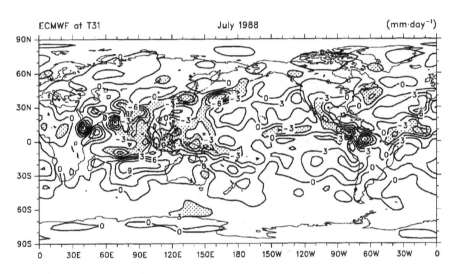

Figure 7.3. For January and July 1988 at T31 resolution, $E-P$ from ECMWF analyses. The contour interval is 3 mm day^{-1} and values less than -3 mm day^{-1} are stippled.

Moisture budget methodology

Use is made of the equation of state, the hydrostatic equation, and the equation of continuity. The following uses standard atmospheric science notation. The vertical mass weighted integral of the specific humidity q is the precipitable water w given by

$$w = \int_0^\infty qp\,\mathrm{d}z = \frac{1}{g}\int_0^{p_s} q\,\mathrm{d}p \qquad (7.1.1)$$

The equation for conservation of water vapor is

$$\frac{\partial q}{\partial t} + \mathbf{v}\cdot\nabla q + \omega\frac{\partial q}{\partial p} = e - c \qquad (7.1.2)$$

where e is the rate of re-evaporation of cloud and rainwater and c is the rate of condensation per unit mass which together produce the precipitation rate, and the role of liquid water in the atmosphere is ignored. Multiplying equation (7.1.2) by L, the latent heat of vaporization for water, gives the latent heat released through evaporation and condensation Q_2 (Yanai et al., 1973): $L(e-c) = -Q_2$ or, using the equation of continuity, (7.1.2) can be expressed in flux form as

$$\frac{\partial q}{\partial t} + \nabla\cdot q\mathbf{v} + \frac{\partial}{\partial p}q\omega = e - c \qquad (7.1.3)$$

Equation (7.1.3) may be vertically integrated to produce

239

E – P

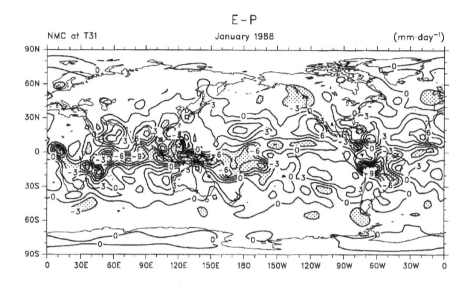

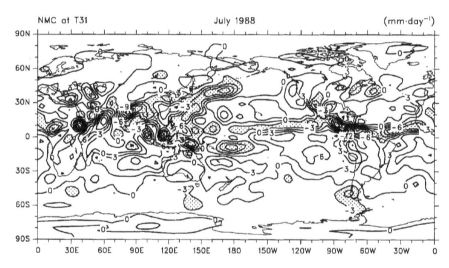

Figure 7.4. For January and July 1988 at T31 resolution, $E-P$ from NMC analyses. The contour interval is 3 mm day^{-1} and values less than -3 mm day^{-1} are stippled.

$$\frac{\partial w}{\partial t} + \nabla \cdot \frac{1}{g}\int_0^{p_s} q\mathbf{v}\mathrm{d}p = E - P \qquad (7.1.4)$$

where E is the evaporation from the surface and P is the precipitation. If the budget of total dry air mass is computed, it generally fails to balance by a substantial amount owing to problems in the analyses of the three-dimensional velocity field. Because this corrupts any budget where quantities are mass weighted, in practice all the velocity fields are adjusted first so that the dry mass budget is satisfied using a method outlined in Trenberth (1991). The details on methods of com-

puting the budgets are given in Trenberth and Guillemot (1995).

Over an annual cycle, the change in storage of moisture in the atmosphere is small. Over land because there is a net runoff in streams, the precipitation must exceed the evaporation, so that $E-P$ should be negative on average. This constraint provides a weak check on any diagnostic results, although one which is typically not satisfied at all well, such as in the results of Peixoto and Oort (1983). More detailed constraints can be included where information exists on the actual runoff (e.g. Oki *et al.*, 1993; Roads *et al.*, 1994); see also Section 1.2.

240

Precipitation

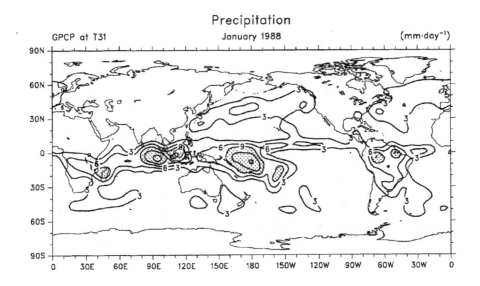

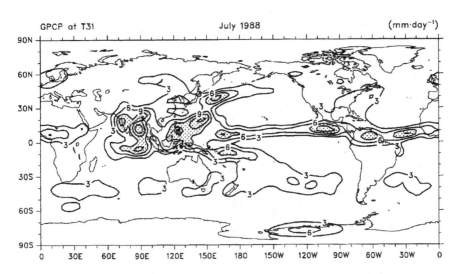

Figure 7.5. For January and July 1988 at T31 resolution, P from the GPCP project. The contour interval is 3 mm day^{-1} and values exceeding 9 mm day^{-1} are stippled.

Evaporation minus precipitation budget comparisons

An examination of the monthly mean E and P fields for 1987 to 1989 from the NASA/Goddard reanalysis reveals that mean values of E exhibit much less spatial structure than P and average about 3 to 5 mm day^{-1} over most of the tropics and subtropics (Figures 7.1 and 7.2). Standard deviations of the monthly means (not shown) for the tropics and subtropics are typically 0.5 mm day^{-1} for E over extensive areas, but several mm day^{-1} for P over only those areas where the tropical convergence zones and monsoon rains are active. It is clear that $E-P$ temporal monthly variability is dominated by variations in location and intensity of rainfall and the spatial structure in $E-P$ is also dominated by the P field.

Figure 7.3 presents $E-P$ from the residual calculation using ECMWF analyses, and Figure 7.4 presents the result from NMC analyses for the same months, January and July 1988, as in Figures 7.1 and 7.2. As a measure of 'truth', Figure 7.5 shows the results for the P field from the GPCP data.

To evaluate the fields for these months, a direct qualitative comparison was made between the GPCP and Goddard P fields and then $E-P$ was compared from Goddard, NMC, and ECMWF. All fields were first truncated at T31. While the GPCP is regarded as an independent measure of 'truth', it is a

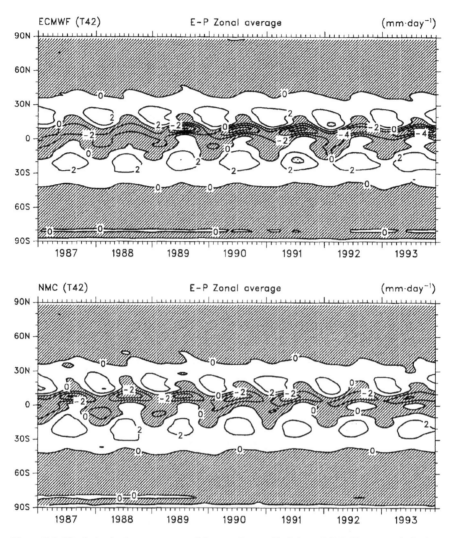

Figure 7.6. The latitude-time sequence of the zonal mean $E-P$ from ECMWF (top) and NMC (bottom). Negative values are hatched and the contour interval is 2 mm day⁻¹. (From Trenberth and Guillemot, 1995.)

product still under development and is least reliable over the high latitudes of the southern hemisphere. For the $E-P$ field in Figures 7.3 and 7.4, the stippled region is selected to correspond very roughly with the stippled regions in the panels for P alone. It is important to note that the characteristics of both ECMWF and NMC analyses are quite different from those in more recent years, as discussed next. Prior to 1989 the ECMWF rainfall was not as intense, while the NMC fields included strong 'bulls-eyes' and stronger features in general.

The wintertime northern hemisphere storm track is very well depicted by the precipitation fields in the Goddard analyses although the storm track is weak in the southern hemisphere. However, in the northern summer, the rainfall maximum off Japan is largely absent in the Goddard analyses (Figure 7.2) although very well represented in the NMC and ECMWF analyses. In the tropics there is too much precipitation implied in all three analyses over land in Central America; this tends to be the case also over southern Asia and over the Ethiopian highlands for NMC and ECMWF. The Goddard Intertropical Convergence Zone (ITCZ) is too weak in both the Atlantic and the Pacific, but the implied P values in the NMC and ECMWF analyses are closer to those from GPCP. In general, the NMC fields in 1987 and 1988 exhibited the strongest features, but with maxima often in the wrong place and too large compared with the GPCP. The NMC fields became quite a bit weaker, closer in magnitude to the GPCP, and with a reduced problem in the 'landlocking' of precipitation after

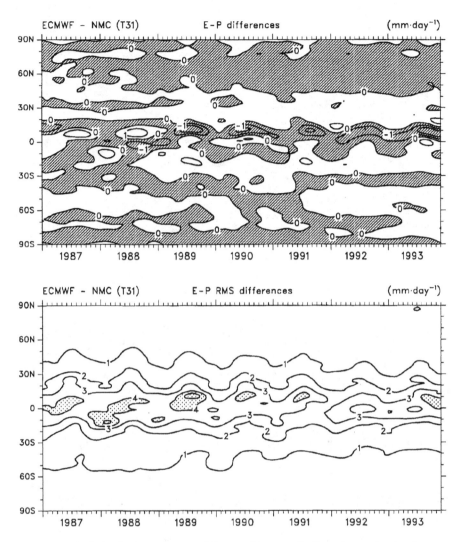

Figure 7.7. The latitude–time sequence of the monthly mean $E-P$ zonal mean (top) and rms differences (bottom) between ECMWF–NMC. Negative values are hatched, the contour interval is 1 mm day^{-1} and values exceeding 4 mm day^{-1} are stippled. (From Trenberth and Guillemot, 1995.)

the introduction of the T126 model (6 March 1991) and Spectral Statistical Interpolation (SSI) analysis system (25 June 1991). Subjectively, for 1987 and 1988, the ECMWF fields look to be most consistent with the GPCP, especially in the northern winter.

Figure 7.6 presents the zonal mean $E-P$ from both ECMWF and NMC as a function of latitude and time from 1987 to 1993. Note the strengthening in the excess of P in the deep tropics after May 1989 for ECMWF, corresponding to the model change at that time, to peak values in excess of 6 mm day^{-1} for $P-E$, values larger than indicated for NMC. The root-mean-square (rms) differences, Figure 7.7, are seen to be up to 3 mm day^{-1} for the zonal means in the tropical

convergence zone and typically 3 to 5 mm day^{-1} for rms differences around each latitude circle.

To shed light on the origin of differences, the total annual mean fields for 1993 are given in Figure 7.8 at T31 resolution, and the mean zonal average and rms differences for January and July 1993 are shown in Figure 7.9. These reveal the tendency for somewhat stronger and more localized values in the ECMWF results, so that the differences tend to be quite spotty. The very broad-scale patterns of the fields are quite similar. However, the localized differences can be very large in places where heavy rain is indicated in one but not the other analysis. The summary (Figure 7.9) also contrasts the mean and rms differences with the rms values of the field

1993 annual mean

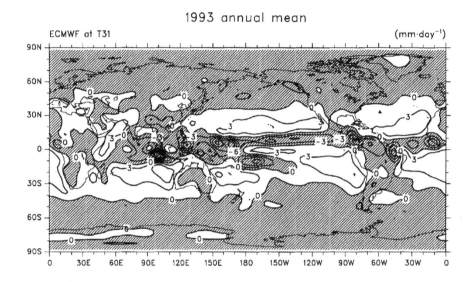

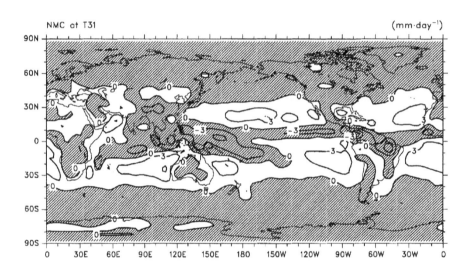

Figure 7.8. Annual mean $E - P$ for 1993 at T31 resolution from ECMWF (top) and NMC (bottom). Negative values are hatched and the contour interval is 3 mm day^{-1}. (From Trenberth and Guillemot, 1995.)

itself from ECMWF. The latter may be taken as a measure of the signal of interest and is typically 4 to 5 mm day^{-1} in the tropics, while the rms difference can be considered as the noise of typically 3 mm day^{-1}. Thus the level of noise is about 60% to 75% of the signal in this field at T31 resolution and has not changed much with time. In spite of this noise, the fields do have a strong large-scale resemblance, and we have explored this by also calculating the rms differences at other resolutions. For example at T15 truncation, the rms differences (Figure 7.9) are remarkably close to being half of the rms differences at T31 resolution, that is ~1.5 mm day^{-1} in the tropics and about 30% of the signal. This result emphasizes

the broad agreement in the large-scale features whereas scales less than about 1000 km do not agree well.

The annual mean for 1993 for each analysis was presented in Figure 7.8 because it is the most recent available. By ignoring annual changes in atmospheric storage, which can be considered negligible, an assessment can be made over land as to whether the constraint that $P > E$ is satisfied in most places (the possible exceptions being places with sources of ground moisture such as lakes or rivers). Clearly there are large areas over Africa and southern Asia, and parts of Australia and South America, where even this relaxed constraint is violated by both analyses. The stronger precipitation in the

244

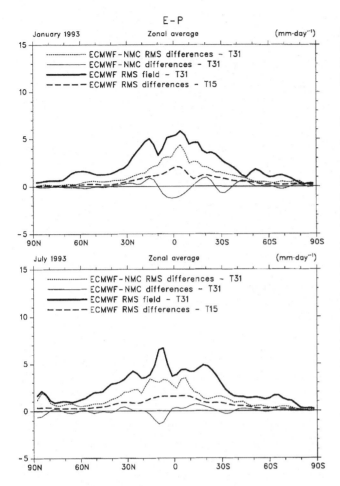

Figure 7.9. $E-P$ meridional profiles of ECMWF rms values computed around each latitude circle at T31 resolution (thick line), ECMWF–NMC zonal means (thin line), and ECMWF–NMC zonal rms differences (dashed) in mm day^{-1} for January (top) and July (bottom) 1993. Also shown are the zonal rms differences at T15 resolution (dotted line).

tropical convergence zones seen in the ECMWF analyses is associated with the stronger low level convergence and divergent circulation that occurred after model changes on 2 May 1989.

Overall correlation coefficients between the monthly mean NMC and ECMWF time series from January 1987 to December 1993 (84 values) with the annual cycle included (Figure 7.10) exceed 0.8 mostly from 30° to 80° N and over the subtropical oceans of the southern hemisphere, but they are often less than 0.3 in the tropics, throughout Africa and South America, and over Antarctica. Correlations are also relatively low over the Rockies and Himalayan regions. Thus the areas of largest discrepancies can be identified as areas of significant topography and regions of poor data coverage.

The ECMWF fields have become much more intense with time (cf. Figure 7.8), implying less agreement with the GPCP after May 1989. In Figure 7.8 it is apparent that the tendency for too much landlocked precipitation over Central America is still present in the most recent analyses at both centers.

Discussion and conclusions

Trenberth and Guillemot (1995) have documented results of tests performed on the sensitivity of w and $E-P$ to vertical resolution, diurnal sampling, and computational algorithms, and the following discussion is based mainly upon that work. In the tropics and subtropics, the atmospheric moisture budget is more sensitive to the divergence field than the moisture amounts, and consequently initialization of the analyses has an impact on the perceived moisture budget. Initialization cuts down on noise but probably also cuts down the magnitude of the signal, so that it impacts on the moisture budget significantly. In the mid-latitudes, quasi-geostrophic dynamics ensure that the divergence field is better known, and uncertainties in the moisture budget stem roughly equally from discrepancies in moisture analyses and the velocity field. The diurnal cycle is not well captured by twice-daily analyses and makes significant differences in $E-P$. This impacts directly on the NMC data used here.

From an evaluation of precipitable water in the global analyses, we conclude that the analyses tend to be dominated by the characteristics of the NWP model climate in the 4DDA and that biases are present. It appears that insufficient account is taken of the data that are available and that the 4DDA system is not capable of exploiting the observations because of the inherent incompatibility with the NWP assimilating model.

Calculations with the full ECMWF 19 model-levels at T106 resolution show that little is lost in $E-P$ on the scales larger than T31 when the fields are first truncated at T42. The horizontal resolution of the data is very important locally in the vicinity of steep orography but not on the large scale in general. The sources of the problems that exist are in the large-scale analyzed fields of moisture and divergence themselves, which are in turn related to the parameterization of moist processes in the NWP model. Trenberth and Guillemot (1995) also showed, by contrasting moisture budgets computed using full vertical resolution model-level data with pressure-level analyses, that the general patterns can be captured quite well with analyses in p coordinates, although high resolution at low levels and proper treatment of the surface via accurate p_s values is vital for reliable results.

There is no acknowledged source of truth for $E-P$. The Goddard reanalyses reveal that variability of E is much less

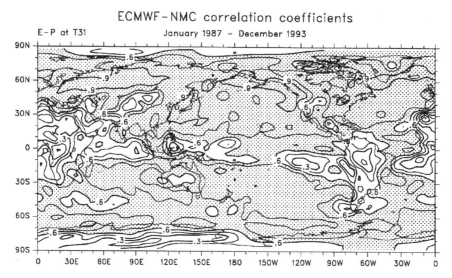

Figure 7.10. Overall correlation coefficients between the monthly mean NMC and ECMWF $E-P$ time series from January 1987 to December 1993 with the annual cycle included. Values greater than 0.6 are stippled.

than that of P so that the main variations in $E-P$ in both space and time arise from the P variations. Under the GPCP project, global fields of precipitation are being assembled and developed, and more months will become available in the future. Nevertheless, indications are that none of the global analyses produces very reliable results for the moisture budget at present. The differences between NMC and ECMWF are substantial and are traceable mostly to the continuing differences in the divergent circulation in the global analyses. Several methods exist for estimating evaporation using space-borne sensors (e.g. Liu, 1988), and it may eventually be worthwhile combining all the estimates of E, P, and $E-P$ to obtain the best estimates of these components of the hydrological cycle, but such an exercise is premature at this point, given that there are rms discrepancies of 60% to 75% in the current estimates of $E-P$ at T31 resolution.

The technique of using annual mean $E-P$ from atmospheric residual calculations and comparing with runoff has been carried out by Oki *et al.* (1993) for each of 35 major river basins for 1985 to 1988 using ECMWF data. They found atmospheric moisture divergence (i.e. the wrong sign) in 7 cases and discrepancies of up to 5656%. For only 23 river basins was the error within 100%, and so the agreement cannot be said to be good. In fact our analysis suggests that the agreement between the atmospheric moisture convergence over complex terrain is poor. In part this may be because of the treatment of the lower boundary in the NWP

models and the use of envelope orography. Somewhat better results are achieved over the United States (Roads *et al.*, 1994) where a very good rawinsonde network helps define moisture transports, although not without substantial corrections to the atmospheric moisture divergence. Over the oceans, Chen *et al.*, (1994) have used the atmospheric moisture divergence to estimate the freshwater budget for the oceans using NMC data for 1979–92. They find some agreement with oceanic estimates of Wijffels *et al.* (1992) but with considerably different magnitudes.

Activities are under way at NCEP and ECMWF to reanalyze retrospectively the observations based upon state-of-the-art 4DDA systems at each center (see Trenberth, 1995). ECMWF will no longer use envelope orography during that exercise. However, it is apparent that the recent analysis systems continue to produce quite different products for moisture and large-scale atmospheric divergent circulation. Prospects may exist for using SSM/I data in the future. ECMWF uses a new variational scheme for assimilation of water vapor radiance information in the 4DDA, and this has also been recently implemented operationally at NCEP but it is not in the NCEP/NCAR reanalyses. Therefore it seems likely that the differences seen above for 1993 may also be present in the reanalysis products.

We conclude that there is enormous scope and good prospects for improving our knowledge of, and ability to model, the global atmospheric hydrologic cycle.

7.2 Factors determining the partitioning of precipitation into evaporation and runoff

P. C. D. Milly

Introduction

Problem definition

Hydrologists have set for themselves the task of quantifying and explaining the distribution and fluxes of water on the Earth, with particular attention to the continental branch of the hydrosphere. By now it must be clear to the reader of this book that the continents exchange water and energy with the atmosphere and the oceans, mainly through three major sets of fluxes, which, for simplicity, will here be termed precipitation, evaporation, and runoff. The term precipitation (P) is meant to include all hydrometeors (raindrops, snowflakes, sleet, etc.) reaching the earth under the influence of gravity, and also includes deposition by dewfall, fogdrip, rime, etc. Evaporation (E) includes direct evaporation from soil, from surface water, or from the outer surface of plant foliage, as well as transpiration through plant tissue. Runoff (Y) includes river discharge, aquifer discharge, and glacier/iceberg discharge. When averaged over the continents and several years, the combined rate of change in water *storage* – as groundwater, soil water, surface water, snowpack, and ice sheets – can generally be assumed to be small compared with these three major *fluxes*; hence,

$$P = E + Y \tag{7.2.1}$$

The relative stability of global-mean sea level is evidence of the validity of this assumption. Even at much smaller spatial scales, available observations of interannual soil-water and groundwater storage tend to support the use of equation (7.2.1) to describe annual means for temperate land areas. From the large-scale perspective of the global water balance, one of the fundamental problems of hydrology is to describe and explain the relation among these three major fluxes, i.e. the splitting of continental precipitation into evaporation and runoff.

Climatic significance of the partitioning

Evaporation from the continents is a major source of atmospheric water vapor. Water vapor, in turn, is (a) the source of all water clouds and precipitation, (b) the major greenhouse gas, and (c) one of the major vectors for large-scale energy transport in the atmosphere. Runoff from the continents affects ocean salinity and, hence, circulation and large-scale energy transport. It follows from these considerations that the traditional viewpoint of the hydrologist (that, in spite of the acknowledged existence of a hydrologic cycle, surface hydrology is passively forced by climate) is incomplete. In fact, continental hydrology exerts a major feedback on climate. From the viewpoint of the climate dynamicist, possibly the most important hydrologic aspect of the land surface is the proportion in which it splits precipitation into evaporation and runoff.

The observed partitioning

According to one estimate (Budyko and Sokolov, 1978), the land areas of the Earth receive an annual precipitation supply of $119\,000\,\mathrm{km^3}$ (or 800 mm average depth), of which $72\,000\,\mathrm{km^3}$ (485 mm) is returned to the atmosphere by evaporation; the other $47\,000\,\mathrm{km^3}$ (315 mm) runs off to the oceans. Globally, then, the continents return about 60% of precipitation directly to the atmosphere, and discharge about 40% to the oceans.

Of course, this time-mean partitioning varies greatly from region to region. In many desert environments, virtually all precipitation eventually evaporates from land; throughflowing rivers and various forms of water-resource development actually make it possible for evaporation to exceed precipitation. In areas of high orographic precipitation, in the monsoon region of southeast Asia, and in the high latitudes of Asia, on the other hand, runoff typically exceeds evaporation. Furthermore, the annual water-balance partitioning changes not only with locale, but also with time, as natural, interannual variability generates alternating periods of relative excess and shortage of precipitation, leading typically to increases and decreases, respectively, of the fraction of annual precipitation that runs off the continents.

Physical factors affecting the partitioning

Theories and observations of processes at the land surface are sufficiently advanced for us to realize that the partitioning of precipitation into evaporation and runoff depends, to some degree, on an extremely large number of complex and interacting factors. These factors include vegetation characteristics, soil characteristics, precipitation, solar and atmospheric radiation, and the physical state of the atmosphere.

The vegetation characteristics that affect the partitioning include density and arrangement of canopy and groundcover, plant structure, root-density profile, leaf size, shape and orientation, leaf-area index, stem-area index, leaf and stem radiative properties, and density and size of stomatal apertures. Some of these vegetation characteristics vary in response to multiple environmental stimuli over a wide range of time scales, with significant signals not only at the astronomical periods of one day and one year, but also on time scales related to effectively random inputs such as those related to weather. The diversity of species and ecosystems on Earth significantly complicates the identification and description of these characteristics.

Water-balance partitioning is also affected by soil characteristics, including radiative properties, surface roughness, porosity, hydraulic conductivity and water potential as hysteretic functions of water content, thermal conductivity, macroporous features, and vapor and liquid thermal diffusion coefficients. Some of these factors are discussed in Section 4.6. Many of them are highly dynamic, and can be affected profoundly by the periodic changes in soil structure that are driven by freeze/thaw cycles, wetting/drying cycles, and cycles of biological activity.

Finally, several atmospheric and radiative variables affect the annual water-balance partitioning. These variables include intensity and form of precipitation, intensity of solar and atmospheric irradiance, ambient air temperature, humidity, windspeed and pressure. Variability of these factors over time (e.g. diurnal, intraseasonal, seasonal) can be just as important as their annual mean values.

Spatial variability adds another layer of complexity to the water-balance problem. Variability of soil, vegetation, weather and topography together give each physical location a unique identity. Resultant horizontal variability in hydrologic response leads inevitably to lateral water and energy fluxes and interactions among different locations. Atmospheric heat and vapor advection, mesoscale atmospheric circulations, lateral subsurface water flows, and overland water flow are all manifestations of this general principle. These horizontal interactions, in principle, necessitate that complete physically based analyses of such systems recognize the full three-dimensional spatial character of the problem.

In the present context, the problem for hydrologists is to identify which, among all of these contributing factors (and possibly others), are the ones most crucial for the determination of the annual water balance. Ideally, this research problem is pursued through an iterative process of observation, hypothesis generation, and hypothesis testing. The hypotheses that survive the tests and, collectively, have the highest explanatory power may be assembled into a model of the water balance.

Physical control of the partitioning: historical background

Asymptotic relations

Brutsaert (1982, pp. 241–3) provides a brief summary of early work on the problem of the annual water balance. Two important hypotheses underlie much of the work – the balance implied by equation (7.2.1), and a presumption of climatic control; specifically, the supposition that annual runoff is, to a great extent, determined by annual precipitation in any given basin, i.e.

$$Y = P - E = f(P) \tag{7.2.2}$$

Also recognized early was a tendency, within a given basin, for evaporation (E) to approach an empirical constant (potential evaporation, E_p) under conditions of high precipitation and to approach P under low precipitation. Collectively, these generalizations were captured by interpolation equations of the form

$$E/P = g(E_p/P) \tag{7.2.3}$$

in which

$$\lim_{x \to 0} g(x) = x \text{ and } \lim_{x \to \infty} g(x) = 1 \tag{7.2.4}$$

Subsequently, it was recognized that E_p is determined, in large measure, by the net radiation balance of the surface, and could be estimated as the amount of water that would be vaporized by the net surface irradiance (Budyko, 1948). In parallel (Penman, 1948) and later (Budyko, 1951) developments, more complete consideration of the entire energy balance led to refinement in the definition of E_p, as did subsequent introduction of the concept of stomatal resistance to transpiration (Monteith, 1965).

The relation among P, Y, E, and E_p implied above is illustrated by curve C in Figure 7.11, which shows Budyko's (1974, p. 325) empirical fit to a large number of experimental points. The asymptotic limits under extremely arid and humid conditions can be understood physically as situations limited by the annual supply of water and energy, respectively. Annual evaporation approaches annual precipitation in regions where the annual input of energy to the surface (as measured by the potential evaporation) greatly exceeds the amount needed to vaporize the annual precipitation (segment A in Figure 7.11). Conversely, where energy input is a small fraction of the necessary amount, the annual evaporation approaches the annual potential evaporation (segment B in Figure 7.11). Evaporation from most land areas is less than both the water and energy limits, as indicated by curve C. The departure from the asymptotes (about 30% of precipitation)

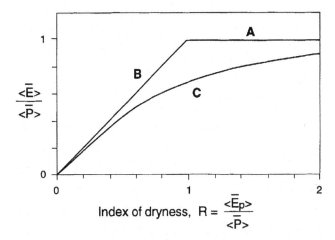

Figure 7.11. Diagram of average annual water balance, showing evaporation ratio as a function of index of dryness. $\langle \bar{E} \rangle$ is evaporation, $\langle \bar{P} \rangle$ is precipitation, and $\langle \bar{E}_p \rangle$ is potential evaporation. Curve C is Budyko's (1974) relation, and segments A and B are its asymptotes.

is large in the region where the index of dryness (ratio of time-space mean of E_p to P) is approximately unity.

Intermediate conditions

A physical theory describing the departure of curve C from asymptotes A and B in Figure 7.11 has not been advanced in the hydrologic literature. One might ask why this departure exists at all, or why it is not larger. Scatter around curves such as C is considerable (Budyko, 1974, p. 326). Are there factors other than the index of dryness that affect the water balance in a systematic way? Indeed, Budyko and Zubenok (1961) have noted that deviations from curve C tend to be positive when precipitation and potential evaporation have seasonal variations that are in phase, and negative when they are out of phase. Quantification and physical explanation of this and other effects on annual water balance have not been given.

Asymptotes A and B define the upper water and energy limits on evaporation and, therefore, the minimum possible runoff from a basin. The task that remains, therefore, is to explain runoff in excess of this minimum. Two related characteristics of land surfaces – finite water storage capacity and finite permeability – may be identified as possible causes of such additional runoff. For example, if the water storage capacity of soil is too small, temporary excesses of water supply will be lost as runoff, even though the index of dryness of the area may exceed 1. Finite-permeability effects enter in two main ways, both of which tend to increase runoff at the expense of evaporation. If precipitation rates exceed the rates

at which water can infiltrate the soil, then runoff will occur, regardless of the long-term water and energy supplies. Secondly, if potential evaporation rates exceed the rates at which water within the root zone can flow the short distances to the plant roots (or to the surface of bare soils), then evaporation may fall below its water- and energy-supply limits.

The relative importance of finite-capacity and finite-permeability factors in determining the annual water balance has not been analyzed in the literature. Eagleson (1978a,b,c) formulated the water-balance problem in terms of the Richards equation for soil water (Section 4.5), allowing the soil permeability to enter the problem. The only allowance for seasonality of climate was the definition of wet and dry seasons, with no water fluxes during the latter; interseasonal water storage was effectively neglected; and intraseasonal (i.e. within-season) storage was implicitly assumed to be unlimited by any capacity. The relation of Eagleson's work to the empirical work described above has not been explored.

Some aspects of the problem of annual water balance, including seasonality, have recently been analyzed in the framework of simple models by Schaake and Liu (1989) and by Dooge (1992). They used single-store water-balance models having a small number of empirical parameters meant to represent unresolved physical processes, including spatial and intraseasonal variabilities. These studies were carried out in a spirit of simplicity and exploration, and it is understandable that some crucial aspects are questionable. In particular, both studies assumed that outflow from the reservoir of water available for evaporation is a linear function of storage and is observable in stream discharge. Physical theory and field evidence suggest, however, that the plant root zone drains in a highly nonlinear way, and that stream baseflow is more closely tied to water storage in the saturated zone, which has a longer time constant. Given the strong nonlinearity of root-zone drainage, it is arguably better described by the threshold concept of a soil-water field capacity.

Other recent work (Milly, 1993, 1994a,b) has explored the possibility that annual water balance may be explained mainly by a finite-capacity model, with no restriction on infiltration. The working hypothesis was that water balance could be described as the simple interaction of water supply, demand, and storage in a finite soil-water reservoir. Milly (1993) treated the special case where storminess (that is, random, intraseasonal variability) of precipitation is the only form of hydrologic variability over time, and Milly (1994a) dealt with the case where seasonality is the only source of variability over time. In neither case did it appear possible to explain all of the observed large-scale runoff. Milly (1994b) included variabilities in time associated with both seasonality and storminess, as well as spatial variability of the storage capacity. The resulting model subsumed the models of Milly (1993, 1994a) as special cases, and appeared capable of explaining most of the large-scale geographic variability of mean water balance over a large study area. By explicitly resolving intraseasonal, interseasonal, and spatial variabilities, the approach avoided the introduction of empirical parameters. The conceptual simplicity of the approach facilitated the non-dimensionalization of the problem even in its most general case. The model and test of Milly (1994b) are summarized in the next section. In the interest of brevity, the summary neglects many of the finer points, for which the reader may refer to the original paper.

A model based on the supply–demand–storage hypothesis

Model formulation

It is hypothesized that the local, annual water balance is controlled by the distributions in time of water supplies (precipitation) and demands (potential evaporation), which are balanced, to the extent possible, by storage of water in the root zone of the soil. It is assumed that the vegetation cover is sufficiently extensive that direct evaporation from the soil need not be considered. The storage capacity of the root zone is characterized using conventional concepts of soil water availability to plants. It is assumed that permeability is sufficiently large that infiltration of water into the soil and uptake of available soil water by roots are unrestricted. The hypothesis can be stated mathematically as a simple water-balance model driven by time series of precipitation and potential evaporation:

$$\frac{dw}{dt} = \begin{cases} 0 & p > e_p \text{ and } w = w_0 \\ 0 & p < e_p \text{ and } w = 0 \\ p - e_p & \text{otherwise} \end{cases} \tag{7.2.5}$$

where p and e_p are local, instantaneous rates of precipitation and potential evaporation, respectively, and w is plant-available soil water (volume per unit area), which can range from a minimum of 0 to a maximum of w_0. The plant-available water-holding capacity, w_0, is a model parameter. Conceptually, it is given by an integral over soil depth,

$$w_0 = \int_0^\infty [\theta_f(z) - \theta_w(z)] r(z) \, dz \tag{7.2.6}$$

where θ_f and θ_w are the volumetric moisture contents of the soil at field capacity and at the wilting point, respectively, and the variable $r(z)$ denotes the fraction of area at depth z that is affected by the root system of the vegetation.

The precipitation is treated as a random process. It is

assumed that precipitation arrives in discrete events that we shall call storms, that the arrival of these storms in time is a Poisson process, and that the amount of precipitation in any storm is governed by the exponential distribution (Benjamin and Cornell, 1970). The expected value of $p(t)$ is given by $P(t)$, and the mean storm arrival rate is $N(t)$. To represent seasonality, it is assumed that $P(t)$, $N(t)$, and $e_p(t)$ can be described adequately by their annual harmonics,

$$P(t) = \bar{P}(1 + \delta_P \sin \omega t) \tag{7.2.7}$$
$$N(t) = \bar{N}(1 + \delta_N \sin \omega t) \tag{7.2.8}$$
$$e_p(t) = \bar{E}_p(1 + \delta_E \sin \omega t) \tag{7.2.9}$$

where δ_P, δ_N and δ_E are the ratios of the amplitudes of the annual harmonics to the annual averages $\bar{P}$, $\bar{N}$ and $\bar{E}_p$, respectively. With $2\pi/\omega$ equal to 1 year, these expressions capture the essential features of the annual land-surface hydrologic forcing outside the tropics. Near the Equator, the noon sun passes overhead twice per year, and these representations may be used with a period of one-half year.

Soil hydraulic characteristics vary greatly at relatively small scales (Warrick and Nielsen, 1980). The nonlinear dependence of water balance on w_0 suggests the need for explicit consideration of spatial variability (Milly and Eagleson, 1987). It is assumed here that the frequency distribution of water-holding capacity within the area of interest, $f_w(w_0)$, is given by the gamma distribution,

$$f_w(w_0) = \frac{\lambda(\lambda w_0)^{\kappa-1} e^{-\lambda w_0}}{\Gamma(\kappa)} \tag{7.2.10}$$

where the mean value of w_0 is given by κ/λ, and its coefficient of variation is given by $\kappa^{-1/2}$. The spatial mean (denoted by angle brackets) of any function $Z(w)$ (such as evaporation or runoff) is found by integrating over the density function (7.2.10), using the relation

$$\langle Z \rangle = \int_0^\infty Z(w) f_w(w) \, dw \tag{7.2.11}$$

Solution

For certain special cases and ignoring spatial variability, the water-balance partitioning was determined analytically using equation (7.2.5), subject to equations (7.2.7), (7.2.8) and (7.2.9), by Milly (1993, 1994a). Milly (1994b) extended those analytic solutions to include spatial variability, and presented a Monte-Carlo solution valid for the general problem. The general solution, in dimensionless form, can be expressed as

$$\frac{\langle \bar{E} \rangle}{\langle \bar{P} \rangle} = G(R, \langle W \rangle, \bar{N}\tau, \delta_P, \delta_N, \delta_E, \kappa) \tag{7.2.12}$$

where τ is $2\pi/\omega$, R is a climatological index of dryness analogous to that of Budyko (1974),

$$R = \overline{E}_p / \overline{P} \qquad (7.2.13)$$

and $\langle W \rangle$ is given by $\langle w_0 \rangle / \overline{P}\tau$. Thus, the time-mean partitioning of precipitation into runoff and evaporation is determined by seven dimensionless numbers. The first of these, the ratio of annual potential evaporation to annual precipitation (index of dryness), has long been recognized as an important determinant of the annual water balance, as seen already in equation (7.2.3). The second factor is the ratio of water-holding capacity to annual mean precipitation amount; large values of the ratio tend to promote evaporation and suppress runoff. The third factor is the mean number of precipitation events per year; for the same amount of annual precipitation, a few heavy rainfalls will produce more runoff than many small ones. Three more factors are the ratios of seasonal fluctuations to annual means of precipitation, storm arrival rate, and potential evaporation. In general, seasonality tends to generate imbalances in water supply and demand, leading to increased runoff, consistent with the empirical conclusion of Budyko and Zubenok (1961). The final factor is a measure of the spatial variability of water-holding capacity.

The general solution (7.2.12) is plotted in the upper panel of Figure 7.12 for representative values of the input parameters. The choice of $\delta_E = 1$ implies a mid-latitude location (where seasonality of E_p is strong), and the choice $\delta_P = \delta_N = 0$ (no precipitation seasonality) is an average for mid-latitudes, where precipitation may have either winter or summer maxima. Representative values for annual-mean storm arrival rate, annual-mean precipitation rate, and water-holding capacity are 100 yr^{-1}, 1000 mm yr^{-1}, and 200 mm, respectively. These give a typical value for $\langle W \rangle$ of 0.2; values of 0.1 and 0.4 are also included for comparison. Corresponding special-case solutions for no seasonality and large storm arrival rates are also plotted in the middle and lower panels, respectively.

In the general case (top panel of Figure 7.12) without spatial variability, a dimensionless storage capacity of 0.4 is almost large enough to absorb all fluctuations over time of water and energy supply; the evaporation curve departs from the asymptotes A and B of Figure 7.11 only slightly near $R = 1$. Halving the value of W from 0.4 to 0.2 (or from 0.2 to 0.1) yields a decrease of evaporation, and increase of runoff, on the order of 10% of precipitation for R greater than about 0.8. The case of strong spatial variability of w_0 is considered with $\kappa = 1$. Because an increase in w_0 enhances evaporation less than an equal decrease of w_0 reduces it, it follows that increasing dispersion of w_0 around some fixed mean decreases the mean evaporation and increases runoff. Budyko's (1974)

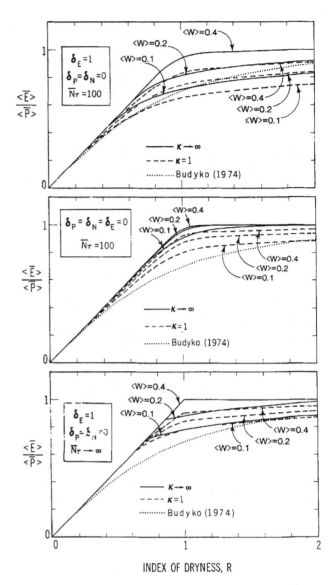

Figure 7.12. Evaporation ratio as a function of index of dryness, for various values of mean dimensionless storage capacity. Solid lines for $\kappa \to \infty$ (no spatial variability of w_0) and dashed lines for $\kappa = 1$ (exponential distribution of w_0). Top: Episodic precipitation and annual cycle of potential evaporation considered. Middle: Same parameters as top, but no annual cycle ($\delta_E = 0$). Bottom: Same as top, but precipitation not episodic ($\overline{N}\tau \to \infty$).

curve lies among the illustrative cases of the general solution in Figure 7.12.

Comparison of the general and special cases illustrated in Figure 7.12 leads to the conclusion, for the chosen parameter values, that both interseasonal and intraseasonal variabilities of forcing contribute to reductions of annual evaporation below its energy- and water-supply limits.

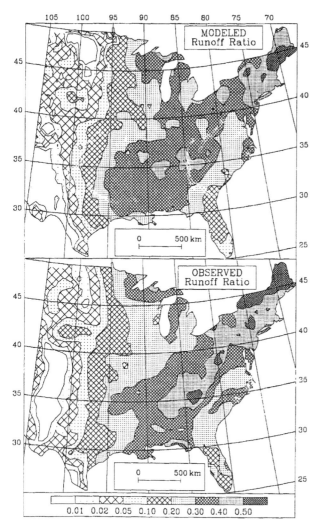

Figure 7.13. Distributions of modeled (top) and observed (bottom) runoff ratios.

A test of the model

The hypothesis of storage-control of the average annual water balance was tested by applying the derived water-balance model to the area of the United States east of the Rocky Mountains; this domain was chosen to provide a wide range of climatic conditions and to avoid the observational problems associated with the complex, mountainous topography west of the region. For each point on a 0.5-degree grid, the seven independent variables in equation (7.2.12) were estimated *a priori* on the basis of published data and methods; there were no free parameters for calibration. The precipitation parameters $\bar{P}$ and δ_P were estimated by Fourier analysis applied to the monthly climatological fields produced by Legates and Willmott (1990a). Storm arrival-rate

parameters $\bar{N}$ and δ_N were estimated from moments of daily precipitation records of first-order weather stations and interpolated to the 0.5-degree grid. Monthly values of potential evaporation were estimated from monthly temperature fields of Legates and Willmott (1990b), using the method of Thornthwaite (1948); resulting values were scaled uniformly by a factor of 1.2, to correct an apparent bias noted by Milly (1994a). Fourier analysis of the scaled monthly values of potential evaporation yielded values of $\bar{E}_p$ and δ_E. The plant-available water-holding capacity of soils, w_0, was estimated with global coverage at 0.5-degree resolution by Patterson (1990), using available global data sets for soils and vegetation. Patterson's (1990) treatment of rooting depths was necessarily highly simplified because data are scarce, and the error of estimation of w_0 is probably large. Detailed soil-survey information from selected areas in the study region were considered in arriving at a nominal value of 10 for κ; sensitivity analyses indicated that the exact value of this parameter was not critical.

Annual runoff was calculated by means of equation (7.2.12), and observed runoff was obtained from the map of Gebert *et al.* (1987). These data are plotted as runoff ratios (ratio of runoff to precipitation) for comparison in Figure 7.13. The geographic distribution of the calculated runoff ratio shares, at least qualitatively, the large-scale features apparent in the observations. In both computations and observations, runoff ratio is the lowest in the western region, where its gradient is predominantly east–west. The computations also reproduce the large-scale band of high observed runoff ratio extending northeastward from the northern coast of the Gulf of Mexico, through the Appalachian Mountains, and into the northeastern United States.

In the west, the major departure from the prevailing east–west gradient in the observations is found in the northern plains, where the westward bulge in the 0.05 runoff-ratio contour marks the area of anomalously high runoff in the Sand Hills of Nebraska. The anomaly is soil-related, rather than climate-determined; the sandy texture of the soil there inhibits water retention. Because the data set of Patterson (1990) recognizes the effect of sandy soil texture on depressing the value of w_0, the observed pattern of runoff ratios in the Sand Hills area is reproduced well by the model.

The runoff error is positive over much of the south-central part of the study area, and negative over much of the Appalachians. Regional errors in potential evaporation and storage capacity are possible causes of these errors. It appears that Patterson's (1990) estimates of w_0 fail to include the inhibiting effect of mountain topography on soil development, at least in the study area, and this may explain the runoff deficit in the Appalachians.

Mean values of P, Y, and E estimated from observations over the study area are 991 mm, 263 mm, and 728 mm; mean modeled values of Y and E are 250 mm and 741 mm. The area-weighted, root-mean-square difference between observed and modeled values of 0.5-degree runoff (and, identically, evaporation) is 78 mm. The correlation between 0.5-degree gridpoint values of observed and modeled runoff was 0.938; the model thus explained 88% of the variance in observed gridpoint runoff. For evaporation, the correlation coefficient was 0.924; the model explained 85% of the variance. On the basis of a simple error analysis, Milly (1994b) concluded that differences between modeled and observed water balances could be explained entirely by input errors and errors in the field of observed runoff. In the absence of more precise estimates of these variables, there is no compelling reason, on the basis of the present comparison alone, to reject the hypothesis underlying the model.

Discussion

The agreement between model and observations in the study just described supports the assumptions underlying the theory, in particular the hypothesis of storage control of the water balance. The differences between model and observations were shown to lie within the range that could be expected given the uncertainties in potential evaporation, precipitation, and especially water-holding capacity of the root zone of the soil. Of course, these uncertainties leave open the possibility that the annual water balance could be affected by factors not included here, such as restriction of rainwater infiltration by limited soil permeability and restriction of transpiration by moisture diffusion toward plant roots. It can only be concluded that it is not necessary to invoke those factors to explain the observational data.

Milly (1994b) used the storage model to examine the factors most important in determining the water-balance partitioning over the study region. In the framework of the storage model, the total annual runoff can, in general, be partitioned into runoff associated with the imbalance between long-term means of precipitation and potential evaporation, runoff caused by differing seasonalities of precipitation and potential evaporation, runoff caused by storminess of precipitation, and runoff caused by the interaction of storminess and seasonality. In the areal mean over the study region, the term associated with annual mean water and energy supplies was the largest, but no term was negligible. The terms associated with fluctuations of the forcing became increasingly important under increasing aridity. The effect of local spatial variability of the soil water-holding capacity on mean runoff (and evaporation) was negligibly small.

If the conceptual model of Milly (1994b) is at all representative of real sensitivities of water balance to its controlling factors, then model-testing with existing large-scale data sets has serious limitations. From a data standpoint, the weak points in the analysis are the values estimated for precipitation, potential evaporation, and storage capacity. It appears that the uncertainties associated with these factors may be at least as large as the differences among conceptual models of water balance, making the rejection of any particular hypothesis an improbable event. Without better estimates of these very basic quantities, it is doubtful whether significant advances can be made in the scientific analysis of the annual water balance at large spatial scales. In particular, it should be noted that the determination of potential evaporation was placed outside the scope of the study of Milly (1994b). Its proper estimation requires consideration of the energy balance at the land surface and of the non-water-stressed resistance of plants to water loss from the surface.

7.3 The water budget of a middle latitude continental region: a modeling and observational study

Peter Rowntree

Introduction

In this section, an attempt is made to use ground-based data to validate the modeled surface water budget in a region of Russia where the UK Meteorological Office Hadley Centre general circulation model simulates too dry a climate in summer. Observed data will be shown to have significant imbalances, and so be of limited value for use in the budgets needed for validation of models. However, they do indicate that model soil moisture is too low in spring, leading to unrealistic drying out in summer. It will be argued that a likely cause is lack of precision in definition of the soil characteristics, together with uncertainties in derivation of the model's soil parameters; the specification of the root depth may also contribute.

An early and important finding from climate models concerning the effects of increasing greenhouse gases was that the middle latitude continents would tend to have lower soil moisture in summer. For example, Mitchell *et al.* (1990) showed widespread decreases in soil moisture in summer with the three high resolution models used for the main assessment. The confidence attached to this prediction in the 1990 IPCC (Intergovernmental Panel on Climate Change) Report was not high (2 stars out of a maximum of 5); one reason for this was that the simulations even of zonally averaged soil moisture differed greatly from each other (Gates *et al.*, 1990). The IPCC Report also included a validation of the soil moisture over part of Russia, for three high resolution models. This validation was against the data of Vinnikov and Yeserkepova (1991) (V&Y hereafter) who summarized seasonal trends in soil moisture, mainly over the period 1972–85, using the large collection of data available in the former USSR. The Hadley Centre (HC) model was shown to be close to the observations in that region (Gates *et al.*, 1990).

The immediate objective of this section is to assess the feasibility of using ground-based data for validating the surface water budget of later versions of the HC model in which the soil moisture in this region is deficient in summer, as it was in the GFDL model assessed by Gates *et al.* (1990). The long-term aim is to improve the realism of the simulation of the surface hydrology in climate models. The model's simulation is compared against observations for part of Russia, using V&Y's data discussed above.

The model

Since the assessment reported by Gates *et al.* (1990), the HC model has been considerably revised; though the physics were largely unchanged, a new dynamics scheme was introduced, giving more realistic levels of eddy activity, and the simulations are now substantially different in such features as eddy activity and temperature and precipitation distributions. The versions of the model used here are similar to those described by Jones *et al.* (1995). Two versions of the model are considered. The second version (M2) differs from the first (M1) in the following respects:

reductions were made in the horizontal diffusion of atmospheric specific humidity, potential temperature and horizontal wind,
modifications (decreases) were made to the evaporation of falling precipitation (correction of the model formulation),
the cloud scheme was changed, leading to increases in summer cloudiness.

The need for a reduction in diffusion was suggested by earlier reductions in moisture diffusion which were found to improve the simulations especially near mountains, with less intense orographic maxima and increased precipitation over adjacent lowlands. The changes in evaporation of precipitation gave some increases in summer rainfall. The cloud scheme was changed because the scheme in M1 was found to give unrealistically small cloud amounts in short-range predictions. This change was not wholly beneficial; in particular it increased top-of-atmosphere albedos particularly over high latitude continents giving substantially higher values than observed (Hall, 1995).

The land surface hydrology parameterization in the model is relevant to the subsequent assessment, so it is useful to describe it in more detail. This version of the model has a single layer, depth equal to the estimated root depth D for the vegetation type specified for that point; these depths range between 0.5 m and 0.8 m in the region of Russia considered here. Surface runoff occurs when the precipitation rate exceeds the saturated hydraulic conductivity of the soil, after allowance is made both for enhancement of the conductivity by vegetation (effects of leaf litter, decayed roots etc.) and also an assumed exponential frequency distribution of the gridbox mean precipitation. Subsurface runoff or percolation through the base of the root zone is calculated using Darcy's equation but with a zero gradient in soil moisture and vertical homogeneity of soil characteristics assumed at the base of the layer so that the soil suction term in Darcy's equation vanishes. The drainage out of the layer is then assumed to be

$$K = K_s \, s^c \tag{7.3.1}$$

where $s = (\theta_A - \theta_{Aw})/(\theta_{As} - \theta_{Aw})$; here θ_A is the available soil moisture equal to $(\theta - \theta_r)$ where θ is total soil moisture and θ_r the soil moisture not available for drainage; θ_{As} and θ_{Aw} are the saturation and wilting values of θ_A, K_s is the saturation value of the hydraulic conductivity and c is an exponent. As discussed by Warrilow *et al.* (1986) who based this fit on a review of the literature, there is a wide range of values for these soil characteristics as well as a number of possible formulations of equation (7.3.1) reported in the literature, no doubt due partly to the difficulty of fitting profiles such as equation (7.3.1) to the observed data.

Two examples of alternative formulations are discussed later in the section. In the Clapp–Hornberger (1978) equations, hydraulic conductivity K is derived using the simple exponential relation

$$K = K_s \, (\theta/\theta_s)^c \tag{7.3.2}$$

This is as equation (7.3.1) except that here θ is the total soil moisture content and θ_s is the saturation value; the exponent c is generally assumed to be related to the slope b of the moisture retention curve by a relation of form

$$c = 2b + 2 + J \tag{7.3.3}$$

in which Clapp and Hornberger (1978) assume $J = 1$.

The second example is the Mualem–van Genuchten relation (van Genuchten, 1980) according to which

$$K = K_s \, S^J \, [1 - \{1 - S^{1/m}\}^m]^2 \tag{7.3.4}$$

where $S = (\theta - \theta_r)/(\theta_s - \theta_r)$. In terms of the parameters used in equations (7.3.2, 7.3.3), assume

$$m = 1/(1 + b) \tag{7.3.5}$$

in which case (7.3.4) becomes of similar form to (7.3.2) and (7.3.3) as S becomes small, albeit with a different definition of S.

Experiments

The two experiments E1 and E2, using respectively M1 and M2, were run for 20 years and for 10 years 3 months respectively, using climatological sea surface temperatures and sea ice, starting from an analysis made using the Met. Office Unified Model and assimilation system for early June. The initial soil moistures were taken from a simulation using the previous version of the Met. Office climate model. The results discussed here are the means for the full duration of the experiments except the first 3 months of E2. Detailed monthly mean data were extracted for gridpoints at 2.5° latitude 3.75° longitude intervals on a section from 60° N 33.75° E to 47.5° N 52.5° E (Figure 7.14e), through an area of Russia relatively well-covered by V&Y's data and with a significant tendency to excessive summer soil dryness in the model. The discussion in this section focuses on the water budget at 55° N 41.25° E, with only limited discussion of the other points.

The surface water budget and availability of observed data

The soil moisture balance may be written:

$$d(SM)/dt = R + M_s - E_w - Y \tag{7.3.6}$$

Here, the terms are the snowmelt (M_s), the rainfall (R), surface runoff and drainage from the soil layer considered (Y), evaporation of water (E_w) and change in soil water content (SM).

All the terms in equation (7.3.6) are available from the model. However, for real data, the situation is less favorable. In particular, the data on rainfall, as distinct from snowfall (S), and for snowmelt are not readily available. Outside the snow season however, equation (7.3.6) can be simplified to:

$$d(SM)/dt = P - E - Y \tag{7.3.7}$$

where E is total evaporation. This equation can also be used for the annual mean water balance since in the average between times with no snow cover,

$$S = M_s + (E - E_w) \tag{7.3.8}$$

An assessment is made below for each term in turn of the data which were readily available for this study.

Soil moisture change (d(SM)/dt)

V&Y have published data on soil moisture in the top meter of soil for about 50 Russian soil moisture stations; these are all west of 105° E and south of 60° N. apart from three north of 60° N west of the Urals. Most are for the period 1972–85. The data are means over the period for the 28th of most months, though with limited data from snow-covered periods of the year. Other Russian data have been used by Robock *et al.* (1995).

Precipitation (P)

The precipitation data for 1972–85 were available for a rather sparse network of stations from the archives maintained in the Hadley Centre. These are based mainly on the data

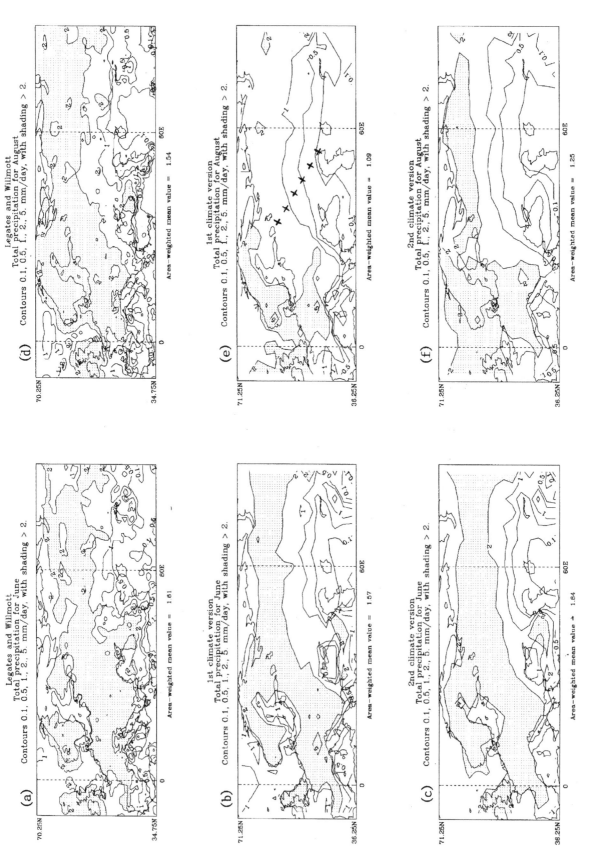

Figure 7.14. Precipitation for June and August as estimated from observations (*a*, *d*) by Legates and Willmott (1990*a*) and simulated by the HC model in experiment E1 (*b*, *e*) and E2 (*c*, *f*). Contours at 0.1, 0.5, 1, 2, 5 mm day⁻¹; stippled above 2 mm day⁻¹. The locations of the model gridpoints studied are shown (X) on (*e*).

received on the GTS (Global Telecommunications System), supplemented where necessary by other sources such as World Weather Records. However, the rainfall stations mostly do not coincide with the soil moisture stations. A further problem is that it is not known to what extent these data have been corrected for errors inherent in the measurement techniques (Groisman *et al.*, 1991). These errors are most serious during snowfall; corrections to the annual mean data can amount to an enhancement of several tens of percent. An alternative, adopted here, is to use corrected data which are available for stations in histogram form, as annual totals and monthly percentages, and also as a mapped analysis of annual totals, in the World Water Balance Atlas (Korzun *et al.*, 1978) for the period 1891–1970.

Surface runoff and drainage (Y)

An analysis of river runoff data is available in Korzun *et al.* (1978). This is in the form of a mapped analysis of annual totals and monthly percentages in histogram form. The histogram data, being for catchments, are not colocated with either the soil moisture or rainfall stations, but do have quite a dense coverage over European Russia. However, as discussed later, the river runoff cannot be expected to equal the runoff Y at the surface and from the root zone, as there are often intermediate storage reservoirs such as groundwater.

Evaporation (E)

Korzun *et al.* (1978) also map annual evapotranspiration; the seasonal variation is shown in histogram form, but only for a few points (in the area of our interest for Moscow, Helsinki and Charkov (50° N, 36° E)).

Discussion

To calculate a soil moisture budget including seasonal variation, knowledge of three of the four terms in equation (7.3.6) would suffice. This is only possible with the available data if we use the data from Korzun *et al.* (1978). However, for validation purposes, we need a balanced annual budget. Korzun *et al.* do not provide this, having made estimates of precipitation, evaporation and runoff which are to some extent independent – although they do state that the evaporation calculations use the Budyko complex method, in which monthly evapotranspiration is computed from potential evaporation, and that precipitation values may be used in the checking process. The lack of balance is evident from examination of, say, runoff at points where precipitation and evaporation contours intersect. Nevertheless, the data have been

Table 7.1. *Annual soil water budget for 1891–1970 from* Korzun et al. *(1978) (mm yr⁻¹) (also gauged precipitation from World Weather Records for 1891–1970)*

Point	P	E	Y	$P-E-Y$	P_{gauge}
60, 33.75	780	400	290	90	556[a]
57.5, 37.5	730	470	220	40	
55, 41.25	705	520	140	45	597[b]
52.5, 45	610	480	105	25	
50, 48.75	370	335	15	20	
47.5, 52.5	250	250	0	0	
55.6, 37.5	700	510	180	10	597[b]

Note: [a] Leningrad (St. Petersburg) (59.9° N 30.3° E)
[b] Moscow (55.6° N 37.5° E)

used, albeit with caution, on the assumption that the data are as good as are available.

Table 7.1 shows the budgets extracted for the model section. Note that there are inevitably uncertainties, and so errors, in reading data from the maps. Assuming smooth variations between contours, the estimated uncertainties are about 20 mm for E and P, 10 mm for Y, in each case 20% of the contour interval. There are substantial discrepancies as expected from the above discussion. All are positive and so could be lessened by reducing P towards the gauged precipitation data such as that shown in Table 7.1 for Leningrad (St Petersburg) against the 60° N point and Moscow (55° N). This suggests that the positive corrections applied to the gauged data may be too large. Also shown in Table 7.1 are estimated values for Moscow itself from the maps in Korzun *et al.* (1978). The balance is better here, and as the monthly distribution is also available for P and E we will use these data in the comparison of modeled and observed climates. The gauged precipitation for Moscow for 1972–85 in the Hadley Centre archives is 699 mm, close to the 700 mm yr⁻¹ shown by the Korzun *et al.* map, and well above the mean for 1891–1970 mean of less than 600 mm yr⁻¹ from World Weather Records; whether this is a real increase or one due to changes in site, measurement methods or application of corrections is beyond the scope of this section (but see Groisman *et al.*, 1991). For Leningrad, the mean for the 1972–85 period was 622 mm, again showing a substantial increase from 1891–1970, but well below the mapped value of 800 mm in Korzun *et al.*

Overall validation of model simulation over Russia

Rainfall distribution

For an assessment of the modeled surface water balance against observations to be useful, we require realistic atmos-

Table 7.2. *Modeled values of the ratio of latent heat flux to the sum of sensible and latent heat flux for 57.5° N 37.5° E and 55° N 41.25° E from May to August*

Experiment	May	June	July	August
57.5° N 37.5° E				
E1	0.727	0.744	0.701	0.547
E2	0.701	0.765	0.796	0.739
55° N 41.25° E				
E1	0.718	0.575	0.277	0.259
E2	0.686	0.740	0.567	0.323

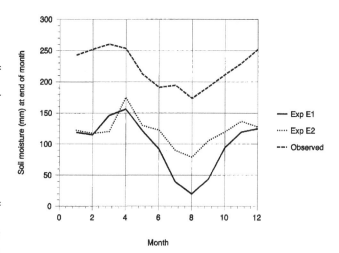

Figure 7.15. Available soil moisture (mm) near end of month (1 = January, 2 = February, etc.) as simulated at 57.5° 37.5° E for Experiments E1 and E2, and as observed (average of Stations 5 and 10 from V&Y – about 58° N 28° E, 57° N 44° E).

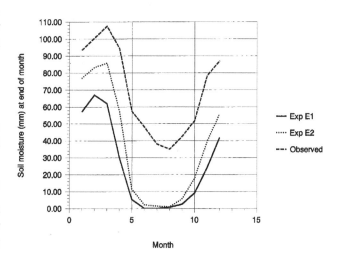

Figure 7.16. As Figure 7.15 for model at 50° N 48.75° E and observed (average of Stations 33 and 35 from V&Y – about 52° N 51° E and 49° N 46° E).

pheric forcing. One of the most important components is the precipitation. However, because of the feedback from evaporation through the atmospheric humidity, errors in the precipitation can be a symptom of faults in the surface parameterization. This is most likely to be the case during summer when the actual or modeled evaporation is limited by soil moisture. Figure 7.14 illustrates the development of errors in modeled precipitation between June and August. Rainfall in June (and July, not shown) is quite realistic near 50–70° N, while in August it is clearly deficient, particularly in E1.

Partitioning between evaporation and sensible heat flux

The modeled rainfall decline in late summer could be due partly to land surface feedbacks. This possibility is indicated by a corresponding decline in the evaporative fraction

$$\mu = (LE/(LE+H)) \tag{7.3.9}$$

(where L is the latent heat coefficient and H is sensible heat flux) at 57.5° N and 55° N in Experiment E1 (Table 7.2). For relatively moist conditions where soil moisture is not limiting evaporation, μ has an asymptotic value which is reached in various limiting conditions including that of large values of the atmospheric resistance (r_A) (Rowntree, 1991). This asymptotic value increases with temperature from about 0.41 at 0 °C to 0.69 at 20 °C and 0.79 at 30 °C. With typical summer temperatures near the surface of 20 °C near 55° N, a value of about 0.7 is associated with soil moistures near and above the level at which they cease to limit evaporation. The decline in μ is less for Experiment E2 than for E1 (Table 7.2), though still marked at the southern point in August.

Soil moisture

Comparison of the modeled soil moisture (Figure 7.15) with observed values from V&Y for near the point at 57.5° N sug-

gests that E1 is also particularly dry in late summer there. This deficiency appears even more serious in similar diagrams for 55° N (shown later in Figure 7.20) and 50° N (Figure 7.16) with available soil moisture almost exhausted by August. To investigate this in more detail, the water balance for the 55° N point is studied in the following subsections.

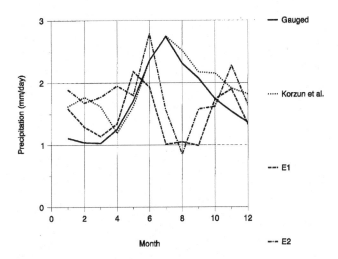

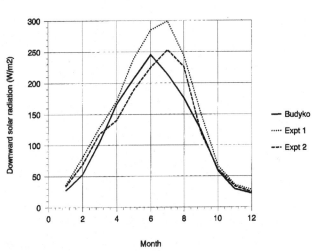

Figure 7.17. Precipitation (mm day⁻¹) as simulated at 55° N 41.25° E for Experiments E1 and E2, and as observed near Moscow (55.7° N 37.6° E) as gauged (1921–1990) and as estimated from Korzun *et al.* (1978).

Figure 7.18. Downward solar radiation (W m⁻²) for model at 55° N 41.25° E and for observed near Moscow from Budyko (1963).

Validation of modeled surface fluxes for 55° N

Precipitation

Figure 7.17 compares the monthly precipitation for the two experiments with the observed for the 55° N point. The precipitation used here is that for Moscow from Korzun *et al.* (1978). There is little difference (< 1%) between the annual precipitation at Moscow and at the 55° N gridpoint. Note that the gauged data, also shown in Figure 7.17, do not differ systematically from the Korzun *et al.* estimates, based on data corrected for exposure, in April to September – i.e. outside the snow season. Figure 7.17 suggests that E1 is too dry relative to the corrected data in the annual mean, with deficient snowfall in winter and too little rainfall from late June to September. E2 has more precipitation in most months but is still too dry in July to September. These differences are mostly consistent with the general character of the differences between the two versions expected from the changes in model formulation.

Radiation

A comparably important component of the forcing is the net radiation which provides the energy for the turbulent surface fluxes of heat and moisture. Observed data on net radiation are less plentiful and less reliable than data for downward solar radiation so we compare the latter with observations (Figure 7.18), using data for Moscow from Budyko (1963).

These relatively early observations, averaging 119 W m⁻², are consistent with an annual mean of 114 W m⁻² quoted by Garratt (1994) from more recent data. Figure 7.18 shows that the observed downward solar radiation lies between the two model versions from April to June but is otherwise less than either. The largest overestimates are in late summer and can be attributed to the drying of the modeled climate at this time, with consequent underestimate of cloudiness in both versions. E1 is sunnier than E2 throughout the year, consistent with the previously noted tendency for less cloudiness; annual means are 147 and 124 W m⁻² respectively. In the comparison of general circulation model and observed data by Garratt (1994), three of the four models reported excessive values for Moscow, ranging from 130 to 146 W m⁻²; the fourth was an early version of the Met. Office model, which gave an underestimate.

Evaporation

The evaporation has been compared with the data in Korzun *et al.* (1978) (World Water Balance, WWB) to provide some assessment of the model simulations (Figure 7.19). E1 agrees with the WWB estimates in spring but falls well below it from late June to September. E2 on the other hand is deficient in spring and, though better than E1, still deficient in summer. This difference between E1 and E2 in spring is consistent with the differences in downward solar radiation (Figure 7.18). Although apparently superior in spring, the large decline in evaporation in June in E1, before there is any serious decline in rainfall, suggests that the soil moisture simulation is not realistic. We consider this below. Also shown in Figure 7.19 are estimates by Budyko (1963) based on a combined heat

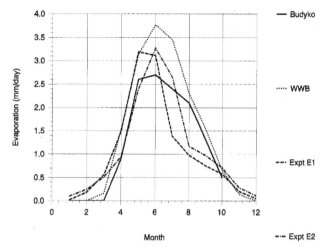

Figure 7.19. Surface evaporation (mm day⁻¹) for model at 55° N 41.25° E and as observed near Moscow from Korzun *et al.* (1978) (WWB) and Budyko (1963).

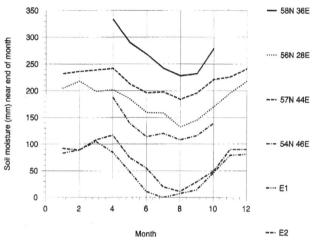

Figure 7.20. Average soil moistures (mm) near the end of each month for the model at 55° N 41.25° E and for the 4 stations from V&Y listed in Table 7.3.

and water balance approach. The annual total is much closer to the model estimates, but the shape of the seasonal variation is similar to the estimate by Korzun *et al.*, showing no evidence of drying of the soil in late summer on the scale found in the model. The lesser annual total evaporation is likely to be linked through the water balance calculation with the use of lower estimates of the precipitation based on gauge data available around 1960 (see earlier).

Discussion of Henning (1989)'s data

Estimates of radiation and turbulent heat flux and runoff quantities could also be extracted from the maps of the heat balance in Henning (1989). For 55° N 40° E, he estimates an annual evaporation of just over 350 mm, lower than Budyko (400 mm) and Korzun *et al.* (510 mm) and also the model (E1: 380; E2: 405). He maps only selected months; compared with Budyko's data, his figures are higher in July (at 2.7 mm day⁻¹), similar in August, and about 40% lower in April and October. He estimates runoff, using a water balance calculation, at about 190 mm yr⁻¹, a similar figure to Korzun *et al.* (1978); this suggests he has used a much lower estimate of precipitation (about 550 mm yr⁻¹), close to the gauged values for Moscow for 1921–50 in World Weather Records.

Soil moisture

The V&Y data do not include a station close either to Moscow or the 55° N 41.25° E point. The following analysis therefore uses an average of the nearest available data to assess the model. Figure 7.20 compares the seasonal variation in soil moisture for the experiments at this point with that in the top meter at four soil moisture stations in V&Y (Table 7.3) which are close to the same latitude belt and reasonably close in longitude.

The shape and magnitude of the model variations are generally similar to the observed, but the mean level is much lower as shown earlier for 57.5° N and 50° N; the soil moisture almost vanishes around August, explaining the low evaporation noted in Figure 7.20. The weakness or absence of peaks associated with spring snowmelt in the observations is not a general feature of the V&Y data, several stations having a mean end of month value more than 20 mm above the adjacent months, in two cases (near 54° N, 70–90° E) about 40 mm above. Such a peak is also noted by Robock *et al.* (1995), in their analysis of Russian soil moisture data. Possibly it is missing because it does not occur near the monthly observation date (28th) at the two stations with full monthly data.

The similarity between modeled and observed soil moisture is perhaps surprising in view of the large differences in precipitation and evaporation; evidently these roughly cancel through the summer as runoff from the top meter is relatively small at this time compared with the other terms. In the experiments, this cancellation follows from the low rainfall and exhaustion of the soil water store – in such arid climates, summer evaporation is obviously constrained by current rainfall. The similarity of precipitation and evaporation for the real data suggests that the opposite control may be operating – restriction of rainfall by evaporation. Such a constraint has been found to be important in a number of model experiments (e.g., Rowntree and Bolton, 1983; Mintz, 1984). Recently, Milly and Dunne (1994) showed the importance,

Table 7.3. *Data on available soil moisture* W *(mm) in the top meter from V&Y used in validation of model at 55° N 41.25° E*

Number	Lat	Long	W_f	W 28 Apr	W 28 Jul	Change	Range
7	58 N	36 E	120	334	242	– 92	106
8	56 N	28 E	175	202	158	– 44	86
10	57 N	44 E	217	242	198	– 44	58
11	54 N	46 E	154	188	120	– 68	80

Note: W_f is the 'field capacity (see text); 'Range' is the difference between the highest and lowest values in the mean annual cycle.

because of this mechanism, of correct specification of the water-holding capacity of the plant-root zone. In their experiments, part of the increased evaporation resulting from the greater water store was used to increase the precipitation and part to reduce the convergence of water from the ocean, the relative increase in each varying geographically.

Causes of limited water capacity

A possible implication of the comparison between the modeled and observed soil moistures is that the HC model has too small a water-holding capacity in the region we are considering.

Likely reasons for the underestimation of water-holding capacity are:

too shallow a root zone;

neglect of the transfer of water from below the root zone during summer;

presence of a shallow water table;

incorrect specification of the soil hydrological parameters – specifically, too low a value of $(\theta - \theta_w)$ at the point where drainage effectively ceases.

These are considered in turn below.

Root depth

The root depth in the model for the 55° N point is 0.67 m, the specified value in the model for the arable cropland and pasture vegetation prescribed for this point in the model. For the natural forest vegetation of the zone in which V&Y state that these stations lie, a deeper root depth of about 0.9 m would be prescribed; the vegetation of the actual plots where the observations were made is grass (K. Ya. Vinnikov, personal communication), as for the plots considered by Robock *et al.* (1995). Robock *et al.* comment that preliminary results

for one site suggest little difference in soil moisture between forest- and grass-covered plots. For the stations they analyzed, the seasonal variation of soil moisture in the 50–100 cm layer was generally less than that in the 0–50 cm layer, suggesting that the roots are mainly drawing water from the upper 50 cm.

Water transfer from below the root zone

Robock *et al.*'s data also show that the mean soil moisture is generally less at 50–100 cm than at 0–50 cm. However, upward transfer of moisture is possible, even with mean soil moisture increasing upward, both because of temporal variations in a highly nonlinear system (e.g., equation (7.3.1) for K, where c is of order 10) and because it is the gradient in soil moisture potential, not that in soil moisture, which determines the transfer. The latter point is relevant only if the soil characteristics, controlling the relation between the potential and soil moisture, vary in the vertical. Thus it is not surprising that an experiment with a new multilayer version of the model in which such transfer could be represented, but with soil characteristics independent of depth, gave only minor improvements in the simulation, with upward transfers into the root zone only up to about 0.1 mm day^{-1}.

Presence of a shallow water table

The data in Table 7.3 include estimates of the 'field capacity' W_f, which V&Y define as 'the maximum water content which can be kept in the soil by the forces counteracting gravitation when there is no contact with groundwater'. The maximum mean values at the end of April all exceed this level, indicating an approximate balance between drainage and recharge by excess of precipitation and snowmelt over evaporation. However, where end of July ('dry season') values exceed W_f, it is possible that the water table is high or the soil impermeable. The model does not include any representation of these possibilities.

Soil properties: specification of soil types

A major possibility to consider is the appropriateness of the soil hydrological properties. There is a number of aspects to this problem. The first is the specification of the soil types in the model. This is in terms of texture (defined in terms of the fraction of sand, silt and clay), thus ignoring the contribution of other soil characteristics to the soil properties. For given 'texture', there is a substantial range of soil characteristics (e.g., Cosby *et al.*, 1984).

Wilson and Henderson-Sellers (1985) define the soils for

most of this region as of 'medium' texture. Their soil texture triangle shows that the effective criteria for this are a clay content of less than about 35%, and a sand content of less than about 65%. Thus there is a large range of sand content. Cosby *et al.* (1984)'s statistical analysis of 11 of the 12 soils in the US Department of Agriculture soils textural triangle, showed that the saturated hydraulic conductivity displayed greatest sensitivity to sand content, with a smaller but still substantial dependence on clay content, such that K_s increased with increasing sand and decreasing clay content. To assess the importance of the consequent uncertainty, 'field capacities' (θ_f), defined here by a drainage rate of 0.01 mm hour^{-1} or about 7.2 mm month^{-1}, have been analyzed for two soils at opposite extremes of this 'medium' range. The names and definitions of these (from Cosby *et al.*, 1984) are given in Table 7.4.

Soil properties: definition of soil hydrological characteristics

The second aspect of the problem is the different characteristics deduced for the soils in different analyses. These analyses have typically been in terms of the Clapp–Hornberger equations (equations 7.3.2, 7.3.3). J is here assumed to be 0.5 rather than $J = 1$ as used by Clapp and Hornberger (1978); the lower value is used here for consistency with the Mualem–Van Genuchten relation discussed below – the effect of this on the present analysis is small.

Table 7.5 shows the values of K_s and b recommended by Cosby *et al.* (1984), together with values for an 'available field capacity' ($\theta_f - \theta_w$) for the two soil types. Note that the problem is simplified here by considering a single-layer soil, so allowing omission of the soil suction or pressure head from consideration. Here θ_w is the wilting value of θ, the value at which evapotranspiration ceases. Although the retained moisture θ_r, the amount not available for drainage, can be less than θ_w, the rate of drainage at θ_w is typically so small using equations (7.3.2) and (7.3.3) that the effective limit to drying out is that set by the wilting point. However, it should be noted that considerable uncertainty is also associated with the value of θ_w: Warrilow *et al.* (1986) noted a range for loam soil of from 0.05 (Marshall and Holmes, 1979) to 0.135 (Rutter, 1975). Even without taking this into account, Table 7.5 shows there are large uncertainties in 'available field capacity' associated with the broad range of soils covered by the medium classification.

Soil properties: alternative formulations for hydraulic conductivity etc.

A further uncertainty in 'available field capacity' as we have

Table 7.4. *Soil types used for analysis of soil hydrological characteristics, with contents of silt, sand and clay from* Cosby et al. *(1984)*

Class	% silt	% sand	% clay
(1) Silty clay loam	56	10	34
(2) Sandy loam	32	58	10

defined it stems from the use of different relations to derive K. Van Genuchten *et al.* (1991) have studied the application of the Mualem–van Genuchten relation (van Genuchten, 1980) (equations 7.3.4, 7.3.5) to the same soils discussed above. Van Genuchten *et al.* (1991) considered two analyses of the soil characteristics, those due to Carsel and Parrish (1988) and Rawls *et al.* (1982). Data for each of these are included in Table 7.5, with b specified according to equation (7.3.5).

The values in the last column under $\theta_f - \theta_w$ are a measure of the amount of water available for evaporation at the end of the wet season, when evaporation becomes the dominant term in the soil water balance. There is clearly a large uncertainty depending not only on where in the Wilson and Henderson-Sellers 'medium' range the soil lies, but also on which particular formulation is used to relate the soil hydraulic conductivity to the soil moisture and also on the value assumed for θ_w or θ_r. The values used by V&Y for the region we are considering are well within the range for 'medium' soils. The Warrilow model values for ($\theta_f - \theta_w$) are near the lower end of the range.

Soil water balance

Observed

To complete this assessment of how the observed data may be used to validate the model's simulation, we need to study the soil water balance. We consider first the observations.

The most convincing results are obtained using the soil moisture data for 1972–85 together with the Moscow precipitation data for the same period; the evaporation and river runoff (Y_R) data from Korzun *et al.* (1978) are used in the absence of an alternative. Equation (7.3.1) shows that this allows us to deduce the sum of the runoff and the change in snowmass ($\mathrm{d}(Sn)/\mathrm{d}t$) as:

$$P - E - \mathrm{d}(SM)/\mathrm{d}t = Y + \mathrm{d}(Sn)/\mathrm{d}t \qquad (7.3.10)$$

Differences between ($P - E$) and Y can thus be divided between changes in the sum of soil moisture, groundwater and snow water content. From October to March they are mainly

Table 7.5. *Soil characteristics for soils (1) and (2) as defined in Table 7.4 according to different analyses*[a]

Analysis	Soil	K_s (m day^{-1})	b	θ_r	θ_s	'S_f'	θ_f	$\theta_f - \theta_w$
Cosby	(1)	0.176	8.72	–	0.464	0.718	0.333	0.228
	(2)	0.452	4.74	–	0.434	0.532	0.231	0.126
Carsel	(1)	0.0168	4.35	0.089	0.43	0.88	0.389	0.284
	(2)	1.061	1.123	0.065	0.41	0.234	0.146	0.041
Rawls	(1)	0.036	6.62	0.040	0.432	0.91	0.317	0.212
	(2)	0.622	3.11	0.041	0.412	0.561	0.249	0.144
Warrilow	medium	0.156	2.75	0.105	0.46	0.445	0.263	0.158

Note: [a] Cosby *et al.* (1984); Carsel and Parrish (1988); Rawls *et al.* (1982) (see text for details); and for medium soils as used in the HC model.
[b] S_f is S as defined in text for $\theta = \theta_f$ for the Carsel and Parrish (1988) and Rawls *et al.* (1982) analyses and for the Warrilow scheme; for Cosby *et al.* (1984), it is θ_f/θ_s. The value of θ at wilting point (θ_w) is assumed to be 0.105.

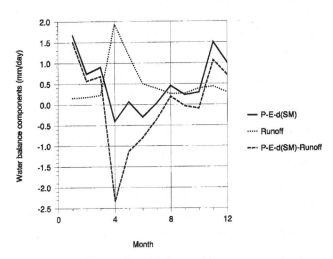

Figure 7.21. Estimates of water balance components (mm day^{-1}) using data for 1972–1985 on soil moisture (based on V&Y's – see text) and on precipitation for Moscow, and estimates of long-term mean evaporation and runoff from Korzun *et al.* (1978).

due to moistening of the soil and accumulation of snow mass. Snow in the Moscow area typically melts around the start of April and covers the ground during late November as shown by ground-based data (Arctic Construction and Frost Effects Laboratory, 1954; Lydolph, 1977) and satellite-based data (Matson *et al.*, 1986), so snow should not affect the data from May to October. We must remember that Y is the loss of water from the top meter, which may be partly to groundwater as well as to rivers, so that even in summer the result is not expected to equal the river runoff. Figure 7.21 shows the components of the budget. The soil moisture data are a simple mean of the data for the four stations in Table 7.3 where available (see Figure 7.20), and a best estimate based on the available data otherwise.

The residual ($P - E -$ d(SM)/dt-river runoff) can be interpreted as dominated by accumulation and loss of snow mass from November to March inclusive, loss of snow in April and then loss of water below 1 meter depth to rivers from May to July. Essentially, the winter precipitation appears as river runoff in spring and early summer. The residual is small from August to October.

Comparison with model

Finally, Figure 7.22 compares the above estimates of ($P - E -$ d(SM)/dt) for reality with similar data from the model experiments. The most important differences occur in spring; for E1, this may just be due to timing of snowmelt – the sum for March and April is the same as for the observed data; for E2, the greater precipitation and lesser evaporation appear responsible. The discrepancy in November appears to be due to the model's soil absorbing more of the precipitation than is observed and so reaching its winter plateau sooner. The fractional error is large in August and September when the model fails to produce the observed small losses from near the surface.

Discussion and conclusions

The observed data are inadequate in several respects. The annual precipitation, evaporation and runoff do not balance

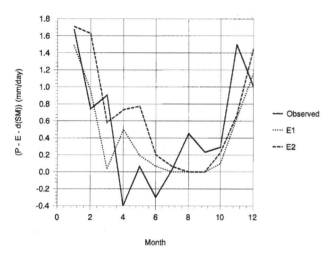

Figure 7.22. Estimates of loss of water from the top soil layer $(P - E - d(SM)/dt)$ (mm day^{-1}) for Moscow and for the model experiments E1 and E2 at 55° N 41.25° E.

in available analyses; the balance could apparently be improved by reducing the corrections for exposure applied to the precipitation data but, as the evaporation data are particularly uncertain, this is not necessarily an appropriate solution. Studies of the annual water balance should be made in catchments with large winter snowfall during field experiments in which evaporation is measured (e.g., GCIP, BOREAS and Mackenzie basin experiments).

Despite these shortcomings, the observations, particularly those of soil moisture, indicate clear problems with the model simulations. One explanation is that soil water-holding capacity is insufficient in the HC model. Use of soil texture categories which encompass too wide a range of soils may contribute to this. Even for a given soil, a wide range of available water-

holding capacities may be calculated depending on the choices made from the literature for the soil characteristics including the equations used to relate them to soil moisture and for the moisture remaining at wilting point. This wide range needs to be reduced by further analysis of soil data in order to give a realistic interpretation of the soil observations (for the soil moisture values of most importance in determining the effective water-holding capacities of the soil layers in those layers from which water is drawn for evaporation).

Another possibility is that the soil moisture observation sites are atypical – e.g., a shallow water table or an impermeable soil layer. V&Y report water table depths of at least 5 m for the first three of the stations in Table 7.3, and 2–2.5 m for the fourth. As noted earlier, for Station 7, the excess of soil moisture above the field capacity in the dry season is difficult to explain otherwise. However, this station is closest in annual range to the modeled behavior, so that its omission would not improve the comparison.

These problems are not easily solved. Soil characteristics are not directly observable from space, so that construction of global data sets must rely either on collation and analysis of existing ground-based data together with further extensive observations in regions with insufficient data, or on the use of indirect methods. The latter may calculate those soil and surface characteristics which optimize the fit to relatively well-observed data such as river runoff and/or remotely sensed surface temperatures (Mahfouf, 1991). Even if the soil moisture were correctly simulated, excessive evaporative demand due to inadequate cloud or a lack of precipitation due to incorrect simulation of the circulation and moisture convergence would prevent a realistic simulation. However, without a sufficient reserve of water before the dry season, a correct simulation is impossible.

7.4 Estimating large-scale runoff

E. Todini and L. Dümenil

Meteorologists are concerned with the vertical flux of water and heat at the land surface, computed using soil and atmospheric column models, i.e., one-dimensional models in the vertical. This ignores horizontal fluxes of water in the soil as well as surface runoff.

Global atmospheric models, known as general circulation models (GCMs) are based upon spatial discretization meshes of one hundred kilometers in size or larger. At this scale, a wide variety of dynamical phenomena cannot be explicitly resolved, leaving an average overall effect to be accounted for.

The extreme variability of land and soil properties, combined with the nonlinearity of processes occurring in the soil, do not allow for a consistent use at the GCM scales, using average parameter values, of the column-type models originally developed for an infinitesimal horizontal dimension. In addition, the need for introducing the representation of horizontal fluxes, both in the soil and on the surface, which is more and more felt by GCM modelers, also requires the extension from the local scale representation to a regional one.

In order to extend the validity of the local-scale results to regional scales one needs to apply an averaging operator, which is called 'expected value' in statistics, and which is generally referred to as 'lumping' when dealing with dynamic systems. If the spatial variability is not substantial, or if one can demonstrate the accuracy of the model formulation at the different scales, one can use lumped parameter values at the larger scales; otherwise lumping should be applied to the physical representation which may result in models that greatly differ from the original smaller scale formulation and which are generally simpler at the new scale.

The rainfall–runoff process

An accurate understanding of the detailed processes acting within a uniform element of soil does not exist. The cost and time involved in first deriving and then applying such detailed knowledge for even a small catchment is likely to remain prohibitive for most applications. Hydrology has therefore been dominated by attempts to simplify those processes, using a variety of assumptions. The main components into which precipitation may be partitioned are evapotranspiration, overland flow, channel flow and saturated and unsaturated subsurface flow.

The literature contains many works that summarize the physics of the complex problem of rainfall–runoff transformation (Dunne, 1978; Freeze, 1980). Many efforts have been made to schematize the whole process in order to develop mathematical models (Dooge, 1973; Todini, 1989). These range from the simple calculation of design discharge to the two-dimensional representation of the various processes based on mass balance, energy and momentum equations (e.g. the SHE model, Abbott *et al.*, 1986; the IHDM model, Beven *et al.*, 1987), and to the three-dimensional representation of all the exchanges (Binley *et al.*, 1989). Taken together, these latter kinds of model comprise the broad category of distributed differential models (Todini, 1988*b*); they are frequently referred to as 'physical models' to highlight the fact that their respective parameters are (or should be) reflected in the field measurements (Beven, 1989). Given their nature, they are mainly used as a mathematical support for the interpretation of physical reality.

Under the pressure of GCM modelers recently there has been a demand for models that are simple in terms of calibration and validation, computationally efficient, and with only a few parameters, corresponding with measurable quantities, to describe the physical processes at the catchment scale. Two main components can be identified in all lumped conceptual rainfall–runoff models that should be considered when deriving the new parameterizations at GCM scales; the first represents the water balance at soil level and the second the transfer to the basin outlet. A third component, the groundwater, although essential in climate impact studies, is not discussed in this section because on the one hand it can easily be modeled as a cascade of linear reservoirs and on the other hand its dynamics are far slower than those of the atmosphere, while its feedback to the atmosphere is mediated by soil moisture content and surface water. The water balance at the soil level is the component that characterizes a rainfall–runoff model and constitutes its most important part. Cordova and Rodriguez-Iturbe (1983) summarize this concept most succinctly, saying 'the problem is not *how* to route but *what* to route'. Possible solutions to the routing part of the models at catchment scale can be found in Todini (1988*b*) where the surface runoff and river discharge are approximated by linear parabolic equations, the analytical solution of which is widely available, and in Naden (1992) who combines the linear parabolic equation solutions with a geomorphological description of the river network which could be extended to GCM scales.

Lastly, an additional component is required in GCM models in order to estimate the major river flows into the seas and oceans. This component must represent the movement of water along the drainage networks from one GCM grid cell to the downstream ones, accounting for two typically observed phenomena, namely the delay and the subsidence of the flood wave.

Vertical lumping

As described above, runoff formation is a complex process which is more or less clearly understood at the hillslope scale, while its extension to catchment or GCM scales must be treated by combining it with a description of the horizontal movement of water within the soil and on the surface. There are three types of horizontal flows which take place under precipitation forcing: groundwater flow, unsaturated zone flow, and surface runoff.

Groundwater movement in general is not rapidly affected by precipitation and its response is in the order of days or months, while both surface runoff and unsaturated zone flow (particularly in the root zone) have much shorter time responses. The root zone horizontal flow, which is mainly driven by soil properties and slope as well as by gravity, also has a strong effect on the spatial distribution of soil moisture available for evapotranspiration, in that it will tend to concentrate saturation in the flatter zones close to the drainage network while at the same time depleting the upper and steeper zones.

Given the presence of strong capillary forces in the unsaturated zone, the description of horizontal flow in unsaturated conditions is generally based upon the knowledge of the vertical profile of moisture content in the soil. Nevertheless, due to the high conductivity, caused by macropores, i.e. wider flow paths present in the coarser texture of the upper soil matrix, (Beven and Germann, 1982), gravity will be the dominant mechanism driving water from the top of the soil to the impermeable or semi-impermeable lower boundary. This boundary will create a perched water table. In this zone, horizontal propagation of water also involving unsaturated flow will occur.

At very small scales, the description of the horizontal component of flow in partly saturated porous media generally requires the use of three-dimensional partial differential equation models that can be reduced to two-dimensional ones when treating problems with a prevailing flow direction, such as for instance in the case of a hillslope. When dealing with larger scales, and in particular at the GCM scale, simplifying assumptions in the physical representation become essential.

At first, in order to simplify the physical representation, it is necessary to acknowledge that the depth of this highly conductive soil (one or two meters at most) will be negligible with respect to the horizontal dimensions even at the hillslope scale; a similar approach was taken by Hurley and Pantelis (1985) in the development of a simplified saturated/unsaturated flow model through a thin porous layer on a hillslope. If this assumption holds, it is then possible to lump the soil moisture profile in the vertical dimension and to describe the horizontal flow as a function of the total soil moisture content. To demonstrate the validity of this hypothesis, two different but well known relationships were used; the first one from Brooks and Corey (1964) and the second one given by van Genuchten (1980), both describing the dependence of the hydraulic conductivity on the reduced soil moisture content, defined as

$$\tilde{\theta} = \frac{\theta - \theta_r}{\theta_s - \theta_r} \qquad (7.4.1)$$

where θ is the soil moisture content, θ_s is the saturation soil moisture content and θ_r is the residual soil moisture content.

Using the Brooks and Corey expression, the hydraulic conductivity was calculated from a number of different vertical profiles having a variation in moisture content with depth while keeping the same integral moisture content and ranging within 0 and 1 (Benning, 1994). The results for a wide range of soil moisture profiles, both increasing and decreasing with depth, are shown in Figure 7.23 where also a flux index, defined as the ratio between the computed flow and the maximum flow corresponding to saturation of the entire soil column, is plotted. This demonstrates that the assumption holds within the range of uncertainty in the soil properties.

A similar analysis was done by using a suitable approximation to allow for analytical integration of the expression for the hydraulic conductivity given by van Genuchten. Table 7.6 shows a comparison of the total reduced soil moisture content given by the original and the approximated expressions, relevant to the same value of what can be defined as a transmissivity for the unsaturated zone, i.e. the integral of the hydraulic conductivity over the vertical depth.

The above-mentioned assumption allows the transient phase of vertical infiltration to be neglected; in other words, when interested in the horizontal movement of water, it is possible to avoid, within the range of reasonable errors, the integration of the unsaturated soil vertical infiltration equation. This can be achieved by assuming a transmissivity which is a function of the total soil moisture content, thus reducing by one the dimensionality of the problem after lumping in the vertical.

Table 7.6. *Values of reduced moisture content producing the same value for transmissivity as a function of depth of water table, calculated with the van Genuchten expression and the approximated one.*

Depth of water table (cm)	Transmissivity ($L^2 S^{-1}$)	Moisture content (van Genuchten) (%)	Moisture content (approximated) (%)
100	0.01	17	17
90	0.2	21	27
80	0.4	28	37
70	0.6	37	46
60	0.8	45	55
50	1.0	54	64
40	1.2	63	72
30	1.4	72	80
20	1.6	81	88
10	1.8	91	95
0	2.0	100	100

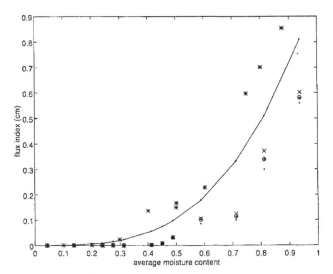

Figure 7.23. The different values of the flux index, calculated both as the integral of the point relative hydraulic conductivity for different soil moisture profiles and as a function of the total soil moisture content (solid line). (From Benning, 1994.)

Horizontal lumping

After lumping the process in the vertical dimension, in order to reach a formulation suitable for GCMs, it is also necessary to reduce the rainfall–runoff problem to a nonlinear reservoir-type model at the GCM grid scale with physical model parameters to be determined by lumping information from topography, soil and vegetation properties. In other words, the objective is to reach an overall water balance equation valid at the GCM scale, by averaging all the microscale effects.

One solution to this is the use of the topographical and physiographical information in the form of a probability distribution instead of explicitly following the real nature of the terrain. Given that the macroscale behavior is characterized by the microscale hillslope processes and by the subgrid drainage and routing processes, in order to substitute the actual processes with an average one, a preliminary proof that the topological succession of slopes will not dramatically modify the shape of the resulting runoff is required. This result has been demonstrated in the literature relating to the Geomorphological Instantaneous Unit Hydrograph (GIUH) (Rodriguez-Iturbe and Valdés, 1979) at the catchment scale, but can also be proved at smaller scales down to the individual hillslope.

On the basis of what was discussed in the previous subsection, if one assumes that the horizontal hydraulic conductivity is very high in a first soil layer and can be derived from the Brooks and Corey expression, namely $k(\tilde{\theta}) = k_s \tilde{\theta}^c$ (where k_s is the saturated conductivity, $\tilde{\theta}$ the reduced moisture content and c a parameter characterizing the soil properties), one can define a transmissivity for the unsaturated zone as

$$T(\tilde{\Theta}) = k_s L \tilde{\Theta}^x \tag{7.4.2}$$

where L is the thickness of the soil layer,

$$\tilde{\Theta} = \frac{1}{L}\int_0^L {}^{L0}\tilde{\theta}(z)\,dz \tag{7.4.3}$$

is the average value along the vertical of the reduced soil

moisture content, and a is a parameter characterizing the soil properties. It is easy to show that the horizontal movement of water in the unsaturated zone, on the same assumptions used in TOPMODEL (Beven and Kirkby, 1979), can be approximated by means of a kinematic wave model (Todini, 1995), which gives

$$(\theta_s - \theta_r) L \frac{\partial \tilde{\Theta}}{\partial t} = p - \frac{\partial q}{\partial x} = p - \alpha\, k_s\, L \tan(\beta) \tilde{\Theta}^{(\alpha-1)} \frac{\partial \tilde{\Theta}}{\partial x} \qquad (7.4.4)$$

where $\tan(\beta)$ represents the terrain slope, q is the horizontal flow, and p is the precipitation.

The model is written in one dimension because it is thought that the movement of water will proceed along the hillslopes towards the closest stream drainage line, and afterwards it will proceed along the drainage network; this allows for the reduction of an additional dimension, which finally reduces the original problem to the solution of a one-dimensional kinematic wave problem.

A higher conductivity in the top layer of limited dimension (the root zone) expressed in TOPMODEL is preserved in this formulation. However, the exponential decay with which this phenomenon is represented in TOPMODEL is replaced by an equation that gives the relationships between the transmissivity and the total moisture content neglecting, after vertical lumping, the dependence of saturated hydraulic conductivity on depth. This means that the kinematic model considers a vertical layer of limited dimensions characterized by a high constant saturated conductivity, but with a corresponding transmissivity varying with the total moisture content.

Kinematic wave equations have been extensively used to model surface runoff. It was the American hydrologist Horton who in 1933 carried out the earliest recorded scientific studies of surface runoff. Later Keulegan (1945) applied the continuity and momentum equations for surface runoff analysis. He investigated the magnitude of the various terms and indicated that a simplified form of the equations, now termed the kinematic equation, would be adequate for surface runoff.

Starting with the formulation of the kinematic wave theory by Lighthill and Whitham (1955), kinematic surface runoff models have been utilized increasingly in hydrologic investigations. The first application of kinematic wave routing to surface runoff and groundwater flow was by Henderson and Wooding (1964). The conditions under which the kinematic flow approximation holds for surface runoff were first investigated by Woolhiser and Liggett (1967); they found it to be an accurate approximation to the full equations for most surface runoff cases.

In an earlier study, Henderson and Wooding (1964) provide solutions for horizontal kinematic subsurface flow through a porous medium of constant permeability both for steady state and a rising water table. They compare the extended Dupuit–Forchheimer equation to a further simplification of the flow equation in which it is assumed that the hydraulic gradient at any point within the saturated zone is equal to the bed slope. Considering a kinematic wave equation valid in a sloping soil mantle of constant saturated hydraulic conductivity overlying a relative impermeable bedrock sloping surface, Beven (1981) stated that this equation was a good approximation to the more correct extended Dupuit–Forchheimer equation. In a second article (Beven, 1982), the vertical propagation in the unsaturated zone was also taken into account, in order to evaluate the time at which the wetting front reaches the bottom of the profile. Assumptions, similar to the ones proposed in this section, were then introduced by Hurley and Pantelis (1985) to model unsaturated and saturated flow through a thin porous layer on a hillslope as a kinematic wave. In addition, a kinematic approximation for subsurface flows (both in the vertical and horizontal directions) has been successfully tested by several other authors (see for instance Stagnitti et al., 1986).

Integrating the partial differential equation (7.4.4) in space on a finite dimension element x, on the assumption that $\tilde{\Theta}$ and p are constant over the integration domain, one obtains:

$$\frac{dV_x}{dt} = px + q_0 - \frac{\alpha k_s L \tan(\beta)}{(\theta_s - \theta_r)^\alpha\, L^\alpha} V_x^\alpha \qquad (7.4.5)$$

where

$$V_x = (\theta_s - \theta_r) L \int_0^x \tilde{\Theta}(\xi)\, d\xi \qquad (7.4.6)$$

and q_0 the upstream inflow to the generic element of size x.

Equation (7.4.5) can be viewed as the following zero-dimensional nonlinear reservoir type equation:

$$\frac{dV_x}{dt} = I_x - O_x \qquad (7.4.7)$$

where $I_x = px + q_0$ is the inflow, $O_x = KV_x^\alpha$ is the outflow and

$$K = \frac{\alpha k_s L \tan(\beta)}{(\theta_s - \theta_r)^\alpha\, L^\alpha} \qquad (7.4.8)$$

is the reservoir constant.

Horizontal lumping can now proceed by linking in series and in parallel, following a tree-shaped drainage network, the nonlinear reservoirs defined by equation (7.4.5). It can be shown (Todini, 1995), subject to a number of simplifying assumptions, that this lumping process does not modify the structure of equation (7.4.5) and that the same nonlinear reservoir structure applies to larger and larger contributing areas up to the catchment scale, with parameters depending upon the surface area, the topography, the topology and the nature of soils. Similarly, when saturation occurs, it is possible to deal with surface runoff since, as mentioned earlier, it can be reproduced as a kinematic wave, leading once again to a nonlinear reservoir type model at the catchment scale.

The above analysis shows that it seems possible to reproduce the overall rainfall–runoff process at the catchment scale as the combination of two simple interconnected nonlinear reservoirs with physically based parameters; the first one representing surface runoff and the second one representing subsurface flow in the saturated/unsaturated soil. Eventually, a third reservoir, essentially linear and with much slower dynamics, might be added, when needed, to represent the groundwater accumulation and flow.

Hydrologic models

A practical demonstration of the validity of the lumping assumptions is given by currently used hydrologic models which implicitly assume vertical as well as, in many cases, horizontal lumping. Two of these models, which have recently raised interest among hydrologists and meteorologists for their simplicity as well as for their notable performances at the catchment scale, will be briefly mentioned in order to clarify the requirements for a lumped runoff production model at the GCM scale.

The first one, the ARNO model (Todini, 1996), derives from the Xinanjiang model developed by Zhao (1977). He expressed the spatial distribution of the soil moisture capacity in the form of a probability distribution function, similar to that advocated by Moore and Clarke (1981). Subsequently, in order to account more effectively for soil depletion due to drainage, the original Xinanjiang model scheme was modified by Todini (1988a), within the framework of a hydrologic study of the river Arno, hence the ARNO model.

The basic concepts expressed in the ARNO model are:

The precipitation input to the soil is considered uniform over the catchment (or sub-catchment) area.

The catchment is composed of an infinite number of elementary contributing areas, each with a different soil moisture capacity and a different soil moisture content, for each of which the continuity of mass can be written and simulated over time.

All the precipitation falling onto the soil infiltrates unless the soil is either impervious or has already reached saturation; the proportion of elementary areas which are saturated is described in space by a distribution function.

The distribution function allows for the description of the dynamics of contributing areas which generate surface runoff.

The total runoff is the spatial integral of the infinitesimal contributions deriving from the different elementary areas.

The soil moisture storage is depleted by the evapotranspiration as well as by lateral subsurface flow (drainage) towards the drainage network and the percolation to deeper layers.

Both drainage and percolation are expressed by simple empirical expressions.

The ARNO model, which in practice behaves like the nonlinear reservoir model discussed above, has been extensively used in hydrologic practice and operationally installed for real-time flow forecasting on several rivers in many countries: the Fuchun in China; the Danube in Germany; the Arno, Tiber, and Reno in Italy. The hydrologic model performances can be assessed by the high value of the explained variance of results obtained in several of these applications (Table 7.7). Because of the simplicity of its formulation it was also successfully included in the Hamburg climate model (Dümenil and Todini, 1992).

The major advantage of the ARNO model is the fact that it is entirely driven by the total soil moisture storage, which is functionally related, by means of simple analytical expressions, to the directly contributing areas, the drainage and the percolation amounts. This makes the model extremely useful in evaluating the total amount of soil moisture available for evapotranspiration, which is one of the major requirements for inclusion in GCMs.

The major disadvantage for its inclusion in GCMs is the lack of physical grounds for establishing some of the parameters, and in particular those relevant to the overall 'drainage' and the horizontal movement in the unsaturated zone, which have to be estimated on the basis of the available precipitation and runoff data. This is not particularly critical in hydrologic applications but it becomes so when the model is used in GCMs, where direct input/output observations are not really available.

Wood et al. (1992) extended the ARNO model concepts to multiple soil layers and they developed an alternative model called VIC for which the same advantages and disadvantages discussed above apply.

Another hydrologic model, the TOPMODEL (Beven and Kirkby, 1979; Sivapalan et al., 1987), is more attractive, in that it focuses on the horizontal movement of water in the un-

Table 7.7. *Basins where the Arno model was applied, their catchment area size and the explained variance of results.*

Catchment	Size (km²)	Explained variance
Fuchun at Lan Xi	18,236	0.96
Tiber at Corbara	6,100	0.98
Arno at Nave di Rosano	4,179	0.98
Danube at Berg	4,037	0.94
Reno at Casalecchio	1,051	0.96

saturated zone as physically related to the soil properties and to topography. TOPMODEL is a variable contributing area conceptual model in which the predominant factors determining the formation of runoff are represented by the topography of the basin and a negative exponential law linking the transmissivity of the soil with the distance of the saturated zone below ground level. Although synthetic, this model is described by Sivapalan *et al.* (1987) as a 'simple physically-based conceptual model' in the sense that its parameters can be measured directly. This definition is somewhat optimistic in view of the considerable simplification inherent in the structure of the model and the doubts and uncertainties encountered in defining the parameters of 'physical models' themselves. But the inclusion of the effects of variability of topography on contributing area dynamics represents a major advance over previous models based on 'point' hydrologic responses assumed to apply at the catchment scale.

It is not simple to summarize the characteristics of the TOPMODEL since it is more a philosophical approach than a specific model. In the perspective of using it for GCM modeling, and without loss of generality, the basic assumptions for a lumped TOPMODEL are described here.

The basic concepts expressed in the TOPMODEL are:

The precipitation is considered uniform over the catchment (or sub-catchment) area.

All the precipitation falling over the soil infiltrates unless the soil has already reached saturation at that particular location.

The total runoff is composed of surface runoff and subsurface flow.

The surface runoff mechanism is driven by the dynamical variation of the saturated contributing areas, which is determined on the basis of the topographic index curve.

The soil moisture storage is depleted by the evapotranspiration as well as by the subsurface flow towards the drainage network and the percolation to deeper layers.

The permeability in a first soil layer is very large at the soil surface and decays exponentially with depth.

Each point in the catchment has a uniquely defined contributing area according to topography.

The downslope movement of water in the capillary fringe of the unsaturated zone is negligible.

The downslope movement of water in the saturated zone is driven by gravity and is a function of the local surface topographic slope.

The subsurface flow is computed for each point, by integrating in space over the relevant contributing area, the basic equation expressing the water movement in saturated porous media, under the assumption of steady state. This reduces the problem to a nonlinear reservoir model for which both a continuity of mass equation and a relation between storage and runoff are available.

The model has recently been used for many catchments where it gives reasonable to good results (Durand *et al.*, 1992; Troch *et al.*, 1993).

The major advantage of TOPMODEL is the possibility of estimating the topographic index curve from the topography of a basin (Quinn *et al.*, 1991). In addition, all the other parameters are physically related to the soil and porous media characteristics, which is also a very useful property of the model when used in GCMs, where no direct calibration of parameters is really possible.

The major disadvantage of TOPMODEL lies in the steady state assumption, which is advocated in order to derive the model integral equations. This assumption, which corresponds to a null travel time from one side to the opposite side of an elementary grid cell (pixel) of the digital terrain model used in the derivation of the Topographic Index Curve, becomes unrealistic for cells of the order of magnitude of hundreds of meters. Thus the horizontal saturated permeability parameter must be artificially increased by orders of magnitude, which in turn makes the infiltrated water immediately drop to the saturated zone leaving the unsaturated zone depleted, with a general underestimation of the actual evapotranspiration and a degree of uncertainty in the actual parameter values to be used in GCM applications.

Application of hydrologic models at the GCM scale

Hydrologic models of the type represented by the ARNO model were implemented and tested in a variety of ways in several GCMs, such as the ECHAM2 model (Dümenil and Todini, 1992; Roeckner *et al.*, 1992), the UK Meteorological Office (UKMO) model (Rowntree and Lean, 1994), the LMD model (Polcher *et al.*, 1996) and the GFDL model (Stamm *et al.*, 1994). In each case, the runoff model is embedded into a parameterization package for the physics describing the radiative and turbulent fluxes at the interface between the atmosphere and land surface. These land surface processes provide the lower boundary condition for the simulation of the atmospheric flow and interact with the remainder of the model in a multitude of ways.

In order to implement the hydrologic scheme into the GCM, therefore, global fields of the variables and parameters used have to be provided on the model grid scale. The most important parameters for the ARNO (Dümenil and Todini, 1992) and VIC models are the form parameter (b), the maximum soil moisture storage capacity (W_{max}), a drainage (and sometimes a baseflow parameter); and for the VIC model (Wood *et al.*, 1992), as we shall see later, the important parameter is the number of vertical layers of soil moisture in the ground.

Polcher *et al.* (1995) describe a test of several schemes which are used in GCM modeling and their sensitivity to the choice of certain parameters. This intercomparison study is part of a larger effort involving more sophisticated hydrologic schemes as well as the more simple GCM applications (Dooge *et al.*, 1994). Each of the schemes was forced by identical atmospheric conditions over an annual cycle which were taken from a GCM simulation of a moist and a semi-arid climate. The introduction of a subgrid-scale representation of soil moisture was found to be important for modeling runoff. Such a scheme yields runoff at a higher frequency than bucket models which assume homogeneity (Figure 7.24). This is an important aspect if the runoff is subsequently used via a river flow model to compute freshwater inflow into the oceans in coupled ocean–atmosphere GCMs. It appears that the formulation of runoff only has an impact on evaporation or other components of the hydrologic cycle when soil moisture is a limiting factor, e.g., in the semi-arid case. However, this study can only give a qualitative picture of the performance of the hydrologic schemes. A more thorough comparison against observed data is given by Rowntree and Lean (1994) and the Project for Intercomparison of Land Surface Parameterization Schemes (PILPS) which is currently under way.

Rowntree and Lean (1994) assessed the effect of the ARNO scheme and the standard UKMO scheme which is based on the solution of the Richards' equation and uses characteristics which are taken as homogeneous over the grid cell. Results from one-column simulations for both schemes were compared with observed river discharge data from several catchments in southern England. After careful selection of the parameters involved, the ARNO scheme was found to be capable of simulating the main seasonal variation of discharge in these catchments, while the UKMO scheme showed deficient runoff during most of the year. A remaining

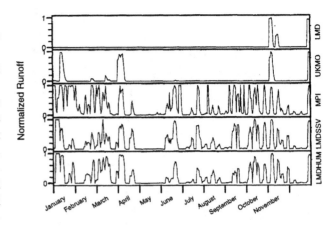

Figure 7.24. Annual cycle of runoff normalized by the soil moisture variations from Polcher *et al.* (1996). From top to bottom: LMD standard model, UKMO standard model, MPI ECHAM ARNO model, LMD model including ARNO and two soil moisture reservoirs, LMD model as in LMDSSV but with saturated values for soil moisture reduced.

problem is the fact that for the ARNO model the runoff in summer is deficient. This may reflect the difficulty of representing a complex hydrologic system with only one moisture variable. In particular, they claim that in order to simulate runoff well, groundwater storage needs to be incorporated. Other aspects include the sensitivity of model results to the parameterization of evaporation, specification of stomatal resistance and radiative effects.

In the ECHAM model of the Max-Planck-Institut the parameters are used following Dümenil and Todini (1992). In particular, the β-form parameter is computed from the variance of the topography so that runoff occurs more easily in regions of steep mountains and rough terrain. Figure 7.25 shows the distribution of this parameter at a horizontal resolution of 0.5 degrees from which lower resolution versions for the standard GCMs can be interpolated. In the current ECHAM4 version of the model this is combined with a drainage component and a single vertical soil moisture reservoir which is, however, of variable size over the continents according to the distribution of total water-holding capacities by Patterson (1990).

Figure 7.26 shows the distribution of W_{max} over the continents at the standard horizontal model resolution of about 2.8 degrees. The drainage component allows for limited drainage between 5% and 90% of soil saturation and higher drainage rates beyond 90%. These model features can be validated against the global estimates of the components of the hydrologic cycle (Baumgartner and Reichel, 1975) as in Figure 7.27, for example, which shows a reasonable representation of the return flow of water from the continents to the

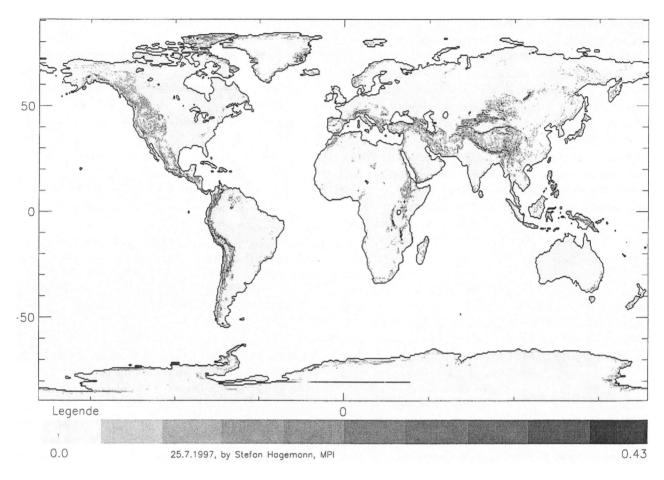

Figure 7.25. Global distribution of the slope parameter β at 0.5 degree resolution as used in the ECHAM model.

oceans. More thorough validation shows that the drainage parameters which are currently used in the model are probably too small in humid regions and that systematic errors in the simulation of the annual cycle of evaporation may be alleviated by using more than one soil reservoir in the vertical.

A similar conclusion is drawn by Stamm *et al.* (1994), who implemented the VIC model in the GFDL model. They use a *b* parameter and values of W_{max} which are both derived from the data by Patterson (1990) using the subgrid-scale information which is available at 0.5 degree resolution to provide boundary fields for the GCM. The results of the VIC model with fixed soil moisture capacities showed that global average soil moisture was lower by about 2.5 cm on average than in a simulation using a standard bucket model. Global evaporation and precipitation were reduced and surface air temperature was increased, particularly over the northern hemisphere in summer. Similar to the ECHAM model, the GFDL-VIC model shows only small sensitivity to the varying soil

water-holding capacities. The authors argue for representation of the surface hydrology in GCMs with two-layer soil models which are capable of representing the cycling of moisture during dry periods by means of surface evaporation, which is generally underestimated by single-layer models, because after rainfall events precipitation is not available for immediate re-evaporation.

River routing in GCMs

One of the requirements of coupled ocean–atmosphere–ice models which are used for climate modeling is the specification of a return flow of freshwater from the continents to the oceans in rivers. This means that the water which becomes available through the rainfall–runoff process at the soil interface needs to be transferred within the grid cell of the GCM and also from grid to grid eventually to reach the river mouth where it enters the oceans. From the hydrologic point of view this requires a representation of river routing at the macro-

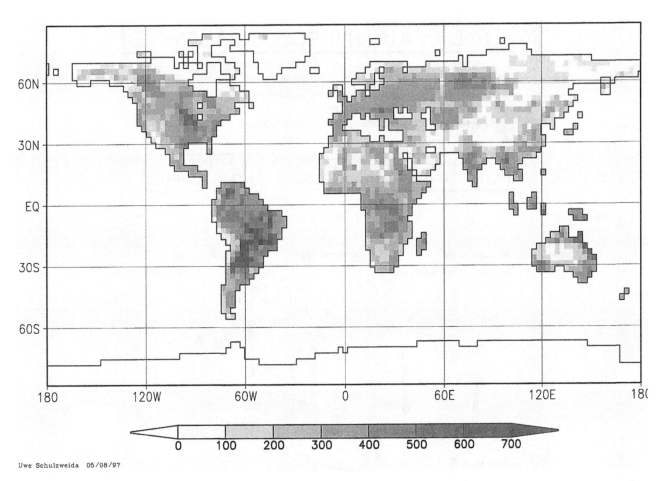

Uwe Schulzweida 05/08/97

Figure 7.26. Global distribution of the soil water-holding capacity W_{max} as used in the ECHAM4 model (interpolated to GCM resolution of 2.8 degrees from the data by Patterson, 1990).

scale for which only a few models have been developed so far. Transferring the water within grid cells of about 2.8 degrees along one side and subsequent transfer from grid to grid in smaller or larger rivers is done on the basis of routing models. In such models the shape of a flood hydrograph is followed downstream from one point (input hydrograph) to another point (output hydrograph) further downstream on the river.

From the phenomenological point of view a flood hydrograph is modified in two ways as the water flows downstream. First, there is a general delay in time of the hydrograph, which will show later at downstream river sections due to its finite travel time, and second, the magnitude of the peak flow rate will be reduced at downstream points mainly due to the successive filling and emptying of the channel storage which produces what is known as subsidence or peak attenuation (Dooge, 1973); in other words the shape of the hydrograph becomes flatter, and the volume of water takes longer to pass a lower section.

At the scale of GCMs and in order to route the overland and subsurface flows produced by the macroscale rainfall–runoff models between cells, there is no need to develop complex channel routing schemes which imply the knowledge of the river geometry and the integration of the full hyperbolic De Saint Venant partial differential equations (Fread, 1985). When the problem is the evaluation of downstream outflows as a function of upstream inflows, even for much smaller scale reaches hydrologists in the past have proposed extremely simplified models such as for instance linear reservoir routing, also known as Nash cascade transfer function (Dooge, 1973), linear parabolic models (Hayami, 1951) or piecewise linearized models such as the piecewise linear Nash cascade (Becker and Glos, 1970) or the Constrained Linear Systems (Natale and Todini, 1977). All these models describe the propagation phenomenon in a lumped form and are generally based upon two parameters in order to describe the delay and the attenuation of the flood waves.

As discussed, for the rainfall–runoff approaches to be used

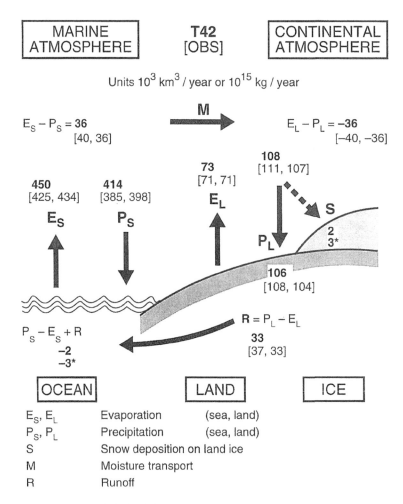

Figure 7.27. Components of the hydrologic cycle as simulated by the ECHAM4 model and according to estimates from observations.

at GCM scales, it is necessary to find lumped parameterizations where the parameters can be derived from their physical meaning. Dooge (1973) as well as Todini and Bossi (1986) showed, under different linearization assumptions, the relationship between the parameters of the Linear Parabolic approximation of the De Saint Venant equations (namely the wave celerity and the diffusivity), with the lumped properties of the river reaches, which makes the Linear Parabolic model extremely attractive for this purpose. On the same lines Na-

den (1992) proposed the use of the Linear Parabolic model in combination with the networks width function for routing runoff along the drainage system of a catchment or of a GCM grid cell.

Unfortunately few of the methods proposed in the hydrologic literature have as yet been used within GCMs. The existing macroscale models fall into two categories: those which have been developed for a specific large catchment such as the CLS model of the Nile (Fahmy et al., 1982) or the

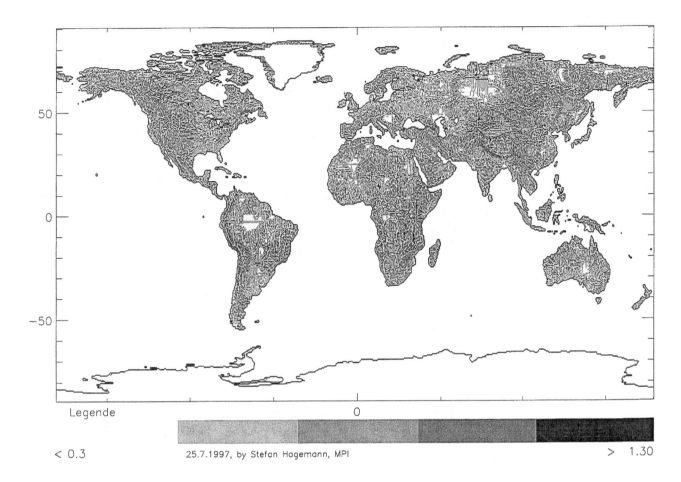

Legende
0
< 0.3
25.7.1997, by Stefan Hagemann, MPI
> 1.30

Figure 7.28. Global distribution of the magnitude of the velocity vector (m s^{-1}) defining the later flow from gridbox to gridbox (Sausen *et al.*, 1991) for a horizontal resolution of 0.5 degrees.

RHINEFLOW model (Kwadijk and van Deursen, 1993), and those which are designed to be applied globally in the GCM context.

A routing model of the Nile was developed for the White and the Blue Niles on a ten-day time scale in order to assess the effects of the reduction in evaporation losses due to the construction of the Jonglei canal in Sudan. The model did not require the integration of full dynamic equations and showed that an extremely simple approach, based on derived unit hydrographs, explained more than 98.8% of the total variance of the downstream Nile flows at Wadi Halfa over a period of more than 50 years, provided that the routing was performed at the appropriate time step (no more than 10 days). The RHINEFLOW model was developed for the river Rhine and its tributaries to study the impact of climate change; its grid size of 3 × 3 km and time resolution of one month are not appropriate for use in GCMs.

The specific requirement for application of routing models

in GCMs is that they should not rely on tuning parameters on the basis of inflow and outflow data, which may be possible in well-studied catchments, but not on the global scale.

The following three schemes have been tested in only a limited domain so far, but may be useful for global applications, because of the horizontal resolution for which they were developed and their simplicity.

In the first scheme, Liston *et al.* (1994) apply their simulation of river discharges to the Mississippi Basin. They use two linear reservoirs per grid cell of 2 × 2.5 degrees which is compatible with current GCM resolutions. The rainfall–runoff process provides the input to the surface reservoir from where outflow enters the groundwater reservoir. Steepness of the orography and river lengths are used as parameters.

The second model (Vorösmarty *et al.*, 1989) was tuned to an application over South America and has later been used for rivers in Africa as well. Again a linear reservoir is used.

The third model (Miller *et al.*, 1994) is applied globally. It

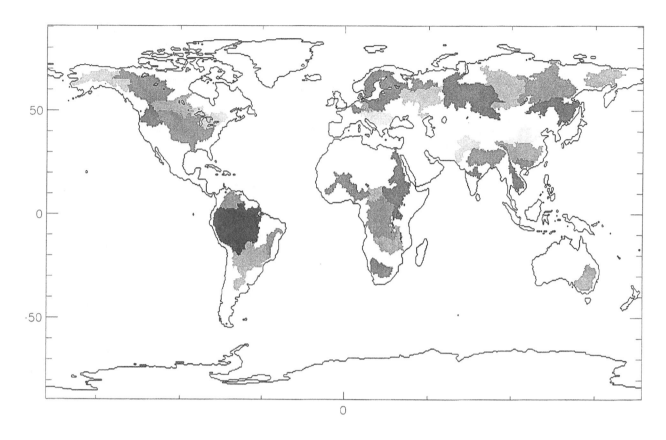

Figure 7.29. Catchments of the 35 largest rivers.

uses information on topography and gridlength, treating a grid-box of 2×2.5 degrees as a linear reservoir. This model is at the same level of complexity as the scheme by Sausen *et al.* (1991) which is currently operational at the Max-Planck-Institute. This scheme treats each gridbox as a two-dimensional linear reservoir which applies different transport coefficients for the fluxes in a north–south and east–west direction. The transport coefficients are a function of the orography in the grid area. Figure 7.28 may serve as an illustration, showing the global distribution of the magnitudes of flow velocities computed from the gridbox coefficients and which clearly reflect the orography. For validation purposes the catchments of the 35 largest rivers were defined (Figure 7.29), and measured discharges for these rivers (Dümenil *et al.*, 1993) are used for comparison purposes provided that they represent conditions which are not anthropogenically influenced.

Although the third scheme gives a reasonably good representation of river discharge for the largest rivers, several errors were identified as illustrated in Figure 7.30. Most of the errors may be attributed to the fact that the scheme does not distinguish between the different flow pathways. Water which is provided by the rainfall–runoff process in the atmospheric part of the model is passed through the channel network too quickly. This is particularly obvious for rivers which exhibit a strong dependence on snow melt processes. For these rivers the lateral transfer to the ocean is overestimated by one month and in winter the simulated discharge is usually much lower than is observed. This is mostly due to the specification of drainage in the ARNO model, which again is too small in the operational model, and experiments using an increased drainage are under way.

Conclusions

The discussion on rainfall runoff and routing processes presented here is aimed at suggesting the future lines of research for developing new parameterizations to be used from the catchment to the GCM scale. The hydrologic representation in GCMs requires in fact the development of three different components. The first is a lumped hydrologic water balance component at the GCM scale, the second is a lumped 'inter-

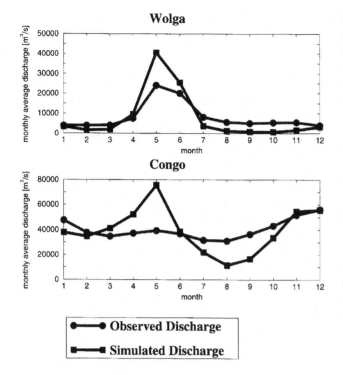

Figure 7.30. Annual cycle of the monthly averaged discharge for a ten-year model simulation using the scheme by Sausen *et al.* (1991). Observations are taken from Dümenil *et al.* (1993).

nal' drainage and routing component, and the third is a lumped 'external' routing component.

At present, hydrologic models used in GCMs do not allow for an easy estimation of their parameters from the physical and geomorphological characteristics of the terrain. Moreover there is a need for internal as well as an external routing of the generated flows which should allow the preservation at GCM scale of the two major flow routing characteristics (the delay and the attenuation) as a function of parameters to be estimated from the topological and the geomorphological river network patterns.

The most important concept for the development of a new hydrologic model is believed to be the concept of lumping the processes by upscaling them from the microscale. The original problem is obviously a three-dimensional problem which can be first reduced to a two-dimensional problem given, as shown above, the possibility of lumping the horizontal flow hydraulic conveyance characteristics in the vertical dimension and of relating them to the total soil moisture content. The problem can be further reduced to zero dimension given the possibility of defining a nonlinear reservoir type model structure valid at the different scales and an aggregation rule for the model parameters (Todini, 1995).

Routing should also be approached in GCMs by means of hydrologic models based on physically meaningful parameters. For this reason, as soon as runoff reaches the drainage network the two-parameter Linear Parabolic approach, with the possibility of relating its parameters to the overall slope and capacity of the major river channels (Todini, 1996), should be used in conjunction with the geomorphological drainage network characteristics either expressed in terms of the GIUH, or in terms of width functions, as proposed by Naden (1992).

References

Abbott, M. B., Bathurst, J. C., Cunge, J. A., O'Connell, P. E. and Rasmussen, J. (1986). An introduction to the European Hydrological System – Système Hydrologique Européen, 'SHE' – 1: History and philosophy of a physically-based, distributed modelling system. *J. Hydrol.*, **87**, 45–59; 2: Structure of physically based, distributed modelling system. *J. Hydrol.*, **7**, 61–77.

Arctic Construction and Frost Effects Laboratory (1954). *Depth of Snow Cover in the Northern Hemisphere*. Boston, Mass: New England Division, Corps of Engineers.

Arkin, P. A. and Xie, P. (1994). The Global Precipitation Climatology Project: first algorithm intercomparison project. *Bull. Am. Meteor. Soc.*, **75**, 401–19.

Baumgartner, A. and Reichel, E. (1975). *The World Water Balance*. Mean Annual Global Continental and Maritime Report n. 9. Hamburg: Meteorologisches Institut der Universität.

Becker, A. and Glos, E. (1970). Stifenmodell zur Hochwasserwellenberechnung in ausufernden Wasserlaufen, *Wasserwirtschaft-Wassertechnik*, 20 Jg, H. 1., Berlin.

Benjamin, J. R. and Cornell, C. A. (1970). *Probability, Statistics and Decision for Civil Engineers*. New York: McGraw-Hill.

Benning, R. G. (1994). Towards a new lumped parameterization at catchment scale. Master Thesis University of Wageningen in collaboration with the University of Bologna.

Beven, K. J. (1981). Kinematic subsurface stormflow. *Water Resour. Res.*, **17**(5), 1419–24.

Beven, K. J. (1982). On subsurface stormflow: Predictions with simple kinematic theory for saturated and unsaturated flows. *Water Resour. Res.*, **18**(6), 1627–33.

Beven, K. J. (1989). Changing ideas in hydrology: the case of physically-based models. *J. Hydrol.*, **105**, 157–72.

Beven, K. J., Calver, A. and Morris, E. M. (1987). *The Institute of Hydrology Distributed Model (IHDM)*. Inst. Hydrol. Rep. no. 98, Wallingford.

Beven, K. J. and Germann, P. (1982). Macropores and water flow in soils. *Water Resour. Res.*, **18**(5), 1311–25.

Beven, K. J. and Kirkby, M. J. (1979). A physically based, variable contributing area model of basin hydrology. *Hydrological Sciences-Bulletin des Sciences Hydrologiques*, **24**, 1–3.

Binley, A., Elgy, J. and Beven, K. J. (1989). A physically based model of heterogeneous hillslopes 1. Runoff production. *Water Resour. Res.*, **25**(6), 1219–26.

Brooks, R. H. and Corey, A. T. (1964). Hydraulic properties of porous media. *Hydrol. Pap.* 3. Fort Collins: Colo. State Univ.

Brutsaert, W. H. (1982). *Evaporation into the Atmosphere*. Boston, Massachusetts: D. Reidel.

Budyko, M. I. (1948). *Evaporation under Natural Conditions* (in Rus.), Gidrometeoizdat, Leningrad (1963, English translation by Isr. Program Sci. Transl., Jerusalem).

Budyko, M. I. (1951). On the influence of reclamation measures upon potential evapotranspiration (in Rus.). *Izv. Akad. Nauk SSSR Ser. Geogr.*, **1**, 16–35.

Budyko, M. I. (ed.), (1963). *Atlas of the Heat Balance of the Earth*. Moscow: Gl. Geofiz. Obs., 69 pp.

Budyko, M. I. (1974). *Climate and Life*. New York: Academic.

Budyko, M. I., and Sokolov, A. A. (1978) Water balance of the Earth. In *World Water Balance and Water Resources of the Earth*. Studies and Reports in Hydrology, 25, ed. V. I. Korzun, pp. 586–91. UNESCO Press.

Budyko, M. I., and Zubenok, L. I. (1961). Determination of evaporation from the land surface (in Rus.). *Izv. Akad. Nauk USSR, Ser. Geogr.*, 3–17.

Carsel, R. F. and Parrish, R. S. (1988). Developing joint probability distributions of soil water retention characteristics. *Water Resour. Res.*, **24**, 755–69.

Chen, T.-C, Pfaendtner, J. and Weng, S.-P. (1994). Aspects of the hydrologic cycle of the ocean-atmosphere system. *J. Phys. Oceanogr.*, **24**, 1827–33.

Clapp, R. B. and Hornberger, G. M. (1978). Empirical equations for some soil hydraulic properties. *Water Resour. Res.*, **14**, 601–4.

Cordova, J. R. and Rodriguez-Iturbe, I. (1983). Geomorphologic estimation of extreme flow probabilities, *J. Hydrol.*, **65**, 159–73.

Cosby, B. J., Hornberger, G. M., Clapp, R. B. and Ginn, T. R. (1984). A statistical exploration of the relationships of soil moisture characteristics to the physical properties of soils. *Water Resour. Res.*, **20**, 682–90.

Dooge, J. C. I. (1973). The linear theory of hydrologic systems. *Tech. Bull. U.S. Dep. Agric.*, No. 1468. Washington, D.C.: U.S. Gov. Print. Off.

Dooge, J. C. I. (1992). Sensitivity of runoff to climate change: a Hortonian approach. *Bull. Am. Meteor. Soc.*, **73**, 2013–24.

Dooge, J. C. I., Bruen, M. and Dowley, A. (1994). Final Report to the European Community of research contract PL8900116 EPOCH.

Dümenil, L., Isele, K., Liebscher, H-J., Schröder, U., Schumacher, M. and Wilke, K. (1993). Discharge data from fifty selected rivers for GCM validation. MPI Report No. 100. Hamburg: Max-Planck-Institut für Meteorologie.

Dümenil, L. and Todini, E. (1992). A rainfall-runoff scheme for use in the Hamburgh climate model. In *Advances in Theoretical Hydrology, a tribute to James Dooge*, ed. J. P. O'Kane, pp. 129–57. European Geophysical Society Series of Hydrological Sciences, 1. Amsterdam: Elsevier.

Dunne, T. (1978). Field studies of hillslope flow processes. In *Hillslope Processes*, ed. M. J. Kirkby, pp. 227–93. New York: John Wiley.

Durand, P., Robson A. and Colin N. (1992). Modelling the hydrology of submediterranean mountain catchments (Mont-Lozère, France) using TOPMODEL: initial results. *J. Hydrol.*, **139** (1992), 1–14.

Eagleson, P. S. (1978*a*). Climate, soil, and vegetation, 1, Introduction to water balance dynamics. *Water Resour. Res.*, **14**, 705–12.

Eagleson, P. S. (1978*b*). Climate, soil, and vegetation, 4, The expected value of annual evaporation. *Water Resour. Res.*, **14**, 731–9.

Eagleson, P. S. (1978*c*). Climate, soil, and vegetation, 6, Dynamics of the annual water balance. *Water Resour. Res.*, **14**, 749–64.

Fahmy, A., Panattoni, L. and Todini, E. (1982). Mathematical model of the river Nile. In *Engineering Application of Computational Hydraulics*, ed. M. B. Abbott and J. Cunge, pp. 111–30. Boston and London: Pitman Advanced Publishing Program.

Fread, D. L. (1985). Channel routing. In *Hydrological Forecasting*, ed. M. G. Anderson and T. P. Burt. New York: John Wiley.

Freeze, R. A. (1980). A stochastic-conceptual analysis of rainfall-runoff processes on a hillslope. *Water Resour. Res.*, **16**(8), 1272–83.

Garratt, J. R. (1994). Incoming shortwave fluxes – a comparison of GCM results with observations. *J. Climate*, **7**, 72–80.

Gates, W. L., Rowntree, P. R. and Zeng, Q-C. (1990). Validation of climate models. In *Climate Change: the IPCC Scientific Assessment*, ed. J. T. Houghton, G. J. Jenkins and J. J. Ephraums, pp. 93–130. Cambridge: Cambridge University Press.

Gates, W. L.+ nine co-authors, (1996). Climate models evaluation. In *Climate Change 1995*, ed. J. T. Houghton, L. G. Meira Filho, B. A. Callander, N. Harris, A. Kattenberg and K. Maskell, pp. 229–84. Cambridge: Cambridge University Press.

Gebert, W. A., Graczyk, D. J. and Krug, W. R. (1987). Average annual runoff in the United States, 1951–80. *Hydrologic Investigations Atlas* HA-710, U. S. Geological Survey.

Groisman, P. Ya., Koknaeva, V. V., Belokrylova, T. A. and Karl, T. R. (1991). Overcoming biases of precipitation measurement: a history of the USSR experience. *Bull. Am. Meteor. Soc.*, **72**, 1725–33.

Hall, C. D. (1995). The U.K. Meteorological office climate model: the AMIP run and recent changes to reduce systematic errors. *Proceedings of the AMIP Scientific Conference*, Monterey, 15–19 May 1995.

Hayami, S. (1951). *On the Propagation of Flood Waves*. Disaster Prevention Research Institute, Kyoto University.

Henderson, F. M. and Wooding, R. A. (1964). Overland flow and groundwater flow from a steady rainfall of finite duration, *J. Geophys. Res.*, **69**(6), 1531–40.

Henning, D. (1989). *Atlas of the Surface Heat Balance of the Continents*. Berlin: Gebruder Borntraeger. 402 pp.

Horton, R. E. (1933). The role of infiltration in the hydrologic cycle. *Trans. Am. Geophys. Union,* **14**, 446–60.

Hurley, D. G. and Pantelis, G. (1985). Unsaturated and saturated flow through a thin porous layer on a hillslope. *Water Resour. Res.*, **21**(6), 821–4.

Jones, R. J., Murphy, J. M. and Noguer, M. (1995). Simulation of climate change over Europe using a nested regional climate model. Part I Assessment of control climate including sensitivity to location of lateral boundaries. *Quart. J. Roy. Meteor. Soc.*, **121**, 1413–49.

Keulegan, G. H. (1945). Spatially varied discharge over a sloping plane. *Amer. Geophys. Union Trans.* Part 6, 956–9.

Korzun, V. I. and others (eds). (1978). World Water Balance and Water Resources of the Earth. UNESCO (English translation; original 1974).

Kwadijk, J. and van Deursen, W. (1993). *CHR/KHR Report: Development and Testing of a GIS-based Water Balance Model for the River Rhine Drainage Basin*. Commission Internationale de l'Hydrologie du Bassin du Rhin.

Legates, D. R. and Willmott, C. J. (1990*a*). Mean seasonal and spatial variability in gauge-corrected, global precipitation. *Internat. J. Climatol.*, **10**, 111–27.

Legates, D. R. and Willmott, C. J. (1990*b*). Mean seasonal and spatial variability in global surface air temperature. *Theor. Appl. Climatol.*, **42**, 11–21.

Lighthill, J. A.. and Whitham, G. B. (1955). On kinematic waves, I, Flood measurements in long rivers. *Proc. Royal Soc. of London*, A, **229**, 281–316.

Liston G. E., Sud, Y. C. and Wood, E. F. (1994). Evaluating GCM land surface hydrology parameterizations by computing river discharges using a runoff routing model: application to the Mississippi Basin. *J. Appl. Met.*, **33**, 394–405.

Liu, W. T. (1988). Moisture and latent heat flux variabilities in the tropical Pacific derived from satellite data. *J. Geophys. Res.*, **93**, 6749–60.

Liu, W. T., Tang, W. & Wentz, F. (1992). Precipitable water and surface humidity over global oceans from Special Sensor Microwave Imager and European Centre for Medium Range Weather Forecasts. *J. Geophys. Res.*, **97**, 2251–64.

Lydolph, P. (1977). Climates of the Soviet Union. *World Survey of Climatology*, **7**. Amsterdam: Elsevier, 443 pp.

Mahfouf, J. F. (1991). Analysis of soil moisture from near-surface parameters: a feasibility study. *J. Appl. Climatol.*, **30**, 1534–47.

Marshall, T. J. and Holmes, J. W. (1979). *Soil Physics*. Cambridge: Cambridge University Press.

Matson, M., Ropelewski, C. F. and Varnadore, M. S. (1986). *An Atlas of Satellite-derived Northern Hemispheric Snow Cover Frequency*. NOAA, U.S. Dept. of Commerce, 75 pp.

Miller, J. R., Russell, G. L. and Caliri, G. (1994). Continental scale river flow in climate models, *J. Climate*, **7**, 914–28.

Milly, P. C. D. (1993). An analytic solution of the stochastic storage problem applicable to soil water. *Water Resour. Res.*, **29**, 3755–8.

Milly, P. C. D. (1994a). Climate, interseasonal storage of soil water, and the annual water balance. *Adv. Water Resour.*, **17**, 19–24.

Milly, P. C. D. (1994*b*). Climate, soil water storage, and the average annual water balance. *Water Resour. Res.*, **30**, 2143–56.

Milly, P. C. D. and Dunne, K. A. (1994). Sensitivity of the global water cycle to the water-holding capacity of land. *J. Climate*, 7, 506–26.

Milly, P. C. D. and Eagleson, P. S. (1987). Effects of spatial variability on annual average water balance. *Water Resour. Res.*, **23**, 2135–43.

Mintz, Y. (1984). The sensitivity of numerically simulated climates to land surface boundary conditions. In *The Global Climate*, ed. J. T. Houghton, pp. 79–105. Cambridge: Cambridge University Press.

Mitchell, J. F. B., Manabe, S., Tokioka, T. and Meleshko, V. (1990). Equilibrium climate change. In *Climate Change: the IPCC Scientific Assessment*, ed. J. T. Houghton, G. J. Jenkins and J. J. Ephraums, pp. 131–64. Cambridge: Cambridge University Press.

Monteith, J. L. (1965). Evaporation and environment. In *The State and Movement of Water in Living Organisms*, pp. 205–34. 19th Symp. Soc. Exp. Biol.

Moore, R. J. and Clarke, R. T. (1981). A distribution function approach to rainfall-runoff modelling. *Water Resour. Res.*, **17**(5), 1367–82.

Naden, P. S. (1992). Spatial variability in flood estimation for large catchments: the exploitation of channel network structure. *J. Hydrolog. Sci.*, **37**, 53–71.

Natale, L. and Todini, E. (1977). A constrained parameter estimation technique for linear models in hydrology. In *Mathematical Models for Surface Water Hydrology*. Chichester: John Wiley.

Oki, T., Musiake, K., Masuda, K. and Matsuyama, H. (1993). Global runoff estimation by atmospheric water balance using ECMWF data set. In *Macroscale Modelling of the Hydrosphere*, Proceedings of the Yokohama Symposium, July 1993. IAHS Publ. no. 214. 163–71.

Patterson, K. (1990). Global distributions of total and total-available soil water holding capacities. M. Sc. thesis. Dept. of Geography, University of Delaware, 118 pp.

Peixoto, J. P. and Oort, A. H. (1983). The atmospheric branch of the hydrological cycle and climate. In *Variations in the Global Water Budget*, ed. A. Street Perrott *et al.*, pp. 5–65. Boston: Reidel.

Penman, H. L. (1948). Natural evaporation from open water, bare soil, and grass. *Proc. R. Soc. London*, Ser. A, **193**, 120–45.

Polcher, J., Laval, K., Dümenil, L., Lean, J. and Rowntree, P. R. (1996). Comparing three land surface schemes used in general circulation models. *J. Hydrol*, **180**, 1963–2000.

Quinn, P., Beven, K., Morris, D. and Moore, R. (1991). *The Use of Digital Terrain Data in Modelling the Response of Hillslopes and Headwaters*. Internal technical report, University of Lancaster.

Rawls, W. J., Brakensiek, D. L. and Saxton, K. E. (1982). Estima-

ting soil water properties. Transactions, ASAE, **25**(5), 1316–20 and 1328.

Roads, J. O., Chen, S.-C., Guetter, A. K. and Georgakakos, K. P. (1994). Large-scale aspects of the United States hydrological cycle. *Bull. Amer. Meteor. Soc.*, **75**, 1589–610.

Roads, J. O., Chen, S.-C., Kao, J., Langley, D. and G. Glatzmaier, G. (1992). Global aspects of the Los Alamos General Circulation Model hydrologic cycle. *J. Geophys. Res.*, **97**, 10,051–68.

Robock, A., Vinnikov, K. Y., Schlosser, C. A., Speranskaya, N. A. and Xue, Y. (1995). Use of midlatitude soil moisture and meteorological observations to validate soil moisture simulations with biosphere and bucket models. *J. Climate*, **8**, 15–35.

Rodriguez-Iturbe, I. and Valdés, J. B. (1979). The geomorphic structure of hydrologic response. *Water Resour. Res.* **15**, 1409–20.

Roeckner, E., Arpe, K., Bengtsson, L., Brinkop, S., Dümenil, L., Esch, M., Kirk, E., Lunkeit, F., Ponater, M., Rochel, B., Sausen, R., Schlese, U., Schubert, S. and Windelband, M. (1992). *Simulation of the Present-day Climate with the ECHAM Model: Impact of Model Physics and Resolution. MPI Tech. Rep.* No. 93. Hamburg: Max-Planck-Institut für Meteorologie.

Rowntree, P. R. (1991). Atmospheric parameterization schemes for evaporation over land: basic concepts and climate modelling aspects. In *Land Surface Evaporation: Measurement and Parametrization*, ed. T. J. Schmugge and J-C. Andre, pp. 5–29. New York: Springer-Verlag.

Rowntree, P. R. and Bolton, J. A. (1983). Simulation of the atmospheric response to soil moisture anomalies over Europe. *Quart. J. Roy. Meteor. Soc.*, **109**, 501–26.

Rowntree, P. R. and Lean, J. (1994). Validation of hydrological schemes for climate models against catchment data. *J. Hydrology*, **155**, 301–23.

Rutter, A. J. (1975). The hydrological cycle in vegetation. In *Vegetation and the Atmosphere*, Vol. 1, ed. J. L. Monteith, pp. 111–54. London: Academic Press.

Sausen, R., Schubert, S. and Dümenil, L. (1991). *A Model for River Runoff for Use in Coupled Atmosphere-ocean Models*. MPI Rep. no. 9. Hamburg: Max-Planck-Institut für Meteorologie.

Savijärvi, H. I. (1988). Global energy and moisture budgets from rawinsonde data. *Mon. Wea. Rev.*, **116**, 417–30.

Schaake, J. C., Jr., and Liu, C. (1989). Development and application of simple water balance models to understand the relationship between climate and water resources. In *New Directions for Surface Water Modeling*, Proceedings of the Baltimore Symposium, May 1989, Intl. Assoc. Hydrol. Sci., Publ. 181.

Schubert, S. D., Rood, R. B. and Pfaendtner, J. (1993). An assimilated dataset for Earth science applications. *Bull. Am. Meteor. Soc.*, **74**, 2331–42.

Sivapalan, M., Beven, K. and Wood, E. F. (1987). On hydrologic similarity 2. A scaled model of storm runoff production.

Water Resour. Res., **23**, 12, 2266–78.

Stagnitti, F., Parlange, M. B., Steenhuis, T. S. and Parlange, J.-Y. (1986). Drainage from a uniform soil layer on a hillslope. *Water Resour. Res.*, **22**(5), 631–4.

Stamm, J. F., Wood, E. F. and Lettenmaier, D. P. (1994). Sensitivity of a GCM simulation of global climate to the representation of land-surface hydrology. *J. Clim.*, **7**, 1218–39.

Thornthwaite, C. W. (1948). An approach toward a rational classification of climate. *Geogr. Rev.*, **38**, 55–94.

Todini, E. (1988*a*). *Un Modello di Previsione di Piena per il Fiume Arno*. Florence: Regione Toscana (in Italian).

Todini, E. (1988*b*). Rainfall-runoff modelling past, present and future. *J. Hydrol.*, **100**, 341–52. Amsterdam: Elsevier.

Todini, E. (1989). Flood forecasting models. *Excerpta*, **4**, 117–62.

Todini, E. (1995). New trends in modelling soil processes from hillslope to GCM scales. In *Role of Water and the Hydrological Cycle in Global Change*, ed. H. R. Oliver and S. A. Oliver, pp. 317–47. New York: Springer.

Todini, E. (1996). The ARNO rainfall-runoff model, *J. Hydrol.*, **175**, 339–82.

Todini, E. and Bossi A. (1986). PAB (Parabolic and Backwater) an unconditionally stable flood routing scheme particularly suited for real-time forecasting and control. *Institute of Hydraulic Construction, Pub.*, **1**, Bologna.

Trenberth, K. E. (1991). Climate diagnostics from global analyses: conservation of mass in ECMWF analyses. *J. Climate*, **4**, 707–22.

Trenberth, K. E. (1992). *Global Analyses from ECMWF and Atlas of 1000 to 10 mb Circulation Statistics*. NCAR Technical Note NCAR/TN-373 + STR, 191 pp plus 24 fiche.

Trenberth, K. E. (1995). Atmospheric circulation climate changes. *Clim. Change*, **31**, 427–53.

Trenberth, K. E. and Guillemot, C. J. (1995). Evaluation of the global atmospheric moisture budget as seen from analyses. *J. Climate*, **8**, 2255–72.

Trenberth, K. E. and Olson, J. G. (1988*a*). *Intercomparison of NMC and ECMWF Global Analyses: 1980–1986*. NCAR Technical Note NCAR/TN-301 + STR, 81 pp.

Trenberth, K. E. and Olson, J. G. (1988*b*). An evaluation and intercomparison of global analyses from NMC and ECMWF. *Bull. Am. Meteor. Soc.*, **69**, 1047–57.

Trenberth, K. E. and Solomon, A. (1993). Implications of global atmospheric spatial spectra for processing and displaying data. *J. Climate*, **6**, 531–45.

Troch, P. A., Mancini, M., Paniconi, C. and Wood, E. F. (1993). Evaluation of a distributed catchment scale water balance model. *Water Resour. Res.*, **29**(6), 1805–17.

van Genuchten, M. Th. (1980). A closed-form equation for predicting the hydraulic conductivity of unsaturated soils. *Soil Sci. Soc. Amer. J.*, **44**, 892–8.

van Genuchten, M. Th., Leij, F. J. and Yates, S. R. (1991). The RETC Code for Quantifying the Hydraulic Functions of Unsaturated Soils. U.S. Environmental Protection Agency Report EPA/600/2-91/065.

Vinnikov K. Ya. and Yeserkepova I. B. (1991). Soil moisture: empirical data and model results. *J. Climate*, **4**, 66–79.

Vorösmarty, C. J., Moore III, B., Grace, A. L., Gileda, M. P., Melillo, J. M., Peterson, B., Jrastetter, E. B. and Steudler, P. A. (1989). Continental scale models of water balance and fluvial transport: an application to South America. *Global Biogeochemical Cycles*, **3**(3), 241–65.

Warrick, A. W. and Nielsen, D. R. (1980). Spatial variability of soil physical properties in the field. In *Applications of Soil Physics*, ed. D. Hillel. New York: Academic.

Warrilow, D. A., Sangster, A. B. and Slingo, A. (1986). *Modelling of Land Surface Processes and their Influence on European Climate*. Meteorological Office. Met O 20 Tech Note DCTN 38.

Wijffels, S. E., Schmitt, R. W., Bryden, H. L. and Stigebrandt, A. (1992). Transport of freshwater by the oceans. *J. Phys. Oceanogr.*, **22**, 155–62.

Wilson, M. F. and Henderson-Sellers, A. (1985). A global archive of land cover and soils data for use in general circulation climate models. *J. Climatol.*, **5**, 119–43.

Wittmeyer, I. L. and Vonder Haar, T. H. (1994). Analysis of the global ISCCP TOVS water vapor climatology. *J. Climate*, **7**, 325–33.

Wood, E. P., Lettenmaier, D. P. and Zartarian, V. G. (1992). A land-surface hydrology parameterization with subgrid variability for general circulation models. *J. Geophys. Res.*, **97** (D3), 2717–28.

Woolhiser, D. A. and Liggett, J. A. (1967). Unsteady one-dimensional flow over a plane: the rising hydrograph. *Water Resour. Res.*, **3**(3), 753–71.

World Weather Records (separate volumes for Before 1920, 1921–30, 1931–40, …, 1971–80), U.S. Department of Commerce.

Yanai, M., Esbensen, S. and Chu, J.-H. (1973). Determination of bulk properties of tropical cloud clusters from large-scale heat and moisture budgets. *J. Atmos. Sci.*, **30**, 611–27.

Zhao, R. J. (1977). *Flood Forecasting Method for Humid Regions of China*. East China College of Hydraulic Engineering.

281

8 The way forward

8.1 Toward an integrated land–atmosphere–ocean hydrology

Moustafa T. Chahine

Complexity is an inherent characteristic of the Earth's climate system because the system's elements are intimately connected and interact in a nonlinear fashion. Both the physical and biological aspects of the system are intrinsically linked to the global hydrologic cycle. Thus, progress in climate research and prediction depends on progress in understanding not only the individual elements of the hydrologic cycle, but also the global interactions between them. Achieving skill in climate prediction based on improved understanding and modeling of the global hydrologic cycle will produce a wide range of social and economic applications in the years ahead.

The first step toward achieving an integrated global hydrologic model of the climate system is to develop coupled land–atmosphere and ocean–atmosphere models and to improve the parameterization of the energy and radiative processes in atmospheric general circulation models. While existing models are very useful for predicting atmospheric dynamics, they are limited in their ability to predict the distribution, phase changes and transport of water in the atmosphere.

Yet, even with improvements in numerical modeling, our understanding of the global hydrologic cycle will remain limited until reliable global data are available on many essential quantities such as clouds, precipitation, evaporation, soil moisture and atmospheric transports. Limited regional and global data sets of key climate parameters, such as rainfall, radiation and clouds, and land-surface parameters are being developed to document climate variability. New data are expected within the coming five years (1998–2002) from more capable satellite observations which will accelerate progress in integrating process studies, models and observations of the hydrologic system.

This section outlines some areas of climate research where studies of the hydrologic cycle are poised to make significant contributions.

Seasonal-to-interannual prediction of the effects of ENSO

Prospects for prediction of the effects of ENSO (El Niño Southern Oscillation) over many regions of the world one or two seasons in advance are greatly improved by progress in modeling ocean–air interactions and in understanding the role of land-surface processes (i.e., soil moisture) in climate models. Prediction skills will be enhanced through better understanding of the patterns of seasonal variability, specifying the mechanisms underlying this variability, and understanding how these mechanisms vary in space and time. Hydrologic studies are being conducted to address these needs, starting with the development of regional and global data sets of key climate parameters (e.g. rainfall, radiation and clouds, and land-surface parameters); conducting continental-scale water budget experiments to identify the mechanisms that control the hydrologic cycle over land; and by improvement of the formulation of energy and freshwater transport and exchanges in atmospheric circulation models.

Improved long-range weather prediction

The Global Atmospheric Research Programme (GARP), conducted in 1979, was a major milestone in weather prediction research. It resulted in improved weather prediction on time scales of several days and provided the research basis for further improvement in forecast skills on the short and medium time scales. The areas of weather prediction most likely to benefit from current and future research in the global hydrologic cycle are the traditional long-range weather forecast and the extended-range forecast of monthly anomalies. In both areas there is dependence on the initial conditions and also on longer time scale forcing from land surface and ocean components.

The current hydrology efforts in atmospheric modeling for weather prediction are focused on the treatment of the energy budget, surface and orographic effects, the vertical distribution of clouds and their optical properties, and the transport of humidity. Processes such as the release of latent heat of freezing (ECMWF, 1996) and the effect of albedo of snow covered areas (Viterbo and Betts, 1997; Betts and Hall, 1997) are being investigated and must be represented correctly.

Prediction of precipitation, water resources and soil moisture over large basin regions

Several large–scale joint meteorological-hydrologic field studies are being conducted in different terrain and climate conditions to develop improved, coupled land-surface and atmospheric models. One consequence has been a revival of interest in boundary-layer physics over vegetated areas.

Land surface processes contribute significantly to the variance of annual precipitation over continents. Soil moisture storage has an integrative nature and may hold the memory of past variations for months. The availability of water in the form of soil moisture is a major factor in determining the relative proportions of sensible and latent heat fluxes. As a result, wet land surfaces tend to correlate with more local rainfall. Betts *et al.* (1993) showed that monthly and seasonal precipitation forecasts over the Midwest USA during the floods of 1993 could have been improved with better formulation of soil moisture and boundary layer conditions in the forecast model. The influence of vegetation in controlling and maintaining soil moisture availability to the atmosphere is an area of active research (Henderson-Sellers *et al.*, 1995). Variability is high among models describing this influence; the differences must be understood, and models must be improved to reflect varying soil and climate conditions properly. Methods of obtaining surface soil wetness from space are being studied and may be implemented within the next 10 years; however, validation of these data will remain a challenging task.

In addition to these local controls on precipitation over land, there are non-local influences. O'Brien and Sittel (1994) showed that, over North America, there is a significant ENSO-related precipitation anomaly except during summer. The lack of a clear ENSO-related signature during summer might be ascribed to the influences of soil moisture and albedo on the persistence of atmospheric circulation regimes. Further understanding of the influence of local vs. non-local factors will require the development of improved, coupled ocean–land–atmosphere models – an effort currently under way.

Prediction of the response of climate to changes in anthropogenic forcing parameters

Prediction of the response of the atmosphere to changes in forcing due to greenhouse gases or aerosols requires accurate determination of the energy fluxes in the atmosphere and at the surface. This is being addressed with improved observations and with better characterization of clouds, precipitation, aerosols and water vapor processes.

Some improvement is likely in the level of accuracy of estimates of interannual variations and regional changes in aerosols from the next series of satellite measurements. In addition, limited studies have been initiated to document changes in the amount and optical properties of tropospheric aerosols, thereby extending the quasi-systematic observa-

tions of stratospheric aerosols already being made by a succession of solar occultation instruments on satellites.

More top-of-the-atmosphere (TOA) radiance data will continue to be available from satellites. To go farther it is necessary to establish the relationship between TOA fluxes and the radiative forcing of cloud systems, and the resulting effect on the general circulation of the atmosphere. This requires greatly improved observations of the vertical structure, optical and microphysical properties of clouds and the vertical profiles of atmospheric temperature, moisture, wind and aerosols.

The current accuracy and reliability of atmospheric moisture determination from both *in situ* measurements and remote sensing does not permit us to assess changes in upper tropospheric water vapor. However, water vapor observations will be considerably improved in the next five years with high spectral resolution soundings. This observational development alone should yield significant advances in determining the vertical transport and mixing of water vapor by a variety of subgrid-scale processes: in particular, deep convective motions, thus pinning down an important and still relatively poorly known feedback for climate change; and in determining the horizontal advection and flux divergence of water vapor in the lower atmosphere, thus improving quantitative model predictions of rainfall.

Coupling of the hydrologic cycle to ocean circulation beyond the tropics, including the polar regions

Ocean models are driven by the fluxes across the ocean surface, including heat, fresh water and momentum. The primary driver of the global thermohaline circulation is thought to be the convective descent of water in the northern Atlantic to levels deep in the ocean. This forcing is largely controlled by the freshwater inflow from higher latitudes, since a decrease of salinity in the upper part of the water column diminishes the potential for vertical mixing and thus deep water formation. To understand this mechanism, climatologies of freshwater flux to the Arctic must be developed, including river runoff and precipitation.

River discharge measurement stations are located mainly on large navigable rivers. For smaller rivers in the Arctic regions this means that estimates of runoff must be derived mainly from improved river routing models. This activity has begun as a part of the two major continental-scale hydrologic experiments under way in Canada and Siberia.

Snow is one of the most difficult processes to model and predict and is one of the most uncertain aspects of polar climatology. The most promising method for determining the freshwater input from atmospheric precipitation over the whole Arctic basin consists in estimating the integrated atmospheric water flux divergence based on upper-air measurements around the basin. New analyses of infrared and microwave data from space will fill gaps in tropical and extratropical rainfall measurements, but measurement of solid precipitation remains based on data from a few special gauges.

Assessment of the rate of acceleration of the hydrologic cycle

It is possible that the rate of recycling of water in the atmosphere may be sensitive to relatively small changes in the dynamics and thermodynamics of meridional transports. In addition, fluctuations in atmospheric humidity may also affect the basic radiation balance by modulating cloudiness and influencing longwave emission. Thus, accurate characterization of the rate of recycling of water in the atmosphere is a sensitive index of the multiple roles of hydrology in the climate system. The recent report from the Intergovernmental Panel on Climate Change (Watson *et al.*, 1996) indicates that climate change may lead to an intensification of the global hydrologic cycle, having major impacts on regional water resources and the possibility of changes in the magnitude and timing of runoff and the intensity of floods and droughts.

One general indicator of the strength of the hydrologic cycle is the mean residence time of water in the atmosphere. The atmosphere recycles its entire water content 33 times per year (total yearly precipitation divided by the total atmospheric precipitable water vapor), giving water vapor a mean global residence time in the atmosphere of about 11 days (Chahine, 1992). A change in this residence time could indicate an acceleration or deceleration of the hydrologic cycle. Current observational data must be assessed for their ability to determine global precipitation, both liquid and solid, and water vapor with sufficient accuracy for regional and seasonal detection of changes in the rate of recycling of water in the Earth's climate system. Analyses must be conducted to estimate the residence time using historical and current observational data and these estimates should be compared to modeled values for various climate change scenarios.

The social and economic outreach

Fluctuations in regional temperature, precipitation and the availability of fresh water, which are linked to fluctuations in the hydrologic cycle, are of significant consequence to human affairs. While basic research in the hydrologic cycle is conceived and conducted within the science community, a collaboration must be initiated between research scientists and the known and expected users of the information who can apply climate variability research (i.e. flood and drought prediction) to issues of agriculture and water resource availability. The type of information that is needed and useful for applications such as these must be defined in parallel with ongoing basic research. Thus, the hydrology research community must initiate contacts with the applications community and find meaningful ways to communicate the results of scientific research.

The way forward

Several of the areas discussed above are being addressed by an international research program in energy and hydrology established by the World Climate Research Programme (WCRP) in 1989 and known as the Global Energy and Water Cycle Experiment (GEWEX) (WCRP, 1990). A central goal of GEWEX has been to engage the cooperation of hydrologic scientists and engineers to foster the interdisciplinary linkages needed to address climate prediction issues. The GEWEX approach is organized into three major categories of activities: hydrologic studies, atmospheric and land-surface radiation studies, and modeling and prediction.

GEWEX has established several large-scale, joint meteorological–hydrologic field studies for different terrain and climate conditions, including Western Europe, the entire Mississippi River basin, the arid landscape of central Asia, the Arctic boreal forest, and the Brazilian rainforest. A major thrust of these studies is to develop coupled atmosphere–land surface models, which characterize the components of the global hydrologic cycle, the transport and precipitation of atmospheric water vapor, the storage of groundwater and evaporation from vegetation and soil, and the runoff from rainfall and river flow.

The GEWEX research effort also aims at determining atmospheric radiative transfer and radiative heating within an air column, from the surface to the top-of-the-atmosphere. Several projects sponsored by GEWEX are contributing to this goal. The first global climatology of monthly-mean cloud amount and optical depth has already been developed from geostationary satellites by the International Satellite Cloud Climatology Project. An accurate determination of the planetary radiation balance and contribution of clouds to this balance is a contribution of the Earth Radiation Budget Experiment. A global climatology of total precipitable water vapor, combining *in situ* measurements with satellite observations is under development by the GEWEX Water Vapor Project. Cloud-radiation process studies and cloud process representation in general circulation models are the focus of the GEWEX Cloud System Study, a major interpretative effort using the most recent cloud-resolving microphysical process models.

GEWEX has assembled and validated several additional climate data sets, including the first global record of monthly mean rainfall over oceans and continents (via the Global Precipitation Climatology Project) by combining rain-gauge data with remote-sensing estimates from geostationary and polar orbiting satellites. And GEWEX is producing the first global climatology of 'soil wetness' (via the International Satellite Land Surface Climatology Project). As more capable satellite observing techniques become available, the GEWEX data sets will be improved and enhanced. Improved temperature and moisture sounders, active lidar and millimeter-wave radars to observe the three-dimensional distribution of clouds, and differential absorption or Raman laser techniques to determine the fine vertical structure of water vapor are among the advanced capabilities expected to be used.

All this has made GEWEX the largest international climate research endeavor at the present time. With the ongoing support of the international research community, major contributions to climate research and prediction are expected to continue.

References

Betts, A. K., Ball, J. H. and Beljaars, A. C. M. (1993). Comparison between the land surface response of the European Centre Model and the FIFE-1987 data. *Quart. J. Roy. Meteor. Soc.*, **119**, 975–1002.

Betts, A. K. and Hall, J. H. (1997). Albedo over the boreal forests. *J. Geophys. Res.*, **102**, 28901–09.

Chahine, M. T. (1992). The hydrologic cycle and its influence on climate. *Nature*, **359**, 373–80.

ECMWF (1996). *Improvements to the 2m Temperature Forecasts.* European Centre for Medium Range Weather Forecasting Newsletter no. **73**, 2–6.

Henderson-Sellers, A., Pitman, A. J., Love P. K., Irannejad, P. and Chen, T. H. (1995). The project for intercomparison of land surface parameterization schemes (PILPS): Phases 2 and 3. *Bull. Amer. Meteor. Soc.*, **76**, 489–503.

O'Brien, J. J. and Sittel, M. C. (1994). *Differences in the Means of ENSO Extremes for Maximum Temperature and Precipitation in the United States.* Florida State University Center for Ocean-Atmospheric Studies Technical Report no. 94–2.Tallahassee, FL: Florida State University.

Viterbo, P. and Betts, A. K. (1997). *The Forecast Impact of Changes to the Snow Albedo of the Boreal Forests.* CAS/JSC Working Group on Numerical Experiments Report No. 25: Research Activities in Atmospheric and Oceanic Modelling. World Meteorological Organization Technical Document No. 792, pp.4.42–4.43. Geneva, Switzerland.

Watson, R. T., Zinyowera, M. C. & Moss, R. H. (eds.) (1996). *Climate Change 1995 – Impacts, Adaptations and Mitigation of Climate Change: Contribution of Working Group II to the Second Assessment Report of the Intergovernmental Panel on Climate Change.* Cambridge: Cambridge University Press.

WCRP (1990). *Scientific Plan for the Global Energy and Water Cycle Experiment, World Climate Research Programme,* WCRP-40 (WMO/TD-No 376), August 1990, 83 pp.

Index